# Geoinformatics for Climate Change Studies

# Geoinformatics for Climate Change Studies

Editors
**P K Joshi • T P Singh**

The Energy and Resources Institute

ISBN 978-81-7993-409-8



**Suggested citation**

P K Joshi and T P Singh (eds). 2011. ***Geoinformatics for Climate Change Studies.*** New Delhi: The Energy and Resources Institute (TERI)

**Published by**
TERI Press
The Energy and Resources Institute (TERI)
Darbari Seth Block
IHC Complex, Lodhi Road
New Delhi – 110 003
India

***Tel.*** 2468 2100 or 4150 4900
***Fax*** 2468 2144 or 2468 2145
India +91 • Delhi (0)11
***E-mail*** teripress@teri.res.in
***Website*** www.teriin.org

Printed in India

Dedicated to

**"Lighting a Billion Lives"**

*– a commitment to enlighten a billion lives with a handful of sunshine*

To displace the use of fossil fuel, thereby mitigating climate change while illuminating the lives of a billion people

# Foreword

"Warming of the climate system is unequivocal!" While climate change debate has been at the fore of international political agenda for the past few decades, the uncertainty surrounding the science, and its mitigation and adaptation aspects have brought the scientific community to the limelight. Anthropologically forced climate change is one of the most significant challenges humankind has ever faced and presents a significant opportunity for increasing the scientific understanding about our planet and about the human activities having major impacts on the global processes governing it.

The science demands a huge amount of spatio-temporal data to dwell with. Handling spatio-temporal data gathered from different sources, for example, the earth observation satellites, weather stations, GPS, and society, requires a very specific skilled and trained cadre in geoinformatics. As the technology is developing by leaps and bounds to collect and make available such data, it places greater responsibility on us to handle it well.

The book *Geoinformatics for Climate Change Studies* is a unique combination of two sciences—climate change and geoinformatics. While the former is the most exigent challenge of the era, the latter brings the knowledge of the most sophisticated technology to handle the wide variety of data required addressing such a challenge. This book provides an in-depth understanding of geoinformatics (remote sensing, GIS, GPS, ICT, and others) as well as an awareness of the causes and consequences of climate change.

A large part of the process of addressing climate change is our ability to learn from the previous experiences. This book brings such experiences in a lucid way so that students, researchers, and professionals interested in climate change and geoinformatics can gain from these. These case studies will also hold immense relevance from an academic standpoint as many of the institutions are coming up with promising programmes to develop a special cadre of trained and skilled manpower to provide solution to make the earth sustainable.

Better understanding of these processes will ultimately give researchers sufficient insights into how to move forward to preserve the ecosystems and environment we have and slow down the negative impact of our past actions. As geoinformatics holds a key role in enhancing our knowledge to address climate change, a thorough understanding of the procedures and techniques used in geoinformatics is a must.

I congratulate the editors in bringing out such a wonderful compendium on the much needed subject. I am sure readers will find the contents of the book interesting and worth applying in their subject areas.

**R K Pachauri**, PhD
Director-General, TERI

# Preface

In recent years there has been an increasing awareness of the effect of climate change on the Earth and its resources. There has also been a dramatic increase in support for research in climate change studies. In particular, fundamental variables of climate such as past, present, and future data on biophysical attributes of the Earth's surface and atmosphere, and tools and methods for processing such data have become increasingly important in a vast array of climate change research and studies. Considerable recent research has focused on global studies and at global scale using such tools and datasets. The growth of research has been marked and is perhaps most apparent with the development of a new field of study—climate change.

Although there has been a long standing interest in global warming and its impact on the Earth's resources, the ability to work at different scales arises in part from advances in technology. Geoinformatics as the primary science and technology for spatio-temporal data analysis is of particular importance and significance. Geoinformatics encompasses remote sensing, geographical information system (GIS), data collected using global positioning system (GPS), information communication technology (ICT), and social interactions. Remote sensing now routinely provides information on the Earth and its environment at the local to global scales. GIS provides a means to store, analyse, and visualize spatial data, including those derived from remote sensing and collected from the field. Together with associated advancements in communication and computational facilities, tools are now often provided to efficiently handle and share a great amount of data of large areas. There are also specialist tools that recognize spatio-temporal characters. The opportunities presented by this situation help to make the area of overlap between climate change and geoinformatics of great interest for research; an area that fits neatly within the core for identifying and communicating climate change research.

Over the last three decades, the scientific evidence for human-induced climate change has become overwhelming. Such research places heavy demands for reliable climate information, which includes spatially distributed information about select climate elements. Similarly, the need for spatially distributed information about weather and climate has increased over the same period, in keeping with the increasing attention given to global warming and climate change by both scientists and policy-makers. Geoinformatics is playing an increasingly important role in climate change research. It is seen as a vehicle for depicting past, present, and future land use, land cover, and climate patterns; for integrating data from various sources with potentially different data models so as to provide a basis for comparison and evaluation; and for visual presentation of the results in a policy-supportive, decision-making environment.

The focus of the book is to fill a gap in the available literature on the subject by bringing together concepts, theories, and experiences of various experts in this field. It is a collection of 16 chapters by scientists who work within the field of climate science and geoinformatics and whose papers reflect the state of research in their respective discipline. All of the chapters are significant contributions to document and understand climate change using geoinformatics, hence the title *Geoinformatics for Climate Change Studies*. The volume also includes chapters by the editors, who felt it appropriate as it reflected continuing contribution to the subject and a hands-on approach.

The diverse chapters in the book include aspects of remote sensing and GIS, but each is mainly focused on climate change research related to different aspects of human beings. The chapters presented vary greatly in conceptual framework, broad concepts, and specific application of new approaches to spatial modelling and applications. Some chapters present grand overview of concepts, while others provide results of large-scale projects and a few report results from pilot studies. All chapters, however, help illustrate ways in which geoinformatics is being or may be used in climate change studies. Most of the representative areas common throughout the globe are touched upon and it is hoped that others may be developed for later use.

Informed, thoughtful science represents humankind's best opportunity for understanding and managing the complexities of climate change. The first two chapters present a common setting of geoinformatics for climate change research and studies. In Chapter 1, Artz and Dangermond propose a GIS-based framework for earth systems modelling and global design that crosses academic, scientific, and industrial domains. A number of existing frameworks and programmes are reviewed to glean best practices and identify gaps. Visualization and exploration including modelling and designing are discussed as useful methods for understanding and responding towards climate change. This chapter is useful in many ways, and is written by top-notch scientists and philosophers of GIS from the Environmental Systems Research Institute (eSRI). The chapter also gives actionable recommendations in the areas of data, education, ethics, organization, models, and design to support the establishment of a GIS-based framework to help provide answers to this global challenge.

In Chapter 2, Joshi, Baruah, and Singh present an overview of the potential of geoinformatics. The chapter provides a series of application scenarios to illustrate the utility of the system, which seems particularly valuable in exploring numerous applications simultaneously. This chapter discusses potential multiple roles of geoinformatics in integrated climate change research. Using case studies, an attempt is made to illustrate geoinformatics in action, from predicting the impacts of climate change as hypothesized future events to presentation of evidence of current climate change.

GIS has become a widespread and popular tool in climate and earth system studies. The next seven chapters on GIS applications present a variety of topics. Bolch and Wheate in Chapter 3 elaborate on glaciers as one of the key indicators for assessment of climate change in remote mountain areas where climate stations are rare or even non-existent. They advocated for multi-temporal remote sensing imagery over the conventional ground-based studies. Geospatial tools provide the opportunity to obtain information on glacier area, width and length, motion, and mass balance as well as other important characteristics such as debris-cover and equilibrium-line altitude. This chapter provides an overview of the advantages, challenges, and limitations of both optical and radar space-borne remote sensing data to map and study glaciers. The examples given refer to different glacier types across the world.

Chapter 4 discusses sea level rise is a coastal hazard, which not only threatens the coast with inundation but is exacerbated by storm surges and wave action. Young highlights the role of geoinformatics in assisting planners by providing the tools necessary to help assess the vulnerability of the coastal environment to sea level rise. The chapter discusses the limitations of current tools and provides high resolution data (example, LIDAR) for profusion of coastal vulnerability mapping.

In Chapter 5, Joshi, Priyanka, and Giriraj discuss the importance of geoinformatics in assessing the forest ecosystems under climate change trajectories. The chapter discusses modelling strategies for predicting the potential impacts of climate change, with examples from different forest ecosystems. The chapter lists some published examples and uses some of the original work to elaborate upon available tools and techniques to understand the impact of the climate change on forest ecosystems.

In Chapter 6, Flügel presents concepts, methods, and toolsets for comprehensive environment and river basin system assessment triggered by ongoing climate change with relevance to Integrated Water Resources Management (IWRM) and sustainable resources management. The synergetic potential of geoinformatics is highlighted in more detail with reference to the regionalized landscape concept of process-oriented response units. Taking examples from South Africa and Germany, the functionality and structural design of the Jena Environment System Analysis Toolset (JESAT) and its different tool components are discussed. In conclusion to the results presented, future research requirements are specified and briefly explained.

Chapter 7 brings a unique setting of climate, population, and energy. Stein and Lemke estimate changes in global future energy demand caused by heating and cooling buildings. Using 64 800 grid cells (180 lines of latitude × 360 lines of longitude), population and climate were calculated for every cell of the grid for 2007 and 2050. For 2050, the calculations were done for two different scenarios (A1FI and B2). This chapter forms a base for the integration of renewable energies in building sustainable planning solutions using geospatial tools.

Diodato, Ceccarelli, and Bellocchi discuss geoprocessing in evapotranspiration data in Chapter 8. This involves transformation of initial dataset derived from primary data sample and topographical information into a larger primary dataset using a topographically weighted radial basis function (TWRBF). In this way, the generated primary large-dataset can be subsequently used in geostatistical analyses for yielding at each unsampled grid-point, instead of a single class, a vector of probabilities of occurrence of different evapotranspiration-threshold data. This was applied to a cropland test site of the Tammaro sub-regional basin (South Italy) where the procedure was used to develop a baseline estimation of evapotranspiration at hilly slope scale incorporating a preliminary analytical GIS. This method can help devise a monitoring scheme for efficiently sampling the most agroclimatically sensitive and uncertain sites.

Chapter 9 introduces the issues of disease and linkage with climate variables. Tripathi, Chaikaew, and Jeefoo discuss the impact of climate on human health. Taking examples from two case studies on diarrhoea and malaria conducted in Thailand, the authors highlight the effect of climate variability on human health. The outcome is expected for better governance in applying strategies to prevent outbreak in future and save thousands of lives.

The next four chapters elaborate on the results of investigations at the national level, pressing contemporary applications in climate change research. Salami and Adepoju demonstrate the use of remote sensing technology for developing a system for land degradation management in Nigeria. Satellite measurements of the Normalised Difference Vegetation Index (NDVI), or greenness, for the period 1982–2006 were used as a proxy for net primary productivity (NPP). This is in contrast with conventional methods of land degradation assessment and monitoring in Nigeria, which are more often localized to specific regions and exist in a disaggregated format, often inaccurate, time consuming and are difficult to obtain in near real-time. Presenting a case from Bangladesh, Salami and Adepoju demonstrate the importance of satellite data to study variations in Land Surface Temperature (LST). Based on MODIS, the annual and seasonal changes in the surface temperature have been analysed based on the satellite data from January 2000 to December 2007 for every 8-day interval and the daily temperature data of last 60 years (1948–2007) collected from 34 stations. Daily temperature is one of most important local phenomena contributing to global climate change.

In Chapter 12, Babqiqi, Messouli, and Kasmi demonstrate a technique to produce high resolution scenarios by using Statistical Downscaling Model (SDSM) and atmospheric general circulation model (HadCM3). The authors elaborate the trends of future temperatures, precipitation, and cold and heat waves over Morocco using SDSM between 1961 and 2099. These scenarios present a reference database of impact assessment models of climate change on socio-economic sectors such as agriculture and water resources.

In Chapter 13, Pradhan discusses the application of GIS technologies to natural hazard detection, modelling, and monitoring. The chapter demonstrates the importance of such technologies over other surveying tools to understand the complexity of the relationship between climate change and its impact on natural hazards. Taking examples from Malaysia, the effective utilization of GIScience tools have been discussed to study natural disasters and address technical requirements. Case studies are also presented to as lessons to adapt in other countries and to help at the time when disaster management is required.

The last three chapters focus on the future use of the technology and elaborate the importance of Spatial Data Infrastructure (SDI) and geoinformatics for climate change, adaptation, and disaster risk reduction. The last chapter focuses on issues related to climate change hazards. In Chapter 14, Paudyal, McDougall, and Apan advocate SDI, a coordinated framework of technologies, standards, and data supported by policies and institutional arrangements that enable sharing and effective usage of spatial information. This chapter discusses differing perspectives of SDI and its components. It proposes improved information management that could help address global issues like climate change at the regional scale. It supports building SDI at catchment scale to improve natural resource management and provide insights into climate variability and change.

In Chapter 16, Joshi, Munsi, and Joshi provide an overview of various adaptation strategies being developed for climate change adaptation (CCA) and disaster risk reduction (DRR), existing institutional frameworks, policies, risk management strategies, and early warning systems. Detailed review has been done on the use and effectiveness of ICT, SDI, and remote sensing and GIS in DRR, which will further help in CCA.

In the last chapter, Marincioni and Abeti discuss the impoatance of ICT, including spatial analysis systems, as an essential tool for the study and management of climate change hazards. However, to become truly effective, ICT innovations must be supported by a culture that favours changes and the sharing of information. This chapter looks into possible applications of ICT to study and manage climate change hazards and discusses known patterns of perception and use of these technologies.

The aim of this book is to provide a clear description, including drawbacks, of the use of geoinformatics in climate change studies and its use for modelling. The editors hope they have provided a timely book that will be a resource for students, researchers, and practitioners. Finally, it should be noted that the chapters in book all benefited from constructive critical comments, notably from colleagues working in these fields and from graduate students and also from discussions with the contributors. This type of input from the community is important and gratefully received.

# Acknowledgements

We greatly appreciate efforts and encouragement of numerous friends and colleagues while preparing this book. First, we would like to thank each of the chapter authors who put up with suggestions and edits; their contributions make up the core of the book. The book presents select original chapters, specifically prepared for it. In the process of assembling material for the book and collecting box items inserted in many chapters, we contacted many people who responded to our emails regarding permissions and requests for high-resolution images. We thank them all for their prompt and encouraging responses. Publication of this book would not have been possible without them.

We are particularly thankful to Dr R K Pachauri, Chancellor, TERI University; Chair, Intergovernmental Panel on Climate Change; and Director-General, The Energy and Resources Institute, for writing the foreword for this book. We both are highly thankful to Dr Pachauri for motivating us and many researchers and scientists all over. Above all, we thank him for informing the importance of climate change vis-à-vis human dimensions to the entire world.

We both have spent some common time at the Indian Institute of Remote Sensing (ISRO/NRSC) while learning and experimenting with geospatial tools. We thank all colleagues and in particular, Dr P S Roy, for his constant guidance and support. He has always been a source of encouragement and motivation.

We are thankful to all our colleagues at the TERI University and Symbiosis Institute of Geoinformatics for their constant support and encouragement.

Dr P K Joshi gratefully acknowledges Gp Capt. (Dr) Rajiv Seth, Registrar, and Dr Bhavik R Bakshi, Vice Chancellor, TERI University, for their support and encouragement on such initiatives. Dr P K Joshi is also thankful to Mr Vinay S P Sinha, a colleague and friend, for his unconditional support throughout the years. Dr P K Joshi is also thankful to his graduate students and research team for valuable inputs and comments on the layout of the book.

We are indebted to Mr M M Joshi for his encouragement to initiate this project and continuous support while developing this book. Dr Richa Sharma of TERI Press deserves special mention for her tireless efforts in reviewing the manuscripts, her endurance with the flooding of confusing mails, and her patient response to each query. We acknowledge

Mr R K Joshi, TERI Press, for his efforts in drawing the graphs and figures and Ms Y S Nongpluh, TERI Press, for her inputs.

Finally, we are thankful to our parents and members of our families for their support. Dr P K Joshi is thankful to Archana and Varun and Dr T P Singh to Deepali, Chiya, and Arnav for their patience; their support is of inestimable value.

# Contents

# A GIS-based Framework for Modelling and Global Design of Earth Systems

**Matt Artz[*] and Jack Dangermond[†]**

[*]*GIS and Science Program Manager, Environmental Systems Research Institute, Inc. (eSRI), Redlands, California, USA*

[†]*President, Environmental Systems Research Institute, Inc. (eSRI), Redlands, California, USA*

*E-mail: martz@esri.com; jdangermond@esri.com*

## INTRODUCTION

From the Medieval Warm Period (10th to 14th century), the Little Ice Age (16th to 19th century) to the Year without a Summer (1816), the earth's climate has undergone radical changes in the distant as well as the recent past, and is almost certain to undergo more radical changes in the not-too-distant future also. With a steady increase in industrialization, population, and urbanization, stressors on the environment, such as pollution, are also on a rise. Such change in climate and environmental quality could have huge implications on the quality of life. Regardless of the stand we take with respect to the often politically charged issue of "global warming" or "global climate change", we owe it to ourselves and our future generation to take an intelligent look at the data available and develop intelligent alternatives.

For decades or longer, efforts have been made to keep careful records of rainfall, temperature, lake level, stream flow, and so on in thousands of locations across the world. Beyond these historical records, additional records of climate change have been kept using dendrochronology (measuring the size of tree rings), which reveals climate data dating back to thousands of years, and palynology (changes in the type and distribution of fossil pollen), which gives us clues about the climate conditions prevailing millions of years ago. The result of all this data collection and related research is the availability of vast array of data describing the conditions at particular locations on the planet at particular points in time. Careful observation and analysis of past records might help us answer questions such as: Are recent weather phenomena a short-term anomaly or a long-term trend? What past climate changes are due to the earth's natural cycle and what changes may have been caused by volcanic eruption, meteorite impact or other cataclysmic disasters?

We live in an age of readily and freely available information. The Internet has generated unprecedented awareness and given access to vast quantities of climate data. Never before have the scientists and non-scientists had such easy and open access to the data and tools needed to study the earth's climate. Both past observations and future predictions are useful in studying climate change. Examining and cross-referencing past and future data can help quantify changes already occurring, as well as help predict patterns and trends that could impact climate in the short and long terms.

Anyone who follows weather forecasts on the Internet or television knows how ambiguous it is to predict the weather. But prediction of short-term, highly localized conditions such as the probability of rain in London on Friday afternoon is quite different from predicting long-term, more general phenomena such as global climate change. For years, scientists have been using sophisticated computer models such as general circulation, atmosphere–ocean interaction, and radiative–convective process models in an attempt to visualize the future of the earth's climate. The output of a particular model can be enlightening, but combining data from multiple models and sources, both past and future, gives a comprehensive and accurate vision of what the future holds for our planet (Artz 2008a).

## TOWARDS A GIS-BASED FRAMEWORK

"The earth is a huge study area. It can be divided into pieces of various sizes and studied at macro or micro scales. For some applications, such as tracking hurricanes, scientists can rely on relatively coarse-grained information but need to update in real time. Conversely, a civil engineer may require centimetre-level precision when constructing a new bridge. The history of geographic information applications has been one of making trade-offs. A person could either study large areas at crude levels of detail or small areas in fine detail. As we approach the end of the first decade of the 21st century, these trade-offs no longer apply" (NGAC 2009). "If we discovered life on another planet, the first thing we would do is conduct a systematic inventory of that planet's life. This is something we have never done on our home planet" (Kelly 2000).

Global climate change is a difficult, complex, politically charged, and vitally important issue. Yet from the knowledge perspective, we are at a distinct disadvantage: at this point in time, we still do not have a clear idea of everything we need to know in order to address the problem in a measured, rational, and above all, scientific manner.

When one thinks about the multitude of issues surrounding climate change science—from root causes to resultant impacts—*geography* is clearly an elemental factor in the equation. Every aspect of climate change affects or is affected by geography, be it at a global, regional, or local level. As a tool for helping us to better understand such geographies, the geographic information system (GIS) is the single most powerful integrating tool for inventorying, analysing, and ultimately managing this extremely complex problem of climate change.

The key to understanding our dynamic climate is creating a framework to take many different pieces of past and future data from a variety of sources and merge them together in a single system. Information technology collates data from many different sources into a common computer database. GIS is a sophisticated technology tool, which is widely used by planners, engineers, and scientists to display and analyse all forms of location-referenced data generated from studies ranging from meteorological studies to studies depicting patterns of human settlement. GIS creates a new framework for studying global climate change by allowing users to conduct an inventory and display large, complex spatial data sets. GIS can also be used to analyse the potential interplay between various factors, bringing us closer to a true understanding of how our dynamic climate may change in the coming decades and centuries.

GIS technology is today used by various professionals for a wide range of applications. For example, the National Oceanic and Atmospheric Administration (NOAA) uses GIS to monitor the world's storm activity; the US Geological Survey uses GIS to collect and analyse data on volcanoes, earthquakes, and tsunamis; and the US Department of Agriculture uses GIS to track the effects of drought on the country's crops. From conservation organizations to international agencies, scientists across the world are using GIS as an integrative platform, which gives those researching and analysing our environment and climate an unprecedented vision and flexibility.

**What is GIS?**

A geographic information system (GIS) integrates hardware, software, and data for capturing, managing, analysing, and displaying all forms of geographically referenced information. GIS allows us to view, understand, question, interpret, and visualize data in many ways, which reveals relationships, patterns, and trends in the form of maps, globes, reports, and charts. A GIS helps address questions and solve problems by looking at the data in a way that is quickly understood and easily shared.

## WHOLE EARTH SYSTEMS SCIENCE

Scientists have long classified various phenomena into logical groupings or "systems". These classifications have helped greatly in advancing the understanding of component physical, biological, and social systems, yet these often create artificial boundaries between disciplines, which can be detrimental to the understanding of larger issues. While advancing the understanding of each of these individual systems is vitally important, ultimately, we need to bring all these systems together to understand how they are interrelated and dependent on one other.

Whole earth systems science offers an opportunity to advance the science and understanding of climate change by providing a framework for a comprehensive, interdisciplinary, and integrated study of our planetary system. Integrating complex physical, biological, and social data and models within a unified framework will give us a single view of the whole earth system and provide us with the tools to manage—and ultimately design—our future in the most effective and efficient way (Artz 2009a).

A GIS-based approach to the whole earth systems science provides a framework for understanding and addressing all the facets of climate change science issues in a holistic manner. Such a framework can provide a base enablement system for managing, visualizing, analysing, modelling, and designing global data. In order to move climate change studies from a massive collection of unrelated or loosely linked endeavours towards an open, integrated framework, six areas need to be addressed, namely, data, education, ethics, models, organization, and design (Artz 2009b).

- *Data* Comprehensive, accurate, and accessible spatial data is at the core of any system designed to analyse and manage our earth. Methods we use to collect, manage, disseminate, and share historical and derived data at a global scale need to be carefully considered.
- *Education* We have been actively engaged in a massive programme of global geoengineering for many decades or centuries. We need to recognize this activity and change the way we think about the earth—not just our impact on the planet, but how (or if) we should manage the situation.
- *Ethics* Geospatial technologies are immensely important in helping us visualize both physical and anthropogenic changes in the earth's climate and related systems, but ultimately, their most valuable contribution lies in analysing that change and supporting decision-making to help shape and design the future of the earth systems in ways that are sustainable while still serving their original purpose. We need to encourage a dialogue on the development of simple and action-oriented guidelines for ethical application of geospatial

technologies in analysing, designing, and implementing purposeful changes in the earth systems.

- *Models* We have reached a high level of sophistication with many individual models, and progress is being made in effecting integration between models. But we need a more comprehensive and open method for consolidating and relating inputs and outputs from all models.
- *Organization* Historically, a government or a non-profit organization or a consortium "adopts" an issue, organizes an approach, and puts resources behind it. Increasingly common today are "open" approaches that are largely self-organizing, relying on "volunteer" resources for most or all of their operations. Much can be learnt from such new approaches for creating organizations to solve complex problems.
- *Design* Once we recognize our environmental imperative to carefully and thoughtfully manage our planet for maintaining the health of all component earth systems and tackle the ethical issues of geoengineering, we can move away from accidental or poorly planned geoengineering into an era of conscious geodesign.

## CURRENT KNOWLEDGE

"...a better world is the common goal all of us—geographers, planners, scientists, and others—have been striving for. Although we have made a lot of progress in building the (technological) infrastructure to help us accomplish this monumental task, we are still not quite there yet" (Dangermond 2009). Various frameworks and programmes already address a number of issues and challenges in establishing a framework for earth systems modelling and global design. Careful review of the approaches towards data collection, education, ethics, models, organization, and design in these frameworks and programmes helps us in identifying concepts and components that can be leveraged as well as gaps that can be filled by a GIS-based framework. The subsequent review presents some interesting and useful examples, and is not meant to present a comprehensive inventory of such frameworks and programmes. Some of the examples are not directly related to the subject of the earth systems modelling and global design, but we can still learn from them.

The OpenStreetMap project leverages volunteers to perform on-the-ground surveys with their personal global positioning system (GPS) and other equipment to create a global base map, which is distributed freely and can be edited by anyone. The non-profit OpenStreetMap Foundation provides support for the project, but does not "control" the project per se.

OpenStreetMap is a model for creating a global data set by citizen volunteers. Organizationally, it provides a good example of a successful structure for managing the creation and distribution of the data, as well as maintaining quality standards.

Global Earth Observation System of Systems (GEOSS) for geospatial information was first suggested by the National Academy of Sciences Mapping Science Committee and was referred to as the National Spatial Data Infrastructure (NSDI). More recently, this architecture has been adopted by the NOAA and others as part of their architecture for the GEOSS. GEOSS serves as a global framework for integrating a large number of global remote-sensing systems into a loosely coupled network and making the data available to many participants, thus providing decision-support tools to a wide variety of users (Dangermond 2005).

Recently, GIS has proven to be an important and reliable tool for the management of spatial information at all geographic levels, from local to global. Over the past 15 years, a number of national, regional, and international organizations have moved towards a vision of building a Global Spatial Data Infrastructure (GSDI) for the sharing of the spatial data. The GSDI association and its membership are responsible for promoting this framework, with a goal of mapping the globe at a resolution of 1 km or better, and including information on a wide variety of geographic features. There are other global data projects in place, most notably, the International Steering Committee for Global Mapping (ISCGM), Global Map project, and the Digital Earth project. Standardized data collection, maintenance, management, and dissemination at the global scale move us closer to the goal of a common operating picture for the planet while ensuring that the widespread participation is key to the success of such global mapping projects.

Critical to the acceptance of the earth systems modelling and global design framework is a new, more enlightened understanding of the relationship of human beings with the environment and their place in it. The relatively new discipline of earth systems engineering and management (ESEM) concerns itself with the design, engineering, analysis, and management of complex earth systems. ESEM takes a holistic view of multiple issues affecting our earth—by not only taking environmental, social, and other considerations upfront in the design process into account, but also considering challenges from an adaptive systems approach, wherein the ongoing analysis feeds back into the continual management of the system (Dangermond 2009).

Brad Allenby, Professor of civil and environmental engineering at Arizona State University and one of ESEM's founders, often emphasizes on the dominant role humans play in the earth systems. "We live in a world that is fundamentally different from anything that we have known in the past," says Allenby. "It is a world dominated by one species, its activities and technologies, its cultures, and the integrated effects of its historical evolution" (Allenby 2008). Ian McHarg already endorsed this concept in the 1960s, and today, we understand that it is even more important to emphasize the anthropogenic elements of earth systems (McHarg 1969). In other words, at this stage of ecological evolution, humans form a significant, if not dominating, component of the natural environment, and all problems need to be addressed and decisions taken keeping the anthropogenic elements in the forefront.

Allenby sees reasoned design and management in the age of the anthropogenic earth as our moral imperative, but the biggest obstacle to our success is that we are not set up to work, or even think, in this way. "We lack solid data and analytical frameworks to make assertions about the costs, benefits, and normative assessments of different ... practices" (Allenby 2005a). And this is why GIS integrated with design is critical to the success of approaches such as ESEM and other logical and rational models for dealing with the environmental and planning problems of ours and those of future generations (Dangermond 2009).

The concept of *anthropogenic biomes* offers a new way forward in understanding our living planet by describing the way humans have reshaped its ecology (Ellis and Ramankutty 2007). Conventional methods of representing biomes on maps are no longer applicable in the age of the Anthropocene, and Ellis and Ramankutty (2007) note that biomes derived from global patterns of human interaction with ecosystems may be a stronger and a better model of global ecological patterns and processes. These provide us with a framework for seamlessly integrating human factors into natural systems, a necessary feature of an all-inclusive whole earth systems modelling framework for our planet.

Numerous papers and articles have been written on the "ethics of climate change", their focus being primarily on humanity's responsibility towards the environment. While such ethical considerations are important to communicate and debate, the focus here is not that *humanity should be doing something to counter climate change* but rather *we need guidelines focused on the responsible and ethical use of technology in shaping decision-making related to climate change* (Artz 2009e).

In his 1942 short story "Runaround" in the book *I, Robot,* Isaac Asimov proposed his well-known three laws of robotics: (1) a robot may not injure a human being or, through inaction, allow a human being to come to harm; (2) a robot must obey orders given to it by human beings, except where such orders would conflict with the First Law, and (3) a robot must protect its own existence as long as such protection does not conflict with the First or Second Law (Asimov 1942). Asimov later abstracted the same concepts and developed a second set of laws governing the design and use of tools; more generally these (1) should be safe to use; (2) must perform their function efficiently, and (iii) must remain intact during the period of use unless their destruction is required for their use or for safety (Asimov 1990).

Asimov's laws transcended the realm of science fiction. While we can apply some of what we learn from Asimov to the development of ethical guidelines for the use of geospatial technology in global design, the overriding theme throughout Asimov's laws pertaining to both robotics specifically and tools generally is a focus on human safety without explicitly taking things like sustainability and the health of our planet into consideration.

We all are familiar with Machines of Loving Grace. The first part of the title is taken from a poem by Richard Brautigan in which he envisioned a future where nature and technology are inextricably linked in a mutually beneficial relationship. Adam Greenfield's essay on ethical guidelines for the user's experience in ubiquitous computing settings provides a useful example of principles that meet the dual requirement of being both useful and humane (Greenfield 2004; Brautigan 1967). Greenfield's five principles are (1) default to harmlessness; (2) be self-disclosing; (3) be conservative of face; (4) be conservative of time; and (5) be deniable (Greenfield 2004). Similar to Asimov's laws regarding tools in general and robotics in particular, Greenfield's focus is towards doing no harm to humans, but still presents us with such valuable concepts as transparency and reversibility.

The GIS Certification Institute (GISCI) has established a Geographic Information Systems Professional (GISP) certification programme for GIS practitioners who have met minimum standards for ethical conduct and professional practice. The GISP Code of Ethics details a number of standards for obligations to society, employers, fund providers, colleagues, professionals, and individuals in society (GISCI n.d. a). Their companion Rules of Conduct for Certified GIS Professionals is "a set of implementing laws of professional practice that seek to express the primary examples of ethical behaviour consistent with the Code of Ethics" (GISCI n.d. b). GISCI's GISP Code of Ethics and Rules of Conduct present a set of detailed, comprehensive, and GIS-centric guidelines on the moral and ethical responsibilities of geospatial practitioners. Responsibility towards the health and welfare of natural systems is not explicitly stated in the GISCI guidelines but should be clearly affirmed in any guideline established for the use of GIS in climate change modelling and global design.

There are a number of models for abstracting, simulating, and understanding complex details of physical, biological, and social systems and subsystems (Goodchild 2005). The domains of the individual modelling packages vary widely—from soils to hydrology, from socio-economics to land use transportation (Wegner 2005; Batty 2005; Maidment 2005). While much progress has been made in recent years to develop models to help us better understand our world in the context of these domains, there is still much more to be done at the macro scale, especially in the area of integration. As we gain more detailed understanding of different granular systems and their components, the challenge in addressing complex issues such as global climate change is bringing these models together to arrive at a more complete picture. The combination of powerful hardware, sophisticated software, and increased human knowledge have all contributed to better models and more accurate simulations, but a GIS-based framework for integrating these disparate representations of past, present, and future states is key to understanding the whole earth (Maguire 2005).

The Earth System Modelling Framework (ESMF) is an open source collaborative project co-sponsored by the US Department of Defense, the National Aeronautics and Space Administration (NASA), the National Science Foundation, and NOAA. The goal of the ESMF project is to build "...high-performance, flexible software infrastructure to increase ease of use, performance portability, interoperability, and reuse in climate, numerical weather prediction, data assimilation, and other earth science applications" (UCAR n.d.). A key component is the definition of architecture for integrating disparate modelling systems, as well as providing support for new, framework-compliant models. A core principle of the ESMF framework is the deconstruction of complex models into small components defined by standards such that they can be quickly and easily assembled in different ways to create new models. One of the key purposes of ESMF is interagency collaboration. The framework streamlines and simplifies dialogue and model/code sharing between analysts and modellers across a wide range of US government agencies. The end result is much more comprehensive model views of climate impacts. However, ESMF is primarily US-centric and is focused on sharing of models and underlying code, not data and workflows.

The Climate Collaboratorium is a project of the Massachusetts Institute of Technology (MIT) Center for Collective Intelligence in the Sloan School of Management. It aims at leveraging new information technology and social media to bring together large numbers of like-minded yet geographically and socially dispersed individuals to collaborate on issues surrounding the global climate change debate. Using what they term *collective intelligence*, Malone and Klein (2007) hope that this framework will "focus ... on a possible use of such a system with a particularly high social return: drawing on the best human and computational resources available to develop government policies about climate change" (Malone and Klein 2007). The Climate Collaboratorium project hopes to show that adopting a framework that is decentralized yet carefully managed can be an effective method to approach large, resource-intensive problems such as global climate change (Malone 2009).

Planet Action is a not-for-profit collaborative initiative launched in June 2007 by Spot Image. Its purpose is to encourage the earth observation industry and professional GIS communities to address climate change by supporting projects that investigate and assess climate change impacts on five focus areas: human dimensions and habitation, drought and water resources, vegetation and ecosystems, oceans, and ice and snow cover. By assisting and funding the projects that will support understanding and action on environmental impacts, the Planet Action initiative hopes to strengthen international cooperation and response to climate change problems. Planet Action projects must meet certain criteria before qualifying for support. Each project must assess climate change related impacts and issues and initiate a course of action. Accepted projects must also incorporate good scientific understanding, resources, and methods. The Planet Action project is a good example of private industry leaders coming together to address global issues usually associated with the governments and NGOs.

Citizen scientists have a strong interest in some facets of science, but they pursue this interest outside of mainstream academic, research, and industrial organizations. These self-directed individuals might very well be using their own resources, working in their garages to develop "the next big thing". But more often, they are networked, working together with fellow citizen scientists. And this is where they become a powerful force to be taken seriously within the scientific community. Scientists and "professionals doing science" are often the ones organizing these citizen science networks; they realize the great value a group of eager volunteers can bring to a project. A good, although somewhat controversial (depending on your belief in intelligent extraterrestrial life), example of a group of volunteers organized to work on a huge project is SETI@home, where volunteers download special software that utilizes the "idle time" on their computers to scan through reams of radio telescope data and search for signs of extraterrestrial intelligence. If nothing else, with more than five million participants worldwide, this project has served as a model for bringing large number of volunteers together to work collectively on a massive project.

Closer home, CPDN and APS@home are two distributed computing projects with an earth science spin. CPDN is investigating how small changes affect climate models. APS@home is looking at atmospheric components of climate change. Although public participation in both CPDN and APS@home is not nearly at the same scale as SETI@home, the potential is certainly there.

There is a huge opportunity for citizen scientists to leverage geospatial technologies in their quest for knowledge, entertainment, and contribution to society. With the relatively recent arrival of powerful (and free) geospatial visualization tools such as Google Earth, ArcGIS Explorer, and NASA World Wind, it is now easier than ever before for citizen scientists to have some fun with maps while making a potentially important scientific contribution. A large number of volunteers are being inducted to work on geospatial problems such as climate change, as evident from the CPDN and APS@home examples. What is needed next is something at a much larger scale, where not just physical, but also biological, social, cultural, economic, and political data and models are integrated to give a more accurate depiction of the complexities inherent in the anthropogenic earth. First, we need to create an environment that successfully brings together a plethora of data sources and modelling systems—a noble vision for GIS technology, but not something to be tackled by citizen scientists. Once the data and technology are in place and a clear framework is established, there comes the opportunity to organize a large group of volunteers who would do the work of tackling this challenge.

Imagine a framework where tens or even hundreds of thousands of citizen scientists log on to a website and download geospatial data sets and work task lists; then using a focused desktop geospatial application, they run different analysis and modelling scenarios as defined in the task list and upload the results of their analysis back on to the website for consolidation. If properly structured and managed, such a project could significantly advance our understanding of the planet. At this scale, it would be difficult, if not impossible, to work if citizen scientists do not participate. They are out there, anxious to help… just waiting for us to create such a framework (Artz 2009c).

In his book *Design with Nature* (1969), Ian McHarg advocated a framework for design that helps humans achieve synergy with nature. Design and planning, which take into consideration both environmental and social issues, help us to ensure that our resources are used appropriately and responsibly, to help us move towards a better future (McHarg 1969). McHarg's pioneering work not only had a fundamental influence on the up-and-coming field of environmental planning, but also simultaneously strengthened the core concepts of the young field of GIS as well.

In his book, McHarg laid out a process by which "environmental data could be incorporated into the planning process." Rejecting the view of a future modelled after some idyllic environmental past, he adopted the view that we should be using our dominance of the earth systems to help in the evolution of the natural world and make it better, rather than "conquer" it. Powerful anthropogenic influence over earth systems represents not just a huge challenge, but an equally huge opportunity—not humans *versus* nature, but humans *with* nature. "While traditional ecological research selected environments with a minimum human influence, I selected arenas of human dominance," McHarg said. Today, more than ever before, it is important to recognize the overwhelming impact of humans on the environment, and that the massive human impacts on the earth are a reality that is not going away, and that we stand at the crossroads. Thus, our challenge is providing designers, engineers, planners, and others with a set of tools and a framework for designing and managing the anthropogenic earth (Dangermond 2009).

Carl Steinitz, an urban planner at Harvard University, conceived many of the early ideas on the application of GIS for landscape analysis and urban planning. He developed a model of landscape change that enables design of alternative futures. Those alternative designs can then be evaluated in terms of their impact on the natural environment as well as their utility to the human population, and the alternative future that achieves the best balance can then be selected for implementation. Landscape architecture and urban and regional planning have taught us to analyse alternative development ideas in a broad environmental context, and GIS tools were a natural outgrowth of this technique; but to date, these design concepts are yet to be fully applied at a truly global scale to help us understand and respond to climate change and other challenges (Steinitz 2000).

### Three Views of a GIS

A geographic information system (GIS) is most often associated with a map. A map, however, is only one way you can work with geographic data in a GIS, and only one type of product is generated by a GIS. A GIS can provide a great deal more problem-solving capabilities than using a simple mapping programme or adding data to an online mapping tool (creating a "mash-up").

A GIS can be viewed in three different ways.

- *The database view* A GIS is a unique kind of database called a geographic database (geodatabase). It is an "information system for geography." Fundamentally, a GIS is based on a structured database that describes the world in geographic terms.
- *The map view* A GIS is a set of intelligent maps and other views that show features and present various relationships on the earth's surface. Maps of the underlying geographic information can be constructed and used as "windows into the database" to support queries, analysis, and editing of the information.
- *The model view* A GIS is a set of information transformation tools that derive new geographic data sets from the existing data sets. These geoprocessing functions take information from the existing data sets, apply analytic functions, and write results into new derived data sets.

By combining data and applying some analytic rules, a model can be created, which helps answer the question that has been posed.

## DISCUSSION

"The long-standing vision of an integrated platform for comprehensive spatial analysis and modelling that has the capabilities for geographic data management, analysis, and visualization of modern GIS is at last beginning to become a reality" (Maguire 2005). A GIS-based framework for earth systems modelling and global design would be an effective platform for visualization and exploration of earth systems data sets. Such a framework would also enable GIS and model integration at a level supporting the most comprehensive spatial analysis and modelling of complex earth systems, and provide sophisticated tools necessary to develop and evaluate a variety of alternative futures in support of global design.

### Visualization and Exploration

Visualization of spatial data is a powerful method for presenting the results of scientific inquiry and analysis. It is also increasingly valuable as a tool used in forming questions at the front end of the process and in the development of the hypotheses that the spatial scientists will be testing. It is even more valuable in an earth systems modelling and global design environment, where the analyst compares and contrasts elements across multiple domains and data sets.

The human eyes and brain represent some of the most sophisticated trend and pattern detection tools available to us. Exploratory visualization of spatial data by scientists and analysts often leads to observations such as: "That looks interesting, I think I see a pattern here...". Identification of spatial patterns with the eye and brain leads to the hypothesis, and spatial analysis combined with the scientific method then confirms or denies the validity of the hypothesis.

The keys to visualization are the availability of data and tools. Combining comprehensive data sets for the study area with simple, easy-to-use visualization tools results in the most productive environment for visualization and exploration. Simplifying the visualization and exploration activity lets scientists and analysts identify patterns and trends more quickly, and helps them formulate the hypotheses they will be testing in the earth systems modelling activity.

### Earth Systems Modelling

GIS itself is an incredibly valuable tool for spatial analysis and modelling, but many stand-alone models are available, which are designed for highly specialized, domain-specific modelling, analysis, and problem solving. Most domain-specific models are not yet, and probably never will be, fully implemented in a GIS framework; however, the spatial display, analysis, and data management capabilities of GIS can still be utilized to greatly streamline almost any modelling workflow (Artz 1989). Figure 1 shows an example of how GIS provides a comprehensive framework for a highway noise modelling workflow.

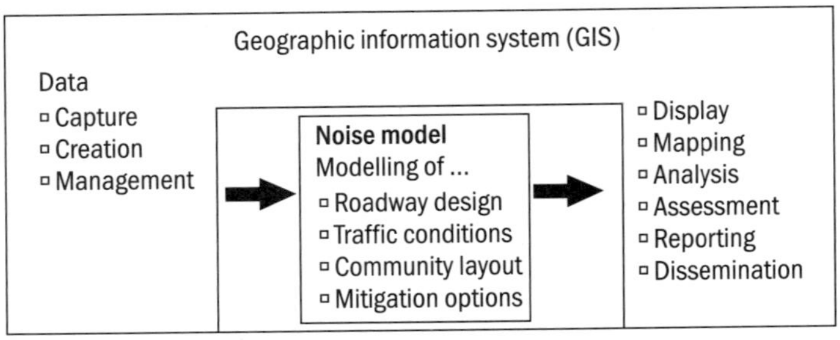


**Figure 1** Using GIS for noise model workflow management and post-modelling support

**Source** Artz (2008b)

Figure 2 shows a more comprehensive modelling framework, wherein GIS is used for workflow management and post-modelling support for multiple domain-specific models; in addition, outputs from multiple models can be compared, analysed, and modelled within the GIS system itself. Such a GIS-based framework offers a comprehensive environment for modelling across complex earth systems.

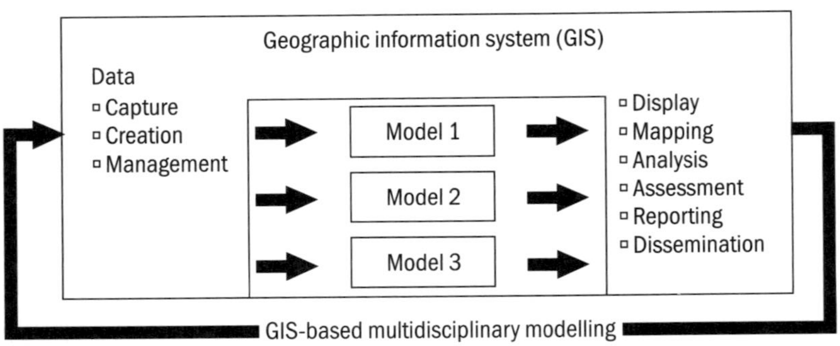


**Figure 2** A GIS-based framework integrating multiple domain-specific models and performing multidisciplinary modelling

Creating a framework that successfully brings together and manages a plethora of data sources and modelling systems to tackle the most pressing environmental issues of our time is surely a huge challenge, but it is a challenge that GIS is well prepared to accept. Once the data and technology framework is in place and a clear workflow is established, the challenge lies in organizing a large group of people to do the work of modelling multiple complex scenarios in order to identify the best out of the possible design futures for the planet (Artz 2009c)

## Global Design

"Man may perish by his own explosive and insidious inventions. For an adjustment to them he leaves himself precious little time, and progressively less as his technological wizardry runs wild and rushes on. If he is to survive at all, it cannot be through slow adjustment. *It will have to be through design* more subtly considered and circumspect, through more cautious planning in advance" (Neutra 1954). "Design is the first signal of human intention … What is our intention as a species and how do we go about thinking about that?" (McDonough 2009).

Michael Batty states that "(a) narrow but suitable definition of design as it pertains to geographic systems … is the process of generating physical artefacts which meet 'agreed' human (social and economic) goals pertaining to specific points or periods in time and space." He goes on to describe the design process as increasingly evolutionary, where human-initiated or -influenced systems grow and evolve in a manner and fashion similar to biological systems. This growth and evolution are in response to ever-changing environments and their associated assemblages of constraints (Batty 2008).

Reasoned design and management in the age of the anthropogenic earth are our moral imperatives, but the biggest obstacle to our success is that we are not currently inclined to work, or even think, in this way. Allenby notes that "(w)e lack solid data and analytical frameworks to make assertions about the costs, benefits, and normative assessments of different ... practices" (Allenby 2005a). GIS and the emerging field of geodesign are critical to the success of approaches such as ESEM and other logical and rational models for dealing with the environmental and planning problems of ours and future generations.

Design considering place was at the core of McHarg's beliefs, and it is the basis for current research and development efforts in the emerging field of geodesign. Geodesign borrows concepts from landscape architecture, environmental studies, geography, planning, sustainability, and integrative studies. Much like GIS and environmental planning preceding its application, geodesign takes an interdisciplinary, synergistic approach to solving critical problems and optimizing location, orientation, and features of projects, both local and global in scale.

Geodesign may be a new term for some people, but GIS and design together have a long history. To a certain extent, geodesign is already practised today by numerous GIS practitioners in fields like urban and regional planning and environmental management. And whether they realize it or not, over the last 40 years, many GIS professionals have been involved in geodesign projects, primarily in the fields of environmental, regional, and urban planning. But geodesign makes this easier by being an integral part of the workflow, both shortening the cycle time of the design process and improving the quality of the results (Artz 2009d). Due to the efforts of Steinitz, the geodesign framework also lets us design and test various alternatives, helping us make the most educated and informed decisions on the best possible future (Steinitz 2000).

When we talk of designing our future, we believe that combining the wealth of data on our world with sophisticated analysis and management tools is the prescription for understanding and shaping the future of our planet—an anthropogenic future wherein advances in human society, technology, and so on are designed in close collaboration with nature, resulting in the best of possible future worlds. It is a huge task and demands a delicate balance, but GIS and geodesign tools make this challenge seem less daunting. Frankly, we have no other choice (Dangermond 2009).

## RECOMMENDATIONS

Longely, Goodchild, Maguire, *et al.* (2001) very clearly emphasized, "Empirical analysis of the real world can be a messy and analytically inconvenient business and so the science of real-world application is the difficult kind—it can rarely refer to apparently universal truths, such as laws of gravity. Rather it is founded on a search for understanding and predictive power in a world where human factors interact with those relating to the physical environment. Social science and natural science are part of what we embrace." In continuation, Masser (2007) emphasized "...it must be recognised that the full potential of GIS can be realized only if the necessary spatial data infrastructures (SDIs) are in place at the local, national, and transnational levels."

Moving forward, we outline some actionable data; educational, ethical, organizational, modelling; and design recommendations for leveraging geospatial technologies in support of climate change studies and managing global change.

### Create a Data Directory and Share Best Practices

Data is at the core of all spatial analysis, and it is the cornerstone of a global modelling and design framework.

- *Share best practices on spatial data collection and management* Much can be shared from a comprehensive review of spatial data collection and management techniques, particularly those related to previous excursions in the realm of climate change modelling and model integration.
- *Create a spatial data directory for climate change studies* GEOSS is taking great strides in creating a global inventory and directory

of remotely sensed data. We propose a similar infrastructure (or even the extension of GEOSS) for historic and derived/modelled data. The aim here is not to build a massive online database of freely available earth systems data, but to create a loosely coupled, decentralized network of data types and locations for use in earth systems modelling and global design projects.

- *Inventory of spatial data models useful in climate change studies* Spatial data models describe the representation of, and relationship between, real world geographic features abstracted as geometric elements. Creating data models across multiple earth systems domains in a manner that is consistent, or at least interoperable, is a challenge. A thorough inventory of climate change related spatial data models and sharing of best practices on interoperability will be of tremendous value to climate change researchers as they move forward in analysing impacts and alternative futures at a comprehensive, global scale.

## Educate the Scientific Community about the Benefits of a GIS-based Framework

Many in the academic, scientific, and commercial communities may lack the conceptual mindset needed for properly modelling and managing the world in the 21st century and beyond.

- *Earth systems thinking* A deep appreciation of earth systems philosophies—especially an understanding of anthropogenic impacts on, and dominance of, these systems—is necessary to ensure the success of an earth systems modelling and global design framework.
- *Geospatial technologies as the primary platform for modelling and managing earth systems* Geospatial technologies create an interoperable environment, offering the best platform for open and integrative studies on climate inputs and outputs across multiple disciplines. We need to promote a framework and workflows based on sharing and collaboration, as opposed to the "stovepipe" isolation and division typical of many disciplines.

## Establish Ethical Guidelines for the Application of Geospatial Technology to Global Design

As the application of GIS technology as a cornerstone of a climate change modelling and global design framework becomes increasingly more obvious, we need a set of guidelines that integrate the nature-centric ethics of climate change with other ethical systems focused on ensuring the health and safety of humanity. The following proposed guidelines should be considered a starting point for initiating a long, meaningful dialogue among the geospatial and scientific communities.

- *Minimize harm to both humans and natural systems* Harmlessness, or what some ethicists refer to as the "harm test", is the common denominator across multiple ethical systems, demanding that actions preclude any type of harm to human beings. At the other end of the spectrum are the "deep greens", the most extreme of whom may even value nature and the health of natural systems more than human life. Even if the goal is environmental sustainability while still meeting the needs of humanity, it is difficult to imagine actions that in all cases would be mutually beneficial to both nature and humans. But one major benefit of a GIS-based framework for climate change modelling and global design is that it provides the power to analyse multiple scenarios and design the best of possible futures, supporting the principle that actions should minimize harm to both humans and natural systems.
- *Complete and comprehensive* Climate change issues are complex and demand an all-inclusive and wide-ranging examination. All relevant aspects of physical, biological, and social systems need to be considered and represented. Multiple data layers describing the intricacies of each relevant system, an array of sophisticated domain-specific models, and use of a GIS-based framework to hold everything together and evaluate multiple future options will ensure that the analysis is both complete and comprehensive.
- *Transparent and defensible* Climate change issues are often ideologically and politically charged, and global design recommendations resulting from spatial analysis and modelling could have massive positive or negative consequences on human health and welfare as well as on earth systems. All research, analysis, and modelling need to be objective, in keeping with high standards of scientific integrity and following the scientific method. Results—and the methods used to obtain them—need to be clearly communicated (GISCI n.d. b). Efforts should be taken to ensure that the process is transparent, and the resulting recommendations are defensible.
- *Adaptable and reversible* Adaptive systems feature feedback loops for sensing and responding to environmental changes (Kosko 1993). Allenby states that "...because the potential outcomes of each action become clear only as the system adjusts, the engineer is behaving unethically if she or he does not monitor the results of the chosen action, and modify them accordingly" (Allenby 2005b). And while the goal of using a GIS-based framework for earth systems modelling and

global design is to get the most comprehensive and complete picture as is possible with current technology, it would be irresponsible to assume that we know everything and thus inextricably lock into an option from which there is no turning back, should things go awry (Artz 2009e).

**What is Design?**

What exactly is *design*? That depends on who you ask. A formal definition might explain how design is the process of planning or sketching the structure or form of something. Other definitions of design are more esoteric, yet much more descriptive. Charles Eames called design "a plan for arranging elements in such a way as to best accomplish a particular purpose." Glen Lowery described design as "a bridge between the abstraction of research and the tangible requirement of real life." And Gavin Heaton defined design simply as "applied imagination".

## Inventory Models and Share Best Practices

Though most domain-specific models are not implemented in a GIS framework, they are instrumental to the success of an earth systems modelling and global design framework, and a complete accounting of available models, how they work, and how they integrate with GIS is essential.

- *Knowledge base of earth systems models* In support of earth systems modelling and global design framework, establish an open, wiki-like knowledge base, cataloguing environmental and earth systems models at all scales.
- *Share best practices on the use of models in a GIS framework* The model knowledge base should include best practices information on how each model integrates with GIS, in terms of data models, data management, display and visualization, and analysis.

## Create an Open Organizational Framework

Organization is critical to the success of a GIS-based framework for addressing climate modelling and design issues. First, we need to create an environment that successfully brings together a plethora of data sources and modelling systems. Once the data and technology are in place, there comes the exercise of organizing the resources needed to face this daunting, but not impossible, challenge.

- *Loose-knit, self-sustaining organizational structure* Although not foreseen as a project that is "owned" by a single entity, such a framework is unlikely to be successful without one or more key sponsors from the academic, government, non-profit, and/or commercial domains stepping forward. Ideally, with support from such sponsors, an online, wiki-like framework could be established to build, store, and share data inventory, modelling best practices, projects and work assignments, and design guidelines and results of various analyses.

- *Framework to manage workflow, projects, and tasks* We imagine a framework wherein large numbers of volunteers could log on to a website to download data sets and work tasks, use free geospatial tools to run different analysis and modelling scenarios as defined in the task list, and then upload the results of analyses back to the website. At this scale, such a project would be difficult, if not impossible, to execute without the participation of volunteers. In fact, merging of online simulation-type and massively multiplayer online role playing game (MMORPG)-type interactive gaming environments with earth systems modelling and global design objectives, if engineered, implemented, and marketed correctly, could potentially garner the participation of huge numbers of people and computers in the largest project of its kind ever.

## Establish Guidelines for Global Design

A geodesign framework will provide a robust set of tools for design professionals to support the design and evaluation of alternative futures for our earth and its systems. And the need for such tools has never been greater. We live in an ever more complex world, where our impact on the natural environment is massive and can no longer be ignored. People are beginning to recognize the importance Neutra placed on the inseparable relationship between humans and nature and also realize McHarg's vision of design with nature, and they want to act. Matt Ball notes that "(t)here is now a growing interest in combining design functionality with the broader geographical context that geospatial tools offer in order to engage more deeply in land-use planning" (Ball 2008). A geodesign framework will finally help us to meet McHarg's challenge, letting us truly "design with nature".

- *Geodesign as a field of study* To what extent are the fundamental spatial concepts associated with GIS relevant in design? To what extent can the fundamental spatial concepts of design be addressed with GIS? Is it possible to devise a curriculum to develop spatial thinking in both GIS and design? To formulate answers to such questions, a specialist meeting on spatial concepts in GIS and design was held in December 2008, in Santa Barbara, California. The purpose of the meeting was to discuss the potential for integrating design more fully into GIS, as well as developing curriculum in spatial thinking. Further discussion is needed to fully develop these concepts and build a curriculum around them.

- *Differentiate between unconscious and conscious global design* Until recently, development projects and other programmes and policies affecting the environment have been mostly short-sighted, project-based, and exclusive. We need to focus on longer-term issues that are global in nature and inclusive of multiple factors. The difference here is *intent*. Once we recognize our environmental imperative to carefully and thoughtfully manage our planet for ensuring the well-being of all components of earth systems and tackle the ethical issues of geoengineering, we can move away from accidental or poorly planned geoengineering into an era of conscious geodesign. Conscious geodesign lets us responsibly control the fate not just of the human race, but of the entire planet and all of its interdependent systems.
- *Robust design tools for GIS environments* Imagine if your initial design concept, scribbled on the back of a cocktail napkin, was backed by the full power of GIS—the sketch goes into the database, becoming a layer that can be compared to all the other layers in the database. Having such a "back of the napkin" design sketches available for immediate analysis and feedback should be a primary area of research and development over the coming years for GIS technology developers. The experience GIS developers have gained with computer-aided design (CAD) integration and sketching tools has led to an appreciation of the power that could be derived by organizing drawing tools, symbology, data models, and process models into one integrated framework for carrying out geodesign.
- *GIS as a foundational design system* Integration of design tools with existing GIS functionality is important, but it is only the first step. Ultimately, our vision is to expand the utility of GIS to the point that it becomes a foundational design system. As we come to grips with our overwhelming impact on the natural world, we are increasingly realizing our inextricable link to nature. And with that, of course, comes an enormous responsibility—a responsibility made all the more gargantuan by the fact that we still have a long way to go towards fully understanding the dynamics of the various systems and developing a robust suite of comprehensive models and other tools to support these activities. As Neutra did with architecture in the 1950s, we need to advance a framework for design and planning, which not just incorporates but also embraces technology, science, and ultimately nature in a system that helps us design and choose the best alternative futures (Artz 2009f).

Only by careful observation of the data, application of scientific principles, and using GIS and other technologies do we have any hope of truly understanding the stressors and impacts on the incredibly complex systems that comprise our anthropogenic earth. A GIS-based framework for earth systems modelling and global design offers the best chance at gaining a true, scientific understanding of earth systems and for taking thoughtful, informed design decisions, along with proposing alternatives that allow human beings and nature to coexist more harmoniously.

## References


Allenby B. 2005a. **Biomass management systems**. In *Reconstructing Earth*. Washington, DC: Island Press

Allenby B. 2005b. **Micro and macro ethics for an anthropogenic earth**. *Professional Ethics Report* **XVIII**(2)

Allenby B. 2008. **The metaphysics of the anthropogenic earth part I: Integrative cognitivism**. Working Paper. Details available at <http://www.lincolncenter.asu.edu/files/documents/centerprg_pacing_paper_metaphysics.pdf>

Artz M. 1989. **CAD and GIS applications in highway noise modeling**. Internoise 89. Proceedings of the 1989 International Conference on Noise Control Engineering: Engineering for Environmental Noise Control. December 1989

Artz M. 2008a. **Science and our changing climate: GIS creates a framework for research and modeling**. In *GIS for Climate Change*. Redlands, California: Environmental Systems Research Institute, Inc. Details available at <http://www.esri.com/library/bestpractices/climate-change.pdf>

Artz M. 2008b. **Using GIS in a modeling workflow**. Details available at <http://gisandscience.com/2008/12/21/using-gis-in-a-modeling-workflow/>

Artz M. 2009a. **Climate change science, GIS, and whole earth systems**. Details available at <http://gisandscience.com/2009/04/29/climate-change-science-gis-and-whole-earth-systems/>

Artz M. 2009b. **Towards a GIS-based framework for climate change studies**. Details available at <http://gisandscience.com/2009/05/19/towards-a-gis-based-framework-for-climate-change-studies/>

Artz M. 2009c. **Geospatial technology and the citizen scientist**. Details available at <http://gisandscience.com/2009/01/07/geospatial-technology-and-the-citizen-scientist/>

Artz M. 2009d. ***GIS Enters the Design Space***. *ArcWatch*. Redlands, California: Environmental Systems Research institute, Inc.

Artz M. 2009e. **Applying GIS technology to global design: ethical considerations**. Details available at <http://gisandscience.com/2009/05/19/towards-a-gis-based-framework-for-climate-change-studies/>

Artz M. 2009f. **GIS and global design**. Details available at <http://gisandscience.com/2009/06/04/gis-and-global-design/>

Asimov I. 1942. **Runaround**. In *I, Robot*. New York: Doubleday and Company

Asimov I. 1990. ***Robot Visions***. Penguin Books Ltd

Ball M. 2008. **What is spatial design, and how can it be applied to sustainability?** *V1 Magazine*

Batty M. 2005. **Socioeconomic applications**. In *GIS, Spatial Analysis, and Modeling,* pp. 147–149, edited by D J Maguire, M Batty, and M F Goodchild. Redlands, California: ESRI Press

Batty M. 2008. **Design as a growth process represented through GIS**. 2008 Specialist Meeting—Spatial Concepts in GIS and Design. Santa Barbara, California, December 2008

Brautigan R. 1967. ***All Watched Over by Machines of Loving Grace***. San Francisco, California: The Communication Company

Dangermond J. 2005. ***GIS Helping Manage Our World***. *ArcNews*. Redlands, California: Environmental Systems Research Institute, Inc.

Dangermond J. 2009. ***GIS: designing our future***. *ArcNews*. Redlands, California: Environmental Systems Research Institute, Inc.

Ellis E C and Ramankutty N. 2007. **Anthropogenic biomes: a framework for earth science and ecology in the 21st century**. American Geophysical Union Fall Meeting, 10–14 December 2007, San Francisco, California. Details available at <http://ecotope.org/people/ellis/presentations/ellis_agu_2007_12_13_for_web.ppt>

GISCI (GIS Certification Institute). n.d. a. ***A GIS Code of Ethics***. GISCI. Details available at <http://www.gisci.org/code_of_ethics.aspx>

GISCI (GIS Certification Institute). n.d. b. **Rules of conduct for certified GIS professionals (GISPs)**. GISCI. Details available at <http://www.gisci.org/Ethics_and_Conduct/rules_of_conduct.aspx>

Goodchild M F. 2005. **GIS and modeling overview**. In *GIS, Spatial Analysis, and Modeling,* pp. 1–18, edited by D J Maguire, M Batty, and M F Goodchild. Redlands, California: ESRI Press

Greenfield A. 2004. **All watched over by machines of loving grace: some ethical guidelines for user experience in ubiquitous-computing settings**. *Boxes and Arrows*. Details available at <http://www.boxesandarrows.com/view/all_watched_over_by_machines_of_loving_grace_some_ethical_guidelines_for_user_experience_in_ubiquitous_computing_settings_1>

Kelly K. 2000. ***All Species Inventory***. Whole Earth

Kosko B. 1993. ***Fuzzy Thinking: the new science of fuzzy logic***. London: Harper Collins

Longely P A, Goodchild M F, Maguire D J, Rhind D W. 2001. ***Geographic Information Systems and Science***. John Wiley & Sons, Ltd

Maguire D J. 2005. **Towards a GIS platform for spatial analysis and modeling**. In *GIS, Spatial Analysis, and Modeling,* pp. 19–39, edited by D J Maguire, M Batty, and M F Goodchild. Redlands, California: ESRI Press

Maidment D R. 2005. **Hydrologic modeling**. In *GIS, Spatial Analysis, and Modeling,* pp. 319–332, edited by D J Maguire, M Batty, and M F Goodchild. Redlands, California: ESRI Press

Malone T W. 2009. **Can collective intelligence save the planet?** Details available at <http://www.climatebiz.com/blog/2009/05/05/can-collective-intelligence-save-planet>

Malone T W and Klein M. 2007. **Harnessing collective intelligence to address global climate change**. *Innovations* **2**(3): 15–26. Details available at <http://www.mitpressjournals.org/doi/abs/10.1162/itgg.2007.2.3.15>

Masser I. 2007. ***Building European SDIs***. Redlands, California: ESRI Press

McDonough W. 2009. **Greener by Design 2009** [Keynote]. San Francisco, California

McHarg I. 1969. ***Design with Nature***. The American Museum of Natural History

Neutra R. 1954. ***Survival through Design***. New York: Oxford University Press

NGAC (National Geospatial Advisory Committee). 2009. ***The Changing Geospatial Landscape***. NGAC

Steinitz C. 2000. **A framework for PLF**. In *Landscape Futures*. Institute for Bioregional Resource Management

UCAR (University Corporation for Atmospheric Research). n.d. About ESMF. Details available at <http://www.esmf.ucar.edu/about_us/index.shtml>

Wegner M. 2005. **Urban land-use transportation models**. In *GIS, Spatial Analysis, and Modeling*, pp. 203–220, edited by D J Maguire, M Batty, and M F Goodchild. Redlands, California: ESRI Press

# Geoinformatics for Climate Change Research

**PK Joshi[*], NG Baruah[#], and TP Singh[†]**
[*]*TERI University, New Delhi, India*
[#]*Department of Geography, University of Cambridge, UK*
[†]*Symbiosis Institute of Geoinformatics, Pune, India*
*E-mail: pkjoshi@teri.res.in; ngb26@cam.ac.uk; tarunsingh@rediffmail.com*

## INTRODUCTION

The earth's environment has always been subject to change. There has been no variation in this trend. However, the pace of this change has varied. The manner and the pace of global environment change and its implications on humankind form one of the most important research bases today. While considering the phenomenon at the human time-scale, changes in weather and vegetation are most noticeable. Human beings are accustomed to daily, seasonal, and inter-annual changes in climatic conditions. Increasing evidence suggests that average climatic conditions, measured over extended time periods, are also changing, over and above the natural variation observed on decadal time-scales. The causes of this climate change are increasingly well understood and have been attributed to anthropogenic activities (McMichael, Campbell-Lendrum, Kovats, *et al.* 2004; IPCC 2007).

To be certain that climatic changes are occurring, a baseline and repeated observations are needed. This requires access to long-term and global data, which is acquired systematically and is calibrated enough to be fully comparable and capable of measuring long-term changes. Since the time immemorial, means of observed environmental processes are very diverse and range from weather probes to sediment fossils. However, Earth Observation Satellites (EOS) are particularly useful in the study of climatic change processes. These provide one of the most systematic ways for collecting data worldwide, in a fully comparable and repeatable way. For this reason, since the past few decades, EOS have found a varied and significant use in global change studies, from weather surveys and forecast to vegetation, ice, and water monitoring (Chuvieco 2008). Also, since the abrupt climate change has been a consequence of human choices, we need spatial modelling tools that represent the world as seen and modified by human beings. This is where geographic information science comes into picture. In order to create such climate change models that include humans, we need geographic information system (GIS).

As there is paucity of detailed experimental evidence on the climate itself and a lack of direct evidence on the extent of anthropogenic damage to the climate, it is extremely important to try to construct models in which we make use of environmental parameters that can be quantified so that the predicted effects of climate can be calculated (Oreskes 2004). Construction of models apart, their usage is wrought with difficulties. This is because the climate is a very complicated system, and extrapolating its behaviour over a long time period is a challenge.

## CLIMATE CHANGE RESEARCH

In 1970, a study called the "Study of critical environmental problems" was carried out by the Massachusetts Institute of Technology (MIT). Its unequivocal conclusion was that global warming was a serious possibility. The session advocated for the aggressive expansion of climate change research. By 1972, more climate change research was undertaken at the Stockholm Conference. Special emphasis was laid on developing numerical general circulation models (GCMs), which successfully simulate the present-day climate and the behaviour of the atmosphere in long time runs. It assumed wide acceptance as the first step in attempts to predict what happens to the atmosphere as a result of anthropogenic activities, considered to be the ultimate goal of the Global Atmospheric Research Programme (GARP), the predecessor to the International Geosphere–Biosphere Programme (IGBP), which was conceived in the early 1980s and is now being implemented. Slowly but surely, climate change made way to the research portfolio of the World Meteorological Organization (WMO), and by the end of the 1970s, climate change was declared an aid issue by the United Nations Environment Programme (UNEP). The First World Climate Conference in 1979 referred to climate change as the process of carrying out an uncontrolled experiment in the earth's atmosphere and called for preventative action. In 1985, a small group of environmental scientists and research managers working on energy and climate met at a conference called "The Second Joint UNEP/ICSU/WMO International Assessment of the Role of Carbon Dioxide and other Greenhouse Gases in Climate Variations and Associated Impacts". They set up the Advisory Group on Greenhouse Gases (AGGG), which successfully launched global warming into world politics and concluded that "it is now believed that in the first half of the next century, a rise of global mean temperature could occur, which is greater than any change in the man's history". Furthermore, they recommended that science-based emission targets should be worked out to limit the rate of change in global mean temperature to a maximum of 0.1°C (WMO 1986, McMichael, Campbell-Lendrum, Kovats, *et al.* 2004).

By 1985, therefore, science leaders formed a network and initiated a dialogue with "policy-makers" on climate change. "The AGGG had succeeded in taking the policy debate into the world of politics, but its institutional base proved too weak to keep the issue out of the hands of big, accountable institutions like governments and the WMO, which had the ability to fund the global change experiment" (Boehmer-Christiansen 1994a, b, 2003). Climate change is, therefore, best understood as a response to the AGGG and its policy advocacy. The Intergovernmental Panel on Climate Change (IPCC) is largely a representative of governmental scientific institutions, especially meteorological offices and atmospheric physics. The challenges being posed by global warming were disseminated by the IPCC in a highly condensed form through the summaries of the "policy-makers" and "scientific findings". Informal links between the IPCC and the IGBP are strong (IGBP 1992). There are programmes to strengthen the scientific capacity of developing countries, and they constitute a common interest with the IPCC and the WMO. The implication is that space science and GIS are deeply involved in these global research efforts (Boehmer-Christiansen 1994a, 1994b, 2003).

The *Third Assessment Report* of the IPCC (IPCC 2001) estimates that globally, the average land and sea surface temperature has increased by 0.6±0.2°C since the mid-19th century, with much of the change occurring since 1976. Warming has been observed in all parts of the world, and precipitation patterns have also changed. In places where precipitation has increased, a disproportionate increase in the frequency of the heaviest precipitation events has been observed (Mason, Goddard, Graham, *et al.* 1999). The level of climatic change that has occurred so far has already had demonstrable effects on a wide variety of ecosystems (Walther, Post, Convey, *et al.* 2002). Climate model simulations have further estimated the likely impacts of past, present, and future greenhouse gas (GHG) emissions on climate. These models are based on heat-retaining properties of gases released into the atmosphere from anthropogenic and natural sources, including the measured climatic effects of various kinds of natural phenomena (Desonie 2007). The models referred by the IPCC have been validated by "back-casting", that is, testing their ability to explain climate variations that have already occurred in the past. The models are able to give approximations of the past patterns only when anthropogenic emissions of non-GHG air pollutants are included along with natural phenomena (IPCC 2001; McMichael, Campbell-Lendrum, Kovats, *et al.* 2004). This brings forward the following facts: (1) the models represent a good approximation of the climate system; (2) natural variations are important contributors to climatic variations; and (3) anthropogenic GHG emissions are an important contributor to climate patterns, and are likely to remain so in the future (IPCC 2001). Considering different combinations of economic development scenarios and model parameterizations, the IPCC concluded that in the absence of specific policy action to reduce GHG emissions, global temperatures would rise between 1.4°C and 5.8°C, from 1990 to 2100 (McMichael, Campbell-Lendrum, Kovats, *et al.* 2004).

### Causes of Climate Change

Throughout the history of the earth, the climate has changed globally and locally during nearly all time periods. Climate change has many natural causes, such as variations in the amount of solar radiation that reaches the earth's system, the position of the earth relative to the sun, and even whether the continents are together or apart. Other factors that are important over shorter time periods are volcanic eruptions and asteroid impacts.

But of late, focus has shifted on anthropogenic causes of climate change since the change is more abrupt and is upsetting the balance of the biosphere. Human activities increase the concentration of greenhouse gas (GHG) in the atmosphere. The dominant anthropogenic source of GHGs is the burning of fossil fuels and biomass (including living matter). Other human activities that alter the earth's climate are deforestation, urbanization, land use land cover and forest change, and air pollution.

**Source** Desonie (2007)

## NEED FOR GEOSPATIAL DATA

Climate change research places heavy demand on reliable climate information, which includes spatially distributed information on selected climate elements. The global and regional climate models are continuously improving and need reliable validation data as well as training data for downscaling scenarios to the spatio-temporal scale desired by the scientists studying the impacts of climate change on nature and society. The need for spatially distributed information on weather and climate has increased similarly over the same period, in keeping with the increasing focus on global warming and climate change by both scientists and policy-makers (Goodchild 1999). Geoinformatics—an umbrella term covering a wide gamut of GIS and related technologies (remote sensing, global positioning system [GPS], image processing, high bandwidth communications)—is widely believed to play an increasingly important role in climate change research. Global climate change and shrinking resources have developed our understanding about the dependence on the dynamic and complex earth system and its environment. Geoinformatics enables to take advantage of the unprecedented amount of digital geodatabase and computational capability through electronic networks. It facilitates multi/interdisciplinary research and capacity to understand the earth as a dynamic and complex system to manage natural resources in a sustainable way (www.earthbyte.org).

## GEOINFORMATICS: AN INTEGRATED SPATIAL RESEARCH TOOLSET

As described above, geoinformatics is a very broad term, which is used to describe activities through GIS and is a synergy of multiple disciplines (remote sensing, GPS, communication technology) (Goodchild 1999). *GIS* is a specialized software package for capturing, storing, manipulating, and displaying geographic information in digital form; *remote sensing* involves the application of digital technologies for imaging the earth from space using satellites or from high altitude using aircraft; in *GPS,* wherein satellite signals are used to determine earth coordinates such as latitude and longitude by direct measurement and communication technology, is a process to store or deliver information using electronic and graphic means, including encoding, transmitting, receiving, storing, retrieving, and decoding. Database management systems (DBMS) is an integral part of geoinformatics, which consists of a collection of interrelated data and a set of programmes to store, access, update, and manage data. Finally, in the electronic communication technologies, the Internet (web services) has notably become an integral part due to its increasing importance in the dissemination of geographic data. In the past, the disciplines of geodesy, photogrammetry, remote sensing, surveying, and cartography worked in an independent fashion to provide hard copy (paper print) maps. In the newer concept of geoinformatics, GPS, remote sensing, digital photogrammetry, and GIS databases are not only being produced as printable maps but are also being displayed on a computer screen and analysed in an interdisciplinary manner. This technology has further extended to mobiles, web, i-pads, and other means of electronic communication.

Remote sensing and GIS are two disparate branches. Remote sensing from aircraft and satellites is the science and art of collecting data on objects located on the earth's surface by using sensors mounted on observation aircraft and satellites, and interpreting the data to provide useful information (Lillesand and Kiefer 2003). On the other hand, GIS comprises a computer system capable of holding digital maps (raster and vector database) called geographical layers or spatial information and their associated statistical and/or descriptive data called attribute data. What geoinformatics has enabled is a framework, wherein remote sensing can be effectively used in combination with GIS. GIS facilitates the analysis of multiple layers of data provided by remote sensing and also allows for statistical analysis of numerous factors without compromising their spatial representation. Data technology suited for the dissemination of large quantities of data on the environment, like the climate system, and the development of related computational mechanism that contribute to the genesis of global information systems form the key ingredient of geoinformatics (Goodchild 1999).

Another most important component to accelerate the success and advancement of geoinformatics is the information and communication technology (ICT). According to Croswell (2005), five key trends have exerted profound influence on the ascent of geoinformatics, namely, pervasive high-performance computing, digital connectivity, geographic data capture and compilation, geographic data management and visualization, and standards and open system. The Internet has been a driver for the overall IT development, forging a trend in digital connectivity. With the development of Internet GIS or web GIS, GIS usage has increased, helping users to access geoinformation at low cost in client–server environment. Thus, geoinformatics may be envisaged as a toolset for assembling, storing, analysing, and simulating environmental information on different scales with respect to time and space.

## CONFRONTING CLIMATE CHANGE
### THE GEOINFORMATICS WAY

Climatological phenomena, such as climate change, are naturally spatially variable, and hence, geoinformatics represents a practical solution to the management of vast spatial climate data sets, which will be derived and applied. Basically, geoinformatics can be thought of as having triple roles in climate change studies—measurement, modelling, and visualization. The measurement of climate change entails derivation of extensive climatic data sets, which may in itself act as evidence for climate change and may also be further used for the derivation of subsequent tertiary outputs. The remote sensing component of geoinformatics is instrumental in the derivation of climatic data sets. The derived climate data may be subjected to analysis or more specifically, modelling in a GIS environment to give tertiary outputs. Finally, the visualization of outputs in a proper perspective, like developing future climate change scenarios, is perhaps the most powerful contribution of geoinformatics to climate change studies. Each of these roles is analysed in detail in the following sections.

### Measurement of Climate Change

In simple terms, this stage is that of data assimilation. The measurement of climate change can be achieved only if the baseline climate data is available for reference. For example, recession of glaciers as evidence of climate change can only be established if any retreat from a certain baseline can be shown. Thus, the derivation of spatial climate data sets is paramount to the measurement of climate change. Over the period, data collection has certainly improved to the point where concerns have shifted from the availability to accessibility and discovery of newer and high-resolution data sources. It is in this context that remote sensing has been envisaged as an invaluable tool. Moreover, data assimilation for models that tackle something as complex as climate change (implying distributed global data management) requires the understanding of the collaborative monitoring of the earth. There is a definite need for technologies and services that allow combining data from various sources to improve capacity of measuring the state of the planet and developing policy-oriented decision tools (Dyras, Dobesch, Gruter, *et al.* 2005). The scaffold provided by GIS is perfect for multi-source data integration. For climate change research, data assimilation basically concerns collecting climatic data for the derivation of baseline climatology and interpolation of climate data for obtaining comprehensive climate data sets.

### *Derivation of Baseline Climatology*

Remote sensing enables the acquisition of large-scale comprehensive climate data sets. An example may be the use of digital terrain models (DTM) for providing baseline climate data sets. Synthetic aperture radar data is used these days to derive these DTMs, replacing land surveying techniques, which were used commonly. The UK LANDMAP Project has used radar data for deriving baseline climate change data (Anon 2001). The generation of comprehensive raster climate data sets is also possible through satellite imagery apart from polar remote sensing. Schadlich, Gottsche, and Olesen (2001) used DTM in conjunction with brightness temperatures derived from thermal infrared channel of METEOSAT to produce land surface temperature maps. The surface ultraviolet maps of Europe and evapotranspiration maps of Tunisia have been developed by Verdebout (2000) and El Garouani, Boussema, and Ennabli (2000), respectively, using similar approaches. In many cases, remote sensing can be used in stand-alone mode without integration with GIS. A classic example of this is reflected in satellite rainfall climatology (Levizzani, Schmetz, Lutz, *et al.* 2001). However, the need for GIS becomes acute while overlaying data from multiple sources. For example, Nichol (1995) combined geographical vector data with surface temperatures derived from Landsat thermal imagery to explore the spatial characteristics of forest canopy temperatures with elevation and land use in Singapore (Chapman and Thornes 2006). Suga, Takeuchi, Kimura, *et al.* (1995) used Landsat data in combination with National Oceanic and Atmospheric Administration (NOAA)/advanced very high resolution radiometer (AVHRR) data to monitor sea surface temperature change in the Sea of Japan, using GIS.

In GIS, climate data can be represented in a variety of ways. If we consider the rainfall data, we observe that it is in point data form when collected from meteorological stations, while it assumes raster form when procured from radar satellite data. The representation of rainfall data in isolines is also not rare. However, climate data is usually in point form. The extrapolation of point climate data across a spatial domain thus attains significance. As mentioned earlier, DTMs can be used to derive baseline climatology. This is done through the process of extrapolation of point data. As estimates from this method are good enough to avoid the consultation of weather records, it has become indispensable in deriving baseline climatology in high and inaccessible mountainous regions, which suffer from the lack of frequent observations due to a sparse network of weather stations. Apart from DTMs, triangular irregular networks (TIN), which are also derived from point climate data, have also been popularly used in microclimate studies (Bernherdson 1992).

The Parameter-elevation Regressions on Independent Slopes Model (PRISM) project in the US is a perfect example of the use of geoinformatics for deriving high-quality spatial climate data sets across the world. PRISM is a knowledge-based empirical two-layered climate analysis system that generates GIS-compatible estimates of climate variables accrued from a variety of sources, including point climate data, DTMs, and other spatial data sets. By using a coordinated set of rules, decisions, and calculations, extensive gridded estimates of precipitation and temperature can be made with respect to different climate regimes (distance, elevation, atmospheric boundary layer, hill slope orientation, and proximity to the coastline) of each DTM facet (Daly, Taylor, Gibson, *et al.* 2000). Using GIS, Agnew and Palutikof (2000) developed a multiple regression model using a 1 km resolution DTM to analyse the variation in geographical and terrain parameters (latitude, altitude, continentality, slope, aspect, and ratio of land to sea) (Chapman and Thornes 2006).

## Gridded Data Sets and Climate Indices

Several European meteorological services publish monthly climatological bulletins, including climate anomaly maps developed by applying geographic information system (GIS) techniques. Sequences of climate anomaly maps might eventually form the basis for long-term gridded data sets. Such data sets provide several options for deriving climatological information for users. Examples of such information are percentile maps for a given month, time series for a given area or pixel point, and other types of statistics based on the gridded data sets. In Norway, such a monthly gridded data set was established covering the period 1901 to present. The elements included are temperature and precipitation.

*Box Contd...*

*Box Contd...*

One of the most common applications of climate data is to derive climate indices for different environmental and societal applications. Gridded climate data sets offer a unique possibility to provide such information as spatially continuous fields based on the gridded data sets themselves, or in combination with other types of data.

**Source** Dyras, Dobesch, Gruter, *et al.* (2005)

The high resolution DTMs allow studying impacts of terrain on climate at meso- or micro-scales. For example, Dubayah (1994) determined solar radiation topoclimatologies and modelled the sensitivity and response of mountainous terrain (del Barrio, Alvera, Puigdefabregas, *et al.* 1997). For example, De Azevedo, Pereira, and Itier (1999) also used advective and radiative sub-models in GIS to extrapolate climate data obtained at sea-level across the entire small volcanic islands (Chapman and Thornes 2006). These studies have great implications for island nations, such as Maldives, in their bid for survival from extinction in the face of sea-level rise brought about by climate change.

While baseline climatology is important, the issue of the use of geostatistical techniques for interpolation of point station data gains precedence when dealing with more spatially comprehensive climate data sets (Tveito, Førland, Alexandersson, *et al.* 2001). Chapman and Thornes (2006) have discussed climate data interpolation with variety of examples. Splining and kriging are common interpolation techniques used in climate change studies. Lennon and Turner (1995) used thin-plate splines determined from a DTM to model the climatic distribution of temperature in the UK. This method has been found to produce accurate results in regions where stations are sparse. For example, Fleming, Chapin, Cramer, *et al.* (2000) used splining in the derivation of an Alaskan baseline climatology from a sparse meteorological station record. Unlike splining, which is a deterministic spatial regression technique, kriging is stochastic, requiring some user input. In contrast to the "rubber sheeting" employed by splining to fit a surface to irregular spaced data, kriging utilizes variograms to analyse the tail structure of variables in given directions to construct a map of spatial variation from a small sample of data points. Kriging was used by Hudson and Wackernagel (1994) to map mean temperatures in Scotland and by Jeffrey, Carter, Moodie, *et al.* (2001) to interpolate daily and monthly rainfall across Australia. With regard to accuracy, splining has been found to perform better than kriging in a pioneering study by Jarvis and Stuart (2001).

Neural networks represent another set of method for data interpolation. They were used by Antonic, Krizan, Marki, *et al.* (2001) to derive an empirical model for seven climatic variables, explaining 98% of the variation in climatic parameters. A feed-forward back-propagation neural network was also used (Rigol, Jarvis, Stuart 2001), which considers both trend and spatial associations of climatic variables. The application of neural networks has the advantage that guiding variables do not need to be linearly related to the interpolation data (Antonic, Krizan, Marki, *et al.* 2001).

### *Evidence of Climate Change*

The retreat of glaciers and sea-level rise represent the most obvious and widely accepted manifestations of climate change evidence. In 1995, the IPCC stated that between one-third and one-half of the existing mountain glaciers could disappear over the next 100 years due to climate change (Gelbspan 1997). Glaciers are regarded as key indicators of climate change because of their proximity to melting conditions, manifestation of which is seen in the form of glacial retreat and thinning (Haeberli and Beniston 1998). Since glaciers respond to regional climate, glacier monitoring provides important evidence of climate change (Dyurgerov and Meier 2000). Time series of glacier changes shows much spatial and temporal variability, and thus, geoinformatics may be used to monitor glacier changes and provide conclusive evidence of climate change (Dyurgerov and Meier 2000).

The state of a glacier is dynamic and subject to constant climate impacts. The monitoring of glacier evolution thus becomes important, which is made possible through multi-temporal satellite image analysis. The use of satellites to monitor glacier area and measure changes in area or position of glacier termini can also reveal the effects of climate change (Li, Sun, and Zheng 1998). Remote sensing and GIS techniques help in establishing the extent of snow and glacier cover for the assessment of possible impacts of global climate change (Krishna 1996, 2005). Glacial retreat or surge can be detected by monitoring and mapping the tongue of glaciers and outlet of glaciers (Aniya and Enomoto 1986). Satellite data from 1984 to 2000 was used by Hall, Klaus, Bindschadler, *et al.* (2001) to assess and measure recession of Pasterze glacier in the Austrian Alps. All of the above constitute important examples of direct evidence of climate change, evaluated using geoinformatics.

## Modelling of Climate Change—Integrated Assessments

While investigators all over the world are conducting research on problems related to climate change, the largest efforts are being directed towards understanding the physical science aspect of this global phenomenon. A lot of emphasis is also being laid on the ecosystem—both managed and unmanaged—responses to climate change. In the last few years, studies concerning the socio-economic implications of climate change have gathered unprecedented pace. There are studies of strategies and technologies that might be used to reduce the future emissions of GHGs. Finally, there are studies on how society and environment might adapt in response to climate change. Integrating the knowledge gathered from natural, physical, and social scientific endeavours is essential for gaining a more complete view of global climate change issues (Martens and Rotmans 2003). Integrated assessments of climate change provide a medium for attaining this goal.

Integrated assessments of climate change are typically carried out using computer models that integrate knowledge from various disciplines. These computer models link representations of the different natural and social science components, including some of the known feedbacks (Rotmans and Dowlatabadi 1998). According to Kandlikar (1999), the objective of integrated assessment is to put the various pieces of the climate problem together and look carefully at the big picture so as to (1) maintain a proper perspective about the problem, since climate change will occur in the presence of ongoing environmental and social changes; (2) develop the understanding necessary to support informed decision-making by the many actors and institutions across the world; and (3) ensure that the type and mix of research undertaken will be as useful as possible to these decisions-makers in both the near- and long-term.

## Climate Change Assessment Tools and Techniques

There is a wide gamut of approaches or methods that can be used in the assessment of climate change impacts, which include quantitative and predictive models, empirical studies, expert judgment, and experimentation.

*Quantitative models* are those models, the variables of which can be expressed in quantitative terms, and the results are very precise. However, the accuracy of such models is dependent on the underlying assumptions. There are broadly three kinds of quantitative models—biophysical, socio-economic, and integrated system models. *Biophysical models* are used to analyse the physical interactions between climatic variables and exposure unit. These are basically of two types: empirical statistical models and process-based models. *Empirical statistical models* are based on the quantitative relationship between climatic variables and a particular system under current climate, while *process-based models* are based on physical laws, principles of the workings of a system, or assumedly universal processes. The estimation of second-order effects, such as those on the production of cereals, on water supply, or on industrial output or socio-economic interaction may require a different group of models, called *economic models*. It is important to distinguish three types of economic models, which depend on the scale of analysis. At the finest scale, *individual models* describe the behaviour of a single actor, such as a farmer or a firm. *Sector (or meso-level) models* describe the behaviour in a sector of the economy. *Economy wide (or macro-level) models* describe economic activity across all sectors. It has been seen that even the individual or sector models fail to address the interactive complexity of climate impact

*Box Contd...*

Box Contd...

phenomena needed for the formulation of tailor-made adaptation and mitigation strategies. For this reason, research efforts are being focused on the development of *integrated systems models.* Empirical assessments of the interactions of climate and society and natural systems can be of great value in anticipating future impacts. This is commonly achieved through regional analogue methods, in which spatio-temporal variations can substitute for future changes. Three kinds of analogues useful for this are historical events, historical trends, and present climate conditions. Empirical studies can also be combined effectively with quantitative model scenarios through remote sensing and GIS.

Each of these approaches has its own advantages and disadvantages, and a good strategy may be to use a combination of approaches in different parts of the assessment or at different stages of the analysis. Many of these models are causal in nature, wherein one or more climatic variables are changed and the consequences predicted and measured. It should be noted that models do not do everything. Models that address only one sector or aspect of a system may simulate that sector or aspect very well. But they may not include interactions from important related sectors or other aspects of the system. Thus, the choice of a model should depend in part on what questions are being asked, that is, whether they are broad or specific questions.

**Source** Tol, Weyant, Davidson, *et al.* (1996)

Integrated assessment models that attempt to provide meaningful input for policy purposes are difficult to construct because of the inherent complexity embedded in the temporal, spatial, and socio-political scales of the climate change problem coupled with the vast uncertainties in relevant scientific and socio-economic variables, which act as impediments in the attempt of integrated assessment models to provide meaningful input to policy. In their current form, integrated assessment models serve an important purpose because they provide an organizing framework for interdisciplinary research and bring together communities from different disciplines working on different aspects of this global problem (Kandlikar and Risbey 1996).

### *Role of GIS in Integrated Assessment Models*

It has been emphasized in the previous section that sectoral studies in climate change can only advance the understanding of the particular components of the climate system. Usually, they are wrought with simplified assumptions, and of late, there has been unequivocal support for a more holistic approach presented by integrated assessment models. An important potential application of GIS systems is in understanding the interactions between sectoral impacts and in providing a more integrated view of resulting societal outcomes, which sectoral studies alone cannot hope to provide (Kandlikar and Risbey 1996).

The contribution of GIS, in a functional sense, for integrated assessments of climate change can be realized broadly in three different ways (Kandlikar 1999), namely, (1) provision of geographical and regional specificity; (2) improvement of data quality and model validation; and (3) capture and display of data along with linkages to analytical and decision support models. Integrated assessment models represent processes at a very high level of aggregation (Nordhaus 1992). More specific results can be obtained when incorporating specific details of a region such as its natural resource characteristics and socio-economic variables. Coupling model experiments with the regional socio-economic data will help illustrate the impacts of climate change on a region which may be particularly important for local communities and institutions (NGOs, regional planning commissions) most vulnerable to ecosystem changes. Thus, the benefits of using GIS to provide regional and geographic specificity to integrated assessment models are critical. From the perspective of scientists and researchers, the integration of ecosystem models and GIS with statistical software will help in managing and evaluating model experiments (Parks 1993). By making specific regional outcomes transparent, GIS systems can aid in incorporating and highlighting the concerns of different stakeholders on a regional or global basis (Mounsey 1991).

The lack of a systematic grounding in historical data has been a limiting factor in attempts to validate integrated assessment models against environmental and societal data (Risbey, Kandlikar, and Patwardhan 1996). One basic problem in integrated assessment models has been the inconsistency of data—the coupling of varied data, often at different scales and in possessing different attributes. For example, while environmental data exhibits continuous spatial variation, social phenomena tend to be more spatially discrete. This discrepancy arises out of differences in purpose and scale, and underlying assumptions in the data collection process. Researchers in the GIS community have encountered these issues in handling and integrating inherent differences in data types and have learnt to address many of them. Besides the need to integrate different kinds of data, integrated assessment models of climate change require the need to represent uncertainty; incorporate dynamics; and address scaling and aggregation issues, because impacts occur in a location-specific fashion, while quantitative estimates of impacts may be required at multiple scales to account for the interests of the various stakeholder groups (Brown 1995). GIS provides a robust structure for addressing these issues of integrated assessment models tackling climate change.

According to Kandlikar (1999), two significant advantages can be derived when coupling GIS with regional and global integrated assessment climate change models. First, GIS can play a role in data integration and quality assurance in the validation of integrated assessment models, and second,

GIS models have a significant role to play in the interpretation of model outcomes and implications for policy. In particular, integrated assessment models, socio-economic and environmental information, and GIS provide a medium for their meaningful synthesis. The uncertainties in the climate problem are so large that any single set of model results has very little heuristic value. However, if the models are exercised over a range of possible outcomes, and GIS is used to envision the possible interactions and outcomes, a more holistic picture of climate change impacts might emerge (Martens and Rotmans 2003). In this regard, it may be useful to combine the multiple outputs from GIS with multi-attribute decision frameworks, which provide a scheme for integrating the utility of different outcomes based on the values and beliefs of different stakeholders.

### *Visualization of Climate Change*

Scientists have generally accepted evidence of climate change and its impacts on society and the environment, but public awareness and policy changes are lagging behind (Sheppard 2005). Communication of the urgency of climate change is a difficult task, and Tickell (2002) has warned that it may require a calamity to induce people and governments to take necessary radical actions. The question researchers often ask is: "What may be the most effective ways to stimulate climate change awareness in society before crises occur?" (Sheppard 2005)

The answer may lie in visual imagery or animated video clips such as model-driven visualizations of climate change (Sheppard 2005). And for this, geoinformatics (geo-visualization) provides the best platform to bring all kinds of spatial and non-spatial data with a muti-media option for browsing various scenarios under multi-criteria evaluation and decision-making. The cognitive advantages of visual representation of data and information over written or verbal information have been widely documented (Tufte 1990). The visual stimuli can trigger innate and instant reflexes and feelings, which can be persistent even in the face of new phenomenon and processes like climate change (Slovic, Finucane, Peters, *et al.* 2002). The other types of visualization tool used in scientific communication are charts, diagrams, maps, graphics, and 3D computer models (Cox 1990) but often are static in nature with limited connectivity with the spatial relationship and linkages. Climate change presents a tricky conundrum. The characteristics of climate change make it difficult to see directly. For example, $CO_2$ and even the carbon source itself are effectively invisible. Authentic source of $CH_4$ and $O_3$ generation and distribution is still a mystery and a daunting research question for many of the scientific communities. Moreover, abstraction, complexity, and scale are another set of challenges to communication of these (Nicholson-Cole 2005).

However, geoinformatics is now a well accepted and a potential tool to recognize quite tangible visual or landscape-related effects of climate change, such as sea-level rise, flooding or drought. In this regard, geo-visualization of landscapes—which is a visual simulation or landscape modelling, representing places of interest and on-the-ground conditions in 3D perspective—comes with fairly high realism (Sheppard and Salter 2004). It assumes a very important role as it is capable of showing past, present, and future worlds, as they would have been seen if the viewers were actually there. Geospatial tools and methods for landscape visualization (Appleton, Lovett, Sunnenberg, *et al.* 2002) are now being applied to depict alternative future scenarios. These are being used for visioning, education, research, public awareness, decision-making, and policy formulation to a great extent (Tress and Tress 2002; Steinitz, Rojo, Bassett, *et al.* 2003; Sheppard and Meitner 2005). For example, Cohen (1997) used GIS and remote sensing imagery with other information to communicate climate change scenarios to Canadian stakeholders. This information on the different scenarios influenced the vision of the future and potential interventions in policy debates. The generic alternative scenarios for future climate change and adaptation, as developed by the IPCC and other organizations, need to be downscaled to the local landscapes. Such information attributed with audio-visual units linked to local specific characterization could be collected in more convincing way (Nakicenovic and Swart 2000), and geoinformatics makes for such a mechanism.

## Climate Change Mitigation–Adaptation

Anthropogenic climate change is associated with various degrees of risks for human society and nature. The two fundamental options for reducing degree of these risks are mitigation of climate change and adaptation to climate change (Fussel 2007a). Both mitigation and adaptation have a spatial dimension. Geoinformatics serves as the tool that enables spatial planning, which is sometimes referred to by climate scientists as the "switchboard" for implementing mitigative and adaptive measures at the local and regional scales (Bulkeley 2006). Mitigation has traditionally received much greater attention in the climate change studies than adaptation, both scientifically and from a policy point of view. In most cases, mitigation strategies are formulated using information based on limited number of scientific disciplines and are thrust upon various policy domains (Roy 2010). A typical approach for mitigation strategies is through institutional arrangements, like financially supported technological development and innovation, or establishing cap-and-trade schemes. In this context, wherein problems are framed and governed by geoinformatics-enabled quantitative modelling, approaches are used to generate highly specialized and often complex knowledge (Pohl 2008; Biesbroek, Swart, van der Knaap 2009).

## Remote Sensing and International Agreements

The global impact of the Kyoto Protocol on technical and scientific issues of relevance to the remote sensing community is considerable and unprecedented. And now REDD/REDD plus has recognized remote sensing as one of the most vital tools for providing measurable, reportable, and verifiable (MRV – measuring, reporting, and verification) baseline estimate. The Protocol further relies on REDD/REDD plus for temporal monitoring and assessment. Remote sensing can play an important role in supporting the requirements of such agreements or protocols by providing systematic observations of land use land cover and forests, supporting establishment of baseline, detecting and quantifying rates of changes, quantifying above-ground biomass stocks and changes in these, and mapping and monitoring certain sources of anthropogenic methane ($CH_4$) and other greenhouse gas (GHG) emissions ($CO_2$, $NO_2$, HFCs, PFCs, SF6, $O_3$, ozone precursors, aerosols, and black carbon).

For establishing a C-stock baseline, optical remote sensing (Landsat TM/ETM+, IRS LISS III, SPOT) as well as non-optical remote sensors synthetic aperture radar (SAR) (Radarsat, Envisat and other) are most useful. Optical remote has been used in different parts of the world but has a limitation of saturation of values and is more related to surface reflectance value. However, non-optical sensors have the greatest potential for quantifying biomass as they can penetrate and provide details about trunk and branches (volume). At more local levels and for a restricted range of environments, airborne light detection and ranging (LIDAR), polarimetric SAR, and very high frequency (VHF) band radar have shown significant potential for biomass estimation.

For detecting anthropogenic sources of GHG, both space-borne optical and non-optical data could be used, but it is not much successful. A wide array of GHG sensors, measurement platforms, monitoring and inventorying systems, and inference methods is needed to meet basic measurement requirements. There are different sensors (like Agriflux, Ameriflux, LIBS and other) in place providing specific information but are limited to ground-based monitoring. Key areas of further research activity to solving the specific needs of the protocol include (1) fusion of optical and SAR data, (2) development of applications using SAR interferometry and polarimetry and low frequency radar (P band), (3) increased synergy between LIDAR and other sensors, (4) development of procedures for data mining, and (5) others exploring new bandwidths (3G or 4G) for data transfer and communication. To ensure the adoption of remote sensing technology in supporting international agreement, access to and greater affordability of remote sensing and other spatio-temporal data are essential. These agreements required ensuring the open exchange of remote sensing data and the provision of appropriate training in the use of technologies.

**Source** Rosenqvist, Milne, Lucas, *et al.* (2003)

In the climate change context, mitigation basically entails limiting global climate change by reducing the emissions of GHGs or enhancing their sinks (McCarthy, Canziani, Leary, *et al.* 2001). In this respect, geoinformatics, more specifically remote sensing technology, could be applied for supporting climate change mitigation measures in the provision of systematic observations of relevant land cover; supporting the establishment of a 1990 carbon stock baseline; detecting and spatially quantifying change in land cover; quantifying above-ground vegetation biomass stocks and associated changes therein; and mapping and monitoring certain sources of anthropogenic methane ($CH_4$) (Rosenqvist, Milne, Lucas, *et al.* 2003). Issues related to mitigation, in particular to afforestation–reforestation–deforestation (ARD) activities, will affect the work of the geospatial scientific community for years to come. Consequently, Rosenqvist, Milne, Lucas, *et al.* (2003) recommend that remote sensing research activities should focus and be aligned to fulfil the specific requirement posed by international agreements and frameworks for broad policy-making and governance. In the present context of climate change, resource depletion, and other factors, the needs relating to carbon accounting and an improved understanding of the terrestrial carbon budget are the major thrust areas (Biesbroek, Swart, van der Knaap 2009).

The climate policy development, scientific initiatives, and political attention were always focused on the mitigation of GHG emission and not on the adaptation strategies, which were more important to handle the changing world and climate (Biesbroek, Swart, van der Knaap 2009). However, the growing recognition that mitigation strategies alone will be insufficient to avoid serious climate change impacts further strengthened the call for adaptation strategies. As a result, scientists and policy-makers recognized that both mitigation and adaptation measures are needed to reduce the impacts of climate change (Klein and Huq 2007; Fussel 2007b; Harrould-Kolieb, Herr Unpublished). In the context of climate change, adaptation refers to the set of actions targeted at the vulnerable systems in response to actual or expected climate stimuli with the objective of moderating degradation and effects (McCarthy, Canziani, Leary, *et al.* 2001; Fussel 2007b). The role of geoinformatics becomes even more pronounced in the case of adaptation to climate change because of two reasons: (1) depending on the specific issue under consideration, adaptation to climate change may have close links with natural resource management, water management, disaster preparedness, urban planning, sustainable development, poverty reduction, and so on, the sectoral areas that are already served by geoinformatics and (2) convenient geographical scaling of adaptation, which means that adaptation can be implemented on a local or regional scale, and its efficacy is less dependent on the actions of others. In four prototypical stages of assessment in climate change adaptation expounded by Fussel and Klein (2006), the geoinformatics approach can be seen to be a natural fit: (1) climate impact assessments, which identify potential (biophysical) impacts of climate change; (2) first- and second-generation vulnerability assessments, which determine the different vulnerabilities of social groups or regions to climate change, considering their adaptive capacity; and (3) adaptation policy assessment, which recommends specific measures for reducing the vulnerability of social groups to climate change and variability.

According to Campbell (2006), spatial planning can be the strategic framework, wherein both mitigation and adaptation measures are positioned in the broader perspective of sustainable development. Not only the scientific community has identified the possibilities to facilitate adaptation and mitigation in spatial planning, but politicians are also aware that the implementation of measures, both adaptation and mitigation, takes place at the local or regional level. Spatial planning can fulfil an important role. The recently published national adaptation strategy of the Netherlands, titled "Make space for climate!", illustrates the growing awareness for spatial planning in climate change (VROM 2007). In spatial planning practice, the differences between time, space, and actors coalesce into one coherent multi-level governance approach, making geoinformatics suitable for an integrative approach to adaptation and mitigation.

## GEOINFORMATICS IN ACTION: SOME APPLICATIONS

The advantage of spatial climate data sets is that they can be compared in a GIS with dissimilar data sourced from many sources. Hence, GIS has enabled the study of the environmental impact of the variation of climate for many applications at a variety of scales. This section outlines some of the many applications of geoinformatics for climate change studies.

### Glaciers

In 1995, the IPCC stated that between one-third and one-half of existing mountain glaciers could disappear over the next 100 years due to the emission of GHGs in the earth's atmosphere (Gelbspan 1997). The International Commission on Glaciers (now the International Commission on Snow and Ice) was established in 1894 to recognize the importance of the relationship between glacier fluctuations and climate variations. "The slightest change in climate can dramatically affect the overall mass balance of glaciers" (Dyurgerov and Meier 2000). Thus, glaciers are key indicators of climate change because of their proximity to melting conditions, manifestation of which is seen in the form of glacial retreat and glacial thinning (Haeberli and Beniston 1998). Since glaciers respond to regional climate, glacier monitoring provides important information on climate variability and change. The best way to infer climate change from glaciers is through monitoring mass balance (Dyurgerov and Meier 2000).

Changes in glacier area and terminus positions, which can be easily extracted using geoinformatics, form two important indicators of glacier response to climate change (Barry 2006). The state of a glacier is dynamic and subject to constant climate impacts. The monitoring of glacier evolution thus becomes important, which is made possible through multi-temporal satellite imagery analysis. Kulkarni, Rathore, Mahajan, *et al.* (2005) performed a study on the retreat of one of the largest glaciers in the Himalayas using geoinformatics. Multi-temporal IRS satellite images were used as evidence of the retreat of Parbati glacier in the Himalayas— confirmed in the field by GPS observations—which were further used to model future changes in retreat using advanced GIS tools. Such studies are very important as the Himalayas have one of the largest concentrations of glaciers outside the Polar regions, and a significant number of its glaciers are shrinking due to climatic variations.

## Changes of Glacier Coverage

Mountain glaciers have declined worldwide as a result of recent global climate change. Recent glacier monitoring uses satellite images as part of a project of the international Global Land Ice Measurements from Space (GLIMS). Such satellite image based studies have permitted widespread, quantitative monitoring of glacier coverage and loss inventories. Narama, Shimamura, Nakayama, *et al.* (2006) used Corona (the US reconnaissance satellite operated from 1960 to 1972) images from 1971 and Landsat 7 Enhanced Thematic Mapper Plus (ETM+) image from 2002 to assess glacier variations in the western Terskey-Alatoo range (a part of the Tien Shan), in the North East of the Kyrgyz Republic. The area of 269 mountain glaciers measured (original combined area of 245 km$^2$) has decreased by 18.4 km$^2$ (8% loss). The results show individual glaciers covering < 1 km$^2$ account for 28% of the combined glacial coverage, but loss from these small glaciers represents 38% of the total decrease in area. The relative change of area is larger for small-scale glaciers, and its contribution to shrinkage of glacier area is high. This shrinkage has been linked to climate changes as recorded at local stations.

The glacier terminus line in 1979 was determined from aerial photographs. The debris-free area of the glacier was determined using the Landsat ETM+ ratio image of bands 3/5. The shrinkage of small-scale glaciers was even higher, a finding that suggests a future decline in water supply, because this area has many small-scale alpine glaciers. There is an obvious link between recent climate change and glacier shrinkage, and this should be assessed in other areas, using satellite images.

*Box Contd...*

*Box Contd...*

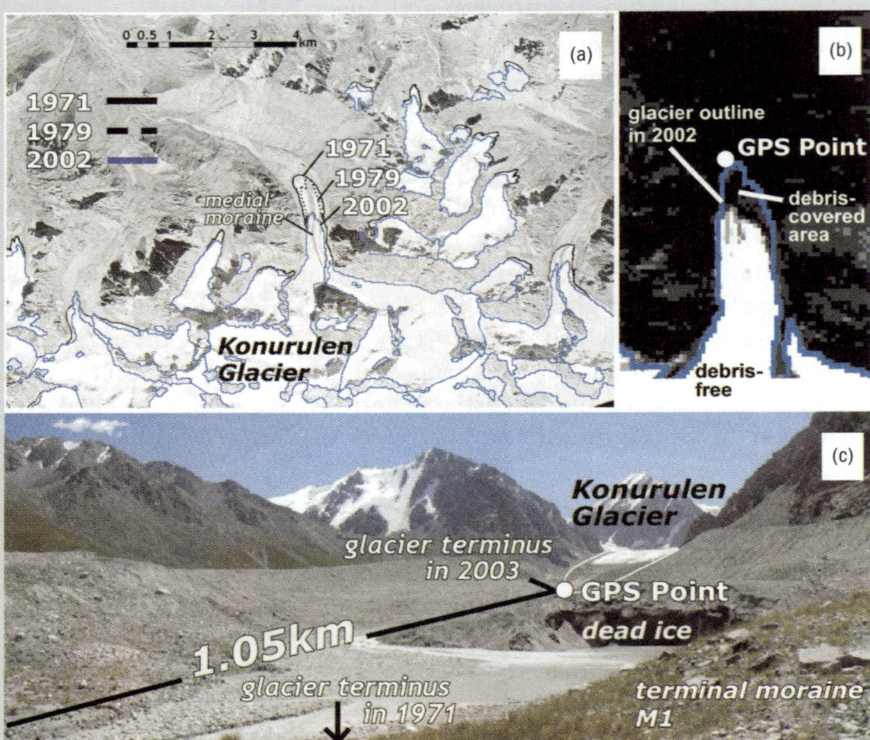


**Figure A** Konur-Olen glacier terminus area: (a) Corona image (glacier loss 0.22 km$^2$ during 1971–79, and 0.44 km$^2$, 1979–2002); (b) debris-free area of the glacier (area 0.12 km$^2$ and length 350 m); (c) dead ice remains

**Source** Narama, Shimamura, Nakayama, *et al.* (2006)

Kulkarni, Bhuguna, Rathor, *et al.* (2007) monitored 466 glaciers in Chenab, Parbati, and Baspa basins, from 1962. Expeditions were organized to identify and map the glacial terminus. The investigation has indicated overall reduction in glacier area from 2077 km$^2$ in 1962 to 1628 km$^2$ at present (21%, overall deglaciation). The number of glaciers has increased due to fragmentation and mean area of glacial extent has reduced from 1.4 km$^2$ to 0.32 km$^2$ between the 1962 and 2001. A total of 127 glaciarates and ice fields less than 1 km$^2$ have shown retreat of 38% from 1962, possibly due to small response time. The study concludes a combination of glacial fragmentation and higher retreat of small glaciers, and climate change is influencing the sustainability of Himalayan glaciers.

*Box Contd...*

*Box Contd...*

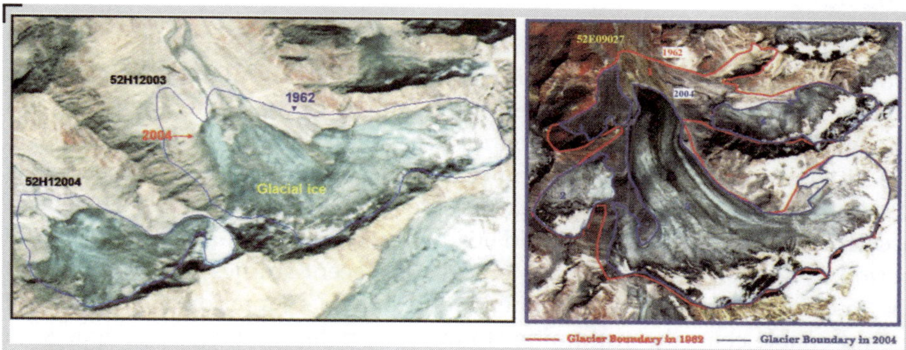


(a) (b)

**Figure B** (a) Resourcesat imagery of LISS IV showing glacial boundary of 1962 and 2004 (these are small mountain glaciers showing negligible accumulation area. These are close to the snow line at the end of ablation season and such glaciers are expected to experience terminal retreat). (b) This glacier split into four glaciers between 1962 and 2004 (however, areal extent is reduced from 7.0 to 5.3 km$^2$)

**Source** Courtesy of Prof. A V Kulkarni (http://dccc.iisc.ernet.in)

Identification of glacier changes is possible only when glacier inventory is available for comparison. In inaccessible areas, where glacier inventory was not complete, remote sensing has been a major tool for filling this void, thus enabling glacier studies in the context of climate change (Aniya, Sato, Naruse, *et al.* 1996). For vast regions like the Himalayas, assessment of glacier cover is crucial in serving as a model for deriving baseline inventory (Krishna 2005). Satellite-derived glacier inventory also aids in building future monitoring strategies (Paul, Kääb, and Haeberli 2007).

Future climate scenarios may be derived from GIS-based modelling, such as determination of surface energy balances of melting glaciers. Hall and Fagre (2003) used spatially explicit models to accomplish GIS-based simulations of climate-induced glacier change in the Glacier National Park for a time period ranging from 1850 to 2100. An approach to estimating sea-level change has been to use GPS to measure elastic deformations resulting from changes in the mass of glaciers (Wahr, DaZhong, and Trupin 1995). GPS data, in conjunction with ASTER DEMs and aerial photographs, has been found to be highly effective in monitoring ice elevation changes and glacier thinning (Rivera, Casassa, Bamber, *et al.* 2005). Continuous satellite monitoring of ice sheets has been deemed necessary for advancing the knowledge of ice dynamics to the point where it is possible to forecast future changes in ice sheets resulting from climatic changes. Proactive steps are required to combat varied glacier response to climate change. Projects like OMEGA—Operational Monitoring system for European Glacial Areas—which aim for continuous regional scale glacier monitoring using geoinformatics, hold the key for future glaciological studies (Pellikka 2003).

## Coastal Zones

One of the more certain consequences of global climate change is accelerated global sea-level rise, which will intensify the stress on many coastal zones. Fluctuations in sea levels often result in changes in coastline. The positional changes of coastline are expressed in terms of shoreline advance or retreat, the latter being the most common result when there is a rise in sea levels (Paskoff 1998). According to climate change research, sea-level rise has been accelerating since 1870 and could lead to a rise of 280–340 mm over the period 1990–2100 (Church and White 2006). Recent studies have focused on the response of coastal environments and on the adaptation measures that would enable coastal communities to mitigate potential negative impacts on natural and man-made systems (Walsh, Betts, Church, *et al.* 2004).

Apart from the risk of coastal flooding, coastal zones also provide a highly significant resource for biodiversity, featuring a wide variety of habitats and key species, and many areas are, therefore, designated as protected conservation sites. Sea-level rise and current coastal management practices threaten the integrity of many coastal sites (Brown 2006). In this context, coastal zone management (CZM) needs to aim at a long-term land use planning to ensure sustainability of resources while providing management flexibility for the future (Kay and Alder 2005; Jolicoeur and O'Carroll 2007). An integrated management to coastal zones might require data set on hydro/morpho-dynamic responses for intensive spatio-temporal modelling. Like any other modelling framework, CZM would be effective with a "bottom-up", which could ensure detailed knowledge of coastal processing and interactions with biotic and abiotic variables at different spatio-temporal scales. However, the coastal regions are always attributed with high degree to variability to calibrate, model, execute, and validate the predictive modelling. Such complicated and diverse landscapes could be taken care of in the GIS environment, which provides an effective framework to integrate and analyse environmental data sources (Argent 2004). These "top-down" approaches would allow evaluating policy interventions and also developing policy-required scenarios at various scales. This brings the scientific modeler and policy-makers together to evaluate and analyse the options at a more strategic scale (Brown 2006).

O'Regan (1996) has reviewed the use of GIS in CZM. Thumerer, Jones, and Brown (2000) used vector GIS for the identification of distinctive flood compartments to assess regional flood risk in eastern England. The grid-based (raster) GIS with an appropriate resolution are relatively efficient for evaluating spatial patterns in flooding. For instance, the Risk Assessment for Strategic Planning (RASP) model developed by Hall, Dawson, Sayers, *et al.* (2003) relates the frequency of flooding in a grid. In this model, each cell with a combined probability of defence types within that compartment failing at critical thresholds is used to generate an expected depth of water. Such models are used at an aggregated level for the national assessment of flood risk. Similarly Lee (2001) identified potential losses and gains in habitats as a result of continuing a similar management policy to the present under future sea-level rise. Nicholls and Wilson (2002) used a GIS-based analysis for a regional level study to explore links between flooding, biodiversity, and agriculture under different future scenarios of climate and socio-economic change.

## Rule-based GIS for Modelling Coastal Floodplains

Brown (2006) developed a large-scale, strategic-level method to understand the implications of climate change and sea level rise for floodplain areas. The rule-based approach allows the evaluation of multiple policy options, to provide relatively transparent and customized response to the end-users using various critical assumptions, and is better than the conventional modelling approach. Thus, it builds a bridge between the policy-makers and future coastland use managers (Figure A).

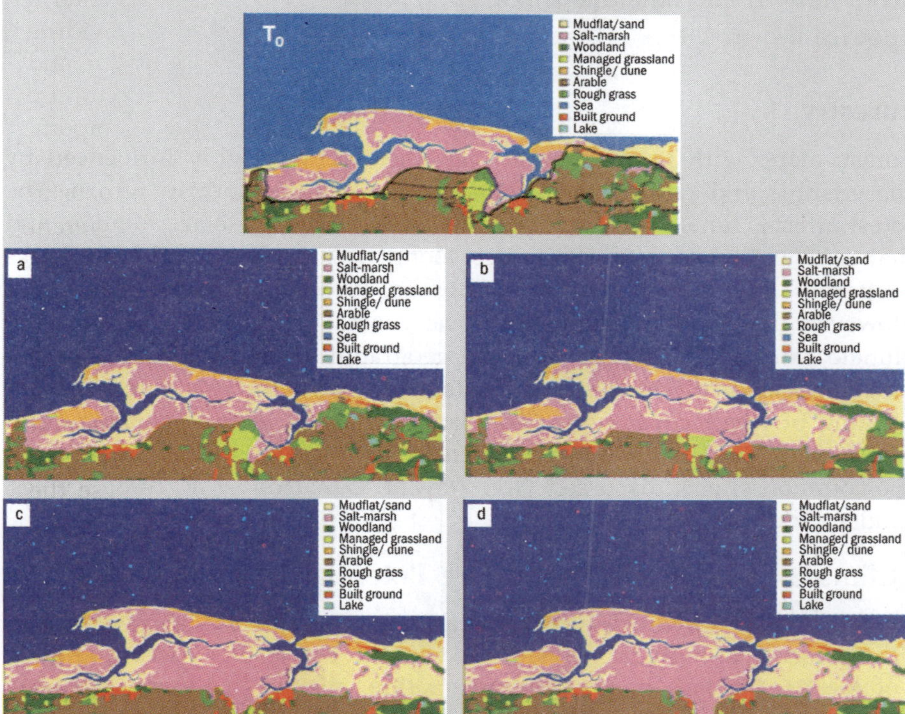


**Figure A** Present data land cover ($T_0$) and simulated future landscapes for the same part of the case study areas under the 2050 high sea-level rise scenario for Policy 1 (a), Policy 2 (b), Policy 3 (c), and Policy 4 (d)

*Box Contd...*

*Box Contd...*

To develop this, impacts from different management policies were explored through their influence on the patterns of flooding and consequent landscape change. A high resolution digital elevation model (DEM) derived from interferometric synthetic aperture radar (IfSAR) was used with tidal and flood frequency data. The method relies on a series of transition rules linking the tidal and flood frequency data to the land-use and land-cover change (LULCC). Using a case study example from the North Norfolk coast in eastern England, the authors demonstrate that the future sea-level rise has much broader policy implications for human and ecological landscape than just flood risk to property.

The rule-based approach incorporated four management policy situations—Policy 1 (maintain existing line), Policy 2 (small-scale realignment), Policy 3 (large-scale inland realignment), and Policy 4 (no defenses). The transition rules and thresholds used in GIS environment can rapidly enable a strategic assessment of future landscape change under different scenarios of climate change and management policy. The study identifies rule-based grid modelling having advantages over the conventional and other modelling approaches.

**Source** Brown (2006)

## Forestry

Forest, along with its resources, is another section highly influenced by the edaphic and climate factors and influences the both in return. The forest affects climate because of its capability to absorb solar radiation, and maintain water and nutrients cycles, along with surface roughness. These properties influence the climate from local to regional and global scales. However, the impact of climate on the vegetation is acute. A change in climate can be expected to shift the geographical distribution of species, its composition, productivity, and interaction with the associated species. Shifting of forest cover types, grasslands, changes in the phenology and distribution, and increase in the intensity of invasive species are some of the common effects discussed (Tol, Weyant, Davidson, *et al.* 1996; Scholes, Linder, Siddiqui 1998).

### Climate Change and Forest Fire Potential

Forest fires are a result of natural activity and are also influenced by the presence of human beings in the vicinity. The results are destruction of forests, increase in forest fragmentation, change in the disturbance regimes and biodiversity, habitat destruction of few species, and many others. Forest fires are expected to increase with the global warming and variations in the climatic parameters. One of the studies taken up by Stocks, Fosberg, Lynham, *et al.* (1998) projected the danger levels of forest fires in Canada and Russia using general circulation models (GCMs). Four

*Box Contd...*

*Box Contd...*

atmospheric GCMs (Canadian Climate Centre GCM, the United Kingdom Hadley Centre GCM, the Max Planck Institute for Meteorology [Germany] GCM and the National Centre for Atmospheric Research [USA]) were chosen for this study, as these models include relatively simple to very complex coupling of the atmosphere, biosphere, and oceans, along with differing parameterizations of sub-grid-scale processes, and provide a robust range of climate predictions.

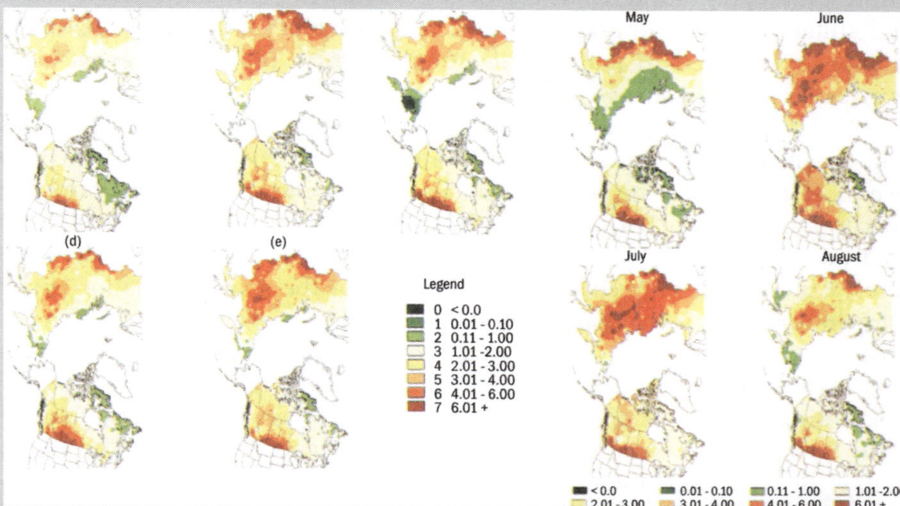


(a) (b)

**Figure A** (a) 2 × $CO_2$ climate scenario for Russia and Canada; (b) monthly analysis using the Canadian General Circulation Model

Temperature and precipitation anomalies between 1 × $CO_2$ and 2 × $CO_2$ runs were combined with baseline observed weather data for both countries for the period 1980–89. Forecast seasonal fire weather severity was similar for the four GCMs, indicating large increases in the aerial extent of extreme fire danger under a 2 × $CO_2$ climate scenario (Figure A[a]). When compared to the monthly 1980s baseline data, the monthly progression analysis (Figure A[b]) showed an earlier start to the fire and a significant increase in the area experiencing high to extremely high fire danger, particularly during June and July. Climate change forecasts have serious implications for forest fire management in both countries. The likely response will be a restructuring of protection priorities to support more intensive protection of smaller, high-value areas, and a return to natural fire regimes over larger areas, with resultant significant impacts on the carbon budget. A strong need is felt to continue modelling future climates, using higher-resolution models as they become available, so that

*Box Contd...*

Box Contd...

the future fire management planning can be accomplished in the most informed manner possible.

**Source** Stocks, Fosberg, Lynham, *et al.* (1998)

In forestry, deforestation and climate change are intrinsically related. One of the most common global warming mitigation strategies involves afforestation programmes. GIS can be used to develop climate zones to select site suitability for afforestation (Ellis, Nair, Linehan, *et al.* 2000). Climate change is expected to manifest itself in the form of extreme events like hurricanes and cyclones. Analysis of extreme events has been accomplished by Pleshikov, Ryzkova, Kaplunov, *et al.* (1998), who modelled the impact of a hurricane on pine stands in Central Siberia, while Foster and Boose (1992) used GIS to analyse the spatial distribution of wind damage and rank the susceptibility of particular forest types. By modelling the stagnation of cold air through a DTM, coupled with sky-view factor data related to the forest canopy, Blennow (1998) explained 89% of the spatial variation of air temperature, which also has implications for climate change. Another major area linked with climate change and forestry is biomass burning. GIS has been extensively used to model and monitor the spread of forest fires by associating climate data with remote sensing imagery (Pew and Larsen 2001; Sunar and Ozkan 2001; Vazquez and Moreno 2001; Zhu, Rong, and Sun, *et al.* 2000). Belda and Melia (2000) modelled forest recovery after fire using an integrated remote sensing and GIS framework. The distribution of the vegetation could also be predicted by analysing the influence of climatic parameter with the regeneration capability of forest types and areas.

## Agriculture

Providing sufficient food for the world's population is becoming more challenging, as agricultural land is progressively degraded through prolonged overuse. Climate change is only going to make matters worse. World food production varies by several percent from year to year, largely as a result of extreme weather conditions such as the El Niño phenomenon and inter-annual climatic variability in many regions. The overall goals of climate change research in agriculture are to define more clearly the ranges of possible impacts, to determine which locations and farming systems may be most vulnerable, to identify their critical thresholds, and to explore adaptation strategies (Tol, Weyant, Davidson, *et al.* 1996).

Climate change, being a spatio-temporal phenomenon, has a direct linkage with agricultural patterns and processes. Knowing and mapping the altered distribution of precipitation and temperature could gather place-specific and tailor-made responses of agriculture. Such attributes along with biophysical parameters are easier to collect using remote sensing data and model GIS. These data sets become handier to the decision- or policy-makers to devise alternatives for the area. Darwin (1995) combined a global computable general equilibrium (CGE) model with a GIS model to analyse the impact of climate change impacts on agriculture. This model accounts for interactions with both non-agricultural sectors and can be extrapolated to other global regions. In this, GIS component describes regional variables such as land, climate, water, agricultural suitability, and others. This approach also provides linkages with projected global climate change. GIS also has massive potential for agro-climatic modelling (Parry, Rosenzweig, Iglesias, *et al.* 1998, Jurisic, Hengl, Duvnjak, *et al.* 1999), which can be combined with climate change models to develop adaptation strategies. By incorporating temperature and aridity thresholds, agro-climatic models can be logically made species-

## Climate Change vis-à-vis Forest Habitat Types

Climate change is bound to affect the number and distribution of species across the world. The mountainous regions are more susceptible to such changes and impacts. Many studies have used remote sensing and geographic information system (GIS)-based modelling to predict the distribution of forest species and shifts as effects of climate change. Such groups of species might not shift rather changes will occur only in some selected species, which will compete with associates and form communities later.

Iverson, Prasad, Matthews, *et al.* (2008) modelled and mapped 134 tree species of eastern USA, individually, to show current and potential future distribution according to two global emission scenarios and three climate models (the Parallel Climate Model, the Hadley CM3 model, and the Geophysical Fluid Dynamics Laboratory Model). Of the 134 species, approximately 66 species would gain and 54 species would lose at least 10% of their suitable habitat under climate change.

The combined forest type maps show all the changes in the northern latitudes, especially the north-eastern US. The results show that only PCM lo (the least harsh scenario) retains any presence of suitable habitat for the spruce-fir type. Habitat for the white-red-jack pine type will be eliminated, whereas the aspen-birch type is severely curtailed in suitable habitat under low emissions, but is eliminated under higher emissions. The oak-hickory type might expand into the north, especially under higher emission scenarios. Though the maple-beech-birch is fairly stable under low emissions, its habitat might get reduced under higher emissions as oak-hickory and oak-pine expand. In this analysis, the southern pine types remain fairly stable even though individual southern pine habitat increases to the north for many pines. The results conclude that climate change would have large impacts on suitable habitat for tree species in the eastern US, especially under a high emissions trajectory (Figure A).

*Box Contd...*

*Box Contd...*

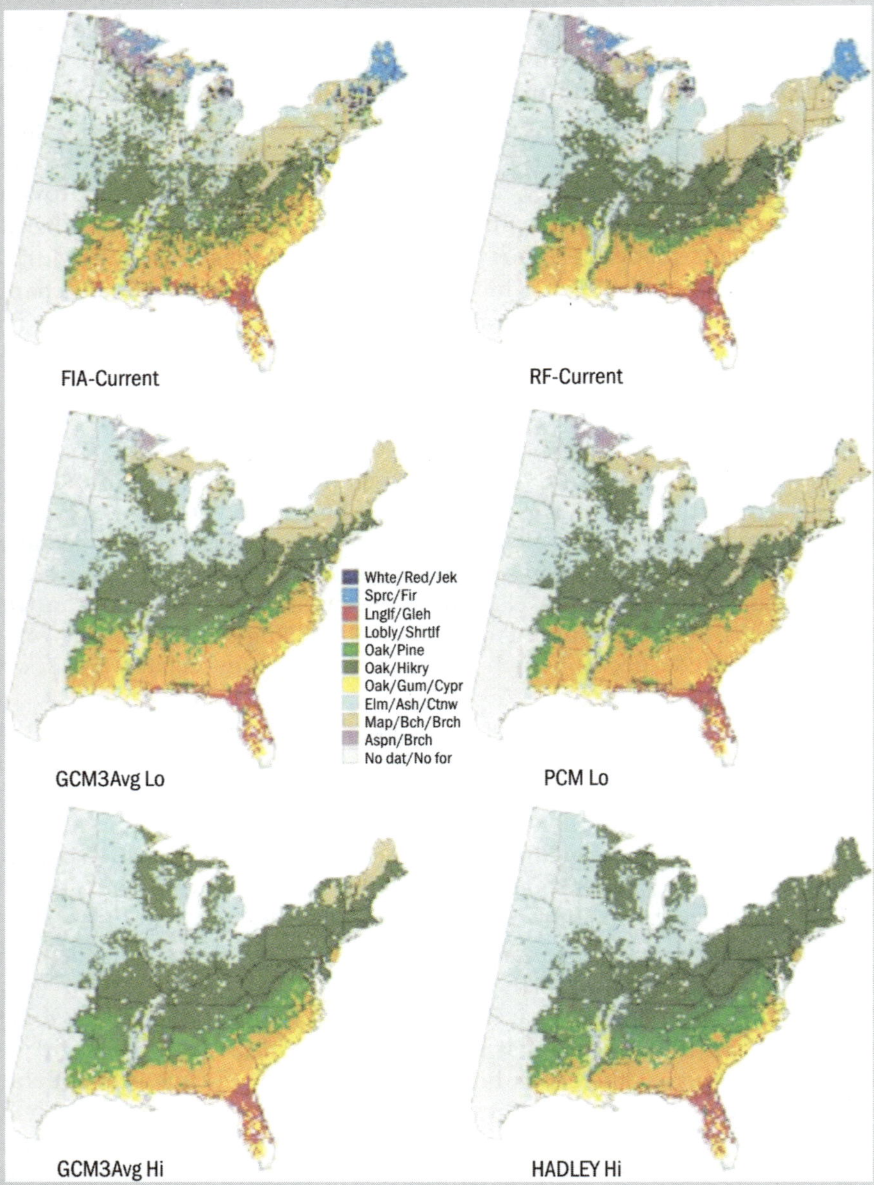


**Figure A** Maps of current and potential future suitable habitat for the forest types, including the FIA estimate of current distribution of abundance, the modelled current map, and scenarios of Parallel Climate Model – low emissions (PCM lo); average of three models – low emissions (Ave lo); average of three models – high emissions (Ave hi); and HadleyCM3 model – high emissions (Had hi)

**Source** Iverson, Prasad, Matthews, *et al.* (2008)

specific as well. The variables used to locate species can also be used to provide estimates of yields using crop simulation models in different climate change scenarios. For example, Priya and Shibaski (2001) use interpolated climate data to drive their model, whereas Kravchenko, Bullock, and Boast (2000) inferred climatic impacts from a DTM. Physiological models are particularly useful to predict yields when crops have been subjected to prolonged stress due to climatic variability (Hijmans, Forbes, and Walker 2000; Chapman and Thornes 2006).

## Mapping Agricultural Vulnerability to Climate Change in India

Understanding the response of a particular sector to climate change paves the pathways for developing mitigation and adaptation strategies. However, in practice, none of the sectors is independent, and often many other factors (social, financial, economic, ecological, and others) also define these responses. Interdisciplinary approaches for understanding the impacts of climate change have been recognized and many studies have contributed to develop a methodology. One such study carried out by O'Brien, Leichenko, Kelkar, *et al.* (2004) presented a systematic methodology for investigating regional vulnerability to climate change in combination with other global stressors using the example of Indian agriculture. The approach relied on local-level case studies for assessing different degrees of vulnerability of various sectors within the country. The study focused on India's agriculture vulnerability to two factors: first, the climate change patterns and second, prediction and economic globalization initiatives. They followed a four tier approach by (1) developing a national vulnerability profile at the district level; (2) inclusion of additional stressor; (3) identification of districts that are "double exposed"; and finally, (4) detailed case studies in selected districts.

The climate change vulnerability was measured as a composite of adaptive capacity and climate sensitivity under exposure to climate change (climate change stressor). The import-sensitivity was measured as a composite of distance to ports, cropping patterns, and crop productivity (globalization stressor). This came up as an innovate approach for mapping vulnerability, which could be used to assess climate impacts in the context of a range of societal changes. The key strength of this approach was evaluation of the relative distribution of vulnerability to multiple stressors at a sub-national level. Moreover, it used both top-down and bottom-up approaches. This was also accomplished by applying Intergovernmental Panel on Climate Change (IPCC) framework to map regional vulnerability as a function of adaptive capacity, sensitivity, and exposure to climate change. Such a framework help in identifying locations for policy intervention is the most critical—both geographically (for example, double exposed districts) and thematically (for example, access to irrigation, credit, alternative crops).

**Source** O'Brien, Leichenko, Kelkar, *et al.* (2004)

*Box Contd...*

*Box Contd...*

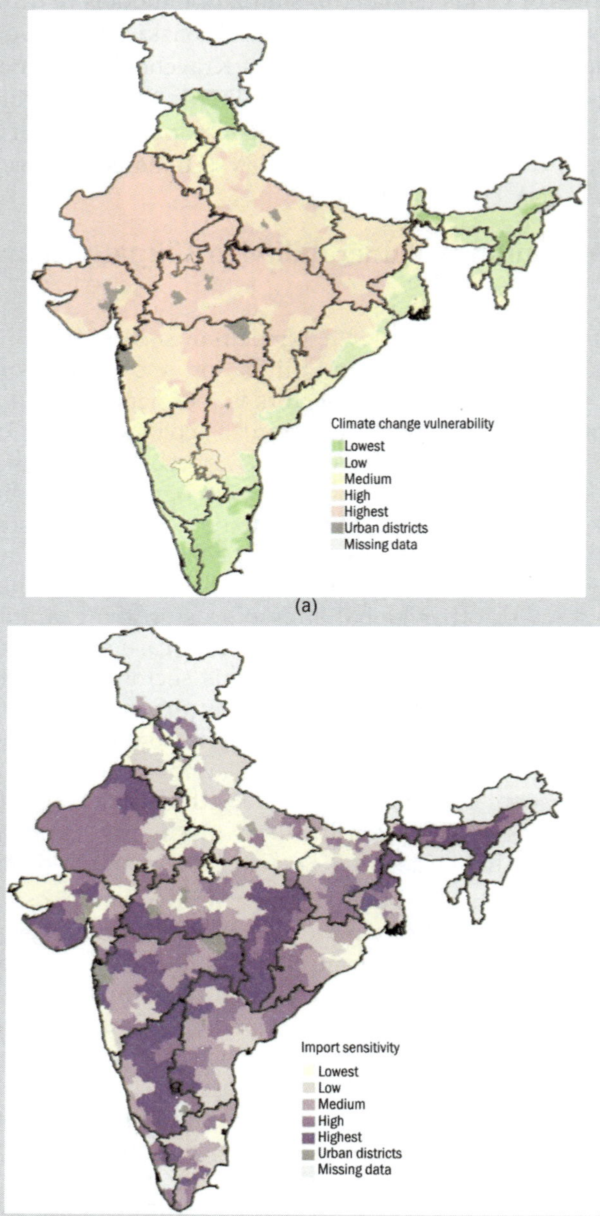


**Figure A** (a) District-level mapping of climate change vulnerability; (b) import-sensitivity map

## Human Health

Health is defined by the constitution of the World Health Organization as "a state of complete physical, mental and social well-being, and not merely the absence of disease or infirmity" (WHO 1946). Climate change will spur environmental degradation—whether chemical contamination or other forms of ecosystem disruption—which can affect both acute and chronic human health problems. The ability to assess the human health impacts of climate change, however, is still at a very early stage of development (Tol, Weyant, Davidson, *et al.* 1996; Balbus, Bouma, Kovats, *et al.* 1998).

Health problems that have special associations to local geography, such as those associated with coastal flooding, especially vector-borne disease, can be effectively analysed using mapping and modelling techniques. This involves mapping of current boundaries of a health problem, followed by identifying and plotting of determinant factors and developing models to project changes in these factors. The changes in the boundary of the health problem are noted with respect to demographic distribution and socio-cultural and economic centres of interest (Kalkstein and Smoyer 1993).

### Malaria Database for Africa—MARA/ARMA Project

Many of the disease outbreaks are frequent in the developing part of the world. Incidentally, these regions also lack a systematically collected disease data or are not equipped to collect spatially geo-referenced disease data sets. In fact, this is one of the greatest weaknesses with the current status of climate change research, which is considered the basic building block for generating models and predicting the scenarios. Because of such gap areas, often the developed information systems remain invalidated. Thus, the climate change impacts will not be assessed because of the absence of baseline data again, to quantify which impacts will be available. Moreover, the availability of climate database with a requisite spatial temporal resolution remains a limitation to accurately defining existing condition (Balbus, Bouma, Kovats, *et al.* 1998).

Overcoming the limitations of adequate surveillance systems for malaria in Africa, the Mapping Malaria Risk in Africa/Atlas du Risque de la Malaria en Afrique (MARA/ARMA) collaboration used parasite ratio (per cent infected) in the surveys of children as a marker of intensity of transmission. The initiative is non-institutional and runs in the spirit of an open collaboration.

To date, over 10 000 independent data points of parasite surveys in children aged 1–9 years have been collected, from published and unpublished sources, through literature searches and country visits. In addition, other associated data (drug resistance, vector distribution, agricultural practice, and so on) has been collected. This data as well as the data defining historical distribution are being used to define the boundaries of malaria transmission within Africa.

*Box Contd...*

*Box Contd...*

Current models are based on long-term mean interpolations and use the raster geographic information system (GIS) package. Continental digital data sets such as the African Data Sampler (ADS), which includes administrative boundaries, populated centres, rivers, roads, and so on, were also acquired. Several malaria risk maps have been produced, both theoretical, based on climate, and statistical, based on the data.

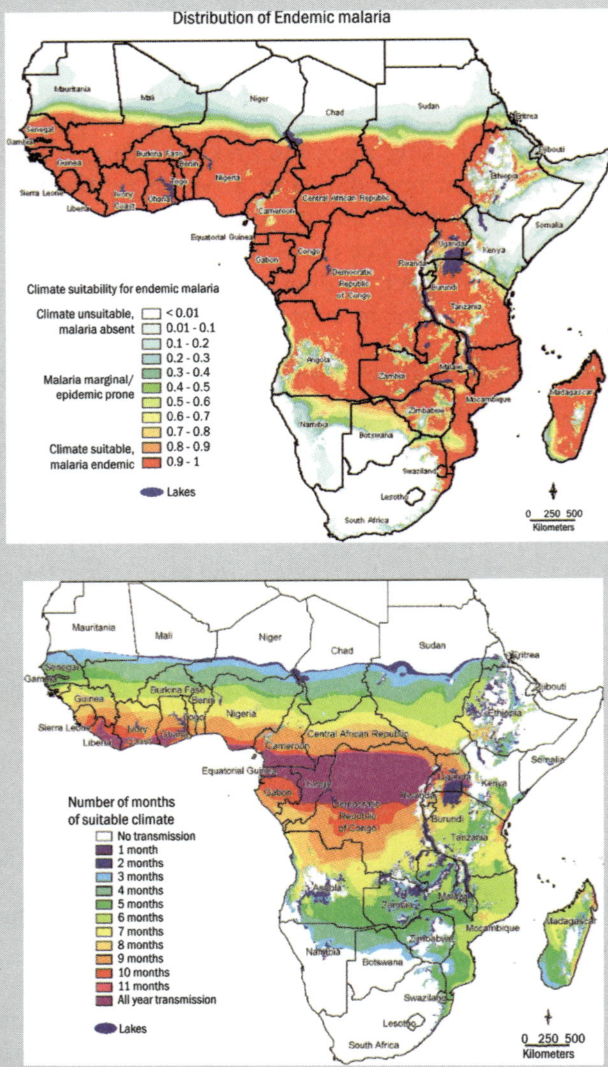


**Figure A** (a) Distribution of endemic malaria. (b) Duration of the malaria transmission season – a climate-based numeric distribution model and existing country-specific malaria maps

*Box Contd...*

*Box Contd...*

Thus, the MARA/ARMA collaboration has demonstrated that despite geographic dispersion, a methodology for creating a database can be successfully instituted. This allows the issue of the impact of climate change on malaria in Africa to move beyond the scenario level. The MARA/ARMA collaboration with its existing skills and methodology also has the potential to serve as a vehicle for collecting important data on other African diseases as well.

**Source** http://www.mara.org.za/ and Craig, Snow, and le Sueur (1999)

Vector-borne diseases have a geographic control and are also restricted by climate and terrain characteristics. These can be modelled effectively using remotely sensed climate data sets with GIS and GPS inputs (Bergquist 2001). Typical examples of such applications from the developing world include malaria (Srivastava, Nagpal, Saxena, *et al.* 2001; Manguin and Boussinesq 1999), lymphatic filariasis (Lindsay and Thomas 2000), and schistosomiasis (Bavia, Malone, Hale, *et al.* 2001; Malone, Bergquist, Huh, *et al.* 2001; Chapman and Thornes 2006).

## CONCLUSION

Climate change is influencing our future. A long period of climatic stabilization lies behind us. The changes are researched, and knowledge on the content of these changes is being developed. Considering the multitude of issues surrounding climate change science—from root causes to resultant impacts—*geography* is clearly an elemental factor in the equation, affecting and being affected by every aspect of climate change, be it at local, regional or global level (Dangermond 2010). Thus, the most dynamic attribute of the era on the earth requires to set a stage by bringing all actors from past, present, and expected future together. The key for this is developing protocols wherein data from different sources, timelines, and spatial and geographical extents can be merged into a single system. Here, geoinformatics appears to be the only solution, which has compatibility to input any kind of data set having geographical linkage, through powerful integration and management of complex decision-making rules. Artz (2010) strongly emphasizes a whole earth systems science-based GIS approach, which provides a plan for understanding and addressing the entire gamut of climate change science issues in a holistic manner, which will serve as a base enablement system for global data management, visualization, analysis, modelling, and design. It is high time to shift from massive data collection approach to integrative frameworks, which could remain an open source platform for the scientific and policy-makers' community.

The scientific community has already demonstrated the incredible power of geoinformatics as a key enabler for advancing our understanding of past and future climate change. The most striking examples of this are the multitude of powerful geo-visualizations that have been developed to communicate the urgency of our current situation by showing the past, present, and possible future of the earth based on the state-of-the-art science (Dangermond and Artz 2010). Geoinformatics is a holistic approach to shape spatial developments, recognizing the interrelations and effects of spatial measures in a long-term perspective. However, there remains some skepticism on the role of spatial data in the climate change debate. But given the degree of complexity, uncertainty, and scale of climate change, geospatial planning might be the only turning point to converge data sets for developing implementable adaptation and mitigation strategies. Above all, it is capable of providing part of the solutions in practice currently: sectoral policies have already formulated their own policy strategies, applying methods according to their own objectives and available funds (Bulkeley 2006). This is where geoinformatics can again play a pivotal role of providing a common integrated framework for the integration of climate change science and policy, ensuring that cross-sector issues are raised and optimal solutions are looked for. Successful coordination to optimize the planning process for combating climate change requires an inter/transdisciplinary approach from scientists, politicians, and society alike, and geoinformatics holds great potential in ironing out the ontological creases that may threaten to derail the process. Moreover, in spatial planning practice, the differences between time, space, and actors coalesce into one coherent multi-level governance approach, making geoinformatics suitable for an integrative approach to adaptation and mitigation.

To summarize, facilitating adaptation to climate change presents many challenges to scientific research and research funding. The climate change research needs to develop methodological developments, monitoring and indicator studies, empirical research, field and experimental research, predictive modelling, scenario development, economic costing, integrated assessment, quantification of the impacts of extreme weather events, modification of existing coping strategies, testing and evaluation of adaptation measures, stakeholder participation and much more. Achieving a concerted, sustained effort in all of these research areas is truly a multidisciplinary exercise in which geoinformatics provides a pivotal unifying role. Continued monitoring with GIS using data from satellite earth observation will ultimately confirm or disprove current thinking and perceptions in climate change research. Either way, future studies are heavily dependent on data-intensive geoinformatics-based spatial analysis.

## References

Agnew M D and Palutikof J P. 2000. **GIS-based construction of baseline climatologies for the Mediterranean using terrain variables.** *Climate Research* **14**: 115–127

Aniya M and Enomoto H. 1986. **Glacier variations and their causes in the northern Patagonia icefield, Chile, since 1944**. *Arctic and Alpine Research* **18:** 307–316

Aniya M, Sato H, Naruse R, Skvarca P, Casassa G. 1996. **The use of satellite and airborne imagery to inventory outlet glaciers of the southern Patagonia icefield, South America.** *Photogrammetric Engineering and Remote Sensing* **62**(12): 1361–1369

Anon. 2001. **Landmap special session.** Proceedings of the First Annual Meeting of the Remote Sensing and Photogrammetry Society, London, UK, September 12–14, 2001. 75 pp.

Antonic O, Krizan J, Marki A, Bukovec D. 2001. **Spatio-temporal interpolation of climatic variables over large region of complex terrain using neural networks**. *Ecological Modelling* **138**: 255–263

Appleton K, Lovett A, Sunnenberg G, Dockerty T. 2002. **Rural landscape visualisation from GIS databases: a comparison of approaches, options, and problems**. *Computers, Environment and Urban Systems* **26:** 201–211

Argent R M. 2004. **An overview of model integration for environmental applications—components, frameworks and semantics**. *Environmental Modelling and Software* **19:** 219–234.

Artz M (ed). 2010. ***GIS for Climate Change,*** 2nd edn. GIS best practices series. ESRI: Redlands, California

Balbus J M, Bouma M, Kovats S, LeSueur D, Martens W V, Patz J. 1998. ***Human Health: Handbook on methods for climate change impact assessment and adaptation strategies,*** Chapter 10, pp. 10-1–10-31, edited by J F Feenstra, I Burton, J B Smith, R S J Tol. UNEP and Institute of Environmental Studies

Barry R G. 2006. **The status of research on glaciers and global glacier recession: a review**. *Progress in Physical Geography* **30**(3): 285–306

Baruah N G, Dutta D, Jeyaram A, Sinha V S P. 2009. **Hyperspectral remote sensing for characterization of wildfire fuel in tropical mixed deciduous forest of West Bengal.** Proceedings at National Symposium of the Indian Society of Remote Sensing, Nagpur, September 2009

Bavia M E, Malone J B, Hale L, Dantas A, Marroni L, Reis R. 2001. **Use of thermal and vegetation index data from earth observing satellites to evaluate the risk of schistosomiasis in Bahia, Brazil**. *Acta Tropica* **79**: 79–85

Belda F and Melia J. 2000. **Relationships between climatic parameters and forest vegetation: application to burned area in Alicante (Spain)**. *Forest Ecology and Management* **135**: 195–204

Bergquist N R. 2001. **Vector-borne parasitic diseases: new trends in data collection and risk assessment**. *Acta Tropica* **79**: 13–20

Bernherdson T. 1992. ***Geographic Information Systems***. Arendal, Norway: VIAK IT

Biesbroek G R, Swart R J, and van der Knaap W G M. 2009. **The mitigation–adaptation dichotomy and the role of spatial planning**. *Habitat International* **33**: 230–237

Blennow K. 1998. **Modelling minimum air temperature in partially and clear felled forests**. *Agricultural and Forest Meteorology* **91**: 223–235

Boehmer-Christianses S. 1994a. **Global climate protection policy: the limits of scientific advice Part 1**. *Global Environmental Change* **4**(2): 140–159

Boehmer-Christianses S. 1994b. **Global climate protection policy: the limits of scientific advice Part 2**. *Global Environmental Change* **4**(3): 185–200

Boehmer-Christianses S. 2003. ***Epilogue: Scientific advice in the world of power politics: climate change – an integrated perspective***, vol. I, Chapter 10, pp. 357–403, edited by P Martens, J Rotmans, D Jansen, K Vrieze. Springer

Brown D G. 1995. **Issues and alternative approaches for the integration and application of societal and environmental data within a GIS.** Working Paper No. 3, Rwanda Society Environment Project, Department of Geography, Michigan State University

Brown I. 2006. **Modelling future landscape change on coastal floodplains using rule-based GIS**. *Environmental Modelling and Software* **21**: 1479–1490

Bulkeley H. 2006. **A changing climate for spatial planning**. *Planning Theory and Practice* **7**(2): 203–214

Campbell H. 2006. **Is the issue of climate change too big for spatial planning?** *Planning Theory and Practice* **7**(2): 201–203

Carter T. 2007. **Adaptation: local climate change impacts, adaptation and vulnerability**. In *The Future Climatic Window: local impacts of climate change*. Austria: Leibnitz

Chapman L and Thornes J E. 2006. **The use of geographical information systems in climatology and meteorology**. *Progress in Physical Geography* **30**: 677-689

Church J A and White N J. 2006. **A 20th century acceleration in global sea-level rise**. *Geophysical Research Letters*: L01602 33

Chuvieco E. 2008. ***Earth Observation of Global Change***. New York: Springer

Cohen S J. 1997. **Scientist–stakeholder collaboration in integrated assessment of climate change: lessons from a case study of Northwest Canada**. *Environmental Modeling and Assessment* **2:** 281–293

Cox D J. 1990. **The art of scientific visualization**. *Academic Computing* **46:** 20–56

Craig M H, Snow R W, and le Sueur D. 1999. **A climate-based distribution model of malaria transmission in sub-Saharan Africa**. *Parasitology Today* **15:** 105–111

Croswell P I. 2005. **Know the past to understand the present and future.** Details available at <http://www.geoplace.com/uploads/OnlineExclusives/techtrends.asp>, last accessed on 29 October 2010

Daly C, Taylor G H, Gibson W P, Parzybok T W, Johnson G L, Pasteris P A. 2000. **High-quality spatial climate data sets for the United States and beyond.** *Transactions of the ASEA* **43**: 1957–1962

Dangermond J. 2010. **The geographic approach to climate change.** *GeoSpatial Today,* April

Darwin R. 1995. ***World Agriculture and Climate Change: economic adaptations.*** Agricultural Economic Report No. 703. Economic Research Service, USDA

de Azevedo E B, Pereira L S, and Itier B. 1999. **Modelling the local climate in island environments: water balance applications.** *Agricultural Water Management* **40**: 393–403

del Barrio G, Alvera B, Puigdefabregas J, Diez C. 1997. **Response of high mountain landscape to topographic variables: Central Pyrenees.** *Landscape Ecology* **12**: 95–115

Desonie D. 2007. ***Climate – causes and effects of climate change.*** New York: Chelsea House

Din A M. 1992. **Global environmental change data and modelling.** *IFIP Transactions A-Computer Science and Technology* **13**: 625–634

Dubayah R C. 1994. **Modelling a solar radiation topoclimatology for the Rio-Grande river basin.** *Journal of Vegetation Science* **5**: 627–640

Dyras I, Dobesch H, Gruter E, Perdigao A, Tveito O E, Thornes J E, Van der Wel F, Bottai L. 2005. **The use of geographic information systems in climatology and meteorology: COST-719.** *Meteorological Applications* **12**(1): 1–5

Dyurgerov M B and Meier M F. 2000. **Twentieth century climate change: evidence from small glaciers.** *Proceedings of the National Academy of Sciences* **97**: 1406–1411

El Garouani A, Boussema M R, and Ennabli H. 2000. **Use of the geographic information system and remote sensing data for the estimation of real evapotranspiration at a regional scale.** *International Journal of Remote Sensing* **21**: 2811–2830

Ellis E A, Nair P K R, Linehan P E, Beck H W, Blanche C A. 2000. **A GIS-based database management application for agroforestry planning and tree selection.** *Computers and Electronics in Agriculture* **27**: 41–55

Fleming M D, Chapin F S, Cramer W, Hufford G L, Serreze M C. 2000. **Geographic patterns and dynamics of Alaskan climate interpolated from a sparse station record.** *Global Change Biology* **6**: 49–58

Foster D R and Boose E R. 1992. **Patterns of forest damage resulting from catastrophic winds in central New-England.** *Journal of Ecology* **80**: 79–98

Fussel H-M and Klein R J T. 2006. **Climate change vulnerability assessments: an evolution of conceptual thinking.** *Climate Change* **75**(3):301–329

Fussel H-M. 2007a. **Vulnerability: a generally applicable conceptual framework for climate change research.** *Global Environmental Change* **17**: 155–167

Fussel H-M. 2007b. **Adpatation planning for climate change – concepts, assessment approaches, and key lessons.** *Sustainability Science* DOI 10.1007/s11625-007-0032-y

Gelbspan R. 1997. ***The Heat is On,*** pp. 7–10, 139–140. Addison-Wesley Publishing

Goodchild M. 1999. **Multiple roles for GIS in global change research**. In *Geographic Information Research*, pp. 257–274, edited by M Craglia and H Onsrud. London: Taylor & Francis

Haeberli W and Beniston M. 1998. **Climate change and its impact on glaciers and permafrost in the Alps**. *Ambio* **27** (4): 258–265

Hall D, Klaus B, Bindschadler R, Schoner W. 2001. **Changes in the Pasterze Glacier, Austria as measured from the ground and space**, pp. 187–193. *The 58th Eastern Snow Conference*

Hall J W, Dawson J, Sayers P B, Rosu C, Chatterton J B, Deacon R. 2003. **A methodology for national-scale flood risk assessment**. *Water and Maritime Engineering* **156:** 235–247

Hall M H P and Fagre D B. 2003. **Modeled climate induced glacier change in Glacier National Park, 1850–2100**. *BioScience* **53**(2)

Harrould-Kolieb E and Herr D. Unpublished. ***Climate Change and Ocean Acidification***. Synergies and Opportunities within the UNFCC Discussion paper, 17 pp.

Hijmans R J, Forbes G A, and Walker T S. 2000. **Estimating the global severity of potato late blight with GIS-linked disease forecast models**. *Plant Pathology* **49**: 697–705

Hudson G and Wackernagel H. 1994. **Mapping temperature using kriging with external drift – theory and an example from Scotland**. *International Journal of Climatology* **14**: 77–91

IGBP (International Geosphere–Biosphere Programme). 1992. ***Global Change. Reducing the Uncertainties***. IGBP, Stockholm, Sweden

IPCC (Intergovernmental Panel on Climate Change). 2001. ***Climate Change 2001: Impacts, adaptation and vulnerability***. Summary for policymakers. Geneva: World Meteorological Organisation

IPCC (Intergovernmental Panel on Climate Change). 2007. ***Climate Change 2007: The physical science basis***. Working Group I Contribution to the IPCC Fourth Assessment Report. New York: Cambridge University Press

Iverson L R, Prasad A M, Matthews S N, Peter M. 2008. **Estimating potential habitat for 134 eastern US tree species under six climate scenarios**. *Forest Ecology and Management* **254**: 390–406

Jarvis C H and Stuart N. 2001. **A comparison among strategies for interpolating maximum and minimum daily air temperatures. Part II: The interaction between number of guiding variables and the type of interpolation method**. *Journal of Applied Meteorology* **40**: 1075–1084

Jeffrey S J, Carter J O, Moodie K B, Beswick A R. 2001. **Using spatial interpolation to construct a comprehensive archive of Australian climate data**. *Environmental Modelling and Software* **16**: 309–330

Jolicoeur S and O'Carroll S. 2007. **Sandy barriers, climate change and long-term planning of strategic coastal infrastructures, ˆIles-de-la-Madeleine, Gulf of St. Lawrence (Qu´ebec, Canada)**. *Landscape and Urban Planning* **81**: 287-298

Jurisic M, Hengl T, Duvnjak V, Martinic I. 1999. **Agro-ecological and land information system.** *Storjarstvo* **41**: 223–231

Kalkstein L S and Smoyer K E. 1993. **The impact of climate change on human health: some international implications.** *Experiencia* **49:** 469–479

Kandlikar M and Risbey J. 1996. **Uses and limitations of insights from integrated assessment modeling.** In *A Usage Guide to Integrated Assessment Models.* Model Visualization and Analysis Project, Center for International Earth Science Information Network (CIESIN). Details available at <http://www.ciesin.org>

Kandlikar M. 1999. **The potential role of GIS in integrated assessments of global change.** In *Geographic Information Research,* pp. 327–336, edited by M Craglia and H Onsrud. London: Taylor & Francis

Kay R and Alder J. 2005. ***Coastal Planning and Management,*** 2nd edn. London: Taylor & Francis. 380 pp.

Klein R J T and Huq S. 2007. **Inter-relationships between adaptation and mitigation.** In *Climate Change 2007: climate change impacts, adaptation, and vulnerability,* Chapter 18. Cambridge: Cambridge University Press

Kravchenko A N, Bullock D G, and Boast C W. 2000. **Joint multifractal analysis of crop yield and terrain slope.** *Agronomy Journal* **92**: 1279–1290

Krishna A P. 1996. **Satellite remote sensing applications for snow cover characterization in the morphogenetic regions of upper Tista river basin, Sikkim Himalaya.** *International Journal of Remote Sensing* **17:** 651–656

Krishna A P. 2005. **Snow and glacier cover assessments in the high mountains of Sikkim Himalaya.** *Hydrological Processes* **19**: 2375–2383

Kulkarni A V, Bahuguna I M, Rathore B P, Singh S K, Randhawa S S, Sood R K and Dhar S. 2007. **Glacial retreat in Himalaya using Indian Remote Sensing Satellite Data.** *Current Science* **92**(1): 69-74.

Kulkarni A V, Rathore B P, Mahajan S, Mathur P. 2005. **Alarming retreat of Parbati glacier, Beas basin, Himachal Pradesh.** *Current Science* **88**(11): 1844–1850

Lee M. 2001. **Coastal defence and the habitats directive: predictions of habitat change in England and Wales.** *Geographical Journal* **167:** 57–71

Lennon J J and Turner J R G. 1995. **Predicting the spatial distribution of climate temperature in Great Britain.** *Journal of Animal Ecology* **64**: 370–392

Levizzani V, Schmetz J, Lutz H J, Kerkmann J, Alberoni P P, Cervino M. 2001. **Precipitation estimations from geostationary orbits and prospects for METEOSAT Second Generation.** *Meteorological Applications* **8**: 23–41

Li Z, Sun W X, and Zheng Q Z. 1998. **Measurements of glacier variation in the Tibetan plateau using Landsat data.** *Remote Sensing of Environment* **63**(3)

Lillesand T M and Kiefer R W. 2000. ***Remote Sensing and Image Interpretation.*** John Wiley: New York

Lindsay S W and Thomas C J. 2000. **Mapping and estimating the population at risk from lymphatic filariasis in Africa.** *Transactions of the Royal Society of Tropical Medicine and Hygiene* **94**: 37–45

Malone J B, Bergquist N R, Huh O K, Bavia M E, Bernardi M, El Bahy M M, Fuentes M V, Kristensen T K, McCarroll J C, Yilma J M, Zhou X N. 2001. **A global network for the control of snail-borne disease using satellite surveillance and geographic information systems.** *Acta Tropica* **79**: 7–12

Manguin S and Boussinesq M. 1999. **Remote sensing in public health: applications to malaria and other diseases.** *Medicene et Maladies Infectieuses* **29**: 318–324

Martens P and Rotmans J. 2003. ***Climate Change – an integrated perspective,*** pp. 1–5. Dordrecht: Kluwer Academic Publishing

Mason S J, Goddard L, Graham N E, Yulaeva E, Sun L, Arkin P A. 1999. **The IRI seasonal climate prediction system and the 1997–1998 El Niño event.** *Bulletin of the American Meteorological Society* **80:** 1853–1873

McCarthy J J, Canziani O F, Leary N A, Dokken D J, White K S (eds). 2001. ***Climate Change 2001: impacts, adaptation and vulnerability.*** Cambridge: Cambridge University Press

McMichael A J, Campbell-Lendrum D, Kovats S,Edwards S, Wilkinson P, Wilson T, Nicholls R, Hales S, Tanser F, Le Sueur D, Schlesinger M, Andronova N. 2004. **Global climate change.** In *Comparative Quantification of Health Risk: global and regional burden of disease attributable to selected major risk factors,* vol. I & II, Chapter 20, pp. 1543–1649, edited by M Ezzati, A D Lopez, A Rodgers, C J L Murray. Geneva: World Health Organization

Mounsey H M. 1991. **Multisource, multinational environmental GIS: lessons learnt from CORINE.** In *Geographical Information Systems: principles and applications,* pp. 185–200, edited by D J Maguire, M F Goodchild, and D W Rhind. London: Longman

Nakicenovic N and Swart R (eds.). 2000. ***Emissions Scenarios 2000.*** Special Report of the Intergovernmental Panel on Climate Change. Cambridge: Cambridge University Press

Narama C, Shimamura Y, Nakayama D, Abdrakhmatov K. 2006. **Recent changes of glacier coverage in the western Terskey-Alatoo range, Kyrgyz Republic, using Corona and Landsat.** *Annals of Glaciology* **43:** 223–229

Nichol J E. 1995. **Monitoring tropical rainforest microclimate.** *Photogrammetric Engineering and Remote Sensing* **61**: 1159–1165

Nicholls R J and Wilson T. 2002. **Integrated impacts on coastal areas and river flooding.** In *REGIS: Regional Climate Change Impact Response Studies in East Anglia and North West England,* edited by I P Holman and P J Loveland. London: DEFRA,

Nicholson-Cole S A. 2005. **Representing climate change futures: a critique on the use of images for visual communication.** *Computers, Environment and Urban Systems* **29**(3): 255–273

Nordhaus W D. 1992. **An optimal transition path for greenhouse gas emissions.** *Science* **258:** 1315–1319

O'Brien K, Leichenko L, Kelkar U, Venema H, Aandahl G, Tompkins H, Javed A, Bhadwal S, Barg S, Nygaard L, West J. 2004. **Mapping vulnerability to multiple stressors: climate change and globalization in India.** *Global Environmental Change* **14**: 303–313

O'Regan P R. 1996. **The use of contemporary information technologies for coastal research and management — a review**. *Journal of Coastal Research* **12**: 192–204

Oreskes N. 2004. **Beyond the ivory tower: the scientific consensus on climate change**. *Science* **306**(5702): 1686

Parks B O. 1993. **The need for integration: environmental modeling with GIS**, In *Environmental Modeling with GIS*, pp. 31–34, edited by M Goodchild, B Parks, and L Steyaert. New York: Oxford University Press

Parry M L, Rosenzweig C, Iglesias A, Erda L. 1998. ***Agriculture: handbook on methods for climate change impact assessment and adaptation strategies***, Chapter 8, pp. 8-1–8-47, edited by J F Feenstra, I Burton, J B Smith, R S J Tol. UNEP and Institute of Environmental Studies.

Paskoff R. 1998. ***Les littoraux. Impact des am´enagements sur leur ´evolution***, 3rd edn. Paris: Colin. 260 pp.

Paul F, Kääb A, and Haeberli W. 2007. **Recent glacier changes in the Alps observed by satellite – consequences for future monitoring strategies**. *Global and Planetary Change* **56**(2007): 111–122

Pellikka P K E. 2003. **OMEGA – an operational glacier monitoring system**. *Geophysical Research Letters* **5**: 02210

Pew K L and Larsen C P S. 2001. **GIS analysis of spatial and temporal patterns of human-caused wildfires in the temperate rain forest of Vancouver Island, Canada**. *Forest Ecology and Management* **140**: 1–18

Pleshikov F I, Ryzkova V A, Kaplunov V Y, Usoltseva J V. 1998. **A computer system for evaluating and predicting hurricane impact on forest**. *Safety Science* **30**: 3–8

Pohl C. 2008. **From science to policy through transdisciplinary research**. *Environmental Science and Policy* **11**(1): 46–53

Priya S and Shibasaki R. 2001. **National spatial crop yield simulation using GIS-based crop production model**. *Ecological Modelling* **136**: 113–129

Rigol J P, Jarvis C H, and Stuart N. 2001. **Artificial neural networks as a tool for spatial interpolation**. *International Journal of Geographical Information Science* **15**: 323–343

Risbey J, Kandlikar M, and Patwardhan A. 1996. **Assessing integrated assessments**, *Climatic Change* **34**: 369–395

Rivera A, Casassa J, Bamber A, Kaab. 2005. **Ice-elevation changes of Glacier Chico, southern Patagonia, using ASTER DEMs, aerial photographs and GPS data**. *Journal of Glaciology* **51**(172)

Roggema R. 2009. ***Adaptation to Climate Change: a spatial challenge***. New York: Springer

Rosenqvist A, Milne A, Lucas R, Imhoff M, Dobson C. 2003. **A review of remote sensing technology in support of the Kyoto Protocol**. *Environmental Science and Policy* **6**: 441–455

Rotmans J and Dowlatabadi H. 1998. **Integrated assessment of climate change: evaluation of methods and strategies**. In *Human Choice and Climate Change: an international social science assessment*, edited by S Rayner and E Malone. Washington, DC: Battelle Press

Roy B. 2010. ***Interdisciplinarity and Climate Change: transforming knowledge and practice for global future***. Taylor & Francis. 263 pp.

Schadlich S, Gottsche F M, and Olesen F S. 2001. **Influence of land surface parameters and atmosphere on METEOSAT brightness temperatures and generation of land surface temperature maps by temporally and spatially interpolating atmospheric correction**. *Remote Sensing of Environment* **75**:39–46

Scholes R J, Linder S, and Siddiqui K M. 1998. ***Forest: handbook on methods for climate change impact assessment and adaptation strategies***, Chapter 12, pp. 12-1–12-30, edited by J F Feenstra, I Burton, J B Smith, R S J Tol. UNEP and Institute of Environmental Studies

Sheppard S R J. 2005. **Validity, reliability, and ethics in visualization**. In *Visualization in Landscape and Environmental Planning*, Chapter 5, pp. 79–97, edited by I Bishop and E Lange. London: Taylor & Francis

Sheppard S R J and Meitner M J. 2005. **Using multi-criteria analysis and visualisation for sustainable forest management planning with stakeholder groups**. *Forest Ecology and Management* **207**(1–2): 171–187

Sheppard S R J and Salter J. 2004. **The role of visualization in forest planning**. In *Encyclopedia of Forest Sciences*, pp. 486–498. Oxford, UK: Academic Press/ Elsevier

Slovic P, Finucane M, Peters E, MacGregor D G. 2002. **The affect heuristic**. In *Heuristics and Biases: the psychology of intuitive judgement*, Chapter 23, pp. 397–420, edited by T Gilovich, D Griffin, D Kahneman. New York, NY: Cambridge University Press

Srivastava A, Nagpal B N, Saxena R, Subbarao S K. 2001. **Predictive habitat modelling for forest malaria vector species *An. dirus* in India – a GIS-based approach**. *Current Science* **80**: 1129–1134

Steinitz C, Rojo H M A, Bassett S, Flaxman M, Goode T, Maddock III T, Mouat D, Peiser R, Shearer A. 2003. ***Alternative Futures for Changing Landscapes: the upper San Pedro river basin in Arizona and Sonora***. Washington, DC: Island Press

Stocks B J, Fosberg M A, Lynham T J, Mearns L, Wotton B M, Yang Q, Zin J-Z, Lawrence K, Hartley G R, Mason J A, McKenney D W. 1998. **Climate change and forest fire potential in Russiona and Canadian boreal forests**. *Climatic Change* **38**: 1–13

Suga Y, Takeuchi S, Kimura H, Inanga A. 1995. **Environmental monitoring of land and sea surface using multisensors**. *Calibration and Application of Satellite Sensors for Environmental Monitoring* **17**: 97–106

Sunar F and Ozkan C. 2001. **Forest fire analysis with remote sensing data**. *International Journal of Remote Sensing* **22**: 2265–2277

Thumerer T, Jones A P, and Brown D. 2000. **A GIS based coastal management system for climate change associated flood risk assessment on the east coast of England**. *International Journal of Geographical Information Science* **14:** 265–281

Tickell C. 2002. **Communicating climate change.** *Science* **297**, 737 (editorial)

Tol R S J, Weyant J, Davidson O, Dowlatabadi H, Edmonds J, Grubb M, Parson E A, Richels R, Rotmans J, Shukla P, Cline W, Frankhauser S. 1996. **Integrated assessment of climate change: an overview and comparison of approaches and results**. In *Economic and Social Dimensions of Climate Change*, edited by J P Bruce, H Lee, and E F Haites. Cambridge: Cambridge University Press

Tress B and Tress G. 2002. **Scenario visualisation for participatory landscape planning: a study from Denmark**. *Landscape and Urban Planning* **982:** 1–18

Tufte E R. 1990. ***Envisioning Information***. Cheshire, CT: Graphics Press

Tveito O E, Førland E J, Alexandersson H, Drebs A, Jónsson T, Tuomenvirta H, Vaarby Laursen E. 2001. **Nordic climate maps**. *DNMI Report 06/01 Klima.*

Vazquez A and Moreno J M. 2001. **Spatial distribution of forest fires in Sierra de Gredos (Central Spain)**. *Forest Ecology and Management* **147**: 55–65

Verdebout J. 2000. **A method to generate surface UV radiation maps over Europe using GOME, Meteosat, and ancillary geophysical data**. *Journal of Geophysical Research – Atmospheres* **105**: 5049–5058

VROM (Ministry of Housing, Spatial Planning and the Environment). 2007. ***Maak Ruimte voor Klimaat! Nationale Adaptatie Strategie,*** De beleidsnotitie. [***Make Space for Climate! National Adaptation Strategy,*** the policy document] (in Dutch), pp. 1–45. The Hague: VROM; Ministry of Transport, Public Works and Water management (V&W); Ministry of Agriculture, Nature and Food Quality (LNV); Ministry of Economic Affairs (EZ); Inter provincial cooperation (IPO); Association of Dutch municipalities (VNG); Union of Water Boards (UvW)

Wahr J, DaZhong H, and Trupin A. 1995. **Predictions of vertical uplift caused by changing polar ice volumes on a viscoelastic earth**. *Geophysical Research Letters* **22**: 977–980

Walsh K J E, Betts H, Church J, Pittock A B, McInnes K L, Jackett D R, McDougall T J. 2004. **Using sea level rise projections for urban planning in Australia**. *Journal of Coastal Research* **20:** 586–598

Walther R, Post E, Convey P, Menzel A, Parmesan C, Beebee T J C, Fromentin M, Hoegh-Gulberg O, Bairlein F. 2002. **Ecological responses to recent climate change**. *Nature* **416**: 389-395

WHO (World Health Organization). 1946. **Constitution of the World Health Organization**. Taken from Final Acts of the International Health Conference, New York, 19 June to 22 July, 116 pp.

WMO (World Meteorological Organisation). 1986. Report of the International Conference on the assessment of carbon dioxide and other greenhouse gases in climate variations and associated impacts, Villach, 9–15 October 1985, WMO no. 661, Geneva

Zhu Q J, Rong T Z, and Sun R. 2000. **A case study on fractal simulation of forest fire spread**. *Science in China Series E-Technological Sciences* **43**: 104


## Websites


http://www.mara.org.za/

http://earthbyte.org

# Reactions of Mountain Glaciers to Climate Change—A Remote Sensing Approach

**Tobias Bolch**[*,†] **and Roger Wheate**[#]
[*]*Universität Zürich, Switzerland*
[†]*University of Northern British Columbia, Canada*
[#]*Technische Universität Dresden, Germany*
*E-mail: tobias.bolch@geo.uzh.ch; wheate@unbc.ca*

## INTRODUCTION

Glacierized areas currently cover about 15 000 000 km$^2$ of land surface, which is more than 10% of the earth's land and corresponds to about 75% of the freshwater resources of the world (Benn and Evans 1998). This fresh water can be critical for agriculture, industrial, and domestic use, and an important source for hydropower generation. Moreover, glaciers in mountains and polar areas have an aesthetic appeal and are a tourist attraction.

A glacier is defined as "a mass of surface ice on land, which flows down under gravity. In general, a glacier is formed and maintained by the accumulation of snow at high altitudes, balanced by melting at low altitudes or discharge into lakes and the sea" (WGMS–UNEP 2008). Glaciers can be classified into three major types.

1 *Mountain glaciers,* which are relatively small glaciers that are formed in high elevations near the top of mountains or mountain ridges. This type can be further subdivided into *valley glaciers* that typically form characteristic glacier tongues and *hanging glaciers,* which cover steep mountain slopes.
2 *Ice caps,* which are small dome-shaped ice sheets with radial flow, usually covering a mountain or a mountain system.
3 *Ice sheets,* which cover large areas and the underlying topography of Greenland and the Antarctica.

Outlet glaciers flow outwards from ice caps and ice sheets channelled by the topography (Figure 1).

**Figure 1** Left: a typical valley glacier (Tschierva Glacier, Swiss Alps); right: ice cap with outlet glacier (Vatnajökull, Iceland)

**Photos** Courtesy T Bolch

Glaciers are sensitive to climate conditions and are, therefore, key indicators of climate change, especially in remote areas (IPCC 2007). Glaciers typically fluctuate in response to climatic conditions, most visible through changes in surface area or length. Lateral and terminal moraines indicate former glacier extents. Glaciers gain moraine rock material from the surrounding mountains and bedrock. *Debris-covered glaciers* occur where large parts of the terminus are covered by debris, if debris supply is high enough, and/or the glacier is melting.

However, length and area changes show a filtered signal of the change in climate as the glacier needs to adjust to the changing situations. This *response time* varies for each glacier and depends mainly on its thickness and the general climatic conditions. For example, the dynamic response time for glaciers in a temperate climate, with a thickness of 150–300 m, is 15–60 years (Paterson 1994; Bahr *et al.* 1998).

The *glacier mass balance* (GMB) shows a direct signal of climate change. Mass balance is concerned with the mass of a glacier and the distribution of its changes in space and time (Paterson 1994). The GMB can be obtained through direct field measurements, or by hydrological and geodetic approaches (Hagg, Braun, Uvarov, *et al.* 2004). Remote sensing offers the best opportunity to estimate glacier mass balance for a large number of glaciers simultaneously through the geodetic method (Paul and Haeberli 2008). Mass gain is mainly through snowfall and avalanches and also through refreezing of meltwater, while ablation and sublimation cause mass loss. The transition between the accumulation and ablation areas is called the equilibrium line. The *equilibrium line altitude* (ELA) is characteristic for each glacier and varies on an annual basis in response to climate. The long-term average, or *steady-state ELA*, is the altitude for which the glacier as a whole has a mass balance of zero, and is said to be in equilibrium with climate. Whether the annual ELA is above or below the steady-state ELA is a key indicator of the state of health of a glacier.

Another typical characteristic of a glacier is its *motion*. This compensates between accumulation and ablation and depends on topographic conditions, mass gain, and climate. Two different types of motion can be distinguished. First, the *glacier flow* is caused by a shear stress, which is determined by the ice thickness and surface slope. Second is the *basal motion*, wherein ice slides over the bedrock, assisted by an existing water film between the ice and the bedrock. As the glacier slides down-slope, the height and width may change. Glacier sliding enables a much faster flow than that resulting from internal deformation. Glacier velocity can vary between several metres and hundreds of metres per year. Some glaciers advance suddenly with much higher velocities. This phenomenon is known as *glacier surge* and is not directly related to climate. Glacier surges can occur in any mountainous area but are especially common in south-central Asia (Hewitt 2007; Kotlyakov, Osipova, and Tsvetkov 2008). Glacier motion can be measured in the field using geodetic methods and can also be computed using remote sensing data.

The recent trends in glacier retreat are mainly caused by the 20th century climate warming. The response of glaciers to climate fluctuations, however, is complex and may also depend on non-climatic factors such as ice dynamics, glacier hypsometry (the distribution of glacier area versus elevation), and topography. Glaciers may even advance due to climate warming if the warming occurs mainly in the winter months, which can lead to more solid precipitation and hence positive mass balances, as experienced in Scandinavia and New Zealand in the 1990s (Andreassen, Elvehøy, Kjøllmoen, *et al.* 2005).

The present contribution describes the advantages and challenges in remote sensing when applied to study mountain glaciers. It is structured as follows: first, a general background of remote sensing is presented, wherein both optical and radar are discussed with a focus on glaciological applications. Following this, remote sensing based methods are described in detail to study glaciers and obtain their characteristic parameters. Sample applications are presented for study areas with different topographic and climatic conditions, and glacier types.

## REMOTE SENSING BASICS

### General and Physical Background

Remote sensing is the survey or investigation and the analysis of objects without approaching them directly. Hence, remote sensing includes photographic methods as well as optical scanners and radar systems.

The origin of remote sensing is directly related to the beginning of photography and, hence, dates back to the 19th century. The first widespread use of aerial photography is related to the development of aircraft and military purposes during the First World War (1914–18). The first (military driven) space-borne mission was the US Corona program, which lasted from 1960 until 1972. These panchromatic images generally have stereo capability and a resolution of ~2 to 8 m (Dashora, Lohani, and Malik 2007; McDonald 1995). A comprehensive and continuous survey of the earth's surface began with the launch of the first Landsat satellite (earlier name: ERTS-1) in 1972. The satellite carried the multispectral scanner (MSS) recording in four wavelengths in the visible and near-infrared spectrum with a resolution of about 80 × 80 m. Recent trends include ultra high resolution imagery with a resolution of 1 m and better (for example, Ikonos, Quickbird) and fast repeat cycles. The US-American SEASAT mission (1978) is regarded as the first satellite mission carrying a radar system (Evans, Alpers, Cazenave, *et al.* 2005). Continuous missions, however, started only in the 1990s (the European ERS-1, the Russian ALMAZ-1, and the Japanese, J-ERS-1). Radar remote sensing is an active system, wherein a beam is emitted and its backscatter is recorded, while optical remote sensing systems are passive and rely on solar radiation and its interaction with the earth's surface. Laser altimetry systems such as the Geoscience Laser Altimetry System (GLAS) onboard the NASA Ice, Cloud, and land Elevation Satellite (ICESat) are also used to study glacier elevation changes from space (Kääb 2008; Muskett,. Lingle, Sauber, *et al.* 2008).

## Optical Remote Sensing

Remote sensing of the earth's surface is based on electromagnetic radiation processes, which allow the identification of objects, structures, and trends. The energy of the radiation depends on the wavelength (λ) and frequency ($f$). Optical (passive) remote sensing systems usually record in the visible (VIS, 0.4–0.75 μm), near-infrared (NIR, 0.75-–.5 μm), short wave infrared (SWIR, 1.5–2.5 μm), and the thermal infrared (TIR, 2.5–12 μm) portions of the spectrum, while radar (active) systems emit and record in the microwave spectrum (1 cm–1 m) (Figures 2 and 3). Three interactions can occur when electromagnetic radiation meets the earth's surface, namely, transmission, reflection, and absorption, wherein reflection can be specular and diffuse. Different materials have typical characteristics with regard to their interaction, leading to the creation of spectral signatures. Thus, different surface types can be visually and digitally distinguished. Most important for remote sensing is the recording of the spectral reflectance of the incoming radiation, which can be described as follows.

$$\rho A = ER(\lambda)/EI(\lambda) \tag{1}$$

where,

$\rho A$ = spectral reflectance

$ER(\lambda)$ = energy of the wavelength, which was reflected from the object

$EI(\lambda)$ = energy of the wavelength, which was incident on the object

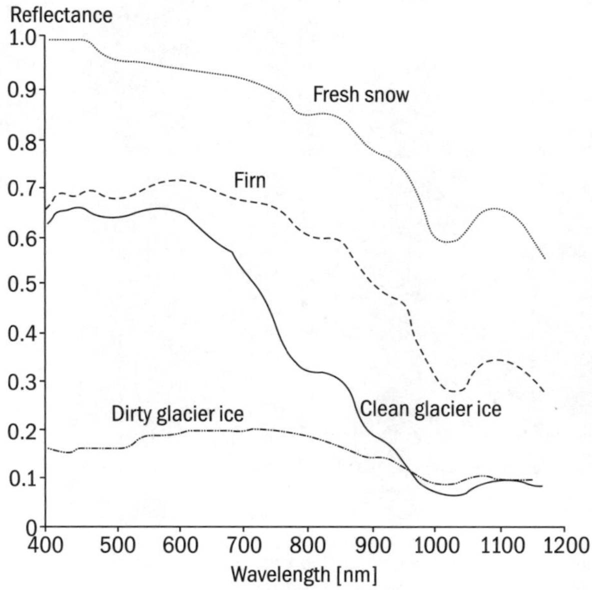


**Figure 2** Reflectance curves for fresh snow, firn, glacier ice, dirty ice

**Source** Hall and Martinec (1985)

In the case of optical passive remote sensing systems, the incoming radiation is the radiation from the sun, which is, therefore, dependent on the atmosphere. The most important atmospheric parameters that determine transmissivity and scattering are water vapour pressure and the content and type of aerosols. There are typical wavelengths in which the incoming solar radiation is highly absorbed; reflectance also depends on latitude and sun angle. Hence, the condition of the atmosphere at the time of acquisition has to be taken into account for interpreting the data acquired at the sensor (Liang, Fang, and Chen 2001; Ouaidrari and Vermote 1999). Atmospheric correction, however, is less important as aerosol content in high mountain area reduces with altitude. In addition, atmospheric correction has limited advantages if only one single scene is utilized (Song, Woodcock, Seto, *et al.* 2001).

A major problem encountered during remote sensing in high mountains is anisotropic reflectance caused by topography (Hugli and Frei 1983), where sun-facing aspects are brighter than areas in shadow (Figure 4).

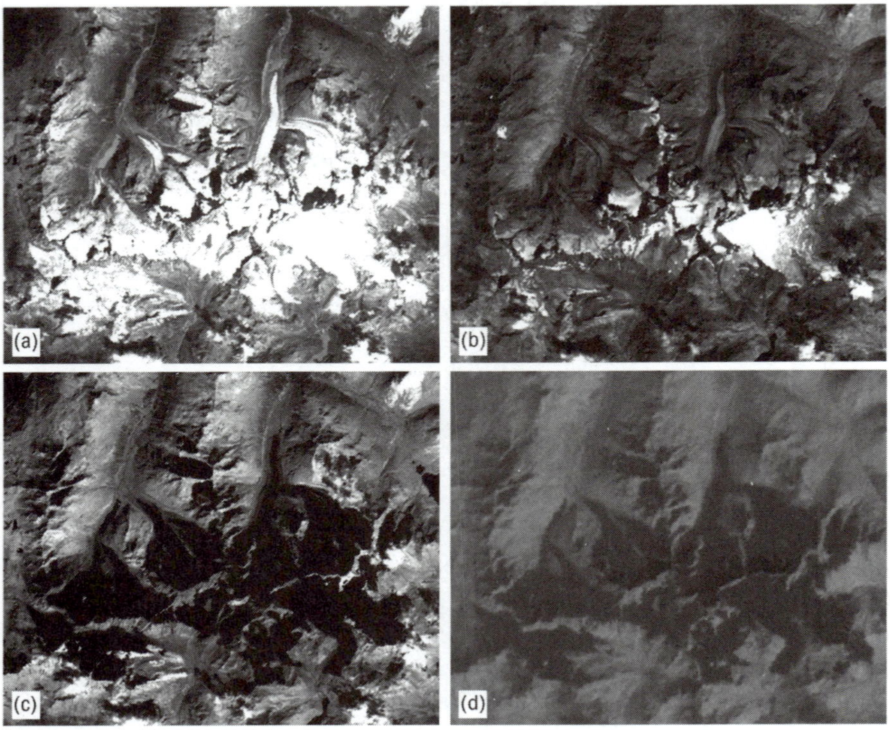


**Figure 3** Reflection from glaciers in different spectral bands based on Landsat ETM. (a) RED (Band 3); (b) near-infrared (NIR, Band 4); (c) short wave infrared (SWIR, Band 5); (d) thermal infrared (TIR, Band 6).

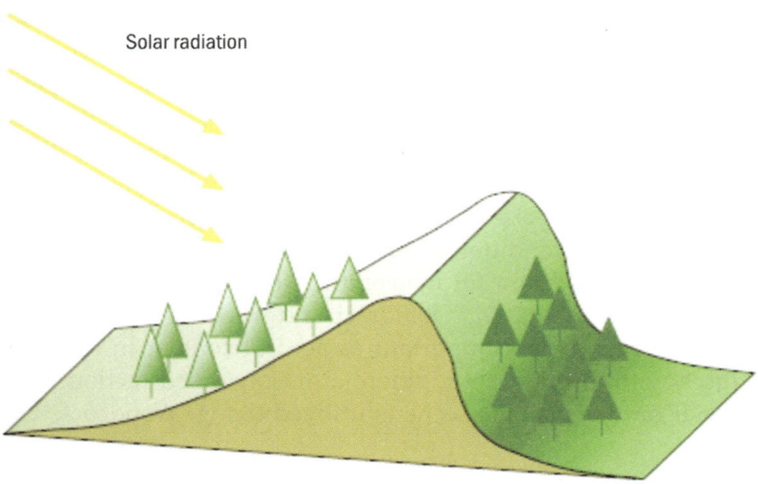


**Figure 4** Different reflectance in the mountains

A simple method to limit this topographic effect is the usage of ratio images. However, these have problems in areas with cast shadow and when only visible bands are utilized (Bishop, Shroder, and Colby 2003; Eckstrand 1996). A common approach for a relief correction is the consideration of reflectance of the surface and local illumination. This requires a detailed digital terrain model (DTM) as well as information about the sun position (angle and azimuth) and the image acquisition geometry. A simple approach is the following cosines correction (Bishop, Colby, Luvall, *et al.* 2005; Teillet, Guindon, and Goodeonugh 1982).

$$Ln = L\lambda/\cos i \tag{2}$$

where,

$Ln$ = normalized radiation

$L\lambda$ = radiation at the surface

$\cos i$ = illuminating factor ($i$), which depends on the sun position and the aspect of the surface

However, this method usually overcorrects steep slopes in the shadow areas because diffuse radiation is not considered (Bishop, Colby, Luvall, *et al.* 2005). An alternative method that accounts for different surface conditions is the Minnaert Correction (Colby 1991; Eckstrand 1996).

$$Ln = L\lambda\ (\cos e)/(\cos^{k} i \cos^{k} e) \tag{3}$$

where,

$Ln$ = normalized radiation

$L\lambda$ = radiation at the surface

$\cos i$ = illuminating factor (i) based on the sun position and aspect of the surface

$e$ = specific irradiation angle

k = Minnaert constant

The Minnaert constant k describes the surface condition. This method is commonly used for topographic correction, although overcorrection remains the main problem when the complex nature of the surface is not sufficiently incorporated. Newer approaches consider the influence of the non-lambert-reflectance or slope (Bishop and Colby 2002; Riano, Chuvieco, Salas, *et al.* 2003). Software packages include the possibility for atmospheric as well as topographic corrections; examples being ATCOR (Richter 2003). Figure 5 indicates the bands of the sensors, illustrated on the earth's incoming electromagnetic spectrum.

Table 1 provides an overview of the selected medium to high resolution satellite data, which is suitable for glaciological applications.

**Table 1** Selection of relevant medium to high resolution satellite data for glaciologic applications

| Satellite/ platform | Sensor/camera | Country | Operation | Spectral resolution | Swath (km) | Spatial resolution (m) | Stereo capability |
|---|---|---|---|---|---|---|---|
| Corona | KH4, KH4-A, KH-4B | USA | 1960–72 | PAN | ~15 × 220 | 2–8 | Yes |
| Hexagon | KH-9 | USA | 1963–80 | PAN | | ~10 | Yes |
| Landsat 1-4 | MSS | USA | 1972–92 | VNIR | 183 × 172 | 80 | No |
| Resurs-F1, Cosmos | KFA-1000 | Russia | 1974–99 | PAN, spectrozonal | 75 × 75 | 5 | Yes |
| Cosmos | KVR-1000 | Russia | 1981-2005 | PAN | 40 × 160 | 2 | No |
| Landsat 4, 5 | TM | USA | Since 1982 | VNIR, SWIR, TIR | 183 × 172 | 30, 120 | No |
| Cosmos | TK-350 | Russia | Since 1984 | PAN | 200 × 300 | 10 | No |
| SPOT 1 | HRV | France | 1986–2002 | PAN, VNIR, SWIR | 60 × 60 | 10, 20 | No |
| Resurs-F2 | MK-4 | Russland | 1988–95 | 6, col, PAN | 144 × 144 | 6–12 | Yes |
| SPOT2, 3 | HRV | France | Since 1990 | PAN, VNIR, SWIR | 60 x 60 | 10, 20 | Across track |
| IRS | LISS-III | India | Since 1996 | VNIR, SWIR | 70 × 70, 140 × 140, 800 × 800 | 5.6, 23, 70, 188 | No |
| SPOT 4 | HRVIR | France | Since 1998 | PAN, VNIR, SWIR | 60 × 60, 2200 × 2200 | 10, 20, 1000 | Across track |
| Landsat 7 | ETM+ | USA | Since 1999 | PAN, VNIR, SWIR, TIR | 183 × 172 | 15, 30, 60 | No |
| Terra | ASTER | Japan | Since 2000 | VNIR, SWIR, TIR | 60 × 60 | 15, 30, 90 | Along track |
| Ikonos | | USA | Since 2000 | PAN, VNIR | 11 × 11 | 1, 4 | Yes |
| Quickbird | | USA | Since 2000 | PAN, VNIR | 1.65 × 16.5 | 0.6, 2.4 | Yes |
| SPOT 5 | HRS/HRG | France | Since 2002 | PAN, VNIR, SWIR | 60 × 60, 1000 × 1000 | 2.5, 5, 10, 1000 | Across track |
| IRS-P6 | LISS-IV | India | Since 2003 | VNIR, SWIR | 24; 70, 140, 740 | 5.8, 23.5, 56–70 | No |
| Cartosat-1 (IRS-P5) | | India | Since 2005 | PAN | 30 × 30 | 2.5 | Along track |
| ALOS | PRISM, | Japan | Since 2006 | PAN | 35 × 35, 70 × 70 | 2.5 | Along track |
| ALOS | AVNIR-2 | Japan | Since 2006 | VNIR | 70 × 70 | 10 | No |
| Worldview-2 | | USA | Since 2009 | PAN, VNIR | 16 × 16 | 0.5, 1.8 | Yes |
| EROS C | | Israel, USA | Since 2009 | PAN, VNIR | 11 × 11 | 0.7, 2.8 | No |
| Sentinel-2 | | ESA/Europe | 2013 (scheduled) | PAN, VNIR, SWIR, TIR | 290 × 290 | 10, 20, 60 | No |

PAN – panchromatic; VNIR – visible and near infrared; SWIR – short wave infrared; TIR – thermal infrared

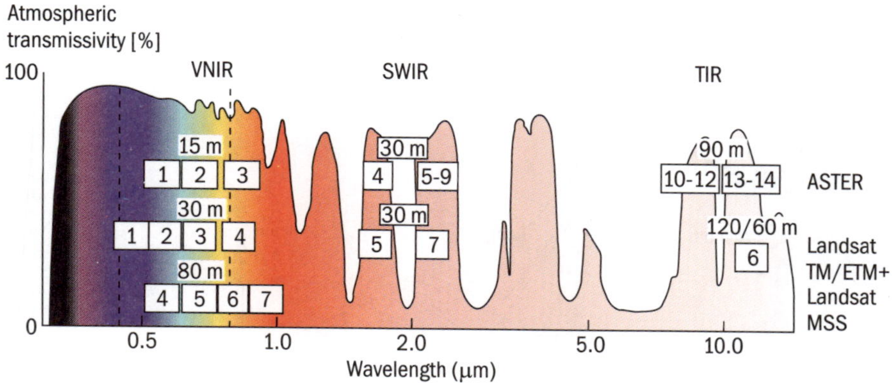


**Figure 5** Bands of the advanced space-borne thermal emission and reflection radiometer (ASTER), thematic mapper (TM), enhanced thematic mapper (ETM+), and multispectral scanner system (MSS) sensors, illustrated on the earth's incoming electromagnetic spectrum

**Source** Modified after Kääb, Huggel, Paul, *et al.* (2003)

## Radar Remote Sensing

Radar is almost weather independent as the beam and its backscatter penetrate through clouds. The utilized frequencies and corresponding wavelengths are usually coded with different letters (Table 2).

**Table 2** Typical wavelengths and the corresponding frequency and its band code

| Band | Frequency (Ghz) | Wavelength (cm) |
|---|---|---|
| Ku | 12–18 | 2.4–1.7 |
| X | 8–12 | 3.8–2.4 |
| C | 4–8 | 7.5–3.8 |
| S | 2–4 | 15–7.5 |
| L | 1–2 | 30–15 |

Most satellite applications utilize synthetic aperture radar (SAR). The main difference between real aperture radar (RAR) and SAR is resolution, which is dependent on the size of the antenna. With RAR, a better azimuth resolution can only be achieved with a longer antenna. As the antenna length is restricted due to technical reasons, it is not sufficient for remote sensing applications. The SAR technique synthesizes a long antenna, achieved using the constant velocity of the radar platform and the impulse that have a certain repetition frequency (Figure 6). The smaller antenna moves along the track, storing all received echoes and simulating a longer antenna. Radar needs to be a side-looking system as objects could not be spatially resolved if the radar beam were emitted in a nadir direction, with the depression angle at 90°.

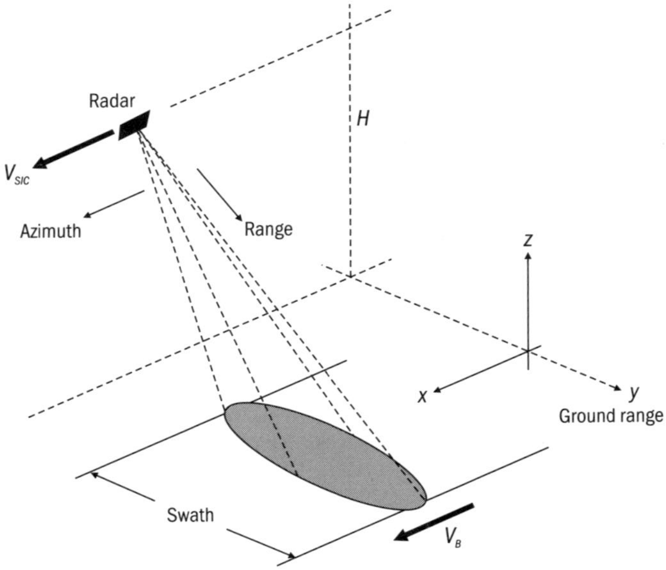


**Figure 6** Synthetic aperture radar (SAR) imaging geometry
**Source** Bamler (1999)

The backscatter signal is influenced by the radar properties, such as wavelength, polarization, and depression angle, and the terrain properties, such as surface roughness, feature orientation, and the dielectric properties of a material. The dielectric constant strongly influences electromagnetic energy and increases if the moisture content of a material increases. High moisture produces a high backscatter coefficient, which leads to a bright signature. Radar could, therefore, be used for the estimation of soil moisture or water content of wet snow and ice. Another influence is the roughness of an object in comparison with the wavelength of the radar beam. The backscatter signal is determined by the average surface roughness of a ground resolution cell. On a smooth surface, the radar beam will be reflected to the opposite direction of the incidence angle. The area will be represented by a dark signature as no energy is returned to the antenna. At a surface with intermediate roughness, part of the signal is reflected and the other part is scattered diffusely. At a rough surface, the radar signal is diffusely scattered at all angles, which causes a strong backscatter signal and thus results in a bright signature. Surfaces do not behave the same for different radar systems as roughness depends on the radar wavelength. A surface could be rough in an X-Band image, have an intermediate roughness in a C-Band image or a smooth surface in an L-Band image if the same depression angle for radar systems in different wavebands and the same vertical relief are assumed (Sabins 1997). Radar waves can also penetrate a surface, depending on the wavelength and moisture, as longer wavelengths have greater penetration depth. Figure 7 illustrates the penetration depth of radar bands for different surfaces; penetration depth decreases as the moisture content of a surface increases.

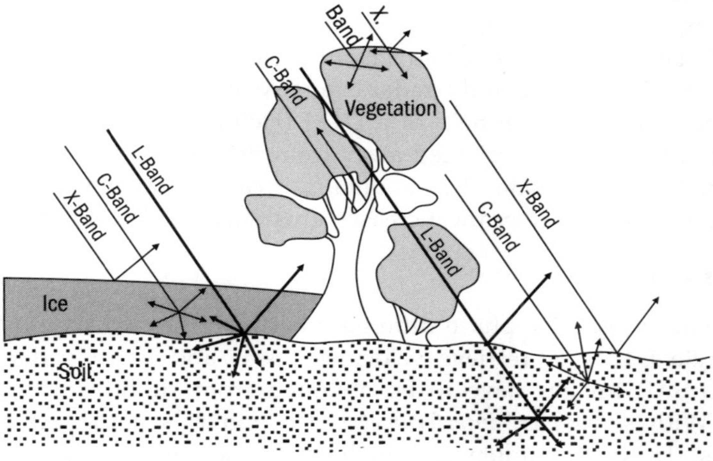


**Figure 7** Penetration depth of different wavelengths
**Source** Modified after Martin (2002)

As SAR is a side-looking system, the topography of the imaged terrain causes specific effects. These are known as foreshortening, layover, and shadow (Figure 8).

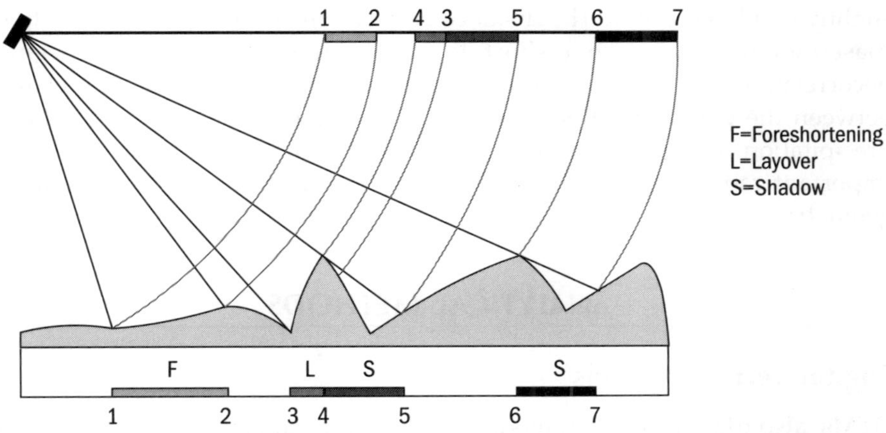


**Figure 8** Problems in radar data caused by the terrain

*Foreshortening* occurs on slopes oriented towards the radar sensor. The near sides of the mountain appear bright and compressed, while slopes oriented away from the sensor appear dark and stretched. Foreshortening depends on the incidence angle of the radar beam, the altitude of the spacecraft, and topography. The foreshortening effect can be compensated with the help of a DTM. The extreme case of the latter effect is called *layover*. Here, the mountain top is imaged before the slope that is oriented towards the sensor, as the radar beam will hit the top of the mountain first and its base only later. This effect occurs when the slope of the mountain is steeper than the incidence angle of the radar beam. As with foreshortening, this can be compensated with the help of a DTM (Hellmann 2000). *Shadow* occurs on steep slopes that are oriented away from the sensor. As the radar beam illuminates the front side of a mountain, it casts a shadow behind. If the look angle of the radar beam increases, this effect will be intensified.

### Interferometry

Interferometry uses the phase difference between the two acquired images to generate DTMs or retrieve velocities (Gens and Van Genderen 1996). The two SAR images have to be obtained from slightly different view angles. With *single-pass interferometry*, the radar is equipped with two antennas. These two antennas can be either arranged perpendicular to the flight direction (across-track) or along the flight direction (along-track). The shuttle radar topography mission (SRTM), for example, was equipped with two antennas to facilitate across-track interferometry, which were used to generate DTMs (Rabus, Eineder, Roth, *et al.* 2003). Along-track interferometry is possible, for example, with TerraSAR-X, as the antenna can be split into two halves. With *repeat-pass interferometry*, the radar visits the same region in two different time periods and with slightly displaced orbits. The distance between the two acquisition positions (baseline) has to be kept short to ensure interferometry. The signals decorrelate if this baseline exceeds a critical distance. The temporal change between the two acquisitions can also worsen the signal correlation due to precipitation, movement, and so on. Table 3 provides an overview of the important radar systems. Figure 9 gives the principles of InSAR imaging geometry.

## ANALYTICAL METHODS

### Digital Terrain Models

DTMs, also digital elevation models (DEMs), are valuable tools for mapping, modelling, analysing, and visualizing glacio-geomorphic phenomena in glacial environment (Bolch, Kamp, and Olsenholler 2005; Etzelmüller, Ødegård, Berthling, *et al.* 2001; Etzelmüller and Sollid 1997; Kamp, Bolch, and Olsenholler 2003). DTMs play an important role in the preparation of ortho-images in high mountain terrain (Finsterwalder 1984), in estimating

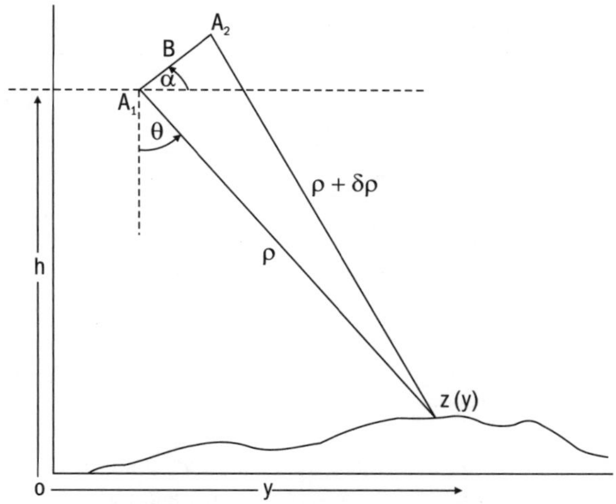


Figure 9 Principles of InSAR imaging geometry

glacier hypsometry (Brocklehurst and Whipple 2004), in the atmospheric correction of satellite imagery (Bishop and Colby 2002), and in derivative parameter maps such as slope, plan, and profile curvature, which can be further used for terrain analysis and debris-covered glacier mapping. Furthermore, DTMs are most important for the estimation of volume change for inaccessible glacier regions (Berthier, Arnaud, Kumar, *et al.* 2007; Bolch Buchroithner, Pieczonka, *et al.* 2008; Rignot, Rivera, and Casassa 2003; Schiefer, Menounos, and Wheate 2007; Surazakov and Aizen 2006).

DTMs can be prepared through different methods, such as space-borne and airborne optical stereo data; interferometric SAR (InSAR) data; space and airborne radar and laser altimetry (for example, light detection and ranging [LiDAR]); and topographic maps. The accuracy of DTMs derived from remote sensing depends on several factors such as spatial resolution of satellite data; positional and vertical accuracy of ground control points (GCPs); as well as contrast and cloud cover in optical satellite images. GCPs can be computed for DTM generation mainly using three methods: precise differential global positioning system (DGPS) measurements during fieldwork; existing rectified topographic maps; and from known satellite position and rotation angles, where DGPS measurements and topographic maps are not available.

The SRTM C-band DTM is a primary DTM source as it is available for free, has a near-global coverage, the available resolution with 30 m/90 m inside/outside the USA is suitable for glacier analysis, and the z-accuracy is of the order of 10 m (Berry, Garlick, and Smith 2007; Falorni, Teles, Vivoni, *et al.* 2005; Jacobsen 2005). The original SRTM DTM contains data gaps in mountainous terrain due to the problems of radar shadow. These gaps,

**Table 3** Important radar systems and their characteristics

| Satellite/ platform | Sensor | Institution/ country | Operation | Band | Pol. | Look angle (°) | Swath width (km) | Resolu-tion (m) |
|---|---|---|---|---|---|---|---|---|
| SEASAT | SAR | NASA/USA | June to September 1978 | L | HH | 20 | 100 | 25 |
| ALMAZ-1 | SAR | NPO/Russia | 1991 | S | HH | 25–60 | 40 | 12–20 |
| ERS-1 | SAR | ESA/Europe | 1991-2000 | C | VV | 23 | 100 | 30 |
| J-ERS-1 | SAR | NASDSA/Japan | 1992 | L | HH | 35 | 75 | 18 |
| Space Shuttle | SIR-C | NASA/USA, DLR/Germany, ASI/Italy | 1994 | L<br>C<br>X | Quad<br>Quad<br>VV | 20–55 | 30–100 | 15–25 |
| ERS-2 | SAR | ESA/ Europe | Since 1995 | C | VV | 23 | 100 | 30 |
| Radarsat-1 | SAR | CSA/Canada | Since 1995 | C | HH | 20–50 | 10–500 | 10–100 |
| SRTM | SAR | NASA-JPL/ USA, DLR/Germany, ASI/Italy | February 2000 | C<br>X | VV+HH<br>VV | 17-65<br>55 | 225<br>50 | 30<br>30 |
| ENVISAT | ASAR | ESA/Europe | Since 2002 | C | Quad | 15-45 | 100<br>450 | 30<br>150 |
| ALOS | PALSAR | JAXA/Japan | Since 2006 | L | Quad | 8-60 | 20-350 | 7-100 |
| Radarsat-2 | SAR | CSA/Canada | Since 2007 | C | Quad | 10-60 | 20-500 | 3-100 |
| TerraSAR-X | SAR | DLR/Germany | Since 2007 | X | Quad | 20-60 | 10-100 | 3.3-17.6 |
| TanDEM-X | SAR | DLR/Germany | Since 2010 | X | Quad | 20-60 | 10-100 | 3.3-17.6 |
| Sentinel-1 | SAR | ESA/Europe | Scheduled for 2013 | C | | | | 5-100 |

SAR – synthetic aperture radar; ASAR – advanced synthetic aperture radar; PALSAR – phased array type L-band synthetic aperture radar; ESA – European Space Agency; NASA – National Aeronautics and Space Administration; NASDA – National Space Development Agency; DLR – Deutsches Zentrum für Luft- und Raumfahrt (German Aerospace Centre); ASI – Agenzia Spaziale Italiana (Italian Space Agency); CSA – Canadian Space Agency; JPL – Jet Propulsion Laboratory; NPO – Nauchno-proizvodstvennoe obyedineniye

however, occur usually at mountain peaks and steep slopes so that glacier snouts in valley beds are not affected. There are newer versions of the SRTM3-DTM, wherein gaps were filled from other data sources. One of these is available from the Consultative Group on International Agricultural Research (CGIAR) Consortium for Spatial Information (http://srtm.csi.cgiar.org/). The main disadvantage is that no data is available north of 60° (for example, for most parts of Alaska) and south of 58° (Antarctica). A new globally available advanced space-borne thermal emission and reflection radiometer (ASTER) DTM (G-DEM) was created in 2009, covering 83°N and 83°S with a resolution of 30 m (http://www.ersdac.or.jp/GDEM/E/index.html). Initial evaluations of the G-DEM are promising (Hayakawa, Oguchi, and Lin 2008). However, spikes and holes are common in areas with rough topography and where few suitable ASTER scenes were acquired. In the near future, more detailed and accurate DTMs of the earth will be available; for example, from the TanDEM-X-mission (Moreira, Krieger, Hajnsek, *et al.* 2004).

SPOT-5, ALOS PRISM, CartoSat I and II, IKONOS, and Quickbird are optical satellites with higher resolutions, which offer stereo capability and can be used to monitor glacier volume changes. The ERS-1 and ERS-2 radar satellites orbited from 1995 until 1999 in a tandem mode and offered the possibility for single pass DTM generation. The data was widely used for DTM generation (Toutin and Gray 2000). Other radar data sources from satellites include Radarsat 1 and 2 and Terra-SAR-X. Figure 10 gives a comparison of an ASTER-derived DTM and a Corona-derived DTM. Table 4 provides an overview of the different methods and data sources for DTM generation and references for further reading.

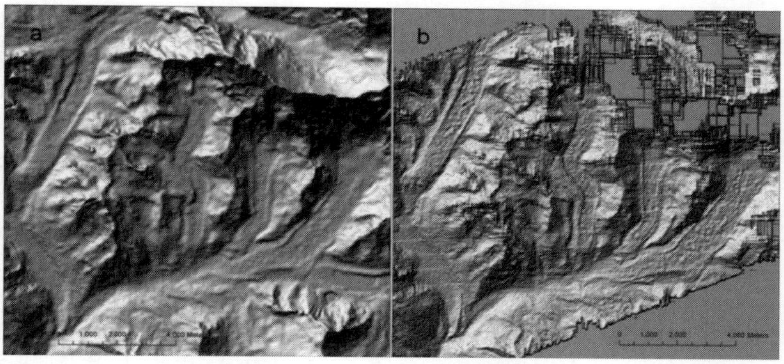


**Figure 10** Comparison of (a) an advanced space-borne thermal emission and reflection radiometer (ASTER)-derived digital terrain model (DTM) from 2002 and (b) a Corona-derived DTM from 1962 of the Mt Everest area, including several debris-covered glaciers

**Source** Bolch, Buchroithner, Pieczonka, *et al.* (2008)

## Glacier Mapping: Optical Imagery

### *Historical Glacier Mapping*

Glacier boundaries and surrounding features were represented only as rough sketches in the first "glacier maps" from the early 17th century. One of the first accurate large-scale maps for glaciological purposes (Venediger Group, Alps) was made by Eduard Richter, the first chairman of the German and Austrian alpine club (Brunner 1987). Historical maps are often the only source on glacier extents, which are available till the mid-20th century, and they can be of great value for change analysis depending on their quality (Brunner 2006). More information on the history and development of glacier mapping can be found in Williams and Ferrigno (2002). Figure 11 presents early glacier maps of the Indian Himalayas.

**Table 4** Comparison of different methods and data sources for DTM generation

| Method | Data source | Suitable DTM resolution/footprint (m) | References |
|---|---|---|---|
| Stereo-photogrammetry | Corona KH4, KH4-A | 20, 10 | Altmaier and Kany (2002); Bolch, *et al.* (2008) |
| | ASTER | 30 | Kääb (2002); Kamp, Bolch, and Olsenholler (2005); Toutin (2008) |
| | SPOT 4, SPOT 5 | 20, 5 | Al-Rousan, Cheng, Petrie, *et al.* (1997); Berthier and Toutin (2008) |
| | ALOS PRISM | 5 | Bignone and Umakawa (2008); Trisatki, Carolita, and Pradana (2009) |
| | CartoSat I | 5 | Ahmed *et al.* (2007); Pieczonka, Bolch, and Buchroithner (2011) |
| | Ikonos | 2 | Toutin (2006); Baltsavias, Favey, |
| | Aerial photo | 1–5 | Bauder, *et al.* (2001) |
| Radar interferometry | SRTM | 30 (90 free) | Berry, Garlick, and Smith, *et al.* (2007); Rabus, Eineder, Roth, *et al.* (2003); Smith and Sandwell (2003) |
| | ERS | 30 | Kervyn (2001); Toutin and Gray (2000) |
| | Radarsat | 30 | Kervyn (2001); Pradhan, Bolch, and Buchroithner (2009); Toutin (2000) |
| | TerraSAR-X | 5 | Yamane, Fujita, Nonaka, *et al.* (2008) |
| Radar altimetry | Envisat-RA | 61–960 | Remy and Parouty (2009) |
| | Cryosat-2 | 250 | |
| Laser altimetry | ICESat | 70 | Kääb (2008); Muskett, Lingle, Sauber, *et al.* (2008); Schutz, Zwally, Shuman *et al.* (2005); Zwally, Schutz, Abdalati, *et al.* (2002) |
| Digitizing | Topographic maps | Variable | Desmet (1997); Racoviteanu, Manley, Arnaud, *et al.* (2007) |

Maps generated from terrestrial photogrammetry can achieve high accuracy for glacier terrain features with scales better than 1:10 000. However, topographic maps in general have to be used with caution as their quality depends heavily on the surveyor and methods, usually involving interpretation of aerial photography. The quality of glacier mapping also depends on the suitability of the photos (for example, seasonal snow cover, shadows) (Bhambri and Bolch 2009). Declassified imagery such as Corona and Hexagon from the 1960s and 1970s are a valuable source for the evaluation of earlier outlines based on maps (Bolch, Yao, Kang, *et al.* 2010).

### *Manual Delineation on Satellite Imagery*

Topographic glacial mapping based on plane table and terrestrial photogrammetry is costly as it requires large amounts of time and human efforts. In comparison, remote sensing data facilitates glacier mapping, as

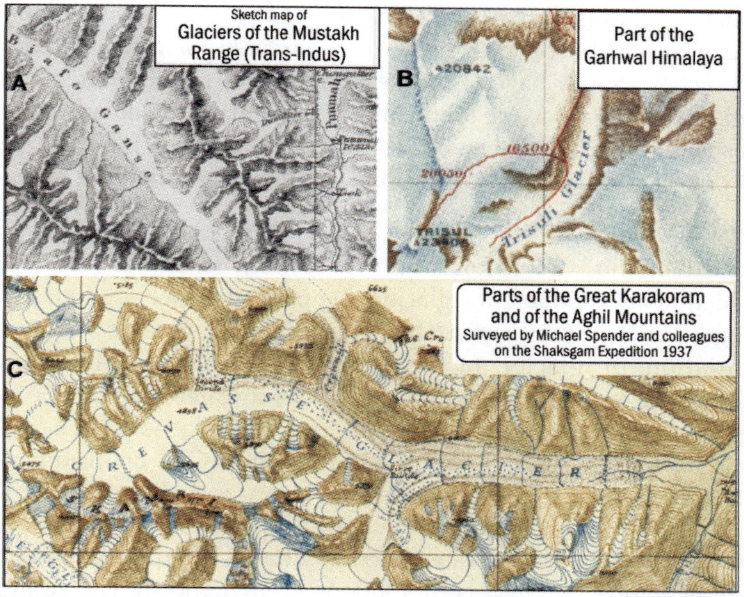


**Figure 11** Early glacier maps from the Indian Himalayas Maps published with the permission of the Royal Geographic Society (with IBG)

fieldwork is only necessary for ground truthing and for obtaining ground control points. The earliest snow and ice studies based on Landsat MSS data were published in the 1970s (Krimmel and Meier 1975; Østrem 1975; Rott 1976). Initially, glacier inventory studies using remote sensing, such as Hall, Williams, and Bayr (1992) or Williams (1986), started with manually digitizing glacier boundaries on false colour composites (FCC) of Landsat MSS and TM images. This method is time consuming for larger areas but is usually highly accurate, subject to resolution and depending on the efficiency of identification and recognition of glacier terrain features.

### *Automated Methods*

Automated methods are required to extract glacier outlines and information for larger areas in a reasonable amount of time. Automated mapping of glaciers involves image processing techniques with multi-spectral data such as simple band mathematics and classification. Automated mapping of snow and ice is based on snow being characterized by high reflectance in the VIS and NIR regions compared to the SWIR region of the solar spectrum. However, these methods fail for debris cover due to similar spectral signals of the surrounding debris, such as lateral and terminal moraines or rocky slopes free of vegetation. Although some efforts have been made to automatically map glaciers based on Landsat MSS data (for example, Della Ventura, Rampini, Rabagliati *et al.* 1987; Gratton, Howarth, and Marceau 1990; Svoboda and Paul 2009), an accurate automated mapping of snow and ice based on multi-spectral remote sensing data is only possible since the launch of Landsat 4 in 1982 with the TM sensor, which contained a short-wave infrared sensor. In order to be suitable for glacier mapping, the utilized scenes should have (1) no seasonal snow, (2) no clouds, (3) a high sun angle, and (iv) an acquisition time at the end of the ablation period (in most cases late summer scenes). Promising approaches for mapping clean ice and debris-covered glaciers are presented in the following sections.

### *Supervised Classification*

A standard method for information extraction based on remote sensing data, which is also suitable for glacier mapping, is supervised classification. Challenges facing the automated mapping of glacier areas include differentiation of glacier facies, identification of debris-covered ice, topographically and cloud-shadowed areas, and waterbodies adjacent to the glaciers. Gratton, Howarth, and Marceau, *et al.* (1990) conducted a supervised classification in a geographic information system (GIS) environment based on MSS and TM data identifying snow, clean and dirty glacier ice, and debris. Manual improvement was necessary, however, in order to achieve good results. Sidjak and Wheate (1999) evaluated algorithms based on different input bands or band combinations from Landsat TM combined in a maximum likelihood classification. They obtained the best results with a combination of the principal components 2–4, a NIR/SWIR ratio image, and the Normalized Difference Snow Index (NDSI ([VIS – SWIR]/[VIS + SWIR])) (Figure 12).

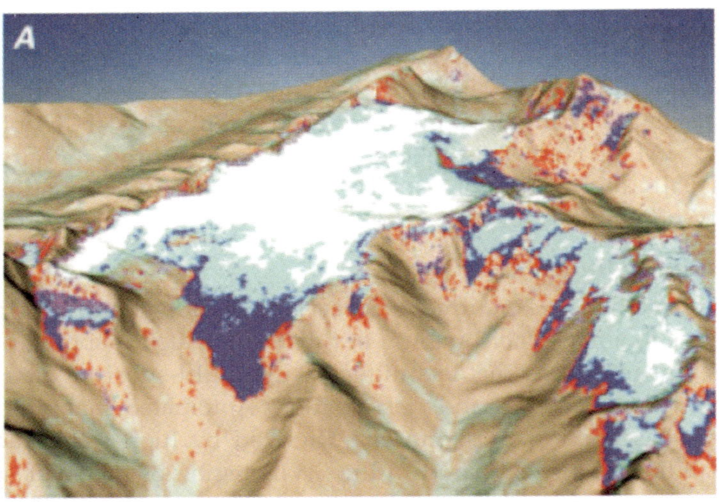


**Figure 12** A perspective view of the Illecillewaet Glacier/Interior Ranges, BC, Canada, facing the southeast. The following glacier facies are shown—accumulation area (snow): white; firn/wet snow: medium blue; bare ice (ablation area): dark blue; debris-covered ice: red; shadowed glacier: purple; meltwater lakes: green. Non-glacier surfaces are shown in grey (forested) and pink salmon (non-forested)

**Source** Wheate, Sidjak, and Wheate (2002)

### *Ratio Images*

Ratio images are simple to calculate. They at least partly eliminate the topographic effect, and are robust and independent of the investigator. The NIR/SWIR ratio has been shown to be effective for differentiating the ice and snow facies, particularly in the areas of shadow (Hall, Ormsby, Bindschadler, *et al.* 1987). Bayr, Hall, and Kovalick 1994 and Rott 1994 proposed a threshold of a ratio image of TM-4/TM-5 (NIR/SWIR) and TM-3/TM-5 (RED/SWIR) to delineate snow and ice areas. Paul (2001) evaluated both ratio image techniques and concluded that the TM-4/TM-5 ratio technique is more appropriate for clean ice glacier mapping. The ratio RED/SWIR, however, performs better in areas with dark shadow and thin debris cover (Andreassen, Paul, Kääb, *et al.* 2008). Paul and Kääb (2005) suggest an additional threshold in the BLUE channel (TM1) to minimize misclassification in shadow. Several inventories used simple and robust ratio methods (Andreassen, Paul, Kääb, *et al.* 2008; Bolch 2007; Bolch, Menounos, and Wheate, *et al.* 2010; Bolch, Yao, Kang, *et al.* 2010; Paul, Kääb, Maisch, *et al.* 2002; Racoviteanu, Paul, Raup, *et al.* 2009). Hall, Riggs, and Salomonson (1995) proposed the NDSI technique for the identification of snow. Racoviteanu, Arnaud, Williams, *et al.* (2008) successfully used the NDSI for glacier mapping of Cordillera Blanca (Northern Peru). The uncertainty of glacier mapping based on ratio images is ~2%–3% (Bolch and Kamp 2006; Paul, Kääb, Maisch, *et al.* 2002), but they are problematic with cast shadows and proglacial lakes. Lakes could be identified using the Normalized Differenced Water Index (NDWI), and in several cases also with the Normalized Differenced Vegetation Index (NDVI) (Bolch and Kamp 2006; Huggel, Kääb, Haeberli, *et al.* 2002). The main problem is to identify turbid lakes, where in addition to areas with cast shadow, manual improvements are usually necessary. Figure 13 illustrates glacier mapping using the ratio method. Repeat analysis with multi-temporal satellite data enables change analysis (Figure 14).

### *Thermal Information*

Several methods exist to map clean ice areas using multi-spectral satellite data. Thermal information would also be suitable, as snow and ice usually have a lower temperature than the surrounding areas, with the air temperatures above 0°C, although the thermal signature is only rarely utilized for mapping clean ice glaciers. Analysing surface temperatures can be a promising method to distinguish between a debris-covered glacier and the surrounding bedrock (Figure 15c) because the underlying ice cools the supraglacial debris or exposed ice cliffs, leading to a cooling effect. This is measurable on the surface with the thermal signature from satellite imagery, provided the debris cover does not exceed a certain density and/or thickness (Ranzi, Grossi, Iacovelli, *et al.* 2004). However, shaded non-glacier areas can

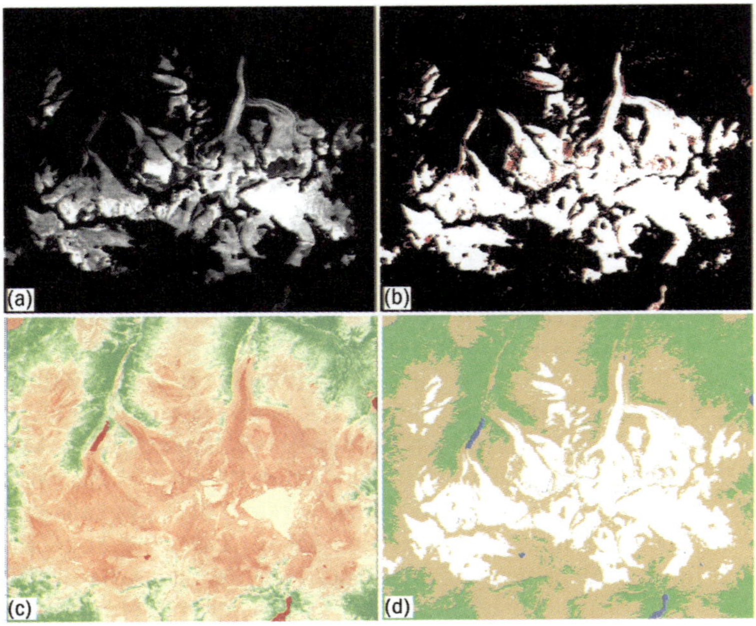


**Figure 13** Example from the Bernina Group/Swiss Alps: (a) near-infrared (NIR)/short wave infrared (SWIR) ratio image; (b) threshold NIR/SWIR>2 (white), RED/SWIR > 2 (white and red); (c) Normalized Differenced Vegetation Index (NDVI); (d) final result: glaciers (white), vegetated areas (green), water (blue), rock and debris (brown)

**Source** Bolch and Kamp (2006)

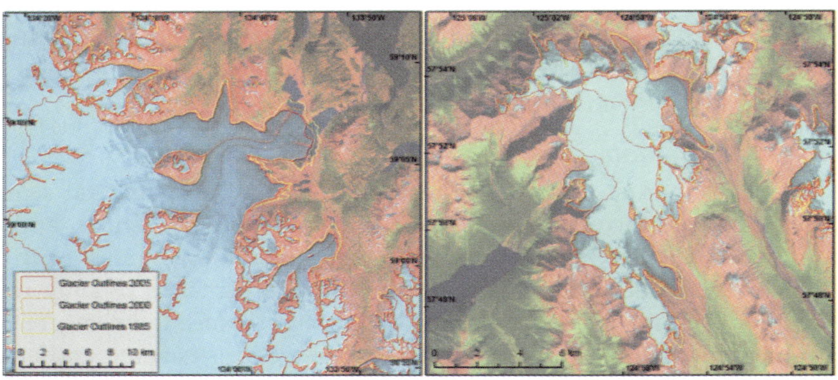

**Figure 14** Example from multi-temporal glacier outlines from Western Canada derived with the ratio method (the outlines indicate glacier recession between 1985 and 2005)

**Source** Bolch, Menounos, and Wheate (2010)

have surface temperatures similar to the clean-ice areas, but debris-covered ice with direct radiation turns out to be warmer. Lithology also has an influence, and thus, significant misclassification can occur when relying on the thermal information alone (Bhambri, Bolch, and Chaujar 2011; Shukla, Gupta, and Arora *et al.* 2010). An additional drawback is the relatively low spatial resolution of the thermal bands (60–120 m).

### *Morphometric Glacier Mapping and Multidimensional Approaches*

Morphometric parameters such as slope and curvature can be used to identify debris-covered glacier tongues (Bishop, Bonk, Kamp, *et al.* 2001; Paul, Huggel, and Kääb 2004) or their margins (Bolch and Kamp 2006; Figure 15a, b, 16a). However, these studies have high inaccuracies or are optimized for one single glacier. The main problems occur at glacier termini and glacier edges, where the transition to non-glaciated terrain is smooth, for example, if the lateral moraine is missing or not represented in the DTM. Multidimensional approaches are more promising, including a combination of different methods; for example, a combination of thermal information and morphometric parameters (Bolch, Buchroithner, Kunert, *et al.* 2007; Shukla, Gupta and Arora 2010; Bhambri, Bolch, and Chaujar 2011). Slope has been identified as the key morphometric parameter (Paul, Huggel, and Kääb 2004; Bhambri, Bolch, and Chaujar 2011). However, the threshold values differ, depending on glacier type and location; for example, it is 12° for Khumbu glacier/Himalayas (Bolch, Buchroithner, Kunert, *et al.* 2007), 24° for Oberaletschgletscher/Swiss Alps (Paul, Huggel, and Kääb 2004), less than 15° for Samudra Tapu glacier/Himachal Himalayas (Shukla, Gupta and Arora 2010), and 18° for Gangotri glacier/Garhwal Himalayas (Bhambri, Bolch, and Chaujar 2011). Using a slope threshold in addition to a supervised classification reduces the misclassification significantly (Figure 16b). Further, processing steps such as the detection of vegetation based on the NDVI, filtering,

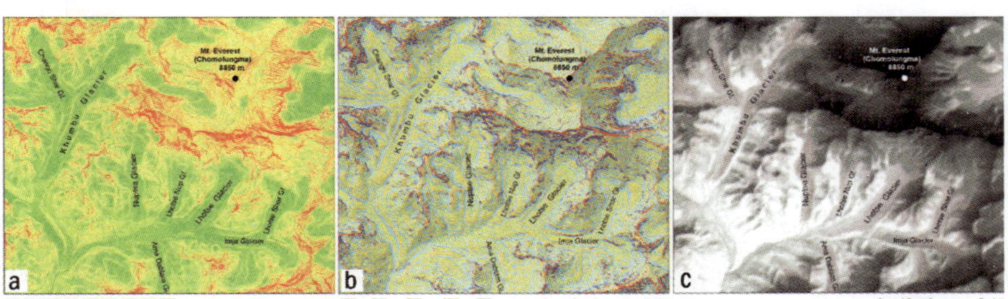


**Figure 15** Morphometric characteristics and thermal information of the Mt Everest area: slope gradient (a), curvature (b), advanced space-borne thermal emission and reflection radiometer (ASTER) thermal bands 13–12–11 (c)

**Source** Bolch, Buchroithner, Kunert, *et al.* (2007)

neighbourhood analysis, and utilizing an area threshold might be included to further reduce misclassification.

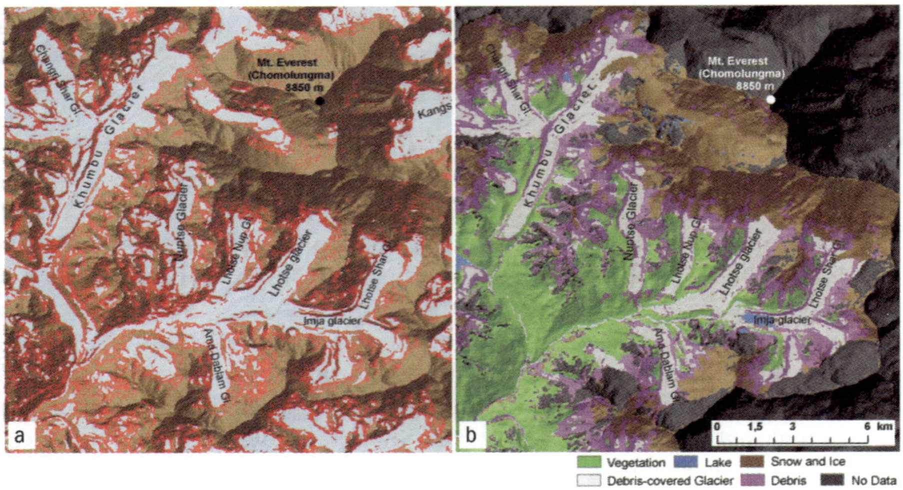


**Figure 16** Comparison of different delineation methods: (a) cluster of slope and curvature (modified after Bishop, Bonk, Kamp, *et al.* 2001); (b) combination of supervised classification (PC 1–3) and slope (<12°)

**Source** Bolch, Buchroithner, Kunert, *et al.* (2007)

## Glacier Mapping: SAR Data

In general, SAR data has the capability to differentiate between snow and ice and the surrounding areas due to the backscatter signal. The main limitation is that space-borne satellite SAR data is usually limited to one frequency, which hampers the correct classification of glaciers (Rott 1994). This is especially true if glaciers have variable surface roughness, for example, due to icefalls, crevasses, and debris cover. Hence, SAR data has played only a minor role in glacier inventories so far. The main advantage of SAR data is the possibility to differentiate different snow facies in a better manner than the optical data. SAR data is, therefore, often used to map temporal changes of snow and ice areas due to the temporal change of backscattering, best observed in the C- and X-Band (Nagler and Rott 2000). The backscattering ratio of a summer and winter image is used to distinguish between the ice and snow in summer, while a winter image is the best to obtain the longer term ELA because the SAR signal integrates over several metres and the boundary between ice and firn is averaged over some years (Floricioiu and Rott 2001). Recent efforts have focused on utilizing SAR data for the detection of glaciers in areas with frequent cloud cover and for debris-covered glaciers where the optical data is limited. One promising approach is the utilization of SAR coherence images (Paul, Strozzi, and Kääb 2010), where areas with changing surfaces (for example, due to glacier motion) decorrelate in contrast to coherent stable areas.

## Glacier Flow

There are two main possibilities to derive glacier flow at the surface from remote sensing data: (1) tracking of features on the glacier surface (*feature tracking*) and (2) differential interferometric SAR (DInSAR). The first method can be applied to both optical and radar data.

### *Feature Tracking*

A common approach to derive the flow velocity of glaciers is feature tracking using multi-temporal imagery. For glaciers, these individual features could be ice pinnacles, crevasses, boulders or a typical pattern of debris/ ice, which can be identified on both images. This is suitable for both optical and radar imagery. With radar data, measurements use the intensity images, but tracking with the coherence image is also possible (Strozzi, Luckman, Murray, *et al.* 2002). The available software incorporate image-to-image cross-correlation in conjunction with a subsequent least squares adjustment in order to precisely estimate sub-pixel accuracy. The feature tracking technique identifies corresponding image blocks on both the images (Figure 17).

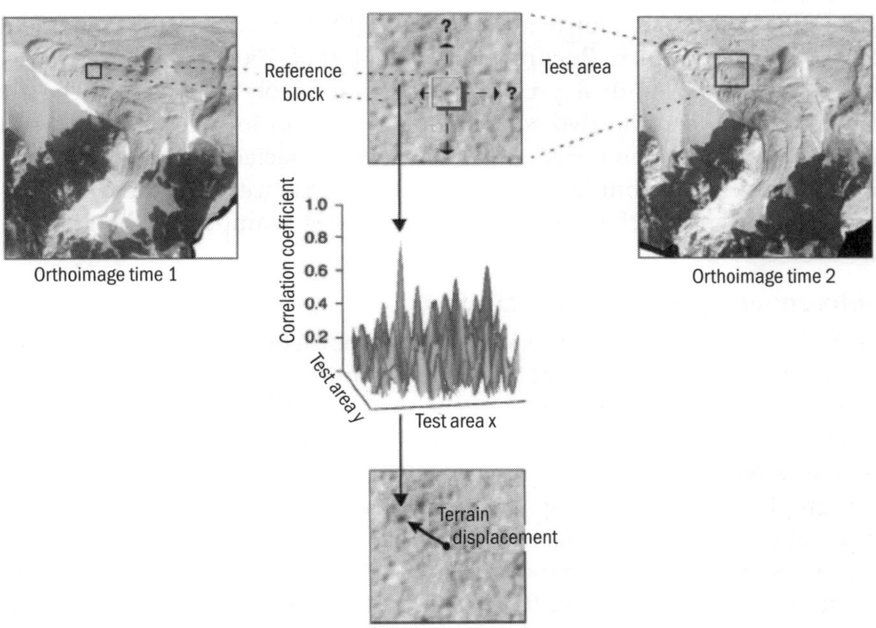


**Figure 17** Scheme of measurement surface displacement based on the identification of image blocks

**Source** Kääb and Vollmer (2000)

The method has been successfully applied to calculate velocity based on optical imagery both for large outlet glaciers (Lucchitta and Ferguson 1986; Scambos, Dutkiewicz, Wilson, *et al.* 1992) and mountain glaciers (Berthier, Vadon, Baratoux, *et al.* 2005; Bolch *et al.* 2008; Kääb 2005; Scherler, Leprince, and Strecker 2008). Table 5 provides an overview of utilized feature tracking software for deriving the velocity of mountain glaciers.

**Table 5** Example of utilized feature tracking software for mountain glaciers

| Name of software | Developer | References |
|---|---|---|
| SAR Software Dresden (SSD) 3.0 | Technische Universität Dresden | Bolch, Buchroithner, Peters, *et al.* (2008) |
| MEDICIS correlator software | Centre National d'Etude Spatiale CNES | Berthier, Vadon, Baratoux, *et al.* (2005) |
| Cosi-Corr | California Institute of Technology | Leprince, Barbot, and Ayoub (2007) Scherler, Leprince, and Strecker (2008) |
| Automatic displacement vector measurement (ADVM) | Technische Universität Graz | Kaufmann and Ladtstädter (2004) |
| Colour Image Analysis Systems (CIAS) correlation image analysis | Universtät Zürich, Universiteit Oslo | Kääb (2005); Kääb and Vollmer (2000) |

In general, all remote sensing image with repeated acquisition of the same area are suitable for feature tracking (Strozzi, Kääb, and Frauenfelder 2004; Kääb 2005; Bolch, Buchroithner, Peters, *et al.* 2008; Luckman, Quincey, and Bevan 2007), with a precise co-registration of the images. The time difference between the two acquisitions can be up to several years, with a longer time period preferred for slow moving glaciers. Feature tracking is most suitable if the aim is to obtain annual glacier velocity, the direction of flow, and long-term changes. Figure 18 provides examples of the glaciers at Mt Everest.

### *Differential Interferometry (DInSAR)*

DInSAR is used for the observation of geometric displacements. Displacements can be measured in centimetres or even millimetres, depending on the time difference between the acquisitions. In total, three overflights are necessary to determine velocities, where the first two images are used as a reference. The resulting interferogram contains information of the topography before the displacement occurred. The image of the third overflight is then used together with the image of the second overflight to retrieve another interferogram. The baseline between the two acquisition positions is optimally small to reduce the effects of topography. The second interferogram contains not only information about the topography but also the displacement. The third image is not necessary if topographic information is available, for example, from the SRTM DEM. DInSAR makes it

**Figure 18** Calculated surface velocities and flow directions for debris-covered glaciers at Mt Everest/Nepal based on advanced space-borne thermal emission and reflection radiometer (ASTER) data (left) and Ikonos data (right)

**Source** Bolch, Buchroithner, Peters, *et al.* (2008)

possible to detect sub-metre changes in images that have a nominal ground resolution of several metres, but a major problem with this technique is coherence. The coherence between two images gets lost if snowfall or melting or snow drift has occurred. To minimize this problem, the time difference between the two acquisitions should be a few days. But with the small time difference, a new problem arises: it is difficult to say how representative the extrapolated data is for annual velocities. Data from the ERS-1/2 mission is the most valuable and widely used data for deriving glacier velocity, as the time difference can be one day with a short baseline (Luckman, Quincey, and Bevan 2007; Mattar, Vachon, Geudtner, *et al.* 1998; Quincey, Richardson, Luckman, *et al.* 2007; Sharov and Etzold 2004).

Luckman, Quincey, and Bevan (2007) compared feature tracking with DInSAR for two glaciers at Mt Everest and concluded that feature tracking yields better results on the lower part of a debris-covered glacier, where the velocity is small, and SAR interferometry achieved better results on the upper part, where velocity is higher.

## Volumetric Glacier Changes/Glacier Mass Balance

The remote sensing based approaches to obtain glacier mass balance include the ELA method (Kulkarni 1992; Rabatel, Dedieu, and Vincent 2005) and the comparison of multitemporal DEMs (also known as the indirect–geodetic approach; Bamber and Kwok 2003; Bamber and Rivera 2007; Etzelmüller and Sollid 1997; Østrem and Haakensen 1995). A complex approach to derive glacier mass balance from remote sensing is the *budget approach*, which is based on ice thickness, surface and depth average velocity, and specific mass balance measurements. These parameters, however, are difficult to obtain using remote sensing alone, especially for mountain glaciers, and are so far only applied to Greenland and Antarctica (Bamber and Kwok 2003).

Due to the availability of the near-global STRM DTM and the possibility to derive more detailed DTMs based on various remote sensing data, the geodetic approach is more often used and offers possibilities for deriving accurate glacier mass balance estimation based on space-borne remote sensing in the near future (Paul and Haeberli 2008). The geodetic approach is especially valuable for debris-covered glaciers, wherein the terminus position can remain relatively stable, and glacier change is mostly recognizable through downwasting (Bolch, Buchroithner, and Benn 2011). The method is based on the assumption that a change in elevation ($dh/dr$) over time from a DTM constructed over the glacier region from various sources can be translated into a change in mass. This method holds true if (1) there is no change in elevation on the bedrock due to neo-tectonic activities, and (2) the density of ice mass has not changed (Bamber and Rivera 2007). Typically, an ice density of about 900 kg/m$^3$ is used (Berthier, Schiefer, Clarke, *et al.* 2010; Zemp, Jannsson, Holmlund, *et al.* 2010). Accurate glacier outlines and DTMs are needed as input for volume and/or mass balance calculations. However, the accuracy of mass balance and volume change estimation derived from geodetic method has several uncertainties due to errors during DTM generation in co-registration (Nuth and Kääb 2011) and ellipsoid and datum conversion, during changes in spatial resolution for the subtraction of various DTMs, along with errors occurring from the assumption of constant value for loss and gain of density of ice/firn.

Thus, the accuracy of volume change/mass balance estimation depends on the accuracy of the DTMs generated from various sources. Therefore, it is essential to estimate errors, for example, elevation inaccuracies using root mean square (RMS) errors in glaciated regions based on existing precise DTMs and global positioning system (GPS) measurements obtained during fieldwork (Figure 19). Table 6 provides a comprehensive overview of selected studies arriving at glacier volume changes based on DTM differencing.

## Extracting Glacier Parameters

Important glacier parameters such as the minimum (elevation of the terminus), maximum and mean/median elevation, slope conditions, aspect, hypsometry, width, and length can be obtained in a GIS if digital outlines and a suitable DTM are available (Schiefer, Menounos, and Wheate 2008; Paul, Barry, Cogley, *et al.* 2009). While information on elevation, slope, and aspect can be extracted directly from the DTM and its derivatives, hypsometry, width, and length require further analysis. GIS techniques

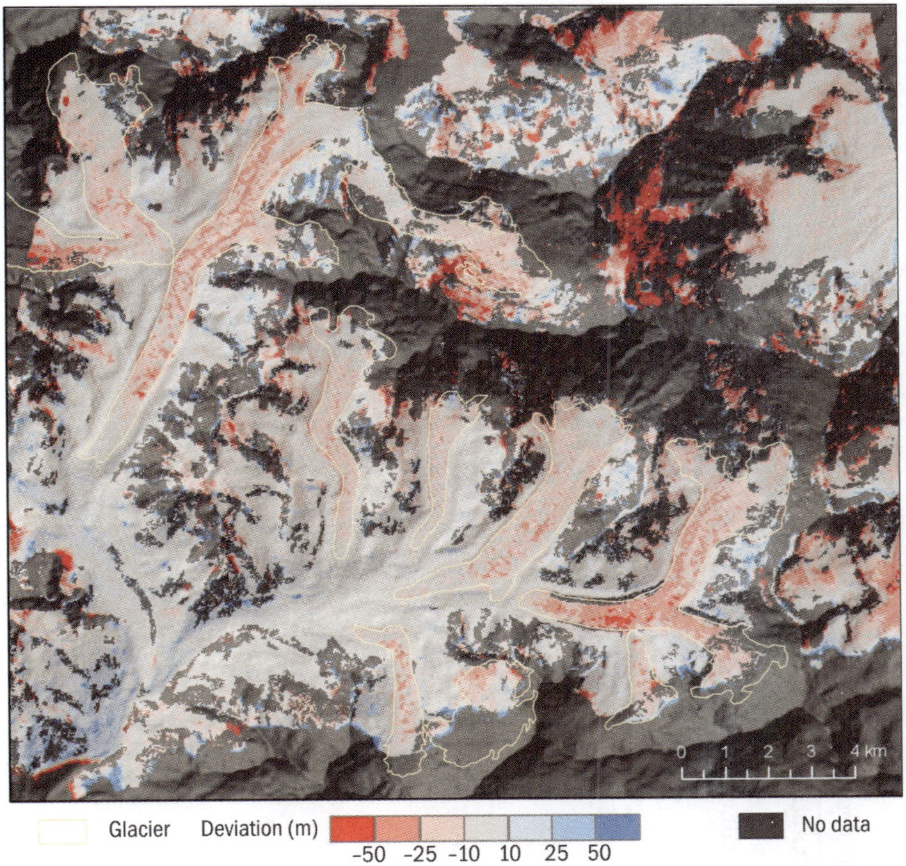


**Figure 19** Example for subtraction of bi-temporal digital terrain models (DTMs) for the calculation of glacier volume changes at Mt Everest/Nepal (the figures show inaccuracies of the utilized DTMs, which have to be taken into account when interpreting the data)

**Source** Bolch, Pieczonka, and Benn (2011)

enable extracting climatic information for the glacier area if additional regionalized climatic data is available or can be modelled using the DTM, such as potential direct solar radiation (Bolch 2007). DTMs combined with glacier outlines are also useful for calculating ice divides in order to split glaciers into their flowsheds based on hydrological analysis (Bolch, Menounos, and Wheate 2010; Manley 2008; Schiefer, Menounos, and Wheate 2008). One main problem is that ice divides are commonly situated in the accumulation area of the glaciers, where optical data is unsuitable for DTM generation due to the low contrast of snow-covered areas. Radar data also has problems due to rapid decorrelation on snow and ice, which limits applications to short repeat pass data.

**Table 6** Selected studies on volume change/mass balance estimations

| Authors | Location | DTM source | Vertical accuracy in m | Brief observations and outcomes |
|---|---|---|---|---|
| Vignon, Arnaud, and Kaser (2003) | Cordillera Blanca, Northern Peru | Topographic maps, ASTER | Topographic maps – ±10<br>ASTER– ±22 | This study estimated the volume change of three small glaciers in Cordillera Blanca and found that a small valley glacier had lost –47.106 $m^3$ water equivalent from 1962 to 2001. |
| Rivera, Casassa, Bamber, *et al.* (2005) | Southern Patàgonia | Aerial photographs, ASTER, GPS data | GPS data – ±0.30<br>Aerial photo – ±8.5<br>ASTER – ±22 | This study found a maximum ice thinning of 54 ± 0.55 m/a, observed at the glacier front from 1975 to 1997, which was supported by GPS data indicating thinning rate of 1.9 ± 0.14 m/a during 1998 to 2001. This study estimated mean net glacier mass balance of –0.29 ± 0.097 $km^3$ w.e.$a^{-1}$ from 1975 to 2001. |
| Berthier, Arnaud, Vincent, *et al.* (2006) | French Alps | SRTM, SPORT5, IGN DTM | SRTM – ±15.7<br>SPORT5 – ±1.1,<br>IGN DTM – ±21.2 | This study evaluated absolute accuracy of SRTM data for the monitoring of glacier volume variations. |
| Keutterling and Thomas (2006) | Gepatschferner glacier, Austria | Aerial photos, topographic map | Aerial photos DTM – ±2.25<br>Topographic map – ±6 | This study estimated a net loss of 26 × $10^6$ $m^3$ glacier volume (approximately 0.9%) from 1990 to 1991 on Northern Tyrol, Austria. |
| Surazakov and Aizen (2006) | Akshiirak glaciers (Tien Shan, Central Asia) | Topographic maps, SRTM | Topographic maps – ±3.3<br>SRTM – ±17 | This study estimated that the volume loss of Akshirak glacier has increased by 2.7 times during 1977 to 2000, compared with historical data from 1943 to 1997; this study found that the errors in surface change increased with the increase in slope. |
| Berthier, Arnaud, Kumar, *et al.* (2007) | Himachal Pradesh (Western Himalayas, India) | SRTM, SPORT5 | SRTM – ±18<br>SPORT5 – ±25 | This study found that the rate of ice loss was twice as higher during 1999 to 2004 than the previous long-term (1977 to 1999) mass balance records for the Himalayas. |

| Authors | Location | DTM source | Vertical accuracy in m | Brief observations and outcomes |
|---|---|---|---|---|
| Schiefer, Menounos, and Wheat *et al.* (2007) | British Columbian glaciers, Canada | Aerial photographs, SRTM, topographic maps | Aerial photographs – ±5<br>SRTM –?<br>Topographic maps – ±25 | This study found significant elevation bias in the SRTM elevation of glaciated region, which can lead to erroneous estimation of volume changes. After the bias-correction, thinning rate was estimated to be –7.8 ± 0.19 m/a in this study. |
| Rivera *et al.* (2007) | Northern Patagonia Icefield, Chile | Topographic maps, ASTER | Topographic maps – ±19<br>ASTER – ±26 | This study mapped glacier boundaries from topographic maps and satellite images and also estimated high thinning rates (up to –4.0 ± 0.97 m/a) at CHN from 1975 to 2001. |
| Racoviteanu, Manley, Arnaud, *et al.* (2007) | Nevado Coropuna, Peruvian Andes | Topographic maps, ASTER, SRTM | SRTM – ±23<br>ASTER– ±61<br>Topographic maps – ±14.7 | This study evaluated the SRTM, ASTER, and topographic DTM using GPS data for glaciological application and found an average thinning of ±5 m at the glacier surface with a significant lowering of the glacier surface at the glacier toes. |
| Bahuguna *et al.* (2007) | Himachal Himalayas | IRS IC and ID PAN | ±15m | This study estimated vertical accuracy of DTM based on the Survey of India (SOI) topographic maps. |
| Bolch *et al.* (2008) | Khumbu Himalayas | ASTER, Corona KH-4 | ASTER – ±30<br>Corona KH-4 – ±56 | Relative error ±10m This study estimated 0.19 $km^3$ volume loss for debris-covered glacier tongues from 1962 to 2002 and found highest downwasting rate >0.5 m/a for the Khumbu glacier from active to stagnant glacier part of the debris-covered glacier tongue spanning about 1 to 3 km. |

**Source** Bhambri and Bolch (2009)

DTM – digital terrain model; ASTER – advanced space-borne thermal emission and reflection radiometer; GPS – global positioning system; SRTM – shuttle radar topography mission

## SUMMARY, AND VISION FOR FUTURE

Remote sensing offers many opportunities for monitoring glaciers and their characteristics and is an important tool to study the reaction of glaciers to climate change. This is especially true as multiple glaciers can be investigated simultaneously in contrast to time-intensive field investigations. This contribution presents the state of remote sensing based analysis of mapping glaciers. Sophisticated methods exist to automatically obtain information about glacier area (supervised classification, ratio images), mass balance (DTM comparisons), motion (feature tracking, DInSAR), as well as other important glacier characteristics such as debris cover or equilibrium-line altitude (supervised classification, radar backscattering). The utilization of further GIS techniques and DTM analysis offers opportunities to derive further important parameters such as length, slope, aspect, and radiation conditions of individual glaciers and, most importantly, change analysis.

However, remote sensing analyses are still limited by many factors such as (1) the availability of suitable optical (no clouds, no seasonal snow, suitable shadow conditions) or radar data (no decorrelation, suitable acquisition angle, and direction), (2) the resolution and the channels of the data (no SWIR channels available at high resolution data), (3) problem of delineating debris-covered glacier parts correctly, (4) limited field measurements for validation and correct orthorectification (ground control points (GCPs) and direct mass balance measurements), and (5) limited data before the 1960s for long-term analysis.

Further problems arise while analysing mountain glaciers, as their location in rough terrain causes additional challenges such as cast shadow, DTM inaccuracies, and correct orthorectification. All these factors can cause errors and have to be taken into account while analysing and interpreting the data. Therefore, an error analysis or at least the presentation of possible errors should be an integrated part of a glacier change analysis. However, glacier fluctuations are significant in most mountains of the world and exceed the typical errors of remote sensing based analysis. Research activities and projects such as Global Land Ice Measurement from Space (GLIMS; Kargel, Abrams, Bishop, *et al.* 2005; Raup, Kääb, Kargel, *et al.* 2007), GlobGlacier (Paul, Kääb, Rott, *et al.* 2009), the Western Canadian Cryospheric Network (WC$^2$N), and other projects are significantly improving the availability of glacier data. Increased accessibility at no or low cost of DTM data (SRTM, Global ASTER DTM) and remote sensing data (free Landsat archive, Corona, ASTER) offer the opportunity to researchers and institutions with limited budgets to study glaciers in detail. The potential accuracy of remote sensing analysis will improve in the future with the availability of more high resolution data from new sensors or missions such as Terra-SAR-X and RadarSat-2 (radar data), Kompsat, Worldview-2 and Eros-C (optical data), TanDTM-X, and SPOT 5 (DTM data).

## References

Al-Rousan N, Cheng P, Petrie G, Toutin T, Valadan Zoej M J. 1997. **Automated DEM extraction and orthoimage generation from SPOT level 1B imagery**. *Photogrammetric Engineering and Remote Sensing* **63**: 965–974

Altmaier A and Kany C. 2002. **Digital surface model generation from CORONA satellite images**. *ISPRS Journal of Photogrammetrie and Remote Sensing* **56**(4): 221

Andreassen L M, Elvehøy H, Kjøllmoen B, Engeset R V, Haakensen N. 2005. **Glacier mass balance and length variations in Norway**. *Annals of Glaciology* **42**: 317–325

Andreassen L M, Paul F, Kääb A, Hausberg J E. 2008. **Landsat-derived glacier inventory for Jotunheimen, Norway, and deduced glacier changes since the 1930s**. *The Cryosphere* **2**(2): 131–145

Baltsavias E, Favey E, Bauder A, Bosch H, Pateraki M. 2001. **Digital surface modelling by airborne laser scanning and digital photogrammetry for glacier monitoring**. *Photogrammetric Record* **17:** 243–273

Bamber J L and Kwok R. 2003. **Remote-sensing techniques**. In *Mass Balance of the Cryosphere: observations and modelling of contemporary and future changes*, edited by J L Bamber and A J Payne. Cambridge: Cambridge University Press

Bamber J L and Rivera A. 2007. **A review of remote sensing methods for glacier mass balance determination**. *Global and Planetary Change* **59**(1–4): 138–148

Bamler R. 1999. *The SRTM Mission: a world-wide 30 m resolution DEM from SAR interferometry in 11 days*, pp. 145–154. Photogrammetric Week 1999. Heidelberg: Wichmann Verlag

Bayr K J, Hall D K, and Kovalick W M. 1994 **Observations on glaciers in the eastern Austrian Alps using satellite data**. *International Journal of Remote Sensing* **15**: 1733–1742

Benn D I and Evans D J. 1998. ***Glaciers and Glaciation***. Hodder Arnold. 760 pp.

Berry P A M, Garlick J D, and Smith R G. 2007. **Near-global validation of the SRTM DEM using satellite radar altimetry**. *Remote Sensing of Environment* **106**(1): 17–27

Berthier E and Toutin T. 2008. **SPOT5-HRS digital elevation models and the monitoring of glacier elevation changes in North-West Canada and South-East Alaska**. *Remote Sensing of Environment* **112**(5): 2443–2454

Berthier E, Vadon H, Baratoux D, Arnaud Y, Vincent C, Feigl K L, Rémy F, Legrésy B. 2005. **Surface motion of mountain glaciers derived from satellite optical imagery**. *Remote Sensing of Environment* **95**(1): 14–28

Berthier E, Arnaud Y, Vincent C, Rémy F. 2006. **Biases of SRTM in high-mountain areas. Implications for the monitoring of glacier volume changes**. *Geophysical Research Letters* **33**: L08502, doi:10.1029/2006GL025862

Berthier E, Arnaud Y, Kumar R, Sarfaraz A, Wagnon P, Chevallier P. 2007. **Remote sensing estimates of glacier mass balances in the Himachal Pradesh (Western Himalaya, India)**. *Remote Sensing of Environment* **108**(3): 327–338

Bhambri R and Bolch T. 2009. **Glacier mapping and glacier change in Indian Himalaya: a review and new results**. *Progress in Physical Geography* **33**(5): 672–704

Bhambri R, Bolch T, and Chaujar R K. 2011. **Automated mapping of debris-covered glaciers in the Garhwal Himalayas using ASTER DEMs and multi-spectral data**. *International Journal of Remote Sensing*

Bignone F and Umakawa H. 2008. **Assessment of ALOS PRISM digital elevation model extraction over Japan**. *The International Archives of the Photogrammetry, Remote Sensing and Spatial Information Sciences.* ***XXXVII*****:** 1135–1138

Bishop M P and Colby J. 2002. **Automated anisotropic reflectance correction of satellite imagery in mountainous terrain**. *International Journal of Remote Sensing* **23**: 2125–2131

Bishop M, Shroder J, and Colby J. 2003. **Remote sensing and geomorphometry for studying relief production in high mountains**. *Geomorphology* **55**: 345–361

Bishop M P, Bonk R, Kamp U, Shroder J F. 2001. **Terrain analysis and data modeling for alpine glacier mapping**. *Polar Geography* **25**: 182–201

Bishop M, Colby J, Luvall J, Quattrochi D, Rickman D. 2005. **Remote-sensing science and technology for studying mountain environments**. In *Geographic Information Science and Mountain Geomorphology*, pp. 147–180, edited by M Bishop and J Shroder. Berlin: Heidelberg; New York: Springer

Bolch T. 2007. **Climate change and glacier retreat in northern Tien Shan (Kazakhstan/Kyrgyzstan) using remote sensing data**. *Global and Planetary Change* **56**: 1–12

Bolch T and Kamp U. 2006. Glacier **Mapping in High Mountains using DEMs, Landsat and ASTER Data,** pp. 13–24. *Grazer Schriften der Geographie und Raumforschung* 41. In Proceedings of the 8th International Symposium on High Mountain Remote Sensing Cartography, 20–27 March 2005, La Paz, Bolivia

Bolch T, Kamp U, and Olsenholler J. 2005. **Using ASTER and SRTM DEMs for studying geomorphology and glaciation in high mountain areas**. In *New Strategies for European Remote Sensing*, pp. 119–127, edited by M Oluic. Rotterdam: Millpress

Bolch T, Menounos B, and Wheate R D. 2010. **Western Canadian Glacier Inventory 1985–2005 and regional glacier recession rates**. *Remote Sensing of Environment* **114**(1): 127–137

Bolch T, Pieczonka T, and Benn D. 2011. **Multi-decadal mass loss of glaciers in the Everest area (Nepal, Himalaya) derived from stereo imagery**. *The Cryosphere* **5**: 349–358

Bolch T, Buchroithner M F, Kunert A, Kamp U. 2007. **Automated delineation of debris-covered glaciers based on ASTER data**. In *GeoInformation in Europe*, pp. 403–410, edited by M A Gomarasca (Proceedings of the 27th EARSeL Symposium, 4–7 June 2007, Bozen, Italy). Millpress

Bolch T, Buchroithner M F, Peters J, Baessler M, Bajracharya S R. 2008. **Identification of glacier motion and potentially dangerous glacier lakes at Mt. Everest area/Nepal using spaceborne imagery**. *Natural Hazards and Earth System Sciences* **8**(6): 1329–1340

Bolch T, Buchroithner M F, Pieczonka T, Kunert A. 2008. **Planimetric and volumetric glacier changes in Khumbu Himalaya since 1962 using Corona, Landsat TM and ASTER data**. *Journal of Glaciology* **54**(187): 592–600

Bolch T, Yao T, Kang S, Buchroithner M F, Scherer D, Maussion F, Huintjes E, Schneider C. 2010. **A glacier inventory for the western Nyainqentanglha Range and Nam Co Basin, Tibet, and glacier changes 1976–2009**. *The Cryosphere* **4**: 419–433

Brocklehurst S H and Whipple K X. 2004. **Hypsometry of glaciated landscapes**. *Earth Surface Processes and Landforms* **29**(7): 907–926

Brunner K. 1987. **Glacier mapping in the Alps**. *Mountain Research and Development*: 375–385

Brunner K. 2006. **Karten dokumentieren den Rückzug der Gletscher seit 1850**. *Wiener Schriften zur Geographie und Kartographie* **17**: 191–200

Colby J. 1991. **Topographic normalization in rugged terrain**. *Photogrammetric Engineering and Remote Sensing* **57**(5): 531–537

Dashora A, Lohani B, and Malik J N. 2007. **A repository of earth resource information – CORONA satellite programme**. *Current Science* **92**(7): 926–932

Della Ventura A, Rampini A, Rabagliati R, Barbero R S. 1987. **Development of a satellite remote sensing technique for the study of alpine glaciers**. *International Journal of Remote Sensing* **8**: 203–215

Desmet P J J. 1997. **Effects of interpolation errors on the analysis of DEMs**. *Earth Surface Processes and Landforms* **22**(6): 563–580

Eckstrand S. 1996. **Landsat TM-based forest damage assessment: correction for topographic effects**. *Photogrammetric Engineering and Remote Sensing* **62**(2): 151–161

Etzelmüller B and Sollid J L. 1997. **Glacier geomorphometry – an approach for analysing long-term glacier surface changes using grid-based digital elevation models**. *Annals of Glaciology* **24**: 135–141

Etzelmüller B, Ødegård R S, Berthling I, Sollid J L. 2001. **Terrain parameters and remote sensing data in the analysis of permafrost distribution and periglacial processes: principles and examples from southern Norway**. *Permafrost and Periglacial Processes* **12**(1): 79–92

Evans D L, Alpers W, Cazenave A, Elachi C, Farr T G, Glackin D, Holt B, Jones L, Liu W T, McCandless W, Menard Y, Moore R, Njoku E. 2005. **Seasat – a 25-year legacy of success**. *Remote Sensing of Environment* **94**(3): 384–404

Falorni G, Teles V, Vivoni E R, Bras R L, Amartunga K S. 2005. **Analysis and characterization of the vertical accuracy of digital elevation models from the Shuttle Radar Topography Mission**. *Journal of Geophysical Research* **110**(F02005)

Finsterwalder R. 1984. **Utilization of orthophotographs in the mapping of high mountain regions**. *Mountain Research and Development*: 315–318

Floricioiu D and Rott H. 2001. **Seasonal and short-term variability of multifrequency, polarimetric backscatter from alpine terrain from SIR-C/X-SAR and AIRSAR data.** *IEEE Transactions on Geoscience and Remote Sensing* **39**: 2638–2638

Gens R and Van Genderen JL. 1996. **Review article SAR interferometry-issues, techniques, applications**. *International Journal of Remote Sensing* **17**(10): 1803–1835

Gratton D J, Howarth P J, and Marceau D J. 1990. **Combining DEM parameters with Landsat MSS and TM imagery in a GIS for mountain glacier characterization**. *IEEE Transactions on Geoscience and Remote Sensing* **28**: 766–769

Hagg W, Braun L, Uvarov V N, Makarevich K G. 2004. **A comparison of three methods of mass balance determination in the Tuyuksu glacier region, Tien Shan**. *Journal of Glaciology* **50**(171): 505–510

Hall D K and Martinec J. 1985. ***Remote Sensing of Ice and Snow***. London and New York: Chapman and Hall. 189 pp.

Hall D K, Williams R S, and Bayr K J. 1992. **Glacier recession in Iceland and Austria**. *EOS Transactions of the American Geophysical Union* **73**(12): 129, 135, 141

Hall D K, Riggs G A, and Salomonson V V. 1995. **Development of methods for mapping global snow cover using Moderate Resolution Imaging Spectroradiometer (MODIS) data.** *Remote Sensing of Environment* **54**: 127–140

Hall D K, Ormsby J P, Bindschadler R A, Siddalingaiah H. 1987. **Characterization of snow and ice reflectance zones on glaciers using Landsat TM data**. *Annals of Glaciology* **9**: 104–108

Hayakawa Y S, Oguchi T, and Lin Z. 2008. **Comparison of new and existing global digital elevation models: ASTER G-DEM and SRTM-3**. *Geophysical Research Letters* **35**: L17404

Hellmann M. 2000. **Classification of fully polarimetric SAR-data for cartographic applications**. Deutsches Zentrum für Luft- und Raumfahrt e.V., Forschungsbericht 2000–19, Oberpfaffenhofen

Hewitt K. 2007. **Tributary glacier surges: an exceptional concentration at Panmah Glacier, Karakoram Himalaya**. *Journal of Glaciology* **53**(181): 181–188

Huggel C, Kääb A, Haeberli W, Teysseire P, Paul F. 2002. **Remote sensing based assessment of hazards from glacier lake outbursts: a case study in the Swiss Alps**. *Canadian Geotechnical Journal* **39**: 316–330

Hugli H and Frei W. 1983. **Understanding anisotropic reflectance in mountainous terrain**. *Photogrammetric Engineering and Remote Sensing* **49**: 671–683

IPCC (Intergovernmental Panel on Climate Change) (ed.). 2007. ***Climate Change 2007 – the physical science basis***. Working Group I: Contribution to the Fourth Assessment Report of the Intergovernmental Panel on Climate Change. Cambridge University Press

Jacobsen K. 2005. **DEMs based on space images versus SRTM height models**. Proceedings of the ASPRS 2005 Annual Conference, Baltimore, Maryland

Kääb A. 2002. **Monitoring high-mountain terrain deformation from repeated air- and spaceborne optical data: examples using digital aerial imagery and ASTER data.** *Journal of Photogrammetry and Remote Sensing* **57**: 39–52

Kääb A. 2005. **Combination of SRTM3 and repeat ASTER data for deriving alpine glacier flow velocities in the Bhutan Himalaya.** *Remote Sensing of Environment* **94**(4): 463–474

Kääb A. 2008. **Glacier volume changes using ASTER satellite stereo and ICESat GLAS laser altimetry. A test study on Edgeøya, Eastern Svalbard.** *IEEE Transactions on Geoscience and Remote Sensing* **46**(10): 2823–2830

Kääb A and Vollmer M. 2000. **Surface geometry, thickness changes and flow fields on Creeping mountain permafrost: automatic extraction by digital image analysis.** *Permafrost and Periglacial Processes* **11**(4): 315–326

Kääb A, Huggel C, Paul F, Wessels R, Raup B, Kieffer H, Kargel J. S. 2003. **Glacier monitoring from ASTER imagery. Accuracy and applications,** pp. 43–53. Proceedings of the EARSeL eProceedings 2 – Observing our cryosphere from space

Kamp U, Bolch T, and Olsenholler J. 2003. **DEM generation from ASTER satellite data for geomorphometric analysis of Cerro Sillajhuay, Chile/ Bolivia.** Proceedings of the Annual Meeting Imaging and Geospatial Information Society (ASPRS), Anchorage. CD-ROM, 9

Kamp U, Bolch T, and Olsenholler J. 2005. **Geomorphometry of Cerro Sillajhuay, Chile/Bolivia: comparison of DEMs derived from ASTER remote sensing data and contour maps.** *Geocarto International* **20**(1): 23–34

Kargel J S, Abrams M J, Bishop M P, Bush A, Hamilton G, Jiskoot H, Kääb A, Kieffer H H, Lee E M, Paul F, Rau F, Raup B, Shroder J, Soltesz D, Stainforth D, Stearns L, Wessels R. 2005. **Multispectral imaging contributions to global land ice measurements from space.** *Remote Sensing of Environment* **99**(1–2): 187–219

Kaufmann V and Ladtstädter R. 2004. **Documentation of the movement of the Hinteres Langtalkar rock glacier,** pp. 893–898. Proceedings of the 20th Congress of the International Society for Photogrammetry and Remote Sensing, 12–23 July 2004, Istanbul, Turkey, IAPRS 35(Part B7)

Kervyn F. 2001. **Modelling topography with SAR interferometry: illustrations of a favourable and less favourable environment.** *Computers and Geosciences* **27**: 1039–1050

Keutterling A and Thomas A. 2006. **Monitoring glacier elevation and volume changes with digital photogrammetry and GIS at Gepatschferner glacier, Austria.** *International Journal of Remote Sensing*: 4371–4380

Kotlyakov V M, Osipova G B, and Tsvetkov D G. 2008. **Monitoring surging glaciers of the Pamirs, central Asia, from space.** *Annals of Glaciology* **48**: 125–134

Krimmel R M and Meier M F. 1975. **Glacier applications of ERTS – 1 images.** *Journal of Glaciology* **15**: 391–402

Kulkarni A V. 1992. **Mass balance of Himalayan glaciers using AAR and ELA methods.** *Journal of Glaciology*: 101–104

Leprince S, Barbot S, and Ayoub F. 2007. **Automatic and precise orthorectification, coregistration, and subpixel correlation of satellite images, application to ground deformation measurements.** *IEEE Transactions on Geoscience and Remote Sensing* **45**(6): 1529–1558

Liang S L, Fang H L, and Chen M Z. 2001. **Atmospheric correction of Landsat ETM+ and surface imagery – part I: methods**. *IEEE Transactions on Geoscience and Remote Sensing* **39**: 2490–2498

Lucchitta B K and Ferguson H M. 1986. **Antarctica: measuring glacier velocity from satellite images**. *Science* **234**(4780): 1105–1108

Luckman A, Quincey D, and Bevan S. 2007. **The potential of satellite radar interferometry and feature tracking for monitoring flow rates of Himalayan glaciers**. *Remote Sensing of Environment* **111**(2–3): 172–181

Manley W F. 2008. **Geospatial inventory and analysis of glaciers: a case study for the eastern Alaska range**. In *Satellite Image Atlas of Glaciers of the World*, edited by R S Williams and J G Ferrigno. USGS Professional Paper 1386-K

Martin C (ed.). 2002. ***Lexikon der Geowissenschaften***. Heidelberg: Spektrum Akademischer Verlag

Mattar K E, Vachon P W, Geudtner D, Gray A L, Cumming I G, Brugman M. 1998. **Validation of alpine glacier velocity measurements using ERS-Tandem-Mission SAR data**. *IEEE Transactions on Geoscience and Remote Sensing* **36**(3): 974–984

McDonald R A. 1995. **CORONA: success for space reconnaissance. A look into the Cold War, and a revolution for intelligence**. *Photogrammetric Engineering and Remote Sensing* **61**: 689–720

Moreira A, Krieger G, Hajnsek I, Hounam D, Werner M, Riegger S, Settelmeyer E. 2004. **TanDEM-X: a TerraSAR-X add-on satellite for single-pass SAR interferometry**. Proceedings of the IGARSS Conference at Alaska, September 2004. 4.

Muskett R R, Lingle C S, Sauber J M, Rabus B T, Tangborn W V. 2008. **Acceleration of surface lowering on the tidewater glaciers of Icy Bay, Alaska, U.S.A. from InSAR DEMs and ICESat altimetry**. *Earth and Planetary Science Letters* **265**(3–4): 345–359

Nagler T and Rott H. 2000. **Retrieval of wet snow by means of multitemporal SAR data**. *IEEE Transactions Geoscience Remote Sensing* **38**(2): 754–765

Nuth C and Kääb A. 2011. **Co-registration and bias corrections of satellite elevation data sets for quantifying glacier thickness change.** *The Cryosphere* **5**(1): 271–290

Østrem G. 1975. **ERTS-1 data in glaciology – an effort to monitor glacier mass balance from satellite imagery**. *Journal of Glaciology* **15**: 403–415

Østrem G and Haakensen N. 1999. **Map comparison or traditional mass-balance measurements: which method is better?** *Geografiska Annaler*: 703–711

Ouaidrari H and Vermote E F. 1999. **Operational atmospheric correction of Landsat TM data**. *International Journal of Remote Sensing* **70**: 4–15

Paterson W. 1994. ***The Physics of Glaciers***. New York: Pergamon

Paul F. 2001. **Evaluation of different methods for glacier mapping using Landsat TM**. *EARSeL eProceedings* (1): 239–245

Paul F and Haeberli W. 2008. **Spatial variability of glacier elevation changes in the Swiss Alps obtained from two digital elevation models**. *Geophysical Research Letters* **35**: L21502

Paul F and Kääb A. 2005. **Perspectives on the production of a glacier inventory from multispectral satellite data in Arctic Canada: Cumberland Peninsula, Baffin Island.** *Annals of Glaciology* **42**: 59–66

Paul F, Huggel C, and Kääb A. 2004. **Combining satellite multispectral image data and a digital elevation model for mapping of debris-covered glaciers.** *Remote Sensing of Environment* **89**(4): 510–518

Paul F, Strozzi T, and Kääb A. 2010. **Mapping clean and debris-covered glaciers from Palsar coherence images.** *Geophysical Research Abstracts* **12**: EGU2010-13091

Paul F, Kääb A, Maisch M, Kellenberger T, Haeberli W. 2002. **The new remote sensing derived Swiss glacier inventory: I. Methods.** *Annals of Glaciology* **34**: 355–361

Paul F, Barry R, Cogley J G, Frey H, Haeberli W, Ohmura A, Ommanney C S L, Raup B, Rivera A, Zemp M. 2009. **Recommendations for the compilation of glacier inventory data from digital sources.** *Annals of Glaciology* **50**: 119–126

Paul F, Kääb A, Rott H, Shepherd A, Strozzi T, Volden E. 2009. **GlobGlacier: a new ESA project to map the World's glaciers and ice caps from space.** *EARSeL eProceedings* **8**(1): 11–25

Pieczonka T, Bolch T, and Buchroithner M F. 2011. **Generation and evaluation of multi-temporal digital terrain models of the Mt. Everest area from different optical sensors.** *ISPRS Journal of Photogrammetry and Remote Sensing* (accepted)

Pradhan B, Bolch T, and Buchroithner M F. 2009. **Elevation modelling using radargrammetry for tropical terrain.** Proceedings of the AGILE Conference, 2–5 June 2009, Hannover, Germany

Quincey D J, Richardson S D, Luckman A, Lucas R M, Reynolds J M, Hambrey M J, Glasser N F. 2007. **Early recognition of glacial lake hazards in the Himalaya using remote sensing datasets.** *Global and Planetary Change* **56**(1–2): 137–152

Rabatel A, Dedieu J-P, and Vincent C. 2005. **Using remote-sensing data to determine equilibrium-line altitude and mass-balance time series: validation on three French glaciers.** *Journal of Glaciology*: 539–546

Rabus B, Eineder M, Roth A, Bamler R. 2003. **The shuttle radar topography mission – a new class of digital elevation models acquired by spaceborne radar.** *Journal of Photogrammetry and Remote Sensing* **57**(4): 241–262

Racoviteanu A E, Manley W F, Arnaud Y, Williams M W. 2007. **Evaluating digital elevation models for glaciologic applications: an example from Nevado Coropuna, Peruvian Andes.** *Global and Planetary Change* **59**(1–4): 110–125

Racoviteanu A E, Arnaud Y, Williams M W, Ordonez J. 2008. **Decadal changes in glacier parameters in the Cordillera Blanca, Peru, derived from remote sensing.** *Journal of Glaciology* **54**(186): 499–510

Racoviteanu A E, Paul F, Raup B, Khalsa S J S, Armstrong R. 2009. **Challenges and recommendations in mapping of glacier parameters from space: results of the 2008 Global Land Ice Measurements from Space (GLIMS) workshop, Boulder, Colorado, USA**. *Annals of Glaciology* **50**: 53–69

Ranzi R, Grossi G, Iacovelli L, Taschner T. 2004. **Use of multispectral ASTER images for mapping debris-covered glaciers within the GLIMS Project**, vol. II, pp. 1144–1147. IEEE International Geoscience and Remote Sensing Symposium

Raup B, Kääb A, Kargel J S, Bishop M P, Hamilton G, Lee E, Paul F, Rau F, Soltesz D, Khalsa S J S, Beedle M, Helm C. 2007. **Remote sensing and GIS technology in the Global Land Ice Measurements from Space (GLIMS) Project**. *Computers and Geosciences* **33**(1): 104–125

Rémy F and Parouty S. 2009. **Antarctic ice sheet and radar altimetry: a review.** *Remote Sensing Letters* **1**: 1212–1239

Riano D, Chuvieco E, Salas J, Aguado I. 2003. **Assessment of different topographic corrections in Landsat-TM data for mapping vegetation types**. *IEEE Transactions on Geoscience and Remote Sensing* **41**(5): 1056–1061

Richter R. 2003. **Atmospheric/topographic correction for satellite imagery (ATCOR-2/3 User Guide, Vers. 5.5)**. Wessling, IMF, DLR

Rignot E, Rivera A, and Casassa G. 2003. **Contribution of the Patagonia icefields of South America to sea level rise**. *Science* **302**(5644): 434–437

Rivera A, Casassa G, Bamber J L, Kääb A. 2005. **Ice elevation changes on Glacier Chico, Southern Patagonian Icefield, Chile**. *Journal of Glaciology*: 105–112

Rott H. 1976. **Analyse der Schneeflächen auf Gletschern der Tiroler Zentralalpen aus Landsat Bildern**. *Zeitschrift für Gletscherkunde und Glazialgeologie* **12**: 1–28

Rott H. 1994. **Thematic studies in alpine areas by means of polarimetric SAR and optical imagery**. *Advances in Space Research* **14**: 217–226

Sabins F. 1997. ***Remote Sensing, Principles and Interpretation***. New York: W H Freeman and Company

Scambos T A, Dutkiewicz M J, Wilson J C, Bindschadler R A. 1992. **Application of image cross-correlation to the measurement of glacier velocity using satellite image data.** *Remote Sensing of Environment* **42**(3): 177–186

Scherler D, Leprince S, and Strecker M R. 2008. **Glacier-surface velocities in alpine terrain from optical satellite imagery – accuracy improvement and quality assessment**. *Remote Sensing of Environment* **112**(10): 3806–3819

Schiefer E, Menounos B, and Wheate R D. 2007. **Recent volume loss of British Columbia glaciers, Canada**. *Geophysical Research Letters* **34**: L16503, doi:10.1029/2007GL030780

Schiefer E, Menounos B, and Wheate R D. 2008. **An inventory and morphometric analysis of British Columbia glaciers, Canada**. *Journal of Glaciology* **54**(186): 551–560

Schutz B E, Zwally H J, Shuman C A, Hancock D, DiMarzio P J. 2005. **Overview of the ICESat mission**. *Geophysical Research Letters* **32**: L21S01

Sharov A I and Etzold S. 2004. **Full-value mapping of glacier rheology using repeat pass SAR interferograms.** ESA SP-550, June 2004. Proceedings of FRINGE Workshop, Frascati, Italy, 1–5 Dec 2003

Sidjak R W and Wheate R D. 1999. **Glacier mapping of the Illecillewaet icefield, British Columbia, Canada, using Landsat TM and digital elevation data.** *International Journal of Remote Sensing* **20**: 273–284

Smith B and Sandwell D. 2003. **Accuracy and resolution of shuttle radar topography mission data.** *Geophysical Research Letters* **30**(9): 1467

Song C, Woodcock C, Seto K, Lenney M P, Macomber S. 2001. **Classification and change detection using Landsat TM data: when and how to correct atmospheric effects?** *Remote Sensing of Environment* **75**: 230–244

Strozzi T, Kääb A, and Frauenfelder R. 2004. **Detecting and quantifying mountain permafrost creep from in situ inventory, space-borne radar interferometry and airborne digital photogrammetry,** *International Journal of Remote Sensing* **25**: 2919–2931

Strozzi T, Luckman A, Murray T, Wegmuller U, Werner C L. 2002. **Glacier motion estimation using SAR offset-tracking procedures.** *IEEE Transactions on Geoscience Remote Sensing* **40**(11): 2384–2391

Surazakov A B and Aizen V B. 2006. **Estimating volume change of mountain glaciers using SRTM and topographic data.** *IEEE Transactions Geoscience Remote Sensing* **44**(10): 2991–2995

Svoboda F and Paul F. 2009. **A new glacier inventory on southern Baffin Island, Canada, from ASTER data: I. Applied methods, challenges and solutions.** *Annals of Glaciology* **50:** 11–21

Teillet P M, Guindon B, and Goodeonugh D G. 1982. **On the slope-aspect correction of multispectral scanner data.** *Canadian Journal of Remote Sensing* **8**: 84–106

Toutin T. 2000. **Evaluation of radargrammetric DEM from RADARSAT images in high relief areas.** *IEEE Transactions on Geoscience and Remote Sensing* **38**(2): 782–789

Toutin T. 2006. **Comparison of 3D physical and empirical models for generating DSMs from stereo HR images.** *Photogrammetric Engineering and Remote Sensing* **72**(5): 597–604

Toutin T. 2008. **ASTER DEMs for geomatic and geoscientific applications: a review.** *International Journal of Remote Sensing* **29**(7): 1855–1875

Toutin T and Gray L. 2000. **State-of-the-art of elevation extraction from satellite SAR data.** *ISPRS Journal of Photogrammetry and Remote Sensing* **55**(1): 13–33

Trisakti B, Caorolita I, and Pradana F. 2009. **Digital elevation model from PRISM-ALOS and ASTER stereoscopic data.** *International Journal of Remote Sensing and Earth Sciences* **6:** 29–38

Vignon F, Arnaud Y, and Kaser G. 2003. **Quantification of glacier volume change using topographic and ASTER DEMs,** pp. 2605–2607. Proceedings of the International Geoscience and Remote Sensing Symposium IGARSS' 03, Toulouse

WGMS–UNEP (World Glacier Monitoring Service–United Nations Environment Programme). 2008. ***Global Glacier Changes: facts and figures***. Chatelaine, Switzerland: DEWA/GRID-Europe

Wheate R D, Sidjak R W, and Whyte G T. 2002. **Mapping glaciers in the Interior Ranges and Rocky Mountains with Landsat data**. Satellite Image Atlas of the World, U.S. Geological Survey. Professional Paper 1386-J-1: J100-110

Williams M W. 1986. **Glacier inventories of Iceland: evaluation and use of sources of data**. *Annals of Glaciology* **8**: 184–191

Williams M W and Ferrigno (eds). 2002. **Satellite image atlas of the glaciers of the world**. USGS Professional Paper 1386

Yamane N, Fujita K, Nonaka T, Shibayama T, Takagishi S. 2008. **Accuracy evaluation of DEM derived by TerraSAR-X data in the Himalayan region**. *The International Archives of the Photogrammetry, Remote Sensing and Spatial Information Sciences* **XXXVII**(Part B7, Commission VII): 203–208

Zemp M, Jansson P, Holmlund P, Gärtner-Roer I, Koblet T, Thee P, Haeberli W. 2010. **Reanalysis of multi-temporal aerial images of Storglaciären, Sweden (1959–99) – Part 2: Comparison of glaciological and volumetric mass balances.** *The Cryosphere* **4:** 345–357

Zwally H J, Schutz B, Abdalati W, Abshire J, Bentley C, Brenner A, Bufton J, Dezio J, Hancock D, Harding D, Herring T, Minster B, Quinn K, Palm S, Spinhirne J, Thomas R. 2002. **ICESat's laser measurements of polar ice, atmosphere, ocean, and land**. *Journal of Geodynamics* **34**(3–4): 405–445

# Geoinformatics and the Mapping of Lands Vulnerable to Sea Level Rise

**Stephen S Young**

*Department of Geography, Salem State University, Salem, MA, USA*

*E-mail: syoung@salemstate.edu*

## INTRODUCTION

Throughout the 20th century and into the 21st century, the global sea levels have been rising after an extended period of little change for thousands of years (Church and White 2006). The main reason for the rising sea levels has been global warming (IPCC 2007). A scientific consensus has emerged that the build-up of atmospheric concentrations of carbon dioxide and other greenhouse gases is the main driving force of a warming global climate (Keeling, Whorf, Wahlen, *et al.* 1995; IPCC 1996, 2007). With focused attention and research on the issue of sea level rise, a clear picture of the current and past rates of sea level rise has emerged, but predictions on future rise still hold tremendous uncertainty, with potential estimates increasing annually as the rate of sea level rise is accelerating (Warrick, Oerlemans, Woodworth, *et al.* 1996; Church and White 2006; Climate Office 2009).

Currently, there are four main reasons for sea level rise: (1) as the ocean warms, the water expands and increases in volume (Church, Gregory, Huybrechts, *et al.* 2001); (2) when ice on land, glaciers, and ice sheets melts, water in the oceans increases (Church, Gregory, Huybrechts, *et al.* 2001); (3) recent modelling of changing currents will make certain regions of the world experience higher sea levels, such as north-eastern US (Yin, Schlesinger, and Stouffer 2009); and (4) the land in many coastal regions is subsiding, leading to a relative rise in sea level (Emery and Aubrey 1991). Although land subsidence is a natural process, it can be enhanced by human activity. For example, the significant subsidence of several metres during the 20th century experienced in Tokyo and Shanghai was due to excessive groundwater withdrawal (Nicholls 1995).

The average rate of sea level rise for the 20th century has been approximately 1.7 mm/year (Church and White 2006). The end of the 20th century has seen an accelerated rate of 0.013±0.006 mm/year (Church and White 2006). Estimates of future sea level rise vary, with the IPCC 2007 report forecasting that sea level will rise between 19 cm and 59 cm from the base period of 1980–99 to the end of the century, that is, 2090–99. In just three years' time, current estimates put the upper range of sea level rise at 100 cm or more (Climate Office 2009). Sea level rise will occur at a faster rate if there is an accelerated melting of the Greenland ice sheet or the collapse of the West Antarctic ice sheet (Nicholls, Wong, Burkett, *et al.* 2007).

Sea level rise is a coastal hazard, which is exacerbated by storm surges and wave action, shoreline erosion, wetland loss, and saltwater intrusion (NRC 2004). Both human infrastructure and the natural environment are vulnerable to sea level rise. Of particular concern are productive coastal wetlands and human settlements located in low elevation areas. Many of the world's coastal settlements are densely populated and are growing rapidly, making future sea level rise even more potentially dangerous. The "low elevation coastal zone" is defined by McGranahan, Balk, and Anderson (2007) as the contiguous area along the coast, which is less than 10 m above sea level, and covers only 2% of the world's land area but contains 10% of the world's population. Most of these settlements are in the developing world, and almost two-thirds of urban areas with more than 5 million people are at least partly in the low elevation coastal zone (McGranahan, Balk, and Anderson, *et al.* 2007). Humans have long been attracted to the coastal environment for settlement, whether for ease of trade or for resources such as seafood. This attraction to the coast continues even today, as demonstrated in China, where the current economic boom is leading to the concentration of humans in the coastal environment (Han and Yan 1999).

In addition to the concern for humans living along the coast, there is a concern for the alteration of the natural environment due to climate change. Coastal ecosystems, both onshore and offshore, are considered some of the most productive ecosystems in the world (Ngoile and Horrill 1993; Agardy, Alder, Dayton, *et al.* 2005). Currently, approximately, one-third of coastal mangroves and about one-fifth of coral reefs have been destroyed (Vitousek, Mooney, Lubchenco, *et al.* 1997; Halpern, Walbridge, Selkoe, *et al.* 2008).

Given the certainty of continued sea level rise and the prospects of an accelerated rise, policy-makers have begun considering how to prepare for the various aspects of sea level rise, including loss of land to inundation, erosion, increased frequency of flooding, loss of coastal wetlands or migration of wetlands, barrier island migration and breaching, salinity increases, and the disruption of human infrastructure (Leatherman 2001; Valiela 2006). In addition, policy-makers are confronting the prospects of adaptation as well as mitigation (Bigio 2003; Klein, Nicholls, and Thomalla 2003; McGranahan, Balk, and Anderson, *et al.* 2007). While weighing the pros and cons of various policies, it is critical to understand what portions of the coast are most vulnerable to sea level rise and to what extent. To determine this, an accurate mapping of the coast needs to be undertaken.

Geoinformatics plays a vital role in assisting planners and policy-makers by providing tools to help assess the vulnerability of the coastal environment to sea level rise. Generating sea level rise vulnerability maps requires a combination of elevation information, current land cover data, and GIS-derived models to predict how various scenarios of different heights of sea level rise will inundate the land as well as influence shoreline erosion and wetland accretion. To date, few maps have been developed at the global scale, though there are some simple, low-resolution, interactive Internet-based global sea level rise maps. Although some broad-scale maps have been developed at the country level, there are only a few detailed maps of some selected regions of the US such as the Chesapeake Bay area (Glick, Clough, and Nunley 2008; Titus and Wang 2008), New Jersey (Titus and Wang 2008; Wu, Yarnal, and Fisher 2002), and the Carolinas (Gesch 2009). In the near future, with the availability of new data sets and techniques, there will be a profusion of coastal mapping.

One of the first major assessments of sea level rise along the US coast included assessments for inundation and shoreline erosion (Kana, Michel, Hayes, *et al.* 1984; Leatherman 1984). The first major nationwide assessment with vulnerability mapping was done by the US Environmental Protection Agency (EPA) in a report to the US Congress on the potential effect of global climate change (Park, Trehan, Mausel, *et al.* 1989). After this report, many studies were conducted throughout the world in the 1990s to estimate the area of land that might be lost to sea level rise (IPCC 1996; Titus and Richman 2001). The vast majority of these assessments, however, failed to publish maps showing the vulnerable areas. More recent mapping projects have shown specific areas vulnerable to sea level rise. These maps, however, tend to be fairly simple and mostly focus on a static coastal environment and depict primarily inundation. Tables 1 and 2 (adapted from CCSP 2009) provide the most up-to-date list of research papers based on sea level rise, depicting sea level rise vulnerability at the regional to global scales.

Changes that will occur along the coast due to sea level rise are driven by complex and interrelated processes. Most reports indicate that inundation will be the main response to sea level rise, but there are more influences, and most studies are yet to fully take into account the complexity of other effects (CCSP 2009). In addition to sea level rise, storm activity, ocean currents, sediment transport, coastal vegetation, and human structures all act in various ways to alter the coastal environment (Carter and Woodroffe 1994; Pilkey and Cooper 2004).

**Table 1** Characteristics of some sea level rise assessments conducted over broad areas

| Study | Study area | Elevation data | Sea level rise scenarios | Elevation accuracy reported | Maps published |
|---|---|---|---|---|---|
| Schneider and Chen (1980) | Conterminous US | 15- and 25-foot contours from USGS maps 1:24000 –scale | 4.6 and 7.6 m | No | Yes |
| USEPA (1989) | Conterminous US | Contours from USGS maps | 0.5,1 and 2 m | No | No |
| Titus, Park, Leatherman, *et al.* (1991) | Conterminous US | Contours from USGS maps, wetland delineation, and tide data | 0.5, 1 and 2 m | No | No |
| FEMA (1991) | US | Coastal floodplain maps | 1 and 3 feet | No | No |
| Small and Nicholls (2003) | Global | GTOPO30 | 5-m land elevation increments | Estimated a 5-m uncertainty for elevation data (no error metric specified) | No |
| Ericson, Vorosmarty, Dingman, *et al.* (2006) | 40 deltas distributed worldwide | GTOPO30 | 0.5–12.5 mm per year for period 2000–50 | No | No |
| Rowley, Kostelnick, Braaten, *et al.* (2007) | Global | GLOBE (GTOPO30) | 1, 2, 3, 4, 5 and 6 m | No | Yes |
| McGranahan, Balk, and Anderson (2007) | Global | SRTM | Land elevations 0 to 10 m (to 10 m elevation defined the "low elevation coastal zone" | No, although increment used in recognition of data limitations | Yes |
| Rowley, Kostelnick, Braaten *et al.* (2007) | Global | GLOBE (GTOPO30) | 1, 2, 3, 4, 5 and 6 m | No | Yes |

*(Table 1 Contd..)*

*(Table 1 Contd..)*

| Study | Study area | Elevation data | Sea level rise scenarios | Elevation accuracy reported | Maps published |
|---|---|---|---|---|---|
| Demirkesen, Evrendilek, Berberoglu, *et al.* (2007) | Izmir, Turkey | SRTM | 2 and 5 m | Yes, but no error metric specified | Yes |
| Marfai and King (2008) | Semarang, Indonesia | Local survey data | 1.2 m and 1.8 m | No | |
| Kafalenos, Leonard, Beagan, *et al.* (2008) | US Gulf Coast | NED | 2 and 4 feet | No | Yes |

SRTM – Shuttle Radar Topography Mission; NED – National Elevation Dataset; USGS – US Geological Survey

GTOPO30 is a global raster digital elevation model (DEM) with a horizontal grid spacing of 30 arc seconds (approximately 1 km).

**Source** CCSP (2009, p. 98, 99)

**Table 2** Characteristics of some sea level rise vulnerability studies conducted over mid-Atlantic locations

| Study | Study area | Elevation data | Sea level rise scenarios | Elevation accuracy reported | Maps published |
|---|---|---|---|---|---|
| Titus and Richman (2001) | US Atlantic and Gulf coasts | USGS DEMs derived from 1:250,000-scale maps | 1.5- and 3.5-m | No | Yes |
| Najjar, Walker, Anderson, *et al.* (2000) | Delaware | 30-m DEMs | 2 feet | No | Yes |
| Kleinosky, Yarnal, and Fisher (2007) | Hampton Roads, Virginia | 10- and 30-m USGS DEMs | 30, 60, and 90 cm | No | Yes |
| Wu, Yarnal, and Fisher *et al.* (2002) | Cape May County, NJ | 30-m USGS DEMs | 60 cm | No | Yes |
| Gornitz, Couch, and Hartig (2002) | New York City area | 30-m USGS DEMs | 5-foot elevation increments | No only qualitative results reported | Yes |
| Larsen, Clark, Guntenspergen, *et al.* (2004) | Blackwater National Wild-life Refuge, Maryland | LIDAR | 30-cm land elevation increment | No | Yes |

*(Table 2 Contd..)*

*(Table 2 Contd..)*

| Study | Study area | Elevation data | Sea level rise scenarios | Elevation accuracy reported | Maps published |
|---|---|---|---|---|---|
| Titus and Wang (2008) | Mid-Atlantic States | Contours from USGS 1:24,000-scale maps LIDAR, local data | 0.5-m land elevation increments | Yes, RMSE vs. LIDAR for a portion of study area | Yes |
| Gesch (2009) | North Carolina | GTOPO30, SRTM, NED, LIDAR | 1 m | Yes with NSSDA error metric (95% confidence) | Yes |

SRTM – Shuttle Radar Topography Mission; NED – National Elevation Dataset; NSSDA – National Standard for Spatial Data Accuracy; LIDAR – Light Detection And Ranging; USGS – United States Geological Survey; RMSE – root mean square error

GTOPO30 is a global raster digital elevation model (DEM) with a horizontal grid spacing of 30 arc seconds (approximately 1 km)

**Source** CCSP (2009, p. 101, 102)

## COASTAL VULNERABILITIES

The concentration of human settlement along coastal regions and economic activity there have had a serious effect on the environment and have made humans more vulnerable to future sea level rise. While bringing more people to the coast, urban development has radically altered the structure of the coastline as well as the flow of water through coastal ecosystems (Rakodi and Treloar 1997). Most of the coastal vulnerability comes from storm surges and high-river flows apart from the higher sea levels. Some of the detrimental environmental effects of coastal urbanization are incorporated into the economic incentives, which drive urban development (McGranahan, Balk, and Anderson, *et al.* 2007). Most urban infrastructure are long-lasting and would be expensive, thus making urban areas less flexible in handling the rising sea level.

Many urban features increase flooding due to increased water flow from impervious surfaces, which drain more rapidly, increasing peak flows and flood risks. Wetlands act as a buffer against tidal floods, as they are able to absorb some of the incoming water. Urban development often uses wetlands for urban expansion or for other reasons such as reducing malaria prevalence. Land compaction and land subsidence due to overextraction of groundwater and a reduction in sediment deposition can exacerbate sea level rise. The risk to humans would be reduced if people and development were encouraged to move away from the coastal areas.

Coastal environments, such as wetlands, beaches, and barrier islands, are integrally related to sea level in that they adjust themselves to the changing sea levels. The current rate of sea level rise is leading to the submersion of low-lying areas, eroding beaches, converting wetlands into open water or forcing their migration horizontally or vertically, increasing coastal flooding as well as the salinity of estuaries and coastal groundwater (CCSP 2009). Natural coastal environments are dynamic and change even without sea level rise. However, sea level rise will add a new dimension to changes along our coasts. It must be emphasized that a rising sea level will not just result in simple inundation but also influence erosion and accretion along the coastal environment. The level and type of erosion will be influenced not only by sea level rise, but also by the geomorphology of the coastline. The most vulnerable to rapid change are sandy-shore environments.

The influences on wetlands are complex and driven by a number of factors, including rates of sea level rise, elevation of surrounding land (horizontal migration can potentially occur on low-lying border regions), changes in sediment supplies assisting vertical accretion (ability to move vertically), and various human activities. These influences make it difficult and complex to model exactly how wetlands will change due to sea level rise. However, due to their critical importance in the environment, there is extensive research into how wetlands will react to sea level rise (CCSP 2009). Sheltered, low-energy areas along the coast, where wetlands are absent and where sediment build-up is minimal, will be unable to grow vertically and, thus, will be submerged by sea level rise. Here elevation and the slope of the shoreline are the determining factors for inundation.

## MAPPING AND MODELLING SEA LEVEL RISE

The vulnerability of coastal areas to sea level rise has been a topic of research in geoinformatics during the past few decades (Schneider and Chen 1980; USEPA 1989; Titus and Richman 2001; CCSP 2009). The most basic aspect of mapping coastal vulnerability has been identifying low-lying lands potentially affected by a given sea level rise scenario (Figure 1). At the heart of these analyses is the use of elevation data. The elevation data for these studies ranges from topographic maps to digital elevation models (DEMs) to remotely sensed Light Detection And Ranging (LIDAR) data. Although considered vital and most important for understanding the vulnerability of coastal regions, elevation-based models alone only provide information on some of the concerns related with sea level rise (FitzGerald, Fenster, Argow, *et al.* 2008; Seiden 2008).

While it is clear that low-lying areas are vulnerable to inundation, the transformation of the coastlines by waves and currents over time adds to the complexity of the situation (Wells 1995). As noted earlier, low-energy coastal areas that transform more slowly are more predictable for inundation. In these cases, the elevation and slope of the land dictate inundation. In higher energy regions, in addition to inundation, coastal erosion will be prominent. The presence of wetlands, beaches, and barrier islands indicates an area of

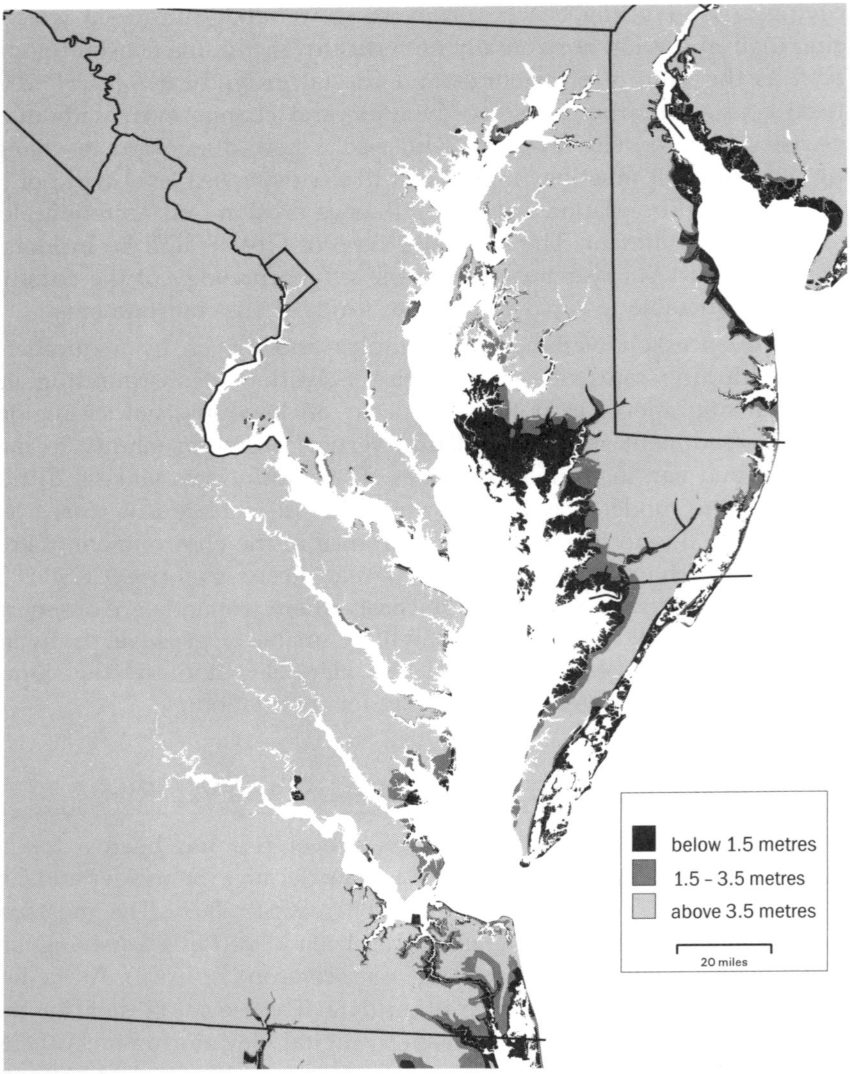


**Figure 1a** Map of the US lands vulnerable to sea level rise – Chesapeake Bay

active sedimentary processes. Irrelevant to coastal energy, with the construction of seawalls, piers, and bulkheads, creating immovable shorelines, hardened coasts will also be directly vulnerable to inundation.

Although there has been decades of geoinformatics research into the mapping and modelling of sea level rise, there is still a great uncertainty associated with these models (CCSP 2009). These earlier studies have been hindered by the quality of availably of data, the simplicity of the models,

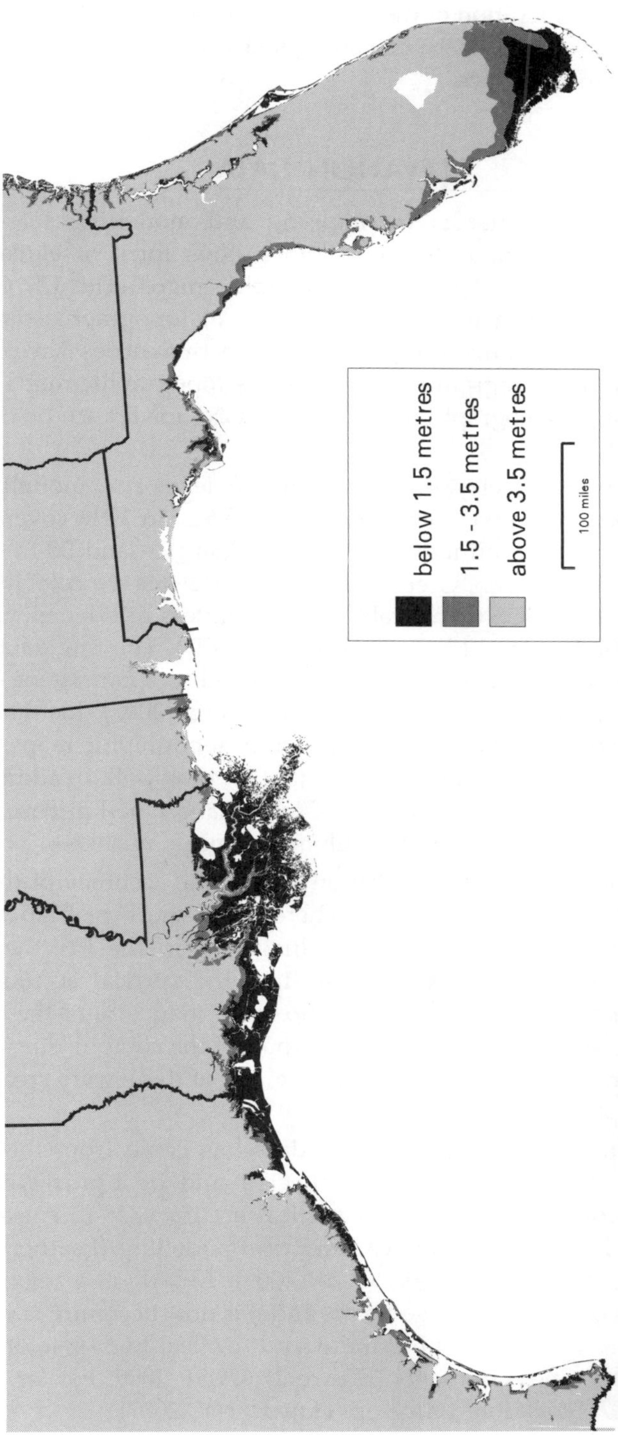


**Figure 1b** Map of the US lands vulnerable to sea level rise – Gulf coast and Florida

**Source** Titus and Richman (2001)

and the lack of consideration of the coastal dynamics. The greatest hindrance has been the poor quality of the elevation data with a poor spatial resolution and poor vertical accuracies.

## ELEVATION DATA

The most important data set in mapping and modelling sea level rise certainly is elevation data. The most ubiquitous form of elevation data has been the one associated with topographic maps. The US Geological Survey (USGS) has been the principal source of topographic data for the US for more than a century (Harvey 1980). Most countries have some sort of national mapping programme with various topographic maps available, such as the Institut Géographique National (IGN) for France or the Survey of India for India.

An improvement in elevation data for sea level rise modelling came when topographic maps were converted into DEMs. In 1999, coverage of the continental US from 7.5-minute DEM source data (10- and 30-m data) was completed (Gesch, Oimoen, Greenlee, *et al.* 2002). After the establishment of the USGS DEM, the USGS then established a seamless gridded elevation data set known as the National Elevation Dataset (NED). The raw data for NED comes from a variety of sources but is processed to a consistent resolution with consistent horizontal and vertical data (Gesch 2007). Just as there are different national agencies producing different topographic maps, there are varied national agencies producing national DEMs as well. In addition, there are now global-scale DEMs such as the Global Land One-km Base Elevation Project (GLOBE) (Hastings, Dunbar, Elphingstone, *et al.* 1999).

An important component of elevation data is the accuracy of the vertical and horizontal aspects of the data. There are variety of map standards that address these issues. For topographic maps in the US, the National Map Accuracy Standards (NMAS) state that the vertical accuracy of the US government maps will be such that no more than 10% of the elevations tested will have an error of more than one-half of the contour interval (USGS 1999). This is important for DEMs because many of these were created based on the topographic maps.

The latest improvement in elevation data has come from the remotely sensed LIDAR data. Currently, the collected and post-processed LIDAR elevation data commonly achieves a vertical accuracy of 15 cm and is the best source of elevation data for sea level rise modelling (Leatherman 2001; Maune, Maitra, and McKay 2007; CCSP 2009). LIDAR is a relatively new developed form of remotely sensed data, and it is now becoming a commonly used data source for the mapping industry to collect high-resolution, high-accuracy elevation measurements (Figure 2) (CCSP 2009; Fowler, Samberg, Flood, *et al.* 2007; Stockdon, Lillycrop, Howd, *et al.* 2007).

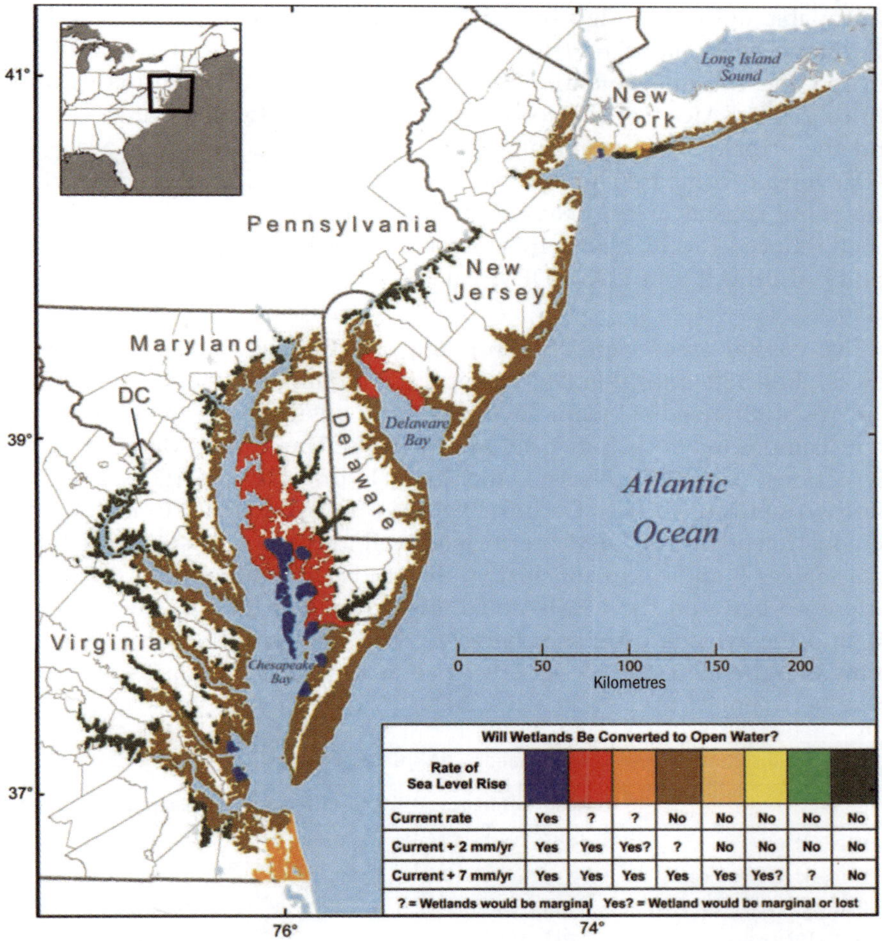


**Figure 2** Wetland survival in response to three sea level rise scenarios
**Source** CCSP (2009 p 221)
**Data source** Reed, Bishara, Cahoon, *et al.* (2008)
**Map source** Titus, Jones. and Streeter (2008)

Because LIDAR is a relatively new data set and can be rather expensive, many researchers will still have to rely on the more traditional DEM data sets for their analysis. Although not as accurate as LIDAR, DEMs might be able to produce reliable results for inundation mapping. Research carried out by the Consortium for Atlantic Regional Assessment (CARA) has shown the ability of 7.5-minute USGS DEMs to perform well in comparison with LIDAR data (CARA 2009). The two main types of elevation data available in the US include USGS DEMs and LIDAR data. The USGS 7.5-minute DEM is available online through the USGS for the entire US. Overall, the USGS 7.5-minute DEM has a root mean squared error (RMSE) of 7 m, which is beyond the extreme potentials for sea level rise in this century (Climate Office 2009; CARA 2009). However, this RMSE is for the entire data set, and the error derived from interpolation will be much less for areas with a flat terrain compared with areas that have a rapid change in elevation. Therefore, coastal areas, because of their general flat terrain, will have a much reduced RMSE. Under this scenario, CARA researchers compared the ability of LIDAR and USGS DEM data to analyse the sea level rise (CARA 2009).

Five coastal areas were chosen along the Atlantic seaboard to compare USGS DEMs (30- and 10-m DEMs) and LIDAR data (up to four time periods). The research used 23 data layers for the five sites, and all data sets were composited to a 5-m grid in Universal Transverse Mercator (UTM) projections using the North American Datum 1983 (NAD 83) and the North American Vertical Datum 1988 (NAVD 88). Results from the study demonstrated that the DEMs performed well in comparison with the LIDAR data (Tables 3 and 4). In conclusion, the study determined that the average difference between the DEMs and LIDAR was less than 0.2 m (RMS: 1.24 m), and the average difference between the DEMs and LIDAR with respect to areas inundated due to sea level rise at difference scenarios was found to be between 4% and 9% (CARA 2009).

**Table 3** Differences between DEM and LIDAR datasets (m)

| | Mean difference | RMS difference | Maximum difference |
|---|---|---|---|
| DEM30 and DEM10 | 0.023 | 0.344 | 8.159 |
| DEM30 and LIDAR mean | 0.192 | 1.243 | 27.528 |
| DEM10 and LIDAR mean | 0.125 | 1.161 | 19.047 |
| Among LIDAR layers | 0.017 | 1.121 | 62.557 |

DEM – digital elevation model; LIDAR – Light Detection And Ranging; RMS – root mean square
**Source** CARA (2009)

Land elevations are recorded relative to sea level. Most elevations are referenced to an orthometric, or sea level referenced, datum. The vertical datum is based on a network of surveyed control benchmarks, much like that for a horizontal datum. The benchmarks are based on local mean sea level from tidal gauges along the coasts. There are specific referenced vertical datums, such as the National Geodetic Vertical Datum of 1929 (NGVD 29) and NAVD 88 (NOAA 2001). The NAVD 88 is a more accurate vertical datum than the NGVD. In addition, NAVD 88 is the official vertical datum on the US, and elevation models in the US should use this datum (CCSP 2009).

**Table 4** Areas inundated by 0.38 m, 0.66 m, and 1 m sea level rise using DEM and LIDAR

| | Inundated area (km$^2$) | | | Difference with LIDAR (%) | |
|---|---|---|---|---|---|
| **Sea level rise** | **DEM 30** | **DEM 10** | **LIDAR mean** | **DEM 30** | **DEM 10** |
| Below 0.38 m | 8.72 | 8.7 | 8.09 | 7.8 | 7.47 |
| Below 0.66 m | 8.87 | 8.88 | 8.54 | 3.95 | 4.04 |
| Below 1.0 m | 10.29 | 10.32 | 9.47 | 8.66 | 9.03 |

DEM – digital elevation model; LIDAR – Light Detection And Ranging

**Source** CARA (2009)

## VULNERABILITY MAPPING

Maps that show the coastal areas that are vulnerable to sea level rise are of great importance to coastal planners and land managers (Coastal States Organization 2007). Map-based analyses of sea level rise not only highlight the natural coastal areas that are vulnerable, but often include summaries of potential human populations, infrastructure, and economic activity at risk. For the past two decades, these analyses have become more common and more detailed with the availability of better data sets, along with digital socio-economic spatial data and the tools to handle the data. In particular, the advancement of geographic information system (GIS) tools has made much of this analysis possible.

One of the first research papers looking at the effects of potential sea level rise in the US was presented by Schneider and Chen (1980). They used USGS 1:24 000 topographic maps and used 15- and 25-foot contours to approximate where inundation would occur. The USEPA (1989) initiated one of the first studies to look at the US coastline with more realistic sea level rise scenarios (0.5–2.0 m). Gesch *et al.*, in their chapter on *Coastal Elevations* in the recent EPA report on sea level rise in the mid-Atlantic region of the US (CCSP 2009), presented the main research articles on broad-scale global sea level rise. They also contributed the main research papers on the Mid-Atlantic region of the US, which is one of the more deeply researched regions in the US because of its great vulnerability (Tables 1 and 2).

In addition to the numerous peer-reviewed research papers and governmental reports, there are numerous non-profit organizations, universities, and local governments, which have developed reports as well (Glick, Clough, and Nunley 2008; Mote, Petersen, Reeder, *et al.* 2008). Although these studies vary in the scale of research, the location of the study area and the accuracy of the data, they are similar in that they have an analogous methodology. Basically, they all use a certain elevation database (DEM-30 m, Shuttle Radar Topography Mission [SRTM], LIDAR, and so on) and then they model the inundation of land based on different sea level rise scenarios. Some studies then go further and estimate the impact on the economy and populations in the areas affected.

Although coastal area vulnerability mapping has been done for decades, the information, data, and tools needed for accurate mapping are still being developed, and so the mapping of coastal areas vulnerable to sea level rise is still in its early stages. Mapping the vulnerability of the coastal environments is only part of the geoinformatics of sea level rise mapping. Planners concerned about the consequences of sea level rise are interested not just in the vulnerability of the coastal environment, but also in the vulnerability of human infrastructure and human population. GIS can provide an integrated format and analysis linking the physical vulnerabilities of our coastal environment with the human vulnerabilities due to sea level rise (Yohe, Neumann, Marshal, *et al.* 1996; Michael 2007). Research is clearly moving in this direction, and geoinformatics will provide the needed tools to researchers and planner across the world.

## References

Agardy T, Alder J, Dayton P, Curran S, Kitchingman A, Wilson M, Catenazzi A, Restrepo J, Birkeland C, Blaber S, Saifullah S, Branch G, Boersma D, Nixon S, Dugan P, Davidson N, Vörösmarty C. 2005. **Coastal systems and coastal communities**. In *Millennium Ecosystem Assessment: conditions and trends*, pp. 513–543, Vol. I. The Millennium Ecosystem Assessment Board. Washington, DC: Island Press

Bigio A G. 2003. **Cities and climate change**. In *Building Safer Cities: the future of disaster risk*, pp. 91–99, edited by A Kreimer, M Arnold, and A Carlin. Washington, DC: World Bank

CARA (The Consortium for Atlantic Regional Assessment). 2009. **Mapping sea-level rise**. Details available at <http://www.cara.psu.edu/about/mappingsealevelrise.asp>, last accessed on 21 April 2009

Carter R W G and Woodroffe C D. 1994. ***Coastal Evolution: late quaternary shoreline morphodynamics***. Cambridge: Cambridge University Press

CCSP (Climate Change Science Program). 2009. ***Synthesis and Assessment Product 4.1: coastal sensitivity to sea level rise: a focus on the Mid-Atlantic region***. US Climate Change Science Program, Washington, DC. 784 pp.

Church J A and White N J. 2006. **A 20th century acceleration in global sea-level rise.** *Geophysical Research Letters* **33**(1): L01602

Church J A, Gregory J M, Huybrechts P, Kuhn M, Lambeck K, Nhuan M T, Qin D, Woodworth P L. 2001. **Changes in sea level**. In *Climate Change 2001: the scientific basis*, pp. 639–693, edited by J T Houghton, Y Ding, D J Griggs, M Noguer, P J van der Linden, D Xiaosu. Cambridge: Cambridge University Press

Climate Office, University of Copenhagen. 2009. **Rising sea levels set to have major impacts around the world.** Press release from *Climate Change Copenhagen Conference,* 10–12 March 2009. Details available at <http://climatecongress.ku.dk/newsroom/rising_sealevels/>

Coastal States Organization. 2007. ***The Role of Coastal Zone Management Programs in Adaptation to Climate Change***. CSO Climate Change Work Group, Washington, DC. 27 pp.

Demirkesen A C, Evrendilek F, Berberoglu S, Kilic S. 2007. **Coastal flood risk analysis using Landsat-7 ETM+ imagery and SRTM DEM: a case study of Izmir, Turkey.** *Environmental Monitoring and Assessment* **131**(1–3): 293–300

Emery K O and Aubrey D G. 1991. ***Sea Levels, Land Levels and Tide Ganges.*** New York: Springer

Ericson J P, Vorosmarty C J, Dingman S L, Ward L G, Meybeck M. 2006. **Effective sea-level rise and deltas: causes of change and human dimension implications**. *Global and Planetary Change* **50**(1–2): 63–82

FEMA (Federal Emergency Management Agency). 1991. ***Projected Impact of Relative Sea Level Rise on the National Flood Insurance Program: report to Congress***. Washington, DC: Federal Insurance Administration. 61 pp.

FitzGerald D M, Fenster M S, Argow B A, Buynevich I V. 2008. **Coastal impacts due to sea-level rise**. *Annual Review of Earth and Planetary Sciences* **36**: 601–647

Fowler R A, Samberg A, Flood M J, Greaves T J. 2007. **Topographic and terrestrial LIDAR**. In *Digital Elevation Model Technologies and Applications: the DEM users manual,* 2nd edn, edited by D Maune. Bethesda, MD: American Society for Photogrammetry and Remote Sensing

Gesch D B. 2007. **The National Elevation Dataset**. In *Digital Elevation Model Technologies and Applications: the DEM users manual,* 2nd edn, pp. 99–118, edited by D Maune. Bethesda, MD: American Society for Photogrammetry and Remote Sensing

Gesch D B. 2009. **Analysis of LIDAR elevation data for improved identification and delineation of lands vulnerable to sea level rise.** *Journal of Coastal Research* (in press)

Gesch D, Oimoen M, Greenlee S, Nelson C, Steuck M, Tyler D. 2002. **The National Elevation Dataset**. *Photogrammetric Engineering and Remote Sensing* **68**(1): 5–11

Glick P, Clough J, and Nunley B. 2008. ***Sea-Level Rise and Coastal Habitats in the Chesapeake Bay Region: technical report***. Washington, DC: National Wildlife Federation

Gornitz V, Couch S, and Hartig E K. 2002. **Impacts of sea level rise in the New York City metropolitan area**. *Global and Planetary Change* **32**(1): 61–88

Halpern B S, Walbridge S, Selkoe K A, Kappel C V, Micheli F, D'Agrosa C, Bruno J F, Casey K S, Ebert C, Fox H E, Fujita R, Heinemann D, Lenihan H S, Madin E M P, Perry M T, Selig E R, Spalding M, Steneck R, Watson R. 2008. **A global map of human impact on marine ecosystems**. *Science* **319**: 948–952

Han S S and Yan Z X. 1999. **China's coastal cities: development, planning and challenges**. *Habitat International* **23**(2): 217–229

Harvey P D A. 1980. ***The History of Topographical Maps: symbols, pictures and surveys.*** London: Thames and Hudson

Hastings D A, Dunbar P K, Elphingstone G M, Bootz M, Murakami H, Maruyama H, Masaharu H, Holland P, Payne J, Bryant N A, Logan T L, Muller J P, Schreier G, MacDonald J S. 1999. ***The Global Land One-kilometer Base Elevation (GLOBE) Digital Elevation Model, Version 1.0***. National Oceanic and Atmospheric Administration, National Geophysical Data Center, Boulder, Colorado, USA. Digital data base available at <http://www.ngdc.noaa.gov/mgg/topo/globe.html>

IPCC (Intergovernmental Panel on Climate Change). 1996. ***Climate Change 1995: the science of climate change***. Cambridge, UK, and New York: Cambridge University Press

IPCC (Intergovernmental Panel on Climate Change). 2007. ***Climate Change 2007: the physical science basis***. Contribution of Working Group I to the Fourth Assessment Report of the IPCC. Cambridge, UK, and New York: Cambridge University Press. 996 pp.

Kafalenos R S, Leonard K J, Beagan D M, Burkett V R, Keim B D, Meyers A, Hunt D T, Hyman R C, Maynard M K, Fritsche B, Henk R H, Seymour E J, Olson L E, Potter J R, Savonis M J. 2008. **What are the implications of climate change and variability for Gulf coast transportation?** In *Impacts of Climate Change and Variability on Transportation Systems and Infrastructure: Gulf Coast Study, Phase I*. A Report by the US Climate Change Science Program and the Subcommittee on Global Change Research, edited by M J Savonis, V R Burkett, and J R Potter. Washington, DC: Department of Transportation. 104 pp.

Kana T W, Michel J, Hayes M O, Jensen J R. 1984. **The physical impact of sea level rise in the area of Charleston, South Carolina**. In *Greenhouse Effect and Sea-level Rise: a challenge for this generation*, pp. 105–150, edited by M C Barth and J G Titus. New York: Van Nostrand Reinhold

Keeling C D, Whorf T P, Wahlen M, van der Plicht J. 1995. **Interannual extremes in the rate of atmospheric carbon dioxide since 1980**. *Nature* **375**: 666–670

Klein R J T, Nicholls R J, and Thomalla F. 2003. **Resilience to natural hazards: how useful is this concept?** *Environmental Hazards* **5**: 35–45

Kleinosky L R, Yarnal B, and Fisher A. 2007. **Vulnerability of Hampton Roads, Virginia to storm-surge flooding and sea-level rise**. *Natural Hazards* **40**(1): 43–70

Larsen C, Clark I, Guntenspergen G R, Cahoon D R, Caruso V, Hupp C, Yanosky T. 2004. ***The Blackwater NWR Inundation Mode. Rising Sea Level on a Low-lying Coast: land use planning for wetlands***. US Geological Survey Open File Report 04–1302. US Geological Survey, Reston, VA

Leatherman S P. 1984. **Shoreline evolution of north Assateague Island, Maryland**. *Shore and Beach* **52**(3):3–10

Leatherman S P. 2001. **Social and economic costs of sea level rise**. In *Sea Level Rise: history and consequences*, pp. 181–223, edited by B C Douglas, M S Kearney, and S P Leatherman. San Diego: Academic Press

Marfai M A and King L. 2008. **Potential vulnerability implications of coastal inundation due to sea level rise for the coastal zone of Semarang city, Indonesia**. *Environmental Geology* **54**(6): 1235–1245

Maune D F, Maitra J B, and McKay E J. 2007. **Accuracy standards and guidelines**. In *Digital Elevation Model Technologies and Applications: the DEM users manual*, 2nd edn, pp. 65–97, edited by D Maune. Bethesda, MD: American Society for Photogrammetry and Remote Sensing

McGranahan G, Balk D, and Anderson B. 2007. **The rising tide: assessing the risks of climate change and human settlements in low elevation coastal zones**. *Environment and Urbanization* **19**(1): 17–37

Michael J A. 2007. **Episodic flooding and the cost of sea-level rise**. *Ecological Economics* **63**: 149–159

Mote P, Petersen A, Reeder S, Shipman H, Binder L W. 2008. ***Sea Level Rise in the Coastal Waters of Washington State***. Seattle, Washington, DC: University of Washington

Najjar R G, Walker H A, Anderson P J, Barron E J, Bord R J, Gibson J R, Kennedy V S, Knight C G, Megonigal J P, O'Connor R E, Polsky C D, Psuty N P, Richards B A, Sorenson L G, Steele E M, Swanson R S. 2000. **The potential impacts of climate change on the mid-Atlantic coastal region**. *Climate Research* **14**(3): 219–233

Ngoile M A K and Horrill C J. 1993. **Coastal ecosystems, productivity and ecosystem protection: coastal ecosystem management**. *Ambio* **22**(7): 461–467

Nicholls R J. 1995. **Coastal mega cities and climate change**. *GeoJournal* **37**: 369–379

Nicholls R J, Wong P P, Burkett V, Codignotto J, Hay J E, McLean R F, Ragoonaden S, Woodroffe C D. 2007. **Coastal systems and low-lying areas**. In *Climate Change 2007: impacts, adaptation and vulnerability*, pp. 315–356, edited by M L Parry, O F Canziani, J P Palutikof, P J van der Linden, C E Hanson. Cambridge, UK: Cambridge University Press

NOAA (National Oceanic and Atmospheric Administration). 2001. ***Tidal Datums and their Applications***. NOAA special publication NOS CO-OPS 1. NOAA National Ocean Service, Silver Spring, MD. 112 pp.

NRC (National Research Council). 2004. ***A Geospatial Framework for the Coastal Zone: national needs for coastal mapping and charting***. Washington, DC: National Academies Press. 151 pp.

Park R A, Trehan M S, Mausel P W, Howe R C. 1989. **The effects of sea level rise on U.S. coastal wetlands**. In *The Potential Effects of Global Climate Change on the United States. Appendix: B, Sea Level Rise*, pp. 1–55. EPA 230-05-89-052. US Environmental Protection Agency, Washington, DC

Pilkey O H and Cooper J A G. 2004. **Society and sea level rise**. *Science* **303:** 1781–1782

Rakodi C and Treloar D. 1997. **Urban development and coastal zone management: an international review**. *Third World Planning Review* **19**(4): 401–424

Reed D J, Bishara D, Cahoon D, Donnelly J, Kearney M, Kolker A, Leonard L, Orson R A, Stevenson J C. 2008. **Site-specific scenarios for wetlands accretion as sea level rises in the mid-Atlantic region**. In *Background Documents Supporting Climate Change Science Program Synthesis and Assessment Product 4.1*, edited by J G Titus and E M Strange. EPA 430R07004. Washington, DC: US Environmental Protection Agency

Rowley R J, Kostelnick J C, Braaten D, Li X, Meisel J. 2007. **Risk of rising sea level to population and land area.** *Eos, Transactions of the American Geophysical Union* **88**(9): 105–116

Schneider S H and Chen R S. 1980. **Carbon dioxide warming and coastline flooding: physical factors and climatic impact**. *Annual Review of Energy* **5**: 107–140

Seiden E. 2008. ***Climate Change: science, education and stewardship for tomorrow's estuaries***. Silver Spring, MD: National Estuarine Research Reserve System. 16 pp.

Small C and Nicholls R J. 2003. **A global analysis of human settlement in coastal zones**. *Journal of Coastal Research* **19**(3): 584–599

Stockdon H F, Lillycrop W J, Howd P A, Wozencraft J A. 2007. **The need for sustained and integrated high-resolution mapping of dynamic coastal landforms**. *Marine Technology Society Journal* **40**(4): 90–99

Titus J G and Richman C. 2001. **Maps of lands vulnerable to sea level rise: modeled elevations along the U.S. Atlantic and Gulf Coasts**. *Climate Research* **18**: 205–228

Titus J G and Wang J. 2008. **Maps of lands close to sea level along the middle Atlantic Coast of the United States: an elevation data set to use while waiting for LIDAR**. In *Background Documents Supporting Climate Change Science Program Synthesis and Assessment Product 4.1: coastal elevations and sensitivity to sea level rise*, pp. 1–44, edited by J G Titus and E M Strange. EPA 430R07004. Washington, DC: US Environmental Protection Agency

Titus J G, Jones R, and Streeter R. 2008. **Maps depicting site-specific scenarios for wetlands accretion as sea level rises in the Mid-Atlantic region**. In *Background Documents Supporting Climate Change Science Program Synthesis and Assessment Product 4.1: coastal elevations and sensitivity to sea level rise*, pp. 176–186, edited by J G Titus and E M Strange. EPA 430R07004. Washington, DC: US Environmental Protection Agency

Titus J G, Park R A, Leatherman S P, Weggel J R, Greene M S, Mausel P W, Brown S, Gaunt G, Threhan M, Yohe G. 1991. **Greenhouse effect and sea level rise: the cost of holding back the sea**. *Coastal Management* **19**(2): 171–204

USEPA (United States Environmental Protection Agency). 1989. ***The Potential Effects of Global Climate Change on the United States: report to Congress***. EPA 230-05-89-050. Washington, DC: US Environmental Protection Agency

USGS (US Geological Survey). 1999. ***Map Accuracy Standards***. US Geological Survey fact sheet FS-171-99. Reston, VA: US Geological Survey. 2 pp.

Valiela I. 2006. ***Global Coastal Change***. Oxford: Blackwell Publishing. 376 pp.

Vitousek P M, Mooney H A, Lubchenco J, Melillo J M. 1997. **Human domination of earth's ecosystems.** *Science* **277**: 494–499

Warrick R A, Oerlemans J, Woodworth P L, Meier M F, le Provosi C. 1996. **Changes in sea level**. In *Climate Change 1995: the science of climate change*, pp. 359–405, edited by J T Houghton, L G Meira Filho, B A Callander. Cambridge, UK: Cambridge University Press

Wells J T. 1995. **Effects of sea level rise on coastal sedimentation and erosion**. In *Climate Change: impact on coastal habitation*, pp. 111–136, edited by D Eisma. Boca Raton, FL: CRC Press

Wu S Y, Yarnal B, and Fisher A. 2002. **Vulnerability of coastal communities to sea-level rise: a case study of Cape May County, New Jersey, USA**. *Climate Research* **22**: 255–270

Yin J J, Schlesinger M E, and Stouffer R J. 2009. **Model projections of rapid sea-level rise on the northeast coast of the United States**. *Nature Geoscience* **2**: 262–266

Yohe C. Neumann J, Marshal P, Ameden H. 1996. **The economic cost of greenhouse induced sea level rise for developed property in the United States**. *Climate Change* **32**: 387–410

# Geospatial Tools to Assess Forest Ecosystems under Climate Change Trajectories

**P K Joshi*, Neena Priyanka*, and A Giriraj†, #**
**TERI University, New Delhi, India*
†*International Centre for Integrated Mountain Development (ICIMOD), Kathmandu, Nepal*
#*International Water Management Institute (IWMI), Colombo, Sri Lanka*
*E-mail: pkjoshi@teri.res.in; neenapriyanka@gmail.com; gudugiri@gmail.com*

## INTRODUCTION

Unsustainable anthropogenic activities have resulted in significant increase in the atmospheric concentrations of greenhouse gases (GHGs) and substantial increase in the global average ambient temperature. The main contributor to increased rates of climate change is the amount of carbon dioxide. Other gases such as nitrogen oxides and methane are also important, but they play a more variable role, depending on the region and type of ecosystem. This situation has resulted in different kind of changes in the climatic regimes at the global scale. Current studies indicate that the increase in the accumulation of atmospheric gases, particularly carbon dioxide, has resulted in positive radiative forcing in climate systems and is the primary cause of global warming (IPCC 2007; Norby, Rustad, Duke, *et al.* 2005; Raupach, Marland, Ciais, *et al.* 2007; Bravo, Jandl, Gadow, *et al.* 2008). This global climate change has significantly affected the natural ecosystems in various regions of the world in multiple ways. These changes range from shifts in different biomes of the earth, increase in disturbances, invasion by exotic species, and high prevalence of wildlife disease, which have already been observed.

Climate change, the phenomenon described as a long-term shift in the regional trends in weather, measured by changes in its rate, range, and magnitude, is plaguing much of the world and is a serious concern at continental, regional, national, and local scales. These long-term changes include variations in temperature and ice cover of Arctic, widespread changes in precipitation amount and trends, temperature regimes, wind patterns, extreme weather events like droughts, heavy precipitation, heat and cold waves, and intensity of tropical cyclones (IPCC 2007). For example, the *Intergovernmental Panel on Climate Change (IPCC) Fourth Assessment Report (AR4)* findings indicate that globally averaged instrumental surface temperature records are likely to increase by 1.1–6.4°C by the end of the 21st century, relative to 1980–90, with a best estimate of 1.8–4.0°C (Figure 1).

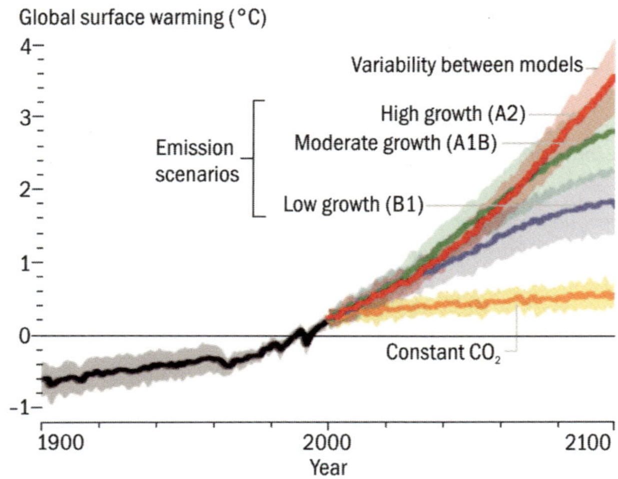


**Figure 1** Temperature projections based on a range of emission scenarios and global climate models ("constant $CO_2$" projects global temperatures with greenhouse gas concentrations stabilized at year 2000 levels)

**Source** http://www.epa.gov/climatechange/science/futuretc.html

This warming is very unusual compared with the past 1000–2000 years of variation, shown by proxy temperature data (Mann and Jones 2003; Briffa, Osborn, and Schweingruber 2004; Cook, Esper, and D'Arrigo 2004; Moberg, Sonechkin, Holmgren, *et al.* 2005), and does not show any sign of ceasing as we enter the 21st century. *IPCC AR4* has reported that the impact of human activities on climate and earth climate systems is unequivocal. It is no longer a scientific enquiry but the concern now rather is the timing and magnitudes of the abrupt changes, which are exceptionally very high. *IPCC AR4* also provides details of wide-ranging implications and adverse impacts of these projections on developing countries, which incidentally also lack capacity to respond to such rapid changes (INCCA 2009, 2010). Unfortunately, the phenomenon is still far from comprehension.

The relationship between forest resources and climate change is even more complex, depending on the dynamics within and between composition, structure, and functions. Deforestation contributes as much as 12%–15% to annual GHGs, which is almost equivalent to the emission from the entire global transportation sector (FAO 2009; WRI 2009). The impacts are already visible, and most of the scientific research has accounted for the accelerated rate of worsening of this problem (World Bank 2010). Some of these include current shifts in forest types (WRI 2008; Houghton, Hall, and Goetz 2009), changes in forest productivity (Dixon, Perry, Vanderklein, *et al.* 1996; Battles, Robards, Das, *et al.* 2005), invasion of exotic species (Martin, Stockton, Allombert, *et al.* 2007; Lockwood, Hoopes, and Marchetti 2007; Bradley, Oppenheimer, and Wilcove 2009), and corridor shifts (Jonzén, Ergon, Lindén, *et al.* 2007; Beier and Brost 2010). Consequently, this could lead to the loss of species, shrinking of the habitat, meta-population formation, and so on. Few of the forest species have looked for newer places to survive, and others have failed to reconstruct with the newer habitat associations. With the prospect of changing climate in the present and future, impacts on forest ecosystems are profound (Guisan and Theurillat 2000; Pearson and Dawson 2003).

In order to improve our understanding of the likely regional impacts on the forest ecosystems, information on current and future projected forest ecosystem scenarios is of utmost importance for sustainable forest management. While considering future scenarios, the uncertainty is taken into account by considering a number of feasible socio-economic scenarios of the future, each with an associated profile of GHG emissions, published by the IPCC (2000). They range from *low emissions* (B1: "global sustainability"—rapid uptake of new technology and adoption of sustainable policies) to *high emissions* (A1F1: "world markets"—an ongoing market-led economy reliant on fossil fuels) in three time slices: "2020s" (2010–39); "2050s" (2040–69); and "2080s" (2070–99). Table 1 gives a brief classification of the IPCC climate change scenarios. Besides scenario-based data, there are various qualitative and quantitative modelling tools that are most commonly adopted for projecting current and future climate scenarios pertaining to forest ecosystems utilizing these climate change scenarios. This chapter intends to discuss some of the approaches to assess forest ecosystems under climate change scenario.

## GEOINFORMATICS FOR FOREST ECOSYSTEMS

As discussed in Chapter 2, *geoinformatics*—an umbrella term, covering a wide gamut of geographic information system (GIS) and related technologies (remote sensing, global positioning system [GPS], image processing, high bandwidth communications)—is widely believed to play an increasingly important role in climate change research. It enables taking advantage of the unprecedented amount of digital geodatabase on forest resource and data sharing and computational capability through electronic networks to model the prospects of forest areas in light of climate change. Given that there are numerous variables involved in a variety of climate change

**Table 1** Intergovernmental Panel on Climate Change (IPCC) scenarios

| Climate change scenarios | Hypothesis |
|---|---|
| A1 | • Very rapid economic growth.<br>• Global population reaches 9 billion in 2050 then decreases thereafter.<br>• Rapid introduction of new and more efficient technologies.<br>• A1F1 (massive use of fossil fuels); A1T (a strong call on non-fossil sources).<br>• A1B (balanced call on various energy sources without heavily relying on one in particular). |
| A2 | • World population reaches 15 billion people in 2100, and rising.<br>• Economic growth and spreading of new efficient technology changes are more fragmented and slower. |
| B1 | • World population reaches 9 billion people in 2050 then decreases.<br>• Economy is dominated by services and informational technologies.<br>• Addressing economic, social, and environmental problems. |
| B2 | • World population reaches more than 2 billion people in 2100 and increases at a slow rate.<br>• Intermediate levels of economic development.<br>• Development and spread of new technologies are uneven and slower than for B1 or A1. |

**Source** SRES, IPCC

scenarios, there are many ways that a forest may respond (Acevedo and Restrepo 2008). Geoinformatics, however, provides an array of tools that may assist an analyst to carry out research on this important topic with unprecedented flexibility. *GIS* has been widely used to study landscape patterns and conservation biology, and carry out bio-geographical analysis (Kameyama, Fukishima, Han, *et al.* 2007; Eigenbrod, Hecnar, and Fahrig 2008). It plays a major role in assisting exploring, modelling, and visualizing of data on species distribution (Lopez-Lopez, Garcia-Ripolles, Aguilar, *et al.* 2006; Vogiatzakis, Mannion, and Griffiths 2006; Zhang, Pennington, and Liu 2007) and studying the effect of climate change, including disturbance events (Pennington 2007). These activities also benefit enormously from innovative methods of data collection, processing, and presentation, which are supported well by information and communication technology (ICT) (Flemons, Guralnick, Krieger, *et al.* 2007; Greene, Minoura, Steiner, *et al.* 2007; Karnatak, Saran, Bhatia, *et al.* 2007; Foody 2008). GIS has played an increasingly important role in biogeography, ecology, climate change, and conservation sciences during the last decade. *Satellite remote sensing* is a significant source of variety of data sets, which can be used to estimate variety, type, cover, and extent of land use and forest cover, meeting a fundamental need that is common to many ecological applications. Remote sensing now routinely provides environmental information at scales from the local to global, and GIS provides, among other things, a means to store, analyse, and visualize spatial data (Kerr and Ostrovsky 2003; Rosenqvist,

Milne, Lucas, *et al.* 2003; Olwoch, Reyers, and van Jaarsveld 2000). *GPS*, in conjunction with remote sensing and GIS techniques, provides vital support to collect, analyse, and store all sorts of geospatial information and expand the scope of information in a timely way while making forest resources assessment and monitoring more efficient.

These tools and techniques play an important role in monitoring forest ecosystems and mapping any change occurring over extended period of time (Coppin and Bauer 1996; Fraser, Abuelgasimb, and Latifovica 2005; Coops, Gillanders, Wulder, *et al.* 2010; Selkowitz 2010). These are useful in carrying out change detection analysis of land use land cover (LULC) patterns of different regions across the world. These technologies with the field experience are converged for multi-criteria decision making (MCDM) or multi objective decision making, as a decision analysis tool or a structured approach that helps decision-makers to determine the overall preferences among alternative options, the different combinations of which can achieve several different objectives. Spatial MCDM is different from the conventional MCDM, because of the inclusion of a geographic component. The challenging problems faced by decision-makers include the involvement of geographic or spatial component. Spatial decision support systems (SDSS) can effectively resolve this class of problem, as it requires information on criterion values and the geographical locations of alternatives in addition to the decision-makers' preferences with respect to a set of evaluation criteria. There are two components of multicriteria SDSS: (1) geodatabase component involving the processes of data acquisition, storage, retrieval, manipulation, and analysis capability and (2) the analytics component like the aggregation of spatial data and the decision-maker's preferences for different alternatives. It has endless applications, as it is cost-effective and repetitive in nature.

## Geospatial Modelling

Spatial models have become an important branch of scientific endeavour. Modelling is a technique that helps researchers and planners to make predictions in time or spatial estimations in a region (Dale, Lowrance, Mulholland, *et al.* 2010). It provides a comprehensive spatio-temporal visualization, which can be useful in understanding the potential change with respect to changing climate and accelerating anthropogenic activities. The model tends to generate approximate regional estimates of environment in response to a given variable. In environmental sciences, these models include weather forecasting models, climate models, ground water models, biological ecosystems models, ecological niche models, and so on (Wegener 2000; Jakeman, Voinov, and Emilio 2008). Spatial models are important for investigating heterogeneous distributions of ecological resources and their consequences (Mladenoff and Baker 1999; Turner, Gardner, and O'Neill 2001). Spatial models are of three categories: scale, conceptual, and mathematical (Steyaert 1993). Another classification was given by Berry (1995): deterministic and stochastic. A third type of classification includes static and dynamic models (Wegener 2000). Other types are also discussed like analytical and simulation, continuous and discrete, mechanistic, process-based, empirical models, and so on (Turner, Gardner, and O'Neill 2001). While studying large and complex landscapes, often scale becomes a challenge, and generally, models are used to understand spatial and ecological patterns and processes (Gustafson, Roberts, and Leefers 2006).

Several types of modelling tools are available for assessing the impact of climate change on forest ecosystems (Sulzman, Poiani, and Kittel 1995). These include the following.

- *Global climate models (GCMs)* These are the freely available models with broad resolution projections of future climate changes. Their relevance to forest assessment is improved by selecting results from transient (not equilibrium) simulations of $CO_2$ increase and from fully coupled ocean–atmosphere models appropriate to the region in question.
- *Regional climate models (RCMs)* These models provide high resolution results on regional and local climate change. Incidentally they are not available for all regions unlike GCMs.
- *Dynamic global vegetation models (DGVMs)* These models (for example, HYBRID, SDGVM, IBIS, TRIFFID, and VECODE) are based on basic principles of photosynthesis, carbon processing, and plant physiology to predict plant functional types (PFTs) (Cramer, Bondeau, Woodward, *et al.* 2001).
- *Species bio-climatic envelope models* These models are the best available tools for studying species-specific information necessary in conservation planning.

Spatial modelling uses present distribution of a species to "train" and "test" the model and predict the future distribution. Envelope construction is performed using GIS or through rule-based techniques such as genetic algorithms or general additive modelling (Peterson, Sanchez-Cordero, Soberon, *et al.* 2001; Berry, Dawson, Harrison, *et al.* 2002; Midgley, Hannah, Millar, *et al.* 2002). These models are incapable of including dynamics transitions, inter-specific competition, and disturbance factors. The latter are based on the correlation between observed species distributions and climatic conditions (Pearson and Dawson 2003; Luoto, Poyry, Heikkinen, *et al.* 2005). For example, Irfan-Ullah, Giriraj, Murthy, *et al.* (2007) used genetic algorithm for rule set prediction (GARP) model to identify the potential distribution of the threatened plant species *Aglaia boudillonni*, using present occurence record and environmental conditions for conservation measures (Figure 2). Once established, the relationship may be used to project the future distribution of species under selected climate change scenarios (Pearson and Dawson 2003; Luoto, Virkkala, and Heikkinen 2007;

Foody 2008). These are successfully applicable at coarse spatial scales and provide valuable initial assessment of climate change (Pearson and Dawson 2003; Luoto, Poyry, Heikkinen, *et al.* 2005; Heikkinen, Luoto, Araujo, *et al.* 2006; Foody 2008). A list of some of the models and their requirements is given in Table 2.

**Table 2** List of species bioclimatic envelope models

| Model | Types of models | Data type | Reference |
|---|---|---|---|
| GLMs | Generalized linear models (logistic regression, auto-logistic regression, others) | P, A | Moning, Werth, Dziock, *et al.* (2010); Marmion, Luoto, Heikkinen, *et al.* (2009) |
| GAMs | Generalized additive models | P, A | Powell, Accad, Austin, *et al.* (2010); Aertsen, Kint, Orshoyen, *et al.* (2010) |
| NPMR | Non-parametric multiplicative regression | P, A | Yost (2008); Davis, Borchert, Meentemeyer, *et al.* (2010) |
| ENFA | Ecological niche factor analysis | P | Hengl *et al.* (2009); Galparsoro, Borja, Bald, *et al.* (2009) |
| GARP | Genetic algorithm for rule set prediction | P, A | Stockwell, Beach, Stewart, *et al.* (2006); Schmidt, König, and Müller (2008) |
| OM-GARP | Open-modeller GARP | P, A | Giovanelli, de Siqueira, Haddad, *et al.* (2010); Guinan, Brown, Dolan, *et al.* (2009) |
| ANNs | Artificial neural networks | P, A | Okunishi, Yamanaka, and Ito (2009); Aertsen, Kint, Orshoyen, *et al.* (2010) |
| CLIMEX | Environmental-threshold-based software | P | Barney and DiTomaso (2010); Lawson, Day, Bowen, *et al.* (2010) |
| DOMAIN | Distance metric | P, A | Phillips, Anderson, and Schapire (2006) |
| MAXENT | Maximum entropy | P, A | Fuller, Morton, and Sarkar (2008); Torres, Brito, Vasconcelos, *et al.* (2010) |
| SVM | Support vector machine | P, A | Kuemmerle, Chaskovskyy, Knorn, *et al.* (2009); Knorn, Rabe, Radeloff, *et al.* (2009) |
| BIOCLIM | Envelope model | P | Beaumont, Hughes, and Poulsen (2005); Santika and Hutchinson (2009) |

P–Presence;
A–absence
**Source** Adapted from Keller, Lodge, and Lewis, *et al.* (2009)

To make the results of these models most applicable to real-world conservation problems, they must be coupled with LULC prediction and

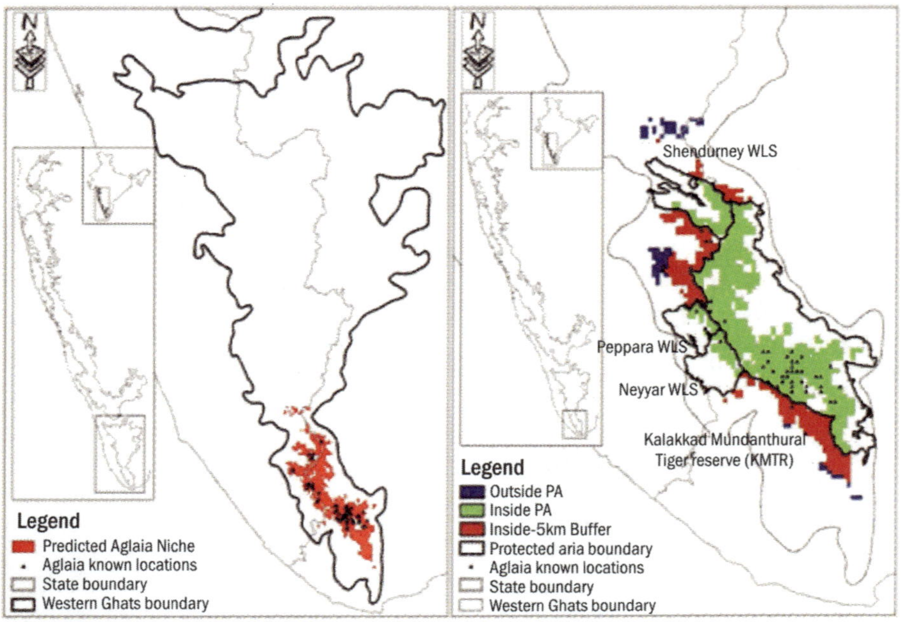


**Figure 2** Known and predicted distribution of *Aglaia bourdillonii* in the entire Western Ghats, India, using GARP-based modelling approach

projection models. The projection models are based on the current pattern of habitat fragmentation, biotic and abiotic derivers, and the future pattern based on the interaction of both (Sala, Chapin, Armesto, *et al.* 2000). The potential shift of a species range approximated by bioclimatic models is then reduced to the available habitat as projected by the LULC model. These are dealt with by using geospatial computational capabilities and could be enhanced in light of climate change. However, most of the time, LULC modelling is the manifestation of the past LULC practices, policy interventions, and consequent impact of land use and demographic control. A variety of LULC change models are available. The detailed discussion is beyond the scope of this chapter. A quick introduction to various models and their brief details are given in Table 3.

Sensitivity analysis based on ground realities of sites, characteristics of individual species, and ecological interaction is an essential supplement to modelling. In fact, it should be an integrative part of the entire modelling exercise, even if it lacks the spatial specificity of the model and results. Models cannot predict species composition at a landscape scale in a dynamic, competitive environment (for example, dynamic vegetation models lack species-specificity; envelope models lack dynamic and competitive elements) (Woodward and Beerling 1997). Strong field observation, legacy data on the

**Table 3** Few illustrations of land-use models employing spatial modeling technique

| Model | Details | Type | Reference |
|---|---|---|---|
| General Ecosystem Model (GEM) | Captures feedback among abiotic and biotic ecosystem components. | Dynamics systems model (DSM) | Fitz, DeBellevue, Costanza, *et al.* (1996) |
| Patuxent Landscape Model (PLM) | Predicts fundamental ecological processes and land use patterns at the watershed level. | DSM | Voinov, Costanza, Wainger, *et al.* (1999) |
| Conversion of Land Use and its Effects (CLUE) Model | Predicts land cover in the future. | DFSM | Veldkamp and Fresco (1996a) |
| Conversion of Land Use and its Effects—Costa Rica (CLUE-CR) Model | Simulated top-down and bottom-up effects of change in Costa Rica. | DFSM | Veldkamp and Fresco (1996b) |
| Area Base Model | Predicts land use proportions at country level. | Area base model, using a modified multinomial logit model | Hardie and Parks (1997) |
| Univariate Spatial Model | Frequency of deforestation. | Univariate spatial models | Mertens and Lambin (1997) |
| Econometric (Multinomial Logit) Model | Predicts land use, aggregated in three classes: natural vegetation, semi-subsistence agriculture, and commercial farming. | Econometric (multinomial logit) model | Chomitz and Gray (1996) |
| Spatial Dynamic Model | Predicts sites used for shifting cultivation in terms of topography and proximity to population centres. | Spatial dynamics model | Gilruth, Marsh, and Itami (1995) |
| Spatial Markov Model | Land use change. | Spatial Markov model | Wood, Lewis, Gray, *et al.* 1997 |
| California Urban Futures (CUF) | Explains land use in a metropolitan setting, in terms of demand (population growth) and supply of land (underdeveloped land available for redevelopment). | Spatial simulation model | Landis (1995) |

*(Table 3 Contd..)*

*(Table 3 Contd..)*

| Model | Details | Type | Reference |
|---|---|---|---|
| Land Use Change Analysis System (LUCAS) | Transition probability matrix (TMP) (of change in land cover), simulates the landscape change and assesses the impact of species habitat. | Spatial stochastic model | Berry, Flamm, Hazen, *et al.* (1996) |
| Simple log weights | Predicts the timberland adjusted for population density. | Simple log weights | Wear, Apt, and Mangold (1998) |
| Logit Model | Predicts the probability of land being classified as potential timberland. | Logit model | Wear, Liu, Foreman, *et al.* (1999) |
| Dynamic model<br>NELUP- Natural Environment Research Council (NERC)—Economic and Social Research Council (ESRC): NERC/ESRC Land Use program (NELUP) | Simulates an optimal harvest sequence.<br>Explains patterns of agricultural and forestry land use under different scenarios. | Dynamic model<br>General systems framework economic component uses a recursive linear planning | Swallow, Talukdar, and Wear (1997)<br>O'Callaghan (1995) |
| NELUP – Extension | Maximizes income, profit is the dependent variable. | Linear planning model | Oglethorpe and O'Callaghan (1995) |
| Forest and Agriculture Sector Optimization Model (FASOM) | Allocation of land in the forest and agriculture sectors, objective function maximizes the discounted economic welfare of producers and consumers in the US agriculture and forest sectors over a nine decade time horizon. | Dynamic, nonlinear, price endogenous, mathematics programming model | Adams, Alig, and McCarl (1996) |
| California Urban and Biodiversity Analysis Model (CURBA) | The interaction among the probabilities of urbanization, its interaction with habitat type and extent, and impacts of policy changes on the two. | Overlay of GIS layers with statistical urban growth projections | Landis, Monzon, Reilly, *et al.* (1998) |
| Cellular automata model | Changes in urban area over time. | Cellular automata model | Clark, Fastie, Hurtt, *et al.* 1998; Kirtland, Gaydos, Clarke, *et al.* (1994) |
| GEOMOD | Predicts land cover future and conservation planning. | Spatial stochastic model | Giriraj, Murthy, Beierkuhnlein, *et al.* (2008) |

resource distribution, calibrated climate data, historical records, interaction with field biologists and experts, and also evidences from palaeo-ecological and palaeo-biogeographical responses to climate change form the central element for modelling. Palaeoclimatic evidence suggests that global climates may be capable of switching rapidly between states (Broecker 1997; Hannah, Midgley, and Millar 2002). The possibility of the reversal of current warming trends within centuries argues that sound conservation plans should be robust to both warming and possible cooling.

## FOREST PHENELOGY

Global climate change significantly alters plant phenology. These processes are dependent on the temperature and precipitation regimes, availability of $CO_2$, and also interactions with factors such as photoperiod, moisture, edaphic conditions, and others (Cleland, Chuine, Menzel, *et al.* 2007). Such changes might vary in tropical and temperate regions, but these would certainly decrease or increase the shifts in leafing and senescence. Satellite remote sensing is a powerful tool to monitor phenology both spatially and temporally. Immediately after the launch of the first operational earth-observing systems in the 1970s (ERTS—Earth Resources Technology Satellite or Landsat—Land Satellite), satellite measurements of surface radiation in the short wavelengths were used to monitor the phenological cycle of forest and agricultural ecosystems. The Normalized Difference Vegetation Index (NDVI = (NIR-R)/(NIR+R)) was developed as an enhancement technique for vegetative areas; later, temporal cycle was used as an indication of the seasonal greenness of the vegetation. Now, it has become important to assess inter-annual variability, temporal trends, and spatial patterns globally, including inaccessible area (Badeck, Bondeau, Bottcher, *et al.* 2004). Since the 1980s, multi-spectral satellite observations from the National Oceanic and Atmospheric Administration's (NOAA) Polar Orbiting Environmental Satellites (POES) (AVHRR—advanced very high resolution radiometer) provide daily global coverage data sets freely. The new generation Systéme Probatoire pour l'Observation de la Terre (SPOT) vegetation sensor has been collecting near-daily surface reflectance data since 1999. The moderate resolution imaging spectroradiometer (MODIS) (onboard Terra and Aqua) and medium resolution imaging spectrometer (MERIS) (onboard the Envisat) sensors are also well suited for phenological studies (Coops, Michaud, Andrwe, *et al.* 2011). Some non-optical sensor like advanced microwave scanning radiometer (AMSR-E) or other radar sensors can be used to detect moisture conditions of the land surface (including vegetation and soil; Doubková and Henebry 2006). Many other medium resolution sensors are available to construct phenology for the climate change studies (Table 4).

**Table 4** List of satellite sensors suitable for phenological studies

| Satellite | Sensor | Operation | Resolution | Frequency |
|---|---|---|---|---|
| Landsat | MSS | 1973–85 | 79 m | 18 days |
| Landsat | TM | 1984–present | 30 m | 16 days |
| Landsat | ETM+ | 1999–present | 30 m | 16 days |
| IRS 1C/P3/1D | WiFS | 1997–2007/1996–2006/ 1997–2010 | 188 m | 5 days |
| IRS P6 | AWiFS | 2003–present | 56 m | 5 days |
| SPOT | Vegetation | 1999–present | 1 km | 1–2 days |
| NOAA | AVHRR | 1982–present | 8 km | Twice monthly |
| NOAA | AVHRR | 1989–present | 1 km | Biweekly |
| Terra | MODIS | 2000–present | 250 m, 500 m, 1 km | 1–2 days |
| Aqua | MODIS | 2000–present | 250 m, 500 m, 1 km | 1–2 days |
| Envisat | MERIS | 2002–present | 300 m | 1–3 days |

## Ecological Effects of Climate Change

Analysis by Thomas, Ohlemüller, Anderson, *et al.* (2008) includes countries and regions that suffer the most severe changes to their natural ecosystems and biodiversity, and explains how the responsibility for those changes is distributed across the world (Figure A). Ecosystems, therefore, respond to changes in temperature and precipitation, which inevitably have an impact on biodiversity. Recent shifts in the distribution of various species towards the poles and to higher altitudes (Parmesan and Yohe 2003; Root, Price, Hall, *et al.* 2003; Walther, Berger, and Sykes 2005; Wilson, Gutiérrez, Gutiérrez, *et al.* 2005; Franco, Hill, Kitschke, *et al.* 2006; Hickler, Prentice, Smith, *et al.* 2006), and the extinction of more than 1% of all amphibian species (Pounds, Bustamante, Coloma, *et al.* 2006) indicate that climate change is already having a major impact on biodiversity. Climatic change is also expected to alter the distribution of most types of vegetation, and there is already evidence of a shift from deciduous woodland to evergreen forest. Such changes will have implications both for biodiversity (Malcolm, Liu, Neilson, *et al.* 2006) and for the humans, rely on natural ecosystems to survive. Thomas, Ohlemüller, Anderson, who *et al.* (2008) used the empirically tested Lund–Potsdam–Jena (LPJ)-dynamic global vegetation model (DGVM) (Sitch, Smith, Prentice, *et al.* 2003) to calculate how temperature, precipitation, and atmospheric $CO_2$ might affect the growth of various plant types between two periods (1945–2045), in a moderate climate-warming scenario (Figure A) and identify major vegetation shift.

*Box Contd...*

*Box Contd...*

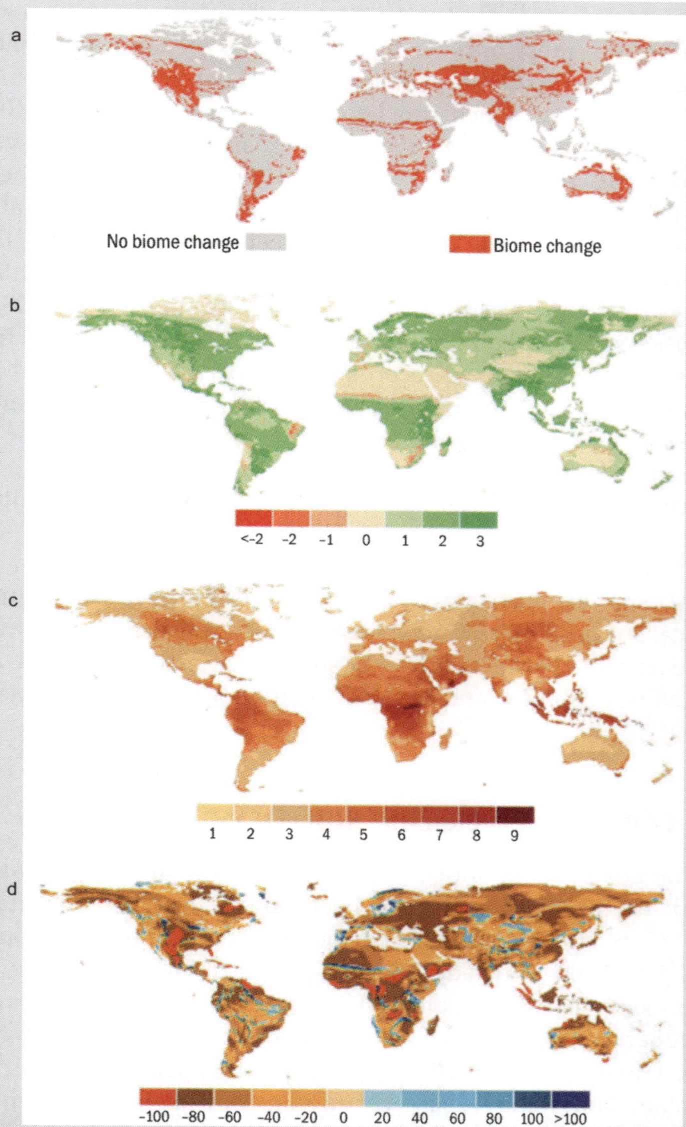


**Figure A** Projected 1945–2045 risks to ecosystems and biodiversity under observed and projected climate change. (a) Locations where vegetation changes from one biome to another are projected, (b) changes in projected leaf area index (cover) of woody plants (declines are shown in brown, increases are shown in green), (c) index of local climate change, measured as standardized Euclidean distance between 2045 and 1945 climates (dark areas indicate large differences), (d) Distributions of shrinking (brown; red denotes complete loss) and expanding (blue) climates, measured as percentage change in area (within 1000 km) between 1945 and 2045 (two grid cells were considered climatically analogous if all climate variables were within ±2°C for the two temperature variables and ±0.2 mm/day for the two precipitation variables)

**Source** Thomas, Ohlemüller, Anderson, *et al.* (2008)

NDVI is empirically related to temporal evolution of the fraction of photosynthetically active radiation (fPAR) absorbed by the green leaves. This is one of the most important biophysical vegetation parameters used for climate modelling. Other parameters are Leaf Area Index (LAI), PAR, greenness, and so on, which can be derived using NDVI. NDVI can also provide a good means to verify trends of ground observed vegetation activity. Myneni, Keeling, Tucker, *et al.* (1997) investigated the 1981–91 AVHRR NDVI record from the GIMMS and the Pathfinder data set for the Northern Hemisphere. They have detected an advance in the active growing season of eight to three days and a delay in the declining phase of four days over this decade. Zhou, Tucker, Kaufmann, *et al.* (2001) looked at the Northern Hemisphere vegetation activity derived from AVHRR NDVI and land surface temperature records for the period 1981–98 and found an increase in mean NDVI for Eurasia and North America. The knowledge of these interactions has led to the assumption that global land use changes, anthropogenic emissions of GHGs ($CO_2$, methane) as well as the observed global warming are possibly enhancing large area biospheric activity (Stockli and Vidale 2004). Figure 3 shows Terra-MODI NDVI images of 1 January 2001 and 1 January 2011. A number of satellite-based sensors with a variety of spectral, spatial, and temporal characteristics have been utilized for phenological studies. Interestingly, many of such sensors are freely accessible to all researchers. For examples, Landsat series, MODIS data and data products, NOVA-AVHRR, and SPOT vegetation are freely available for such studies.

## FOREST COVER DISTRIBUTION AND SHIFTS

Climate change impact on forest land use, especially in terms of change from forest land to other land cover, has been one of the most important issues in the LULC dynamics (Mayaux, Bartholome, Fritz, *et al.* 2004; Butler 2005; Brink and Eva 2009). Considering changes in forest cover, it is important to make projections on the potential forest cover in the present and future climate scenarios, to account for the magnitude and direction of change (Goparaju, Tripathi, and Jha 2005; Giriraj, Murthy, Beierkuhnlein, *et al.* 2008; Giriraj, Babar, and Reddy 2008; Giriraj, Beierkuhnlein, and Murthy 2010; Sharma, Lahkar, Ghosh, *et al.* 2008; Joshi 2010a). Many studies have been carried out at global, regional, and national levels, to provide evidence for the shift in forest types in different climate change scenarios as predicted by the IPCC (IPCC 2007). A list of projected impacts of climate change on forest ecosystems is given in Table 5.

For such studies, species bioclimatic envelope models (S-BEMs) are generally used to predict the distribution of species under current, past, and future climatic conditions by inferring the species' environmental requirements from native localities. S-BEMs can be evaluated for their ability to predict current species distributions, but it is unclear whether models

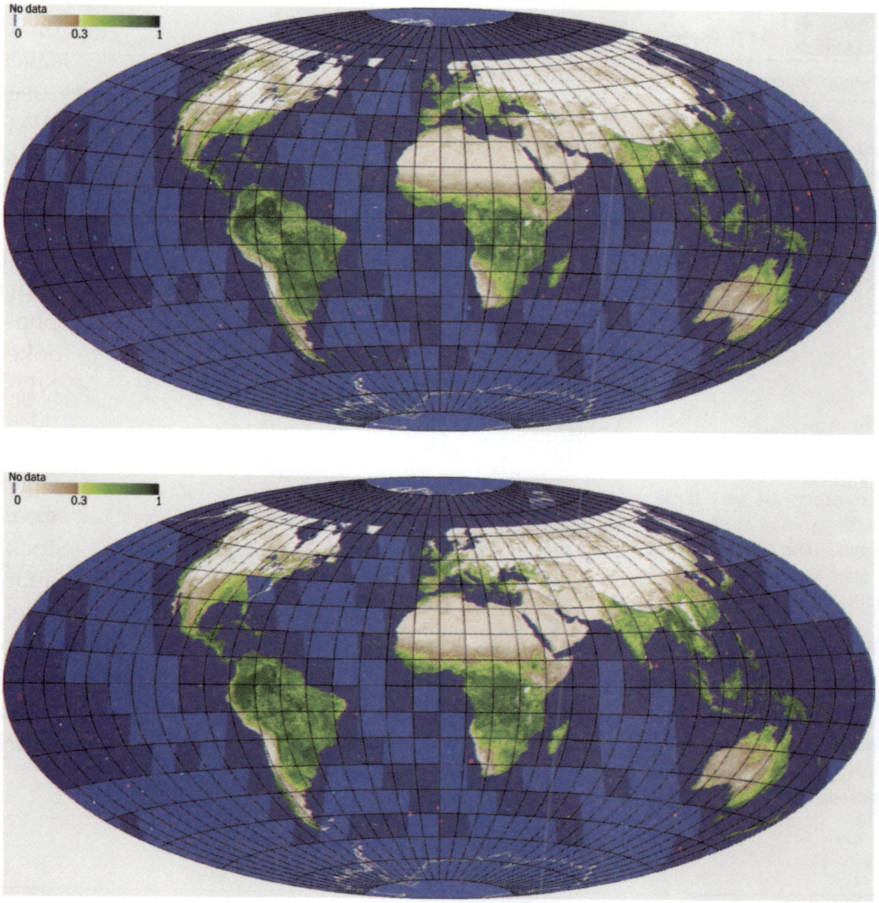


**Figure 3** Terra-MODIS MOD13.NDVI of 01/01/2001 and 01/01/2011

that are successful in predicting current distributions are equally successful in predicting distribution under different climates (for example, different regions or time periods) (Hijmans and Graham 2006). However, many of the studies rely on the geospatial modelling tools to generate rules for predicting the distribution. For example, Lasch, Lindner, Erhard, *et al.* (2002) developed a forest simulation model (FORSKA-M) to assess the sensitivity of the species composition to climate change. The simulation indicated that climatic warming would lead to a shift of natural species towards more drought-tolerant species and a decrease in the species diversity, productivity, and groundwater recharge.

With the occurrence of climate change and global warming, large shifts in the vegetation distribution are expected at very unprecedented rates. Such shifts will be most rapid and extreme at the ecotones, the boundaries between two ecosystems, and among the associates and mixed communities. Present models do not adequately take into account such rapid effects

**Table 5** Main projected impacts on forest ecosystems

| Change in conditions | Impacts in ecosystems |
|---|---|
| Changes in the structure and functioning of terrestrial and marine ecosystems (global warming of 2°C to 3°C)[a] | • Increase of 2°C: 15% ecosystems affected.<br>• Increase of 3.5°C: 40% ecosystems affected.<br>• Disturbance regimes: wildfire and insects.<br>• Major biome changes, including emergence of novel biomes.<br>• Amazon forests, China's taiga, Siberian and Canadian tundra are very likely to show major changes (~3°C)[a].<br>• Forest expansions are projected in North America and Eurasia. With temperature <2°C, tropical forests are likely to experience severe impacts, including biodiversity losses.<br>• Increasing high risk of extinction for roughly 20% to 30% (of species assessed so far varying from 1% to 80%).<br>• Changes in species' ecological interactions, with predominantly negative consequences for goods and services.<br>• Shifts in species distribution. |
| Ecosystems are very likely to be exposed to atmospheric $CO_2$ levels much higher than in the past 650 000 years | • Accelerated release of C from vulnerable C stocks (peatlands, tundra frozen loess, permafrost soils, and soils of boreal and tropical forests).<br>• Projected increase in C sequestration by poleward taiga expansion.<br>• C emission increase from albedo changes, wildfire, forest declines at taiga's equatorial limit, and methane losses from tundra will be likely greater than the increase in C sequestration by taiga.<br>• Tropical forest sequestration by 2100 is likely to be dominated by climate change impacts, especially in drier regions. |
| The resilience of many ecosystems is likely to be exceeded by 2100 | • Key ecosystem properties (for example, biodiversity) or regulating services (for example, carbon sequestration) are very likely to be impaired.<br>• Extractive use from and fragmentation of wild habitats are very likely to impair species adaptation.<br>• Disruption of species' ecological interactions. |
| By 2100, ocean pH is very likely to be lower than during the last 20 million years (ocean acidification) | • Negative impacts on marine shell-forming organisms (for example, corals, crabs, squids, marine snails, clams and oysters) and their dependent species.<br>• Likely to impair aragonite-based shell formation in a wide range of planktonic and shallow benthic marine organisms. |
| Impacts on the oceans for global mean temperature increases of about 1.5°C to 3°C[a] | • Shift in the productivity of the oceans.<br>• The low-productivity zones in sub-tropical oceans are likely to expand by about 5% in Northern Hemisphere and by about 10% in Southern Hemisphere.<br>• The productive polar sea-ice biomes are very likely to contract by about 40% (Northern Hemisphere) and about 20% (Southern Hemisphere).<br>• Loss of corals due to bleaching (especially for the Great Barrier Reef).<br>• Climate change and direct anthropogenic impacts (such as pollution and harvesting) are expected to cause annual bleaching (about 2030 to 2050) followed by mass mortality.<br>• Slow down of the Meridional Overturning Circulation, with potentially serious consequences for fisheries. |

*(Table 5 Contd..)*

(Table 5 Contd..)

| Change in conditions | Impacts in ecosystems |
|---|---|
| | • Polar species, including predators such as penguins, seals and polar bears, are very likely to experience habitat degradation and losses as sea-ice biomes shrink.<br>• Likely regional changes in the distribution and productivity of particular fish species.<br>• Local extinctions of particular fish species are expected at edges of ranges. |
| Forestry production will be affected by the projected changes in the frequency and severity of extreme climate events | • Food and forestry trade is projected to increase in response to climate change, with increased food-import dependence of most developing countries.<br>• Increment of wildfires.<br>• Production increase will shift from low-latitude regions in the short term, to high-latitude regions in the long term. |
| Greater rainfall variability | • Likely to compromise inland and coastal wetland species through shifts in the timing, duration, and depth of water levels. |

**Source** Adapted from Parry, Canziani, Palutikof, *et al.* (2007) and Easterling, Aggarwal, Batima, *et al.* (2007)

[a]Global mean annual temperature change relative to 1880–1999 (°C).

because of multiple forces altering such composition and structure, for example, anthropocentric initiatives, mortality rates, and other biotic and abiotic stresses. This is also because of the lack of structured field studies and calibrated climatic data. For example, studies by Allen and Breshears (1998) reported rapid shifts (2 km or more in <5 years) between semi-arid ponderosa pine forest and pine–juniper woodland through mortality of ponderosa pines in response to a severe drought (Allen, Macalady, Chenchouni, *et al.* 2010).

### Biome (*Quercus ilex*) Shift in Montseny Mountain (NE Spain)

Shifts in the distribution of biomes and plant species in response to global warming have been described as an impact of climate change worldwide. Beach forests and heather (*Calluna vulgaris*) are being replaced by holm oak (*Quercus ilex*) forest at medium altitude (~ 1400 m). This replacement has been observed to occur through a progressive isolation and degradation of beech stands. A study (Penuelas and Boada 2003) compared vegetation distribution of 1945 with the current situation and reported the changes that occurred between 1969 and 2001 in the distribution of vegetation patches on the south-facing slope of one of these Montseny mountains between 1200 m and 1700 m of altitude. The Erica heathland keeps a similar area, whereas the fern lands are also retiring to high altitudes. Figure A shows the replacement of heathlands and grasslands by holm oak forest at medium altitude (1200–1400 m) in Montseny mountains (Catalonia, NE Spain), from 1969 to 2001. The progressively warmer

Box Contd...

*Box Contd...*

conditions, complemented by the land use changes (mainly the cessation of rational land management, shepherd firing, which was practised before 1980s), are the apparent causes, providing a paradigmatic example of global change affecting distribution of plant species and biomes.

**Figure A** Replacement of heathlands and grasslands by holm oak forest at medium altitude (1200–1400 m) in Montseny mountains (Catalonia, NE Spain), from 1969 to 2001

**Source** Penuelas and Boada (2003)

Iverson and Prasad (2001) evaluated the potential changes in tree species richness and forest community types for the eastern USA. They used five scenarios of future climate changes resulting from the doubling of atmospheric $CO_2$. Using DISTRIB (an empirical model that uses a regression tree analysis approach), in combination with individual species (~80) and forest types, a slight increase in average species richness and larger increase with temperature rise are reported. In another similar initiative, Iverson, Prasad, and Matthews (2008) modelled and mapped 134 tree species using predictive data-mining tool. They found that 66 species would gain and 54 species would lose at least 10% of their suitable habitat under climate change.

Honnay, Verhaeghe, and Hermy (2001) reported that 85% of all forest plant species had an extremely low success of colonizing in fragmented landscapes in western Belgium. Belotelov, Bogatyrev, Kirilenko, *et al.* (1994) used Holdridge Life Zone (HLZ) (Holdridge 1947) to investigate geographical distribution of main biomes under potential climate changes at a global scale. They considered two processes that determine delays of biome shift—destruction of an old biome after the death of dominant vegetation and formation of a new biome with new dominant species. Harrison, Berry, Butt, *et al.* (2006), using SPECIES neural network model for modelling the distribution of 45 species, reported that distribution might be affected due to climate change, but the magnitude of change will vary for different species. Similarly, Soja, Tchebakova, French, *et al.* (2007) provided an evidence of migration of keystone ecosystems in upland and lowland tree line of mountainous regions. Nitschke and Innes (2008) developed tree and climate assessment (TACA) to model the response of individual species and suggested that the studied ecosystem could evolve into a new ecosystem over the next 100 years. Lenoir, Gégout, Marquet, *et al.* (2008) compared the altitudinal distribution of 171 forest plant species between two periods—1905–85 and 1986–2005—in western Europe. They concluded that global warming has resulted in a significant upward shift (29 m per decade) in species.

Horikawa, Tsuyama, Matsui, *et al.* (2009) found a decrease in the suitability of Japanese stone pine (*Pinus pumila*) in coming years. While studying tree migration in eastern USA, Woodall, Oswalt, Westfall, *et al.* (2009) reported that the northern set of species provided an evidence of northward migration, which will continue in future, and southern species had no shift, while general species had shown southern expansion. Another study, conducted for *Pinus cembra* in Alps and Carpathian mountains, provided evidence for an upslope shift (Casalegno, Amatulli, Camia, *et al.* In press). Dirnböck, Dullinger, and Grabherr (2003) used ordinal regression models for 85 alpine plant species and predicted the loss of potential habitat by 2050. More profound changes would bring a severe contraction of the alpine, non-forest zone, because of the expansion of the tree line conifer *Pinus mugo*

Turra, and many alpine species will lose major parts of their habitat. Doyle, Krauss, Conner, *et al.* (2010) predicted the retreat and migration of tidal forests along the northern Gulf of Mexico due to sea-level rise. Studying the French Mediterranean forests, Vennetier and Ripert (2009) developed a new bioclimatic model using both local and global variables, and concluded that species-rich areas might disappear by 2050. Hanewinkel, Hummel, and Cullmann (2010) modelled forest biome shifts in southwest Germany and evaluated these economically.

In India, Ravindranath and Sukumar (1996) have studied the potential impacts of climate change on forest cover in India. They reported that due to climate change, productivity could increase and forest type boundaries could shift along the altitudinal and rainfall gradients. The shift might occur from lower to higher elevations, and the drier forest types might transform to moist forest types. In another study, Ravindranath, Joshi, Sukumar, *et al.* (2006) used climate projections of RCM of Hadley Centre (A2 and B2 scenarios) and BIOME4 vegetation response model. They found that in 2085, 77% (A2) and 68% (B2) of the forest types in India might shift towards wetter forest types in north-east region and drier forest types in the north-west region, respectively, in the absence of anthropogenic disturbances. They also found that the productivity might double due to an increase in $CO_2$ concentrations in A2 scenario, and 70% increase in productivity might take place in B2 scenario. Panigrahy, Anitha, Kimothi, *et al.* (2010) reported timberline change detection using topographic map and satellite imagery and reported vegetation ingression and land cover changes in Nanda Devi Biosphere Reserve in the Central Himalayas. Gopalakrishnan, Jayaraman, Swarnim, *et al.* (2011), using the model output on climate projections, suggested by Ravindranath, Joshi, Sukumar, *et al.* (2006), modelled the teak distribution (*Tectona grandis*) and explained that 30% of teak grids in India are vulnerable to climate change under both A2 and B2 scenarios because the future climate may not be optimal for teak at these grids (Figure 4).

Roy, Joshi, Singh, *et al.* (2006) used HLZ model (Holdridge 1947) to classify land over India according to the HLZ. Geographical analysis showed that there are 19 life zones. HLZs are not only basic bioclimatic units but also describe similar human activities in a region. But these show only broad life zone responses to the transient dynamics of vegetation. Earlier studies have found 20 HLZs in Africa (Valarde, Mlahi, Moran, *et al.* 2005) and 26 in China (Yue, Liu, Jorgenses, *et al.* 2001). HLZ model is widely used globally due to its simplicity and convenience for calculation of indices and better correspondence with vegetation types (Chang 1993; Pan, Li, Gong, *et al.* 2003) and climate parameters. Roy, Joshi, Singh, *et al.* (2006) have further used this along vegetation cover type/land use type map (depicting 35 classes, including 21 non-forest classes) for biome classification in India, which is perceived to be the precursor to climate change modelling. The biome

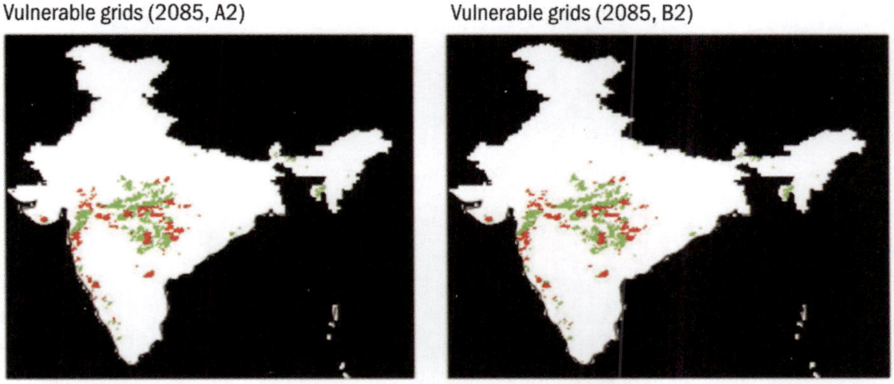


**Figure 4** Distribution of vulnerable teak grids in A2 and B2 scenario (2085) (grids marked red are vulnerable, while those marked green are not. In A2 scenario, 1110 [33.3%] of total 3329 teak grids are vulnerable, while 1009 [30.3%] teak grids are vulnerable in B2 scenario)

classification map depicts 7 biomes and 19 sub-biomes. It can be easily used in the study of the response of Indian biomes to global change, regionally and globally (Figure 5).

As the extension of above study, Joshi (2010b) used the rules for preparing potential vegetation cover type map of India. The study used different predicted climate change scenarios to map the dynamics of vegetation cover type. Using the broad phenological classes, it could be assessed that deciduous and very dry deciduous forests are the most dynamic one, which will see a wide interchange in 2020 and gradual change in 2050 and 2080. The evergreen and semi-evergreen forests will also have transitions in the Himalayan region. The semi-evergreen forest might increase in both western and eastern Himalayas, while interaction in the Western Ghats will be more dominated by the deciduous forest types (Figure 6).

Joshi (2010b) also used GARP with the climate change scenario (2020, 2050, and 2080) to predict the potential (in the absence of human interference) distribution of dominant vegetation cover types over India. Species occurence data (geographic space) combined with suite of eco-geographical variables (for example, precipitation, temperature, evapo-transpiration—for current and future climate scenarios) were projected on the ecological space. The current and future models were then evaluated in the context of change in distribution pattern and shifts if any. Correspondence analysis indicated a general shift towards wetter forest types in the North East region and drier forest types in the North West region. Though the study is at the infancy stage, it can be envisaged as the first evidences of the forest vulnerability, using geospatial tools at 1 km resolution. Approach is innovative and simple to extrapolate at higher resolutions, and the generated database can be upscaled for global analysis and downscaled for regional/local planning.

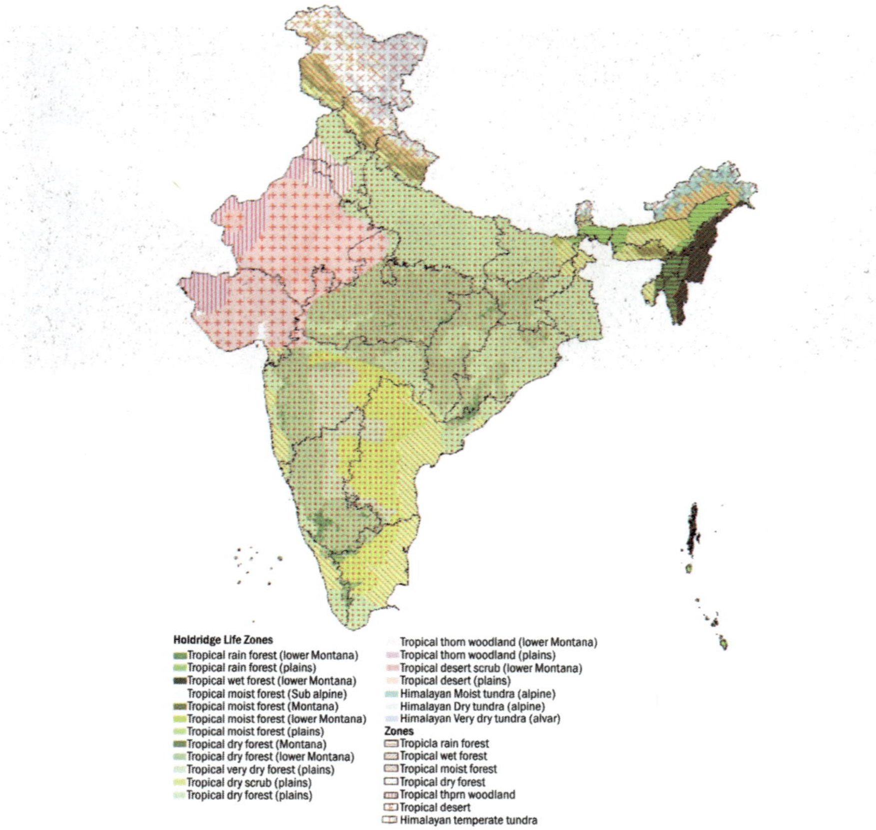


**Figure 5** Holdridge Life Zone classification of India (using bio-temperature, precipitation, and potential evapotranspiration)—precursor to biome classification

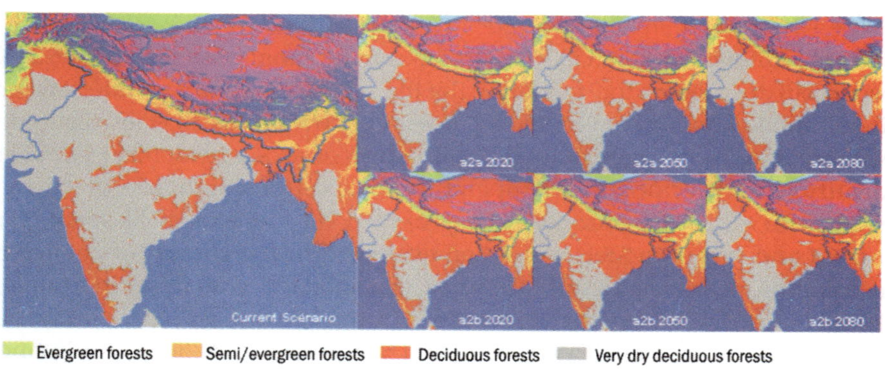


**Figure 6** Vegetation cover dynamics over India, using Holdridge Life Zone classification (using bio-temperature, precipitation, and potential evapotranspiration) for current and future climatic conditions (a2a: 2020, 2050, 2080; a2b: 2020, 2050, 2080)

A summary of the predicted change by 2080 is given in Table 6. The forest types are arranged in the decreasing order of carbon sequestration potential and net primary productivity (NPP). For example, major part of (≥75%) temperature conifer (TCF) would change to tropical evergreen forest (TEF), which is a relatively high NPP zone. At the same time, some part (≤25%) of this would change to tropical moist deciduous forests (TMDF), temperate forests (TF), and tropical dry deciduous forests (TDDF). This is one of the comprehensive databases generated on the dynamics of forest cover types in India, in the light of climate change.

**Table 6** Summary of predicted changes in the vegetation cover types, by 2080

| | TEF | TCF | TMDF | TF | TDDF | GL/SL | TMF | CF | TTF |
|---|---|---|---|---|---|---|---|---|---|
| TEF | | | | | | | | | |
| TCF | ✓✓ | | ✓ | ✓ | ✓ | | | | |
| TMDF | ✓✓✓ | | | | ✓ | | | | |
| TF | ✓ | ✓ | ✓ | | ✓ | | | | |
| TDDF | ✓✓ | | | | | ✓ | | | ✓ |
| GL/SL | ✓ | | | | ✓✓ | | | | |
| TMF | | | | ✓✓✓ | | | | | |
| CF | | | | ✓✓✓ | | | | | |
| TTF | | | | | ✓✓✓ | ✓✓ | | | |

TEF – tropical evergreen forests; TCF – temperate conifer forests; TMDF – tropical moist deciduous forests; TF – temperate forests; TDDF – tropical dry deciduous forests; GL/SL –grassland/scrublands; TMF – temperate mixed forests; CF – confer forests; TTF – tropical thorn forests

Towards high productive zone

Towards low productive zone

Extent of change – ✓✓✓ ≥ 75%, ✓✓ ~ 50%, ✓ ≤ 25%

The projected changes indicate that majority of shifts will result in high productivity of forests. Such a shift towards high productive zone could be linked to increase in $CO_2$ and also temperature. There are also instances where forest types would change to low productive zone. This is used to identify the vulnerable areas where the impact of climate change is negative in terms of productivity. As per one of the estimates (Ojea, Ghosh, Agrawal, *et al.* 2009), the costs for adaptation to such changes in forest distribution in India are in the range of $239.78–1486.72 million per year, depending on the adaptation scenario.

A more focused study in the Western Himalayan Eco-region (Joshi, Rawat, Narula, *et al.* 2011) assessed the shifts in forest cover types of between 700 m and 4500 m altitude. GARP and climatic envelopes were used to assess the distribution of five prominent forest cover types in four different scenarios. The summary of the shifts observed is given in Table 7.

On increasing precipitation, a downward shift in the temperate evergreen and tropical semi-evergreen was observed, while sub-tropical conifer and tropical moist-deciduous forests showed a slight upward shift, and temperate conifer showed no shift. On increasing temperature, an upward shift in all forest types was observed except sub-tropical conifer forests, which showed no significant changes. When both temperature and precipitation were changed, the actual distribution was maintained and slight upward shift was observed in all the forest types except sub-tropical conifer.

**Table 7** Shifts in forest cover type distribution in the western Himalayas under different climate change scenario

| Forest cover type | Current distribution (altitude in m) | Increased precipitation | Increased temperature | Increased precipitation + temperature |
|---|---|---|---|---|
| Temperate evergreen | >4500 | ↓ | ↑ | ↓ |
| Tropical semi-evergreen | 3000–4500 and ≥ 4500 | ↑ | o | ~ |
| Temperature conifer | 3000–4500 and ≥ 4500 | ~ | ↑ | ~ |
| Sub-tropical conifer | 3000–45 000 | ↑ | ~ | ~ |
| Tropical moist-deciduous | 900–1800 and 1800–3000 | ↑ | ↑ | ↑ |

↓ downward shift, ↑ upward shift, ~ no significant change, o cluttered distribution

## VEGETATION VULNERABILITY: EASTERN HIMALAYAS

The Eastern Himalayan region covers a broad spectrum of ecological zones and three global biodiversity hotspots in eastern Nepal, north-eastern India, Bhutan, the Tibet Autonomous Region and Yunnan of China, and northern Myanmar. The welfare of approximately 400 million people living downstream is inextricably linked with the natural resources of the Eastern Himalayas. The spatial distribution of trends in annual and seasonal temperature (using HadRM2 and PRECIS models) is illustrated in Figure 7, which clearly explains that the major parts of the region are showing warming trends. HadRM2 simulation over the region projects the highest mean temperature increase of 3.2°C during DJF and SON, the lowest increase of 2°C during MAM, and a 2.9°C increase in the annual mean temperature by the 2050s. Similarly, B2 (A2) scenarios from PRECIS simulation over the same region project the highest mean temperature increase of 3.5°C (5.3°C) during DJF, the lowest increase of 2.5°C (3.8°C) during SON (MAM and JJA), and a 2.9°C (4.3°C) increase in annual mean temperature by the 2080s. The HadRM2 model projects an increase of monsoon precipitation by 7% of the current simulation; the B2 (A2) scenarios of the PRECIS model predict an increase of monsoon precipitation by 17% (28%) of the baseline simulation. The greenhouse warming scenarios of both models predict an enhancement of monsoon precipitation of up to 30% over most parts of this region and even up to 40% over a few high-altitude pockets. Reduction in HadRM2 simulated summer precipitation over northern high-altitude regions, however, is lower in the PRECIS simulation (Figure 7).

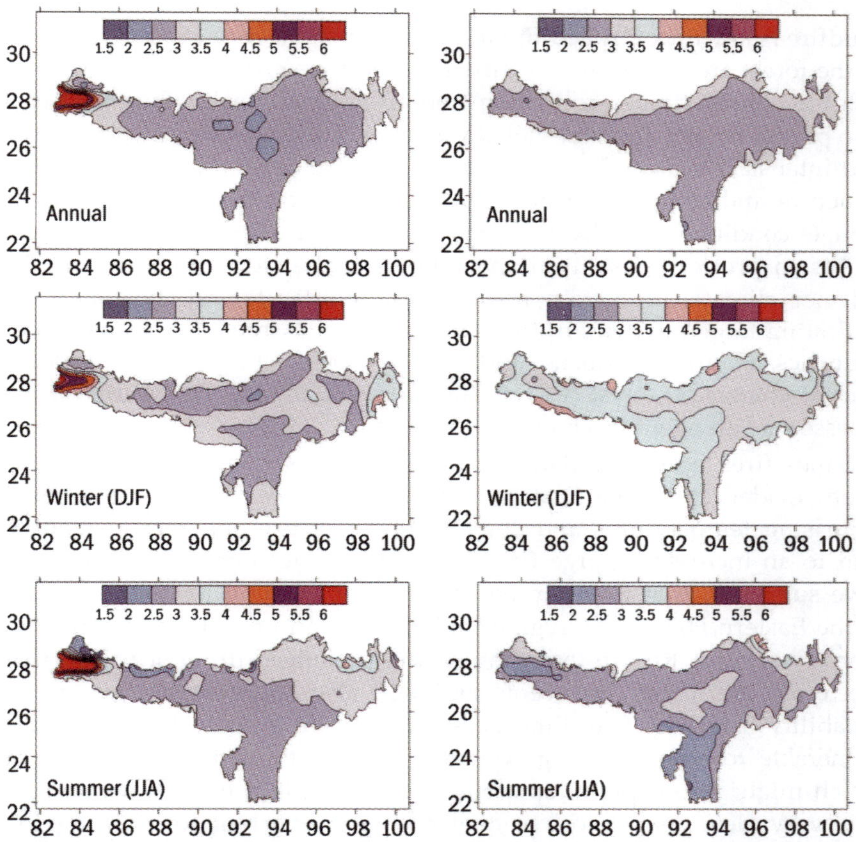


**Figure 7** Spatial distribution of scenarios (°C) of the simulated annual, winter, and summer mean temperature over the eastern Himalayan region; left: HadRM2 (2041–60); right: PRECIS (2071–2100)

**Source** Sharma, Chettri, Tse-ring, *et al.* (2009)

## FOREST DISTURBANCE

Understanding of how each disturbance in the forest landscapes is influenced by climate and affects forests, and might be exacerbated by climate change, providing a background for examining ways to cope with the impacts of climate change (Dale, Joyce, McNulty, *et al.* 2001). Though it is very complicated to evaluate if the changes occurred are the result of past disturbance and/or the results of the climate change, it is equally important to identify if climate change alters the distribution, extent, frequency, or intensity of any of these disturbances due to which large impacts (such as loss of species regeneration) are likely to be expected.

## Wildfires

Wildfire is any uncontrolled fire in the combustible vegetation, which occurs in the forest areas. Fire affects different forest resources and values, namely, wildlife habitat, timber, recreation, and also human health. Beyond this, it is also part of the equation for global warming. The frequency, magnitude (size and intensity), seasonality, and type of wildfire depend primarily on forest structure and composition but are defined and controlled by weather and climate conditions. Stocks, Fosberg, Lynham, *et al.* (1998) used four current GCMs to project forest fire danger levels in Canada and Russia under a warmer climate. Prediction results were similar for all the four GCMs, indicating large increases in the aerial extent of extreme fire danger in both countries. Flannigan, Stocks, and Wotton (2000) addressed the impacts of climate change on forest fires on the forests of the US. Forest fire has been assessed as an agent of change for forests.

Small fires occur regularly in the Himalayas (Figure 8), and there is strong evidence from the Eastern Himalayan climate change projections that increasing temperatures and changes in forest use and management will lead to an increase in large fire events at low altitudes. Modelling studies have suggested that fire may become more important for forest dynamics in the Eastern Himalayan region than changes in growth conditions due to global warming. Rise in temperature and prolonged drought period would accelerate the forest fire events and will also suppress the regeneration capability after the fire. Pine ecosystems have already been considered vulnerable to climate change because of their limited drought tolerance, which might also explain replacement and shift in the forest stands. Thus, it is very clear that wildfires may become important for shaping forest ecosystems in the future even in areas where fire disturbance is currently low (Moser, Temperli, Schneiter, *et al.* 2010).

## Invasive Species

Invasive species pose serious ecological and economic threats to the forest ecosystem, which once established in the native ecosystems are difficult to eradicate (Bradley, Blumenthal, Wilcove, *et al.* 2009; Bradley, Oppenheimer, and Wilcove 2009). Geospatial tools can be instrumental in determining areas that are of concern with regard to the invasion for better mitigation measures, prioritization, and management (Soberón and Peterson 2005; Peterson, Soberón, and Sánchez-Cordero 1999; Peterson, Allen, and Holling 2008). Priyanka and Joshi (unpublished) assessed the preliminary niche models of the potential ecological and geographical invasion of *Lantana camara,* a long-established invader with a wide distribution, in the

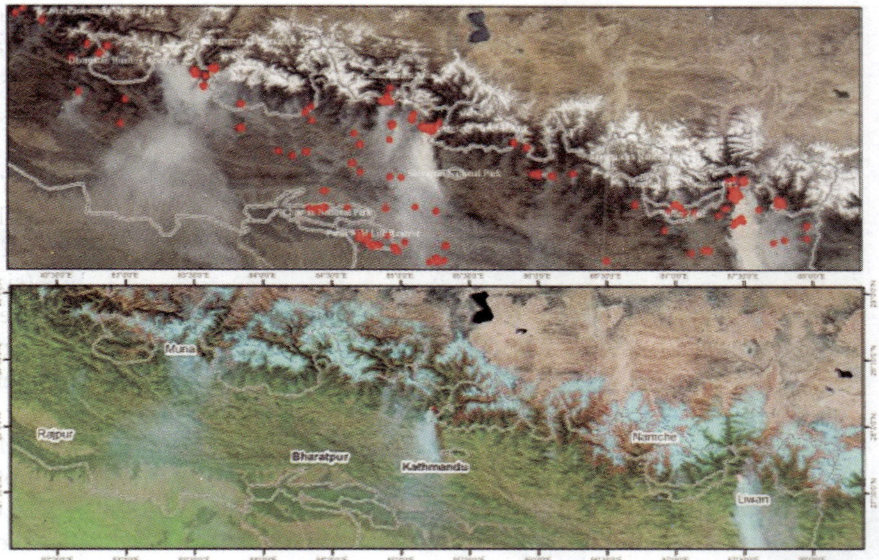


**Figure 8** Distribution of forest fire over the central Himalayas

parts of Western Himalayan region. The research work emphasizes the importance of modelling techniques in invasion studies through evaluation of the multi-model dimensions of current and future invasion mapping potential using ecological niche models. The current and future (2020, 2050, and 2080) predicted maps established the fact that under given locality records, the invasion potential of the species will be altered with the aerial range changes driven mainly by shifting largely of low to medium and medium to high suitability range in future (Figure 9). The expansionary trend revealed by the species is in concurrence with the IPCC report (2007) that projected greater increase in the proportion of plant invasion in response to amplified warming.

In another initiative, using occurrence records of *Eupatorium adenophorum* Spreng., a native of Mexico, introduced in India as an ornamental plant (in 1920s) in Uttarakhand in India, and 22 environmental variables, the ecological niche of the species was modelled using GARP and Maxent. The models were validated, and the variables influencing the distribution of *E. adenophorum* were identified. Potential distribution was then estimated for the current as well as future climatic conditions. It was found that precipitation and minimum temperature are important factors that limit the distribution of the species. Two future climatic scenarios were evaluated—A2a and B2a. It was found that the species would expand its range in both the scenarios, though more in the B2a scenario rather than A2a, probably because the species prefers the cooler climate of B2a scenario (Table 8). Such a study is significant because it can help in detecting potential regions, where an invasive species may spread in the future, and thus can aid in

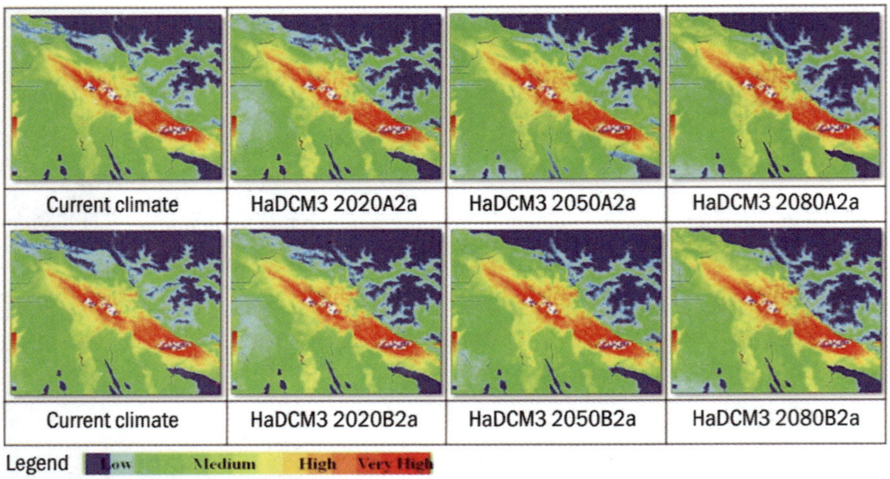


**Figure 9** The current and future ranges for Lantana camara in the parts of the Himalayas (the area covers Rajaji National Park and Jim Corbett National Park, two important protected areas of India, known for tiger and elephant)

**Table 8** Direction of spread of *Eupatorium adenophorum* in the future relative to the current distribution ("m" indicates the direction in which the maximum range shift is occurring)

| Scenario | Mode | Year | Direction of range shift | | |
|---|---|---|---|---|---|
| | | | Lateral | Upward | Downward |
| A2a | Maxent | 2020 | ↔m | ↑ | |
| | | 2050 | ↔ | ↑m | ↓ |
| | | 2080 | ↔ | | ↑m |
| | GARP | 2020 | ↔m | ↑ | ↓ |
| | | 2050 | | ↑ | |
| | | 2080 | | ↑ | |
| B2a | Maxent | 2020 | ↔ | ↑m | ↓ |
| | | 2050 | | ↑m | ↓ |
| | | 2080 | ↔ | | ↑m |
| | GARP | 2020 | ↔ | | |
| | | 2050 | ↔ | ↑m | |
| | | 2080 | ↔ | ↑m | |

**Source** Sheel (2010)

formulating better management plans to help contain the spread of such a species.

## Insects

Climate is an important factor in defining the insect species and their interaction with plants in the forest ecosystems. Climate patterns (temperature, precipitation, moisture, and others) are useful in explaining favourability of these within the ecosystems like coexistence, dominance, outbreaks, and even extinction. Williams, Long, Wargo, *et al.* (2000) provide a detailed review of the effect of climate change on forest insect and disease outbreak. The magnitude and patterns of changes will vary among the pathogens, insects, forest types, and climatic variables. Such an understanding needs to be developed at the ecosystem level, considering the host and parasite relationship and type of infestation. However, many of the studies carried out in the US, Europe, Africa, and also few countries in Asia have determined greater challenges of insect infestation in different forest ecosystems. Understanding and mapping such changes in space and time are the prerequisites to sustainable disease reduction and prioritization at specific areas based on risk maps.

## Mountain Pine Beetle

The mountain pine beetle (*Dendroctonus ponderosae* Hopkins, Coleoptera: Curculionidae, Scolytinae), a native insect of the pine forests of western

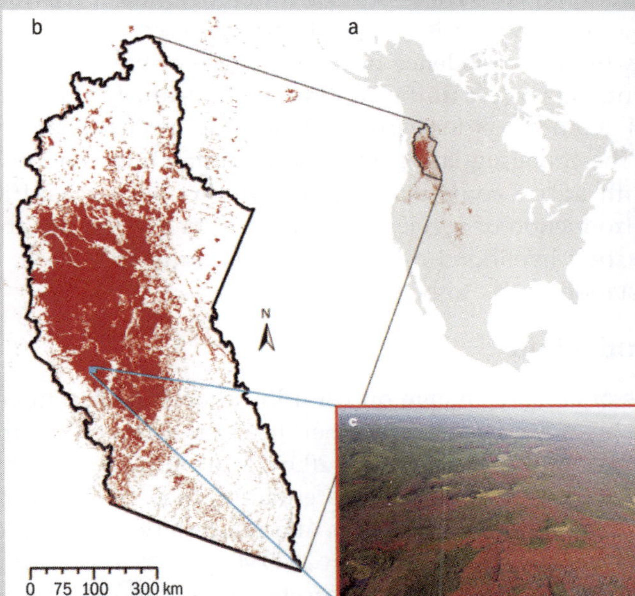


**Figure A** (a) Extent (dark red) of mountain pine beetle. (b) The study area includes 98% of the current outbreak area. (c) A photograph taken in 2006 showing an example of recent mortality: pine trees turn red in the first year after beetle kill, and grey in subsequent years

Photo credit: Joan Westfall, Entopath Management Ltd

**Source** Kurz, Dymond, Stinson, *et al.* (2008)



*Box Contd...*

Box Contd...

North America, periodically erupts. It causes tree mortality, reduces forest carbon uptake, and increases emission from the decay of dead trees. As per a study, the cumulative impact of the beetle outbreak in the affected region of British Columbia, Canada, during 2000–20 will be 270 million tonnes of carbon (or 36 g carbonm22 yr21 on average over 374 000 km$^2$ of forest). Climate change has contributed to the unprecedented extent and severity of this outbreak. Insect outbreaks such as this represent an important mechanism by which climate change may undermine the ability of the northern forests to take up and store atmospheric carbon.

## MITIGATION OPTIONS

Forests have tremendous potential to mitigate the impact of climate change and combat the challenges. Forest and forest ecosystems are important in their own way for providing unique services to the living world. They are the largest terrestrial sinks of carbon, and their removal brings the carbon back to the environment. These provide different kinds of ecosystem services, ranging from photosynthesis, the fundamental process of life, to surrogating biodiversity, the basic evidence of the existence of life. Such services could only be maintained with a unified approach of sustainable forest management (SFM). SFM should look forward to identifying mechanisms and measures to confront the changing landscape due to human interventions. This could be done with social constructive interventions with scientific basis. For example, introduction of species that are eco-friendly and are able to protect people and their livelihood options. Also checking the rate of degradation and deforestation needs to be stressed.

### Plant Introduction

The impacts of climate change on rural livelihoods in developing countries are likely to be more severe due to their heavy reliance on natural resources (Adger 2003; David and Twyman 2005; Morton, DeFries, Shimabukuro, *et al.* 2007). Thus, there is a need to enhance and strengthen the adaptive capacities of rural poor. Afforestation and reforestation (A/R) projects under the Clean Development Mechanism (CDM) (Kyoto Protocol) are already functional and take care of any reforestation activity on a barren piece of land. Low-cost input technology should be introduced for cultivating hardy perennial crops that are able to withstand the vagaries of climate and give assured returns. For example, *Simarouba glauca,* a native of South America, is a hardy multipurpose oil tree species that grows under a wide range of hostile ecological conditions (Joshi and Hiremath 2000). This paradise tree has the ability to improve rural livelihoods and generate substantial incomes for farmers dependent on marginal lands. Malviya (2009) used niche modelling approach to model current and future potential distribution of *S. glauca* and looked into the prospects for its introduction/cultivation by using multi criteria evaluation (MCE). This included using a range of multiple criteria like Human Development Index (HDI), rural and marginal worker population, and finally, area under wastelands. Results indicate that under current climate scenarios, 31 out of the 34 Indian provinces were found to be suitable to varying degree for the introduction of the species (Figure 10). However, under the future climatic conditions, the potential distribution of the species is expected to increase drastically. Such studies have the potential to assist decision-makers in taking informed decisions by providing visualization of scenarios in spatio-temporal context and evaluating options that are scientifically sound (Ascough II, McMastera, Dunna *et al.* 2002; Robertson, Peter, Villet, *et al.* 2003).

### *REDD/REDD+*

Deforestation also contributes 20% of the total GHG emissions in the tropical countries (IPCC 2007), the second largest source of GHG emissions in the world (Parker, Mitchell, Trivedi, *et al.* 2009). Avoided deforestation (conserving forests) was discussed as a CDM activity for the commitment period of 2008–12. But it was rejected due to apprehensions about diluting the effect of fossil fuel reduction, methods to measure and reduce emissions (Gibbs, Brown, Niles, *et al.* 2007), leakage, and so on. Reducing emissions from deforestation and forest degradation (REDD and now REDD+) in developing nations was introduced for the first time in 2005 at the COP-13 of United Nations Framework Convention on Climate Change (UNFCCC). REDD is a concept to assist voluntarily reducing emissions by providing incentives for discouraging deforestation and forest degradation (Chandrasekharan, Kumar, Ravindranath, *et al.* 2010). It seems to provide benefits to those countries that already have large amounts of depleted forests and countries that have maintained their forests in pristine forms (Mollicone, Freibauer, Schulze, *et al.* 2007; Grassi, Monni, Federici, *et al.* 2008; Rawat 2010a, b). COP-13 at Bali reconsidered to work on methodological issues, and COP-15 at Copenhagen gave a clear intent to get REDD going without any delay. CIFOR's Global Comparative Study (GCS) on REDD+ is a major step in this learning process. Over four years, the project aims at identifying what works in REDD+ initiatives at national and sub-national levels. This web-enabled GIS-based system will be able to provide finer details of the initiatives and developments (Figure 11).

For measuring REDD, knowing (1) the aerial extent of deforestation and forest degradation, (2) spatial location of the deforestation or forest degradation (forest types), (3) forest biomass lost, (4) C content of each forest type (tonne/area), and (5) process of forest loss, which affects the rate and timing of emissions (Olander, Gibbs, Steininger, *et al.* 2008), are the most important criteria. For this, remote sensing instruments mounted

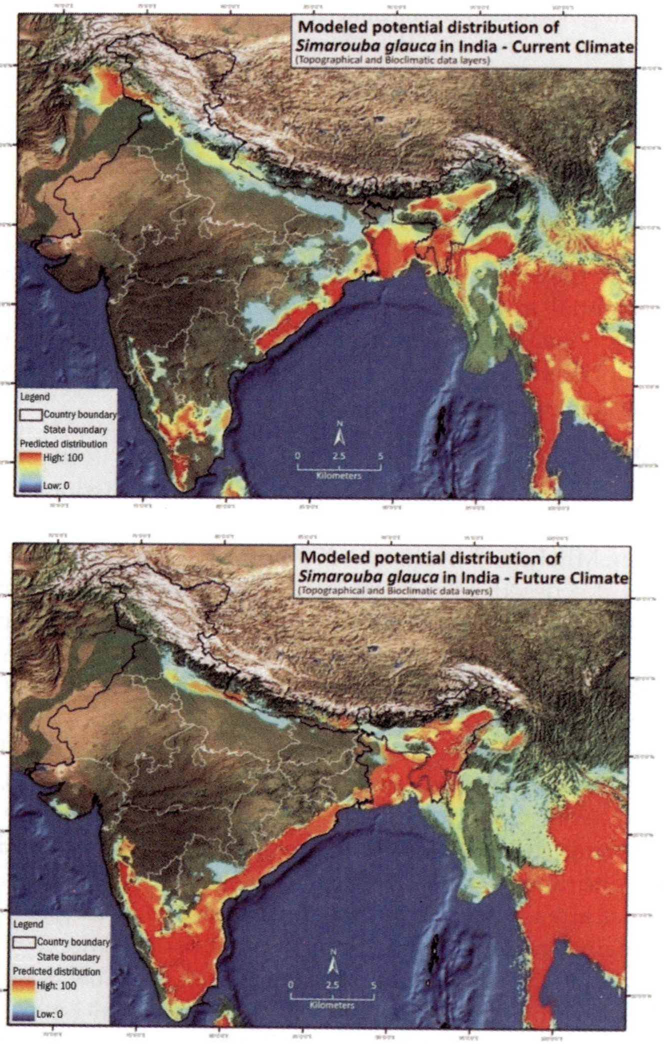


**Figure 10** Map of potential distribution of *Simarouba glauca* (Paradise tree) under current and future climate conditions in India

**Source** Malviya (2009)

on satellites provide images of the earth's surface (since 1972) and are generally the most reliable data sources for accurately estimating changes in large areas (Coppin and Bauer 1996; Lu, Mausel, Brondizio, *et al.* 2004; Olander, Gibbs, Steininger, *et al.* 2008). Change detection techniques could be used to estimate the changes in forest areas over the years. The biophysical modelling can be used for estimating biomass, forest carbon stocks, and dynamics over the time period. The spatially detailed sensor (as fine as

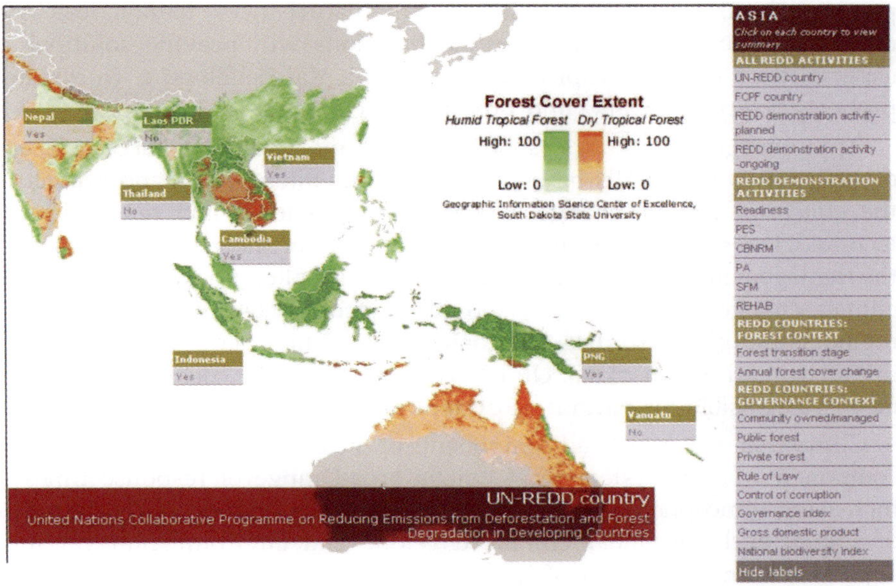


**Figure 11** Map of Asia (other areas include Africa and Latin America) showing early GCS on REDD survey results (at least nine countries in Latin America, Africa, and Asia will be included in the study over the four years)

~10 cm pixels) can be used to directly collect measurements of tree height and crown area or diameter. LIDAR, or light detection and ranging, directly estimates the height and vertical structure of forests (Dubayah and Drake 2000).

## CONCLUSION

Climate change is globally evident and can affect forests and forest ecosystems all over the world to varying extent. This might result in the loss of many important species, primarily habitat specialists species, that are restricted to narrow geographical ranges and these showing signs of phenological variation. The scope of climate change research targeting forests and forest ecosystems should be broadened with the inclusion of less-researched topic. For instance, issues related to health of the species or forests, wildlife distribution, impact on natural habitat, and protected area network appear to be missing from the research agenda. Further, addressing responses of ecosystems, its structure and functions, and communities in face of global warming is imperative. In addition, climate change should not be considered as a single risk factor; instead, it should be considered as one of the several horsemen that synergistically imperil forest ecosystem, particularly, in conjunction with human disturbances, habitat loss, and overexploitation. Adapting these combinations will provide solutions to especially of the the tropical forest region, the developing world, and probably the most diverse and productive ecosystems in the world.

Applying geoinformatics (such as GIS, remote sensing, GPS, ICT, and others) should essentially be used in analysing, interpreting, and predicting the effects of climate change. Use of modern science and renovations of technological solutions are no more options; rather these are necessities for the management of forest ecosystem to minimize alterations in face of global climate change. Such initiatives should be incorporated into local, national, regional, and global studies and conservation action plans. And in all such initiatives and action plans, climate change should be considered as a key factor responsible for natural ecosystem degradation and alteration.

Climate change affects all components of the earth system. Yet, individual components of the system react with a broad range of response times to increasing atmospheric GHGs concentrations. The combined consideration of spatial and non-linear climate envelope and environmental variables applied within the context of species interactions or habitat dynamics may be useful in assessing the potential consequences of climate change for forest ecosystem. However, lack of data, complexity of potential current and future projection models, and the propagation of uncertainty are inherent and remain great challenges to be overcome in near future. The discussed work provides a repository of research studies that are undertaken to showcase the importance of modelling tools in assessing forest ecosystems under climate change trajectories. Using such studies while drafting policy or decision-making could provide better insights for taking remedial measures under anticipated climate change, especially in the context of forest ecosystem sustenance and management. A major challenge, and thus, a step forward, is to implement regional-level priorities, so that decision-making on landscape change can take place at the regional level. Thus, an interdisciplinary research agenda integrated with monitoring networks and projection models is needed to provide information at all levels of decision-making, from developing policy to managing units.

## References

Acevedo M A and Restrepo C. 2008. **Land-cover and land-use change and its contribution to the large-scale organization of Puerto Rico's bird assemblages.** *Diversity and Distributions* **14**: 114–122

Adams D M, Alig R J, and McCarl B A. 1996. **An analysis of the impacts of public timber harvest policies on private forest management in the US.** *Forest Science* **42**(3): 343–358

Adger W N. 2003. **Social capital collective action and adaptation to climate change.** *Economic Geography*.**79**:387–404

Aertsen W, Kint V, Orshoyen J, Ozkan K, Muys B. 2010. **Comparison and ranking of different modelling techniques for prediction of site index in Mediterranean mountain forests.** *Ecological Modelling* **221**: 1119–1130

Allen C D and Breshears D D. 1998. **Drought-induced shift of a forest/woodland ecotone: rapid landscape response to climate variation**. *Proceedings of the National Academy of Sciences of the United States of America* **95**: 14839–14842

Allen C D, Macalady A K, Chenchouni H, Bachelet D, McDowell N, Vennetier M, Kitzberger T, Rigling A, Breshears D D, Hogg E H, Gonzalez P, Fensham R, Zhangm Z, Castro J, Demidova N, Lim J, Allard G, Running S W, Semerci A, Cobb N. 2010. **A global overview of drought and heat-induced tree mortality reveals emerging climate change risks for forests.** *Forest Ecology and Management* **259**: 660–684

Ascough II J C, McMastera G S, Dunna G H, Andalesb A A. 2002. **The GPFARM DSS for agroecosystem sustainability: past future and lessons learned,** pp. 1–8. In *International Environmental Modelling and Software Society (iEMSs) 2010.* International Congress on Environmental Modelling and Software Modelling for Environment's Sake, Fifth Biennial Meeting, Ottawa, Canada

Badeck F-W, Bondeau A, Bottcher K, Doktor D, Lucht W, Schaber J, Sitch S. 2004. **Responses of spring phenology to climate change**. *New Phytologist* **162**(2): 295–309

Barney J N and DiTomaso J M. 2010. **Bioclimatic predictions of habitat suitability for the biofuel switchgrass in North America under current and future climate scenarios**. *Biomass and Bioenergy* **34:** 124–133

Battles J J, Robards T, Das A, Waring K, Gilless J K, Biging G, Schurr F. 2005. **Climate change impacts on forest growth and tree mortality: a data-driven modeling study in a mixed-conifer forest of the Sierra Nevada**. *Climatic Change* **87**: S193–S213

Beaumont L J, Hughes L, and Poulsen M. 2005. **Predicting species distributions: use of climatic parameters in BIOCLIM and its impact on predictions of species' current and future distributions**. *Ecological Modelling* **186**: 251–270

Beier P and Brost B. 2010. **Use of land facets to plan for climate change: conserving the arenas not the actors**. *Conservation Biology* **21**(10): 234–244

Belotelov N V, Bogatyrev B G, Kirilenko A P, Venevsky S V. 1996. **Modelling of time-dependent biomes shifts under global climate changes**. *Ecological Modelling* **87**: 29–40

Berry J K. 1995. **What's in a model**? *GIS World* **8**(1): 26–28

Berry M, Flamm R, Hazen B, Macintyre R. 1996. **LUCAS: a system for modeling land use change**. *IEEE Computational Science and Engineering* **3**:24–35

Berry P M, Dawson T P, Harrison P A, Pearson R G. 2002. **Modelling potential impacts of climate change on the bioclimatic envelope of species in Britain and Ireland**. *Global Ecology and Biogeography* **11**: 453–462

Bradley B A, Oppenheimer M, and Wilcove D S. 2009. **Climate change and plant invasion: restoration opportunities ahead**? *Global Change Biology* **15**: 1511–1521

Bradley B A, Oppenheimer M, and Wilcove D S. 2009. **Climate change increases risk of plant invasion in the eastern United States**. *Biological Invasions* **12**(6): 1855–1872

Bradley B A, Blumenthal D M, Wilcove D S, Ziska L H. 2009. **Predicting plant invasions in an era of global change**. *Trends in Ecology and Evolution* **25**(5): 310–318

Bravo F, Jandl R, Gadow K V, LeMay V. 2008. **Introduction**. In *Managing Forest Ecosystems: the challenge of climate change,* edited by F Bravo, V LeMay, R Jandl, K von Gadow. Springer

Briffa K R, Osborn T J, and Schweingruber F H. 2004. **Large-scale temperature inferences from tree rings: a review**. *Global Planet Change* **40**: 11-26

Brink A B and Eva H D. 2009. **Monitoring 25 years of land cover change dynamics in Africa: a sample based remote sensing approach**. *Applied Geography* **29**: 501–512

Broecker W S. 1997. **Thermohaline circulation: the Achilles heel of our climate system: will man-made $CO_2$ upset the current balance**? Science **278**: 1582–1588

Butler R A. 2005. **A place out of time: tropical rainforests and the perils they face**. Details available at <http://rainforestsmongabaycom/20burkinahtm>, last accessed on 20 January 2011

Casalegno S, Amatulli G, Camia A, Nelson A, Pekkarinen A (In press). **Vulnerability of *Pinus cembra* in the Alps and Caparthian mountains under present and future climates**. *Forest Ecology and Management*

Chandrasekharan I, Kumar R S, Ravindranath M, Kushwah D J. 2010. **Incentive mechanism for preventing deforestation and promoting conservation of forest ecosystem**. *Current Science* **98**(4): 489–498

Chang D S. 1993. **The potential evapotranspiration (PE) index for vegetation and vegetation–climate classification (II)—an introduction of main methods and PEP program**. *Acta phytoecol Geobot Sinca* **13**: 197–207

Chomitz K M and Gray D A. 1996. **Roads, land use and deforestation: a spatial model applied to Belize**. *World Bank Economic Review* **10**(3): 487–512

Clark J S, Fastie C, Hurtt G, Jackson S T, Johnson C, King G A, Lewis M, Lynch J, Pacala S, Prentice I C, Schupp E W, Webb T III, Wyckoff P. 1998. **Reid's paradox of rapid plant migration: dispersal theory and interpretation of paleo-ecological records**. *Bioscience* **48**: 13–24

Cleland E E, Chuine I, Menzel A, Mooney H A, Schwartz M D. 2007. **Shifting plant phenology in response to global change**. *Trends in Ecology and Evolution* **22**(7): 357–365

Cook E R, Esper J, and D'Arrigo R. 2004. **Extra-tropical northern hemisphere temperature variability over the past 1000 years**. *Quaternary Science Reviews* **23**: 2063–2074

Coops N C, Gillanders S N, Wulder M A, Gergel S E, Nelson T, Goodwin N R. 2010. **Assessing changes in forest fragmentation following infestation using time series Landsat imagery**. *Forest Ecology and Management* **259**: 2355–2365

Coops N C, Michaud J, Andrwe M E, Wulder M A. 2011. **Comparison of a regional level habitat index derived from MERIS and MODIS estimates of canopy-absorbed photosynthetically active radiation**. *Remote Sensing Letters* **2**(4): 327–326

Coppin P R and Bauer M E. 1996. **Change detection in forest ecosystems with remote sensing digital imagery**. *Remote Sensing Reviews* **13**: 207–223

Cramer W, Bondeau A, Woodward F I, Prentice I C, Betts R A, Brovkin V, Cox P M, Fischer V, Foley J A, Friend A D, Kucharik C, Lomas M R, Ramankutty N, Sitch S, Smith B, White A, Young-Molling C. 2001. **Global response of terrestrial ecosystem structure and function to $CO_2$ and climate change: results from six dynamic global vegetation models**. *Global Change Biology* **7**: 357–373

Dale V H, Joyce L A, McNulty S, Neilson R P, Ayres M P, Flannigan M D, Hanson P J, Irland L C, Lugo A E, Peterson C J, Simberloff D, Swanson F J, Stocks B J, Wotton B M. 2001. **Climate change and forest disturbances**. *Bioscience* **51**(9): 723–734

Dale V H, Lowrance R, Mulholland R, Robertson G P. 2010. **Bioenergy sustainability at the regional scale**. *Ecology and Society* **15**(4): 23

David T S G and Twyman C. 2005. **Adaptation and equity in resource dependent societies**. In *Fairness in Adaptation to Climate Change*, pp. 223–237, edited by W N Adger, J Paavola, S Huq, M J Mace. Cambridge: The MIT Press

Davis F W, Borchert M, Meentemeyer R K, Flint A, Rizzo D M. 2010. **Pre-impact forest composition and ongoing tree mortality associated with sudden oak death in the Big Sur region, California**. *Forest Ecology and Management* **259**: 2342–2354

Dirnböck T, Dullinger S, and Grabherr G. 2003. **A regional impact assessment of climate and land use change on alpine vegetation**. *Journal of Biogeography* **30**(3): 401–417

Dixon R K, Perry J A, Vanderklein E L, Hiol F H. 1996. **Vulnerability of forest resources to global climate change: case study of Cameroon and Ghana**. *Climate Research* **6**: 127–133

Doubková M and Henebry G M. 2006. **Synergistic use of AMSR-E and MODIS data for understanding land surface phenology: a case study from grasslands of Great Plains**. Digest of IGARSS 2006. Denver, CO

Doyle T W, Krauss K W, Conner W H, From A S. 2010. **Predicting the retreat and migration of tidal forests along the northern Gulf of Mexico under sea-level rise**. *Forest Ecology and Management* **259**: 770–777

Dubayah R and Drake J B. 2000. **LIDAR remote sensing for forestry applications**. *Journal of Forestry* **98**: 44–46

Easterling W E, Aggarwal P K, Batima P, Brander K M, Erda L, Howden S M, Kirilenko A, Morton J, Soussana J-F, Schmidhuber J, Tubiello F N. 2007. **Food fibre and forest products**. In *Climate Change 2007: Impacts Adaptation and Vulnerability*. Contribution of Working Group II to the Fourth Assessment Report of the Intergovernmental Panel on Climate Change, pp. 273–313, edited by M L Parry, O F Canziani, J P Palutikof, P J van der Linden, C E Hanson. Cambridge: Cambridge University Press

Eigenbrod F, Hecnar S J, and Fahrig L. 2008. **Accessible habitat: an improved measure of the effects of habitat loss and roads on wildlife populations.** *Landscape Ecology* **23**: 159–168

FAO (Food and Agricultural Organization). 2009 (Submission to UNFCCC AWG LCA). **Enabling agriculture to contribute to climate change.** Details available at <http://unfcccint/resource/docs/2008/smsn/igo/036pdf>

Fitz H C, DeBellevue E B, Costanza R, Boumans R, Maxwell T, Wainger L, Sklar F H. 1996. **Development of a general ecosystem model (GEM) for a range of scales and ecosystems.** *Ecological Modelling* **88**:263–297

Flannigan M D, Stocks B J, and Wotton B M. 2000. **Climate change and forest fires.** *Science of the Total Environment* **262**: 221–229

Flemons P, Guralnick R, Krieger J, Ranipeta A, Neufeld D. 2007. **A web-based GIS tool for exploring the world's biodiversity: the global biodiversity information facility mapping and analysis portal application (GBIF-MAPA).** *Ecological Informatics* **2**: 49–60

Foody G M. 2008. **GIS – biodiversity applications.** *Progress in Physical Geography* **32**(2): 223–235

Franco A M A, Hill J K, Kitschke C, Collingham Y C, Roy D B, Fox R, Huntley B. Thomas C D. 2006. **Impacts of climate warming and habitat loss on extinctions at species' low-latitude range boundaries.** *Global Change Biology* **12**: 1545–1553

Fraser R, Abuelgasimb T, and Latifovica R. 2005. **A method for detecting large-scale forest cover change using coarse spatial resolution imagery.** *Remote Sensing of Environment* **95**: 414–427

Fuller T, Morton D P, and Sarkar S. 2008. **Incorporating uncertainty about species' potential distributions under climate change into the selection of conservation areas with a case study from the Arctic Coastal Plain of Alaska.** *Biological Conservation* **141**: 1547–1559

Galparsoro I, Borja A, Bald J, Liria P, Chust G. 2009. **Predicting suitable habitat for the European lobster (*Homarus gammarus*) on the Basque continental shelf (Bay of Biscay) using ecological-niche factor analysis.** *Ecological Modelling* **220**: 556–567

Gibbs H K, Brown S, Niles J O, Foley J A. 2007. **Monitoring and estimating tropical forest carbon stocks: making REDD a reality.** *Environmental Research Letters* **2**

Gilruth P T, Marsh S E, and Itami R. 1995. **A dynamic spatial model of shifting cultivation in the highlands of Guinea West Africa.** *Ecological Modelling* **79**(1–3): 179–197

Giovanelli J G R, de Siqueira M F, Haddad C F, Alexandrino J. 2010. **Modeling a spatially restricted distribution in the Neotropics: how the size of calibration area affects the performance of five presence-only methods.** *Ecological Modelling* **221**: 215–224

Giriraj A, Babar S, and Reddy C S. 2008. **Monitoring of forest cover change in Pranahita Wildlife Sanctuary, Andhra Pradesh, India, using remote sensing and GIS.** *Journal of Environmental Science and Technology* **1**(2): 73–79

Giriraj A, Beierkuhnlein C, and Murthy M S R. 2010. **Evaluating forest fragmentation and its tree community composition in the tropical rain forest of southern Western Ghats (India) from 1973 to 2004**. *Environmental Monitoring and Assessment* **161**: 29–44

Giriraj A, Murthy M S R, Beierkuhnlein C, Irfan-Ullah M. 2008. **Modelling spatial and temporal forest cover change patterns (1973–2020): a case study from South Western Ghats (India)**. *Sensors* **8**(10): 6132–6153

Gopalakrishnan R, Jayaraman M, Swarnim S, Chaturvedi R K, Bala G, Ravindranath N H. 2011. **Impact of climate change at species level: a case study of teak in India**. *Mitigation and Adaptation Strategies for Global Change* **16**(2): 199–209

Goparaju L, Tripathi A, and Jha C S. 2005. **Forest fragmentation impacts on phytodiversity – an analysis using remote sensing and GIS**. *Current Science* **88**(8): 1264–1274

Grassi G, Monni S, Federici S, Achard F, Mollicone D. 2008. **Applying the conservativeness principle to REDD to deal with the uncertainties of the estimates**. *Environmental Research Letters* **3**(3): 1–12

Greene S L, Minoura T, Steiner J J, Pentacost G. 2007. **WebGRMS: prototype software for web-based mapping of biological collections**. *Biodiversity and Conservation* **16**:2611–2625

Guinan J, Brown C, Dolan M F J, Grehan A J. 2009. **Ecological niche modelling of the distribution of cold-water coral habitat using underwater remote sensing data**. *Ecological Informatics* **4**: 83–92

Guisan A and Theurillat J P. 2000. **Assessing alpine plant vulnerability to climate change: a modeling perspective**. *Integrated Assessment* **1**: 307–320

Gustafson E J, Roberts L J, and Leefers L A. 2006. **Linking linear programming and spatial simulation models to predict landscape effects of forest management alternatives**. *Journal of Environmental Management* **81**: 339–350

Hanewinkel M, Hummel S, and Cullmann D A. 2010. **Modelling and economic evaluation of forest biome shifts under climate change in Southwest Germany**. *Forest Ecology and Environment* **259**: 710–719

Hannah K, Midgley G F, and Millar D. 2002. **Climate change-integrated conservation strategies**. *Global Ecology and Biogeography* **11**(6): 485–495

Hardie I W and Parks P J. 1997. **Land use with heterogeneous land quality: an application of an area-base model**. *American Journal of Agricultural Economics* **79**(2): 299–310

Harrison P A, Berry P M, Butt N, New M. 2006. **Modelling climate change impacts on species' distributions at the European scale: implications for conservation policy**. *Environmental Science and Policy* **9**: 116–128

Heikkinen R K, Luoto M, Araujo M B, Virkkala R, Thuiller W, Sykes M T. 2006. **Methods and uncertainties in bioclimatic envelope modelling under climate change**. *Progress in Physical Geography* **30**: 751–777

Hengl T, Sierdsema H, Radovic A, Dilo A. 2009. **Spatial prediction of species' distributions from occurrence-only records: combining point pattern analysis, ENFA and regression-kriging**. *Ecological Modelling* **220**: 3499–3511

Hickler T, Prentice I C, Smith B, Sykes M T, Zaehle S. 2006. **Implementing plant hydraulic architecture within the LPJ dynamic global vegetation model.** *Global Ecology Biogeography* **15**: 567–577

Hijmans R J and Graham C H. 2006. **The ability of climate envelope models to predict the effect of climate change on species distributions**. *Global Change Biology* **12**: 2272–2281

Holdridge L R. 1947. **Determination of world plant formations from simple climatic data.** *Science* **105**: 367–368

Honnay O, Verhaeghe W, and Hermy M. 2001. **Plant community assembly along dendritic networks of small forest streams**. *Ecology* **82**: 1691–1702

Horikawa M, Tsuyama I, Matsui T, Kominami Y, Tanaka N. 2009. **Assessing the potential impacts of climate change on the alpine habitat suitability of Japanese stone Pine (*Pinus pumila*)**. *Landscape Ecology* **24**: 115–128

Houghton R A, Hall F, and Goetz S J. 2009. **Importance of biomass in the global carbon cycle**. *Journal of Geophysical Research* **114**: G00E03. Details available at <http://www.epagov/climatechange/science/futuretchtml>

INCCA (Indian Network of Climate Change Assessment). 2009. **Climate change and India – towards preparation of a comprehensive climate change assessment.** Ministry of Environment and Forests, New Delhi. 24 pp.

INCCA (Indian Network of Climate Change Assessment). 2010. **Climate change and India: a 4×4 assessment – a sectoral and regional analysis for 2030s.** INCSA Report #2, Ministry of Environment and Forests, New Delhi. 160 pp.

IPCC (Intergovernmental Panel on Climate Change). 2000. IPCC Special Report on Emission Scenarios. Summary for Policy Makers. Geneva. Switzerland

IPCC (Intergovernmental Panel on Climate Change). 2007. ***Climate Change 2007: Synthesis Report.*** Contribution of Working Groups I, II, and III to the *Fourth Assessment Report of the IPCC.* Geneva, Switzerland

Irfan-Ullah M, Giriraj A, Murthy M S R, Peterson A T. 2007. **Mapping the geographic distribution of *Aglaia bourdillonii* Gamble (Meliaceae) an endemic and threatened plant using ecological niche modeling**. *Biodiversity and Conservation* **16:** 1917–1925

Iverson L R and Prasad A M. 2001. **Potential tree species shifts with five climate change scenarios in the Eastern United States**. *Forest Ecology and Management* **155**(1–3)

Iverson L, Prasad A, and Matthews S. 2008. **Modelling potential climate change impacts on the trees of the northwestern United States**. *Mitigation and Adaptation Strategies for Global Change* **13**: 517–540

Jakeman A J, Voinov A, and Emilio A. 2008. ***Environmental Modelling Software and Decision Support,*** pp. 1–11. Elsevier

Jonzén N, Ergon T, Lindén A, Stenseth N C. 2007. **Introduction (bird migration and climate, Special Issue 17)**. *Climate Research* **35**: 1–3

Joshi P K. 2010a. **Assessing vulnerability of vegetation types to human dimensions vis-à-vis climate change in Northeastern Himalaya – a precursor to adaptation strategies and management**. 9th CTWF International

Workshop on Climate and Environmental Change – Challenges for Developing Countries, November 17–19, 2010. Beijing

Joshi P K. 2010b. **Geospatial tools for climate change studies in forestry sector**. Symposium on Climate Change and Disaster Management. 13th Annual International Conference and Exhibition on Geospatial Information Technology and Applications. Map India 2010, January 19–21, 2010, Gurgaon, India

Joshi J and Hiremath S. 2000. **Simarouba: a potential oilseed tree**. *Current Science* **78**(6): 694–697

Joshi P K, Rawat A, Narula S, Sinha V S P. 2011. **Assessing impact of climate change on forest cover type shifts in Western Himalayan Eco-region**. *Journal of Forestry Research* (accepted)

Kameyama S, Fukishima M, Han M, Kaneki M. 2007. **Spatio-temporal changes on habitat potential of endangered freshwater fish in Japan**. *Ecological Informatics* **2**: 318–327

Karnatak H C, Saran S, Bhatia K, Roy P S. 2007. **Multicriteria spatial decision analysis in web GIS environment**. *Geoinformatica* **11**: 407–429

Keller R P, Lodge D M, and Lewis M A. 2009. ***Bioeconomics of Invasive Species: integrating ecology economics policy and management***, pp. 63–69. New York: Oxford University Press

Kerr J T and Ostrovsky M. 2003. **From space to species: ecological applications for remote sensing**. *Trends in Ecology and Evolution* **18**: 299–305

Kirtland D, Gaydos L, Clarke K, DeCola L, Acevedo W, Bell C. 1994. **An analysis of human-induced land transformations in the San Francisco Bay/Sacramento area**. *World Resources Review* **6**(2): 206–217

Knorn J, Rabe A, Radeloff V C, Kuemmerle T, Kozak J, Hostert P. 2009. **Land cover mapping of large areas using chain classification of neighboring Landsat satellite images**. *Remote Sensing of Environment* **113**: 957–964

Kuemmerle T, Chaskovskyy O, Knorn J, Radeloff V C, Kruhlov I, Keeton W S, Hostert P. 2009. **Forest cover change and illegal logging in the Ukrainian Carpathians in the transition period from 1988 to 2007**. *Remote Sensing of Environment* **113**: 1194–1207

Kurz W A, Dymond C C, Stinson G, Rampley G J, Neilson E T, Carroll A L, Ebata T, Safranyi L. 2008. **Mountain pine beetle and forest carbon feedback to climate change**. *Nature* **452**: 987–990

Landis J D. 1995. **Imagining land use futures: applying the California urban futures model**. *Journal of the American Planning Association* **61**(4): 438

Landis J D, Monzon J P, Reilly M, Cogan C. 1998. **Development and pilot application of the California Urban and Biodiversity Analysis (CURBA) Model**. Presented at the 1998 ESRI International User Conference, October 7–9. Details available at <http://wwwesricom/library/usercoc98/PROCEED/TO600/PAP571/P571HTM>

Lasch P, Lindner M, Erhard M, Suckow F, Wenzel A. 2002. **Regional impact assessment on forest structure and functions under climate change — the Brandenburg case study**. *Forest Ecology and Management* **162**: 73–86

Lawson B E, Day M D, Bowen M, Klinken R, Zalucki M P. 2010. **The effect of data sources and quality on the predictive capacity of CLIMEX models: an assessment of *Teleonemia scrupulosa* and *Octotoma scabripennis* for the biocontrol of *Lantana camara* in Australia**. *Biological Control* **52**: 68–76

Lenoir J, Gégout J C, Marquet P A, de Ruffray P, Brisse H. 2008. **A significant upward shift in plant species optimum elevation during the 20th century.** *Science* **320**: 1768–1771

Lockwood J L, Hoopes M F, and Marchetti M P. 2007. ***Invasion Ecology***. Blackwell Publishing. 304 pp.

Lopez-Lopez P, Garcia-Ripolles C, Aguilar J, Garcia-Lopez F, Verdejo J. 2006. **Modelling breeding habitat preferences of Bonelli's eagle (*Hieraaetus fasciatus*) in relation to topography disturbance climate and land use at different spatial scales**. *Journal of Ornithology* **147**: 97–106

Lu D, Mausel P, Brondizio E, Moran E. 2004. **Change detection techniques**. *International Journal of Remote Sensing* **25**(12): 2365–2407

Luoto M, Poyry J, Heikkinen R K, Saarinen K. 2005. **Uncertainty of bioclimate envelope models based on the geographical distribution of species**. *Global Ecology and Biogeography* **14**: 575–584

Luoto M, Virkkala R, and Heikkinen R K. 2007. **The role of land cover in bioclimatic models depends on spatial resolution**. *Global Ecology and Biogeography* **16**: 34–42

Malcolm J R, Liu C R, Neilson R P, Hansen L, Hannah L. 2006. **Global warming and extinctions of endemic species from biodiversity hotspots**. *Conservation Biology* **20**: 538–548

Malviya S. 2009. **Spatio temporal modeling and evaluation of the distribution of *Simarouba glauca* in India – strategising rural livelihood adaptation under climate change**. MSc Thesis submitted to TERI University, New Delhi

Mann M E and Jones P D. 2003. **Global surface temperatures over the past two millennia**. *Geophysical Research Letters* **30**(15): 1820–1825

Marmion M, Luoto M, Heikkinen R K, Thuiller W. 2009. **The performance of state-of-the-art modelling techniques depends on geographical distribution of species**. *Ecological Modelling* **220**:3512–3520

Martin J L, Stockton S A, Allombert S, Gaston A J. 2007. **Top down and bottom-up consequences of unchecked ungulate browsing on plant and animal diversity in temperate forests: lessons from a deer introduction.** *Biological Invasions*

Mayaux P, Bartholome E, Fritz S, Belward A. 2004. **A new land-cover map of Africa for the year 2000**. *Journal of Biogeography* **31**: 861–877

Mertens B and Lambin E. 1997. **Spatial modelling of deforestation in southern Cameroon: spatial disaggregation of diverse deforestation processes**. *Applied Geography* **17**(2): 143–162

Midgley G F, Hannah L, Millar D, Rutherford M C, Powerie L W. 2002. **Assessing the vulnerability of species richness to anthropogenic climate change in a biodiversity hotspot**. *Global Ecology and Biogeography* **11**: 445–451

Mladenoff D J and Baker W L. 1999. ***Spatial Modeling Of Forest Landscape Change: approaches and applications,*** pp. 1–14. Cambridge, UK: Cambridge University Press

Moberg A, Sonechkin D M, Holmgren K, Datsenko N M, Karlen W. 2005. **Highly variable northern hemisphere temperatures from low- and high-resolution proxy data**. *Nature* **433**: 613–617

Mollicone D, Freibauer A, Schulze E D, Braatz S, Grassi G, Federici S. 2007. **Elements for the expected mechanisms on 'reduced emissions from deforestation and degradation' REDD under UNFCCC**. *Environmental Research Letters* **2**: 1–7

Moning C, Werth S, Dziock F, Bassler C, Bradtka J, Hothorn T, Muller J. 2010. **Lichen diversity in temperate montane forests is influenced by forest structure more than climate**. *Forest Ecology and Management* **258**(5): 745–751

Morton D C, DeFries R S, Shimabukuro Y E, Anderson L O, Arai E, del Bon Espirito-Santo F, Freitas R, Morisette J. 2007. **Cropland expansion changes deforestation dynamics in the southern Brazilian Amazon**. *Proceedings of the National Academy of Sciences of the USA* **103**: 14637–14641

Moser B, Temperli C, Schneiter G, Wohlgemuth T. 2010. **Potential shift in tree species composition after interaction of fire and drought in the Central Alps**. *European Journal of Forest Research* **129**(4): 625–633

Myneni R B, Keeling C D, Tucker C J, Asrar G, Nemani R R. 1997. **Increased plant growth in the northern high latitudes from 1981–1991**. *Nature* **386**: 698–702

Nitschke C R and Innes J L. 2008. **A tree and climate assessment tool for modelling ecosystem response to climate change**. *Ecological Modelling* **210**: 263–277

Norby R J, Rustad L E, Dukes J S, Ojima D S, Parton W J, Del Grosso S J, McMurtie R E, Pepper D A. 2005. **Ecosystem responses to warming and interacting global change factors.** In *Terrestrial Ecosystems in a Changing World,* edited by J G Canadell, E E Pataki, and L F Pitelka. Heidelberg: The IGBP Series Springer

O'Callaghan J R. 1995. **NELUP: an introduction**. *Journal of Environmental Planning and Management* **38**(1): 5–20

Oglethorpe D R and O'Callaghan J R. 1995. **Farmlevel economic modelling within a river catchment decision support system**. *Journal of Environmental Planning and Management* **38**(1): 93–106

Ojea E, Ghosh R, Agrawal B B, Joshi P K. 2009. **The costs of ecosystem adaptation: methodology and estimates for Indian forests.** Basque Centre for Climate Change (BC3) Working Paper Series 2009–10, BC3, Bilbao, Spain. Details available at <http://wwwbc3researchorg/lits_publicationshtml>

Okunishi T, Yamanaka Y, and Ito S. 2009. **A simulation model for Japanese sardine (*Sardinops melanostictus*) migrations in the western North Pacific**. *Ecological Modelling* **220**: 462–479

Olander L P, Gibbs H K, Steininger M, Swenson J J, Murray B C. 2008. **Reference scenarios for deforestation and forest degradation in support of REDD: a review of data and methods**. *Environmental Research Letters* **3**: 1–11

Olwoch J M, Reyers B, and van Jaarsveld A S. 2009. **Host–parasite distribution patterns under simulated climate: implications for tick-borne diseases**. *International Journal of Climatology* **29**: 993–1000

Pan Y, Li X, Gong P, He C, Shi P, Pu R. 2003. **An integrative classification of vegetation in China based on NOAA AVHRR and vegetation–climate indices of the Holdridge Life Zone**. *International Journal of Remote Sensing* **25**(5): 1009–1027

Panigrahy S, Anitha D, Kimothi N M, Singh S P. 2010. **Timberline change detection using topographic map and satellite Imagery**. *Tropical Ecology* **51**: 87–91

Parker C, Mitchell A, Trivedi M, Mardas N, Sosis K. 2009. **The little REDD+**. Details available at <http://wwwamazonconservationorg/pdf/redd_the_little_redd_book_dec_08pdf>, last accessed on January 24, 2011

Parmesan C and Yohe G. 2003. **A globally coherent fingerprint of climate change impacts across natural systems**. *Nature* **421**: 37–42

Parry M L, Canziani O F, Palutikof J P, van der Linden P J, Hanson C E (eds). 2007. **IPCC 2007: climate change 2007: impacts adaptation and vulnerability**. Contribution of Working Group II to the *Fourth Assessment Report of the Intergovernmental Panel on Climate Change*. Cambridge: Cambridge University Press. 976 pp.

Pearson R G and Dawson T P. 2003. **Predicting the impacts of climate change on the distribution of species: are bioclimate envelope models useful?** *Global Ecology and Biogeography* **12**: 361–371

Pennington D D. 2007. **Exploratory modelling of forest disturbance scenarios in central Oregon using computational experiments in GIS**. *Ecological Informatics* **2**: 387–403

Penuelas J and Boada M. 2003. **A global change-induced biome shift in the Montseny mountains (NE Spain)**. *Global Change Biology* **9**: 131–140

Peterson G, Allen C R, and Holling C S. 1998. **Ecological resilience, biodiversity, and scale**. *Ecosystems* **1**: 6–18

Peterson A T, Soberón J, and Sánchez-Cordero V. 1999. **Conservatism of ecological niches in evolutionary time**. *Science* **285**: 1265–1267

Peterson A T, Sanchez-Cordero V, Soberon J, Bartley J, Buddemeier R W, Navarro-Singuenza A G. 2001. **Effects of global climate change on geographic distributions of Mexican Cracidae**. *Ecological Modelling* **144**: 21–30

Phillips S J, Anderson R P, and Schapire R E. 2006. **Maximum entropy modeling of species geographic distribution**. *Ecological Modelling* **190**:231–259

Pounds J A, Bustamante M R, Coloma L A, Consuegra J A, Fogden M p L, Foster P N, La Marca E, Masters K L, Merino-Viteri A, Puschendorf R, Ron S R, Sánchez-Azofeifa G A, Still C J, Young B E. 2006. **Widespread amphibian extinctions from epidemic disease driven by global warming**. *Nature* **439**: 161–167

Powell M, Accad A, Austin M P, Choy S L, Williams K J, Shapcott A. 2010. **Predicting loss and fragmentation of habitat of the vulnerable subtropical rainforest tree *Macadamia integrifolia* with models developed from compiled ecological data**. *Biological Conservation* **143**: 1385–1396

Raupach M R, Marland G, Ciais P, Le Quéré C, Canadell J P, Klepper G, Field C B. 2007. **Global and regional drivers of accelerating $CO_2$ emissions.** *Proceedings of the National Academy of Sciences USA* **104**(24): 10288–10293

Ravindranath N H and Sukumar R. 1996. **Impacts of climate change on forest cover of India.** *Commonwealth Forestry Review* **75**: 76–79

Ravindranath N H, Joshi N V, Sukumar R, Saxena A. 2006. **Impact of climate change on forests in India.** *Current Science* **90**(3): 354–361

Rawat V R S. 2010a. **Reducing emissions from deforestation in developing countries (REDD) and REDD plus under the UNFCCC negotiations.** *The Indian Forester* **136**(1): 129–133

Rawat V R S. 2010b. **REDD-plus at Copenhagen: a win situation for all participating nations.** *The Indian Forester* **136**(2)

Robertson M P, Peter C I, Villet M H, Ripley B S. 2003. **Comparing models for predicting species' potential distributions: a case study using correlative and mechanistic predictive modelling techniques.** *Ecological Modelling* **164**: 153–167

Root T L, Price J T, Hall R K, Schneider S H, Rosenzweig C, Pounds J A 2003. **Fingerprints of global warming on wild animals and plants.** *Nature* **421**: 57–60

Rosenqvist A, Milne A, Lucas R, Imhoff M, Dobson C. 2003. **A review of remote sensing technology in support of the Kyoto Protocol.** *Environmental Science and Policy* **6**: 441–455

Roy P S, Joshi P K, Singh S, Agarwal S, Yadav D, Jegannathan C. 2006. **Biome mapping in India using vegetation type map derived using temporal satellite data and environmental parameters.** *Ecological Modelling* **197**: 148–158

Sala O E, Chapin F S, Armesto J J, Berlow E, Bloomfield J, Dirzo R, Huber-Sanwald E, Huenneke L F, Jackson R B, Kinzig A, Leemans R, Lodge D M, Mooney H A, Oesterheld M, Poff N L, Sykes M T, Walker B H, Walker M, Wall DH. 2000. **Global biodiversity scenarios for the year 2100.** *Science* **287**: 1770–1774

Santika T and Hutchinson M F. 2009. **The effect of species response form on species distribution model prediction and inference.** *Ecological Modelling* **220**: 2365–2379

Schmidt M, König K, and Müller J V. 2008. **Modelling species richness and life form composition in Sahelian Burkina Faso with remote sensing data.** *Journal of Arid Environments* **72**: 1506–1517

Selkowitz D J. 2010. **A comparison of multi-spectral multi-angular and multi-temporal remote sensing datasets for fractional shrub canopy mapping in Arctic Alaska.** *Remote Sensing of Environment* **114**: 1338–1352

Sharma P K, Lahkar B P, Ghosh S, Rabha A, Das J P, Nath N K, Dey S, Brahma N. 2008. **Land use and land cover change and future implication analysis in Manas National Park, India, using multi-temporal satellite data.** *Current Science* **95**(2): 23–227

Sharma E, Chettri N, Tse-ring K, Shrestha A B, Fang Jing, Mool P K, Eriksson M. 2009. ***Climate Change Impacts and vulnerability in the Eastern Himalayas.*** ICIMOD, Kathmandu

Sheel V. 2010. **Assessing potential distribution of *Eupatorium adenophorum* in middle Himalaya, India**. MSc Thesis submitted to TERI University, New Delhi

Sitch S, Smith B, Prentice I C, Arneth A, Bondeau A, Cramer W, Kaplan J O, Levis S, Lucht W, Sykes M T, Thonicke K, Venevsky S. 2003. **Evaluation of ecosystem dynamics plant geography and terrestrial carbon cycling in the LPJ dynamic global vegetation model**. *Global Change Biology* **9**: 161–185

Soberón J amd Peterson A T. 2005. **Interpretation of models of fundamental ecological niches and species' distributional areas**. *Biodiversity Informatics* **2**: 1–10

Soja A J, Tchebakova N M, French N H F, Flannigan M D, Shugart H H, Stocks B J, Sukhinin A I, Parfenova E I, Chapin III F S, Stackhouse Jr P W. 2007. **Climate-induced boreal forest change: predictions versus current observations**. *Global and Planetary Change* **56**: 274–296

Steyaert L T. 1993. **The state of environmental simulation modelling**. In *Environmental Modelling with GIS*, pp. 16–30, edited by M F Goodchild and L T Steyaert. New York: Oxford University Press

Stockli R and Vidale P L. 2004. **European plant phenology and climate as seen in a 20-year AVHRR land surface parameter dataset**. *International Journal of Remote Sensing* **25**(17): 3303–3330

Stocks B J, Fosberg M A, Lynham T J, Mearns L, Wotton B M, Yang Q, Jin J Z, Lawrence K, Hartley G R, Mason J A, Mckenney D W. 1998. **Climate change and forest fire potential in Russian and Canadian boreal forests.** *Climatic Change* **38**: 1–13

Stockwell D R B, Beach J H, Stewart A, Vorontsov G, Vieglais D, Pereira R S. 2006. **The use of GARP genetic algorithm and internet grid computing in the Life mapper world atlas of species biodiversity**. *Ecological Modelling* **195**: 139–145

Sulzman E W, Poiani K A, and Kittel T G F. 1995. **Modeling human-induced climatic change: a summary for environmental managers**. *Environmental Management* **19**: 197–224

Swallow S K, Talukdar P, and Wear D N. 1997. **Spatial and temporal specialization in forest ecosystem management under sole ownership**. *American Journal of Agricultural Economics* **79**: 311–326

Thomas C D, Ohlemüller R, Anderson B, Hickler T, Miller P A, Sykes M T, Williams J W. 2008. **Exporting the effects of climate change**. *EMBO Reports* **9**: S28–S33

Torres J, Brito J C, Vasconcelos M J, Catarino L, Goncalyes J, Honrado J. 2010. **Ensemble models of habitat suitability relate chimpanzee (*Pan troglodytes*) conservation to forest and landscape dynamics in Western Africa**. *Biological Conservation* **143**: 416–425

Turner M G, Gardner R H, and O'Neill R V. 2001. ***Landscape Ecology – in theory and practice***. New York: Springer-Verlag

Valarde S J, Mlahi Y, Moran D, Wright H, Hussain S. 2005. **Valuing the impacts of climatic change on protected areas in Africa**. *Ecological Economics* **53**: 21–23

Veldkamp A and Fresco L O. 1996a. **CLUE: a conceptual model to study the conversion of land use and its effects**. *Ecological Modelling* **85**: 253–270

Veldkamp A and Fresco L O. 1996b. **CLUE-CR: an integrated multi-scale model to simulate land use change scenarios in Costa Rica**. *Ecological Modelling* **91**: 231–248

Vennetier M and Ripert C, 2009. **Forest flora turnover with climate change in the Mediterranean region: a case study in Southeastern France**. *Forest Ecology and Management* **2585**: S56–S63

Vogiatzakis I N, Mannion A M, and Griffiths G H. 2006. **Mediterranean ecosystems: problems and tools for conservation**. *Progress in Physical Geography* **30**: 175–200

Voinov A, Costanza R, Wainger L, Boumans R, Villa F, Maxwell T, Voinov H. 1999. **Patuxent landscape model: integrated ecological economic modeling of a watershed**. *Journal of Ecosystem Modelling and Software* **14**: 473–491

Walther G R, Berger S, and Sykes M T. 2005. **An ecological 'footprint' of climate change**. *Proceedings of the Royal Society London* B **272**: 1427–1432

Wear D N, Apt R, and Mangold R. 1998. **People space time: factors that will govern forest sustainability,** pp. 348–361. Transactions of the 63rd North American Wildlife and Natural Resources Conference, Washington, DC, Wildlife Management Institute

Wear D N, Liu R, Foreman J M, Sheffield R M. 1999. **The Effects of population growth on timber management and inventories in Virginia**. *Forest Ecology and Management* **118**: 107–115

Wegener M. 2000. **Spatial models and GIS**. In *Spatial Models and GIS: new potential and new models,* pp. 3–18, edited by Fotheringham and Wegener. London: Taylor and Francis Inc.

Williams D W, Long R P, Wargo P M, Liebhold A M. 2000. **Effects of climate change on forest insect and disease outbreaks**. In *Responses of Northern US Forests to Environmental Change Ecological Studies,* pp. 455–494, edited by R A Mickler, R A Birdsey, and J Hom. New York: Springer-Verlag

Wilson R J, Gutiérrez D, Gutiérrez J, Martínez D, Agudo R, Monserrat V J. 2005. **Changes to the elevational limits and extent of species ranges associated with climate change**. *Ecological Letters* **8**: 1138–1146

Wood E C, Lewis J E, Gray T G, Lietzow R W. 1997. **The development of a land cover change model for southern Senegal**. Presented at the Land Use Modeling Workshop June 1997. Sponsored by the USGS and NCGIA. Details available at <http://ncgiancgiaucsbedu/conf/landuse97/papers/wood eric/pecdochtml>, last accessed on June 2010

Woodall C W, Oswalt C M, Westfall J A, Perry C H, Nelson M D, Finley A O. 2009. **An indicator of tree migration in forests of the Eastern United States**. *Forest Ecology and Management* **257**: 1434–1444

Woodward F I and Beerling D J. 1997. **The dynamics of vegetation change: health warnings for equilibrium 'dodo' models**. *Global Ecology Biogeography Letters* **6**: 413–418

World Bank. 2010. **Environmental licensing for hydroelectric projects in Brazil: a contribution to the debate**. Brazil Country Management Unit Report 40995-BR

World Resources Institute (WRI). 2008. **Earth trends: the environmental information portal**. Details available at <http://earthtrendswriorg>, last accessed on 30 January 2011

World Resources Institute (WRI). 2009. **Guidelines for reporting information on public climate finance**. Details available at <wwwwriorg/publication/>, last accessed on 30 January 2011

Yost A C. 2008. **Probabilistic modeling and mapping of plant indicator species in a Northeast Oregon industrial forest USA**. *Ecological Indicators* **8**: 46–56

Yue T, Liu J, Jorgenses S E, Gao Z, Zhang S, Deng X. 2001. **Changes of Holdridge life zone diversity in all of China over half a century**. *Ecological Modelling* **144**: 153–162

Zhang J T, Pennington DD, and Liu XH. 2007. **GBD-Explorer: extending open source java GIS for exploring ecoregion-based biodiversity data**. *Ecological Informatics* **2**: 94–102

Zhou L, Tucker C J, Kaufmann R K, Slayback D, Shabanov N V, Myneni R B. 2001. **Variations in northern vegetation activity inferred from satellite data of vegetation index during 1981 to 1999**. *Journal of Geophysical Research* **106**: 20 069–20 083

# Geoinformatics for Comprehensive Impact Assessment and Analysis of Climate Change for Integrated Water Resources Management

**Wolfgang-Albert Flügel**

*Department of Geoinformatics, Hydrology and Modelling (DGHM), Friedrich-Schiller University, Jena (FSU-Jena), Löbdergraben 32, Germany*

*E-mail: c5WAFL@uni-jena.de*

## INTRODUCTION

Climate change is a global challenge (Alcamo and Henrichs 2002; IPCC 2007a, b, c), and throughout the world, there is a rising concern that global warming and climate change in general will influence a wide spectrum of drivers that are likely to impact and alter our continental and global hydrological domains in terms of natural and socio-economic processes (IPCC 2000; IUCN 2003). In response to this challenge, the Intergovernmental Panel on Climate Change (IPCC) has been established by the World Meteorological Organization (WMO) and United Nations Environment Programme (UNEP) to assess scientific, technical, and socio-economic information relevant for the understanding of the potential impacts and vulnerabilities, and to develop conceptual options for mitigation and technological as well as socio-economic adaptation.

Since the reduction in greenhouse gas emissions has proved to be difficult (Querner 2002; IUCN 2003), it is becoming ever more evident that mitigation of the impact from this component will not be sufficient to protect societies from the effects of climate change. Instead, adaptation strategies have to be developed, and integrated water resources management (IWRM), as defined by the GWP–TAC (2000), has been identified internationally as an appropriate strategy to cope with likely climate change induced water resources related impacts on people, businesses, societies, economies, and environment at large (IUCN 2003; GWP–TEC 2004).

The IWRM process is driven by a system-oriented, integrated, and "bottom-up" management approach, which is extensively supported by a volume of literature (Heathcote 1998; Jain and Singh 2003; GWP 2009). Integration is promoted across sectors and groups in society, and is realized by identifying an agreed set of water management guiding principles by applying a participatory approach, bringing together representatives of key stakeholders, water managers, and decision-makers (Giupponi, Mysiak, Fasio, *et al.* 2004; Giupponi, Sgobbi, Mysiak, *et al.* 2008). This process, however, is not easy to implement because of water demands from agriculture, rural and urban households, and industries; differing transboundary political interests; water-related legislation and governance structures; and missing hydrometeorological infrastructure and landscape information, as the majority of river basins in this world are ungauged basins, as defined by the International Association of Hydrological Sciences–Prediction in Ungauged Basins (IAHS-PUB) initiative (Young and Romanowicz 2004).

On a global scale, semi-arid river basins that already have severe water stress at present, show a consistent population growth and widely suffer from land degradation, floods, and droughts, causing severe socio-economic problems, decrease in livelihood opportunity, increase in poverty, and even loss of human lives (Wallace and Gregory 2002). Because of the stressed environment, the human population is already highly vulnerable to the impacts of climate change (Fischer, Shah, and van Velthuizen 2002; Querner 2002). Scenario-based simulations on a global scale (Alcamo and Henrichs 2002) indicate that the scarce water resources are likely to come under even more stress in the forthcoming decades (IPCC 2007a, b, c; IUCN 2003).

The system-oriented IWRM process and the lack of sufficient hydrometeorological and landscape information, along with socio-economic data, require the integrated application of geoinformatics. Geoinformatics generates synergy by deploying and developing IT components of geo-computation, namely, remote sensing, geographic information system (GIS), Web-based data information systems, modelling tools, and state-of-the-art strategies to access, analyse, and model challenging problems related to sustainable and environmentally sound land and water resources management. As a self-contained scientific discipline, geoinformatics develops innovative scientific concepts, and develops, applies, and enhances methods and techniques for environment system assessment, analysis, and integrated modelling, and thereby, providing needed information and know-how for decision support. Applying a holistic systems approach (Flügel 2000) and considering scale-related system process dynamics, geoinformatics toolsets like the Jena Environment System Analysis Toolset (JESAT) have been developed (Flügel, Michl, Krause, *et al.* 2009). The system integrates software components of remote sensing, GIS, global positioning systems (GPS), landscape assessment and terrain analysis, process studies from field studies, distributed modelling of environment system dynamics, object-related information knowledge systems, and networking technologies.

## CONCEPTUAL BACKGROUND AND GEOINFORMATICS

### Framework of Climate Change

Present global state-of-the-art climate modelling is presented in the *Fourth Assessment Report* of the IPCC (2007a, b, c), which have compiled the results from 21 different atmosphere–ocean global circulation models (AOGCM) for the IPCC scenario A1B (IPCC 2000), evaluated as multi-model data sets (MMD-A1B) in the Program for Climate Model Diagnosis and Intercomparison (PCMDI). According to the *Fourth Assessment Report*, the temperature projections based on the MMD-A1B models represent a strong warming during the 21st century. Warming greater than the global mean is projected for South Asia (3.3°C) and East Asia (3.3°C), and significantly higher warming than the global mean is reported in the continental interior of Asia (3.7°C in central Asia, 3.8°C in Tibet, and 4.3°C in northern Asia).

The challenge of downscaling from the global macro-scale resolution to a regional basin scale must be met to refer the global climate model (GCM) results to a scale that is relevant for IWRM. This is done by means of regional climate models (RCM) based on climate processes (Ahrens 2003; Klein, Frei, Gurtz, *et al.* 2005; Kumar, Sahai, Kumar, *et al.* 2006) and further by means of geo-statistics using climate and land surface parameters (Lapin, Damborská, and Melo 2001; Schmidli and Frei 2005; Wilby, Charles, Zorita, *et al.* 2004) complemented by deploying satellite-based techniques. The downscaling procedures generate terra bytes of modelled hydro-meteorological data time series, which require sophisticated software tools to extract the spatial information required for multi-scale river basin system analysis.

### Participatory IWRM Process

The IWRM process concerns coordinated decision-making on land and water resources management and requires a holistic system approach that accounts for scale-related cross-sectoral issues together with participatory and adaptive strategies towards sustainable management of land and water resources. The key action is collaboration between stakeholders when dealing with competing and sometimes conflicting water demands for the limited river basin water resources. This must be complemented by interdisciplinary research integrating scientific disciplines from natural sciences, engineering, and socio-economy, and accounting for governance regulations implemented in different political domains.

Involving key stakeholders in a participatory approach strives to define commonly agreed decision principles to achieve sustainable water resources management in terms of quantity and quality and requires proper communication interfaces (Horlitz 2006). The NetSyMod approach presented by Giupponi, Camera, Fassio, *et al.* (2006) and Giupponi, Sgobbi, Mysiak, *et al.* (2008) has been tested successfully in EU projects. It applies the Delphi method (Giupponi, Camera, Fassio, *et al.* 2006; Nevo and Yolande 2007), which attempts to obtain knowledge-based consensus from a group of experts, using repeated responses of questionnaires and controlled feedback related to the issues of IWRM decision-making.

## Adapting IWRM to Climate Change

Adapting IWRM strategies to climate change is based on a set of emission scenarios (IPCC 2000), which are treated as driving forces in the driver pressures state impact response (DPSIR) approach proposed by the European Environmental Agency (EEA 1999). This approach requires, to a large extent, geo-spatial information provided by geoinformatics comprising a comprehensive system assessment and scenario-based modelling analysis. Results are applied to decision support systems (DSS) like the MULINO-DSS (Giupponi, Mysiak, Fasio, *et al.* 2004; Giupponi 2007), which integrate expert knowledge and stakeholder views obtained from Delphi exercises. The DPSIR approach has been applied for the analysis of adaptation strategies for climate change in Europe, and its methodical concept offers great potential for application internationally (EEA 2005, 2008b). The decision-making process, to a large extent, is based on indicators that in some way express the status of the natural, socio-economic, and governance domains, and therefore, must be readily available to decision-makers by means of user-friendly decision information support tools (DIST).

## Application of Geoinformatics

Within the framework covered by the ongoing adaptation of our environment to changing land use and land cover (LULC), partly triggered by climate change on the one hand and its gradual impacts on IWRM on the other, geoinformatics as a synergetic discipline provides innovative concepts, methods, and tools to carry out the assessment and analysis of these impacts on sustainable land and water resources management. This is shown schematically in Figure 1 and can be described as follows.

- River basin systems comprise three different core components: the abiotic and biotic resources and the exploiting socio-economic development. These three system components are interactively linked with each other by means of process-based ecosystem services (ESS) and ecosystem functions (ESF) as feedback interactions.
- All components are either directly or indirectly influenced by governance regulations and policies, that is, by the Water Framework Directive of the European Community (EC) that strives to implement sustainable land and water resources management.

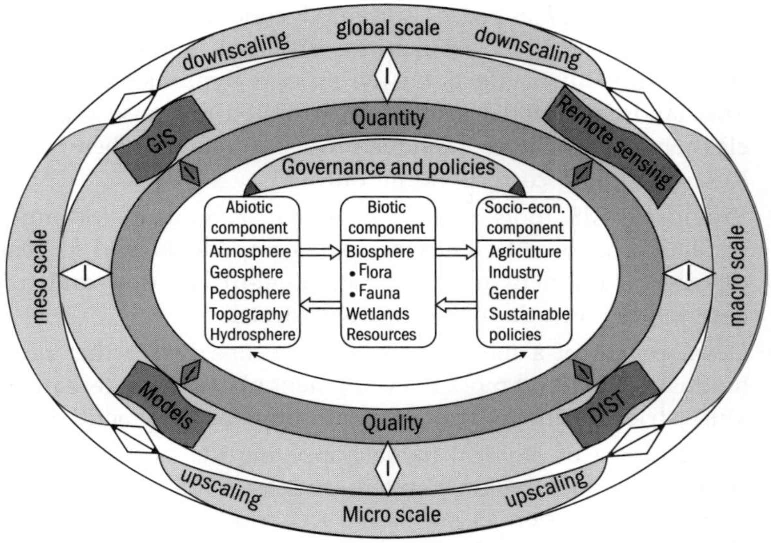


**Figure 1** Integration of geoinformatics tools for multi-scale integrated water resources management

- The interactive system process dynamics among the three core components and their management by governance structures impact land and water resources in terms of quantity and quality.
- For the spatial system assessment of the core components, geoinformatics developed remote sensing techniques, GIS, process models, and Web-based DIST for component interaction analysis and management decision support. By means of respective methodical interfaces, they relate the quantity and quality issues of the river basin to the three core system components, accounting for the scale of interest.
- From the scale-specific integrated system assessment and analysis, enhanced knowledge and insight are obtained, which by means of conceptual and software interfaces can by upscaled to higher scales or downscaled to lower scales, thus generating the synergy required to understand integrated process system dynamics.

Ready-to-use toolsets like JESAT (Flügel, Michl, Krause, *et al.* 2009) or Integrated Land Management Systems (ILMS) (Fink, Beisecker, Kralisch, *et al.* 2007) have been developed to address and elaborate on the following research challenges that address the issues of sustainable IWRM and adaptation to climate change.

- Application of a holistic approach for multi-scale system assessment, and modelling analysis of river basins and their nested tributary networks.

- Development of a process-oriented conceptual landscape model that synergistically classifies the different landscape components according to their effects on the process dynamics under research. This landscape model provides distributed response units (RUs) as class entities that have a unique process dynamics when transferring RU input into a corresponding output.
- Providing GIS tools to delineate RU accounting for multi-scale resolution of the different landscape components, and to generate a topological RU network for the gradient-driven flow and transport between RU and the receiving river stretch.
- Development of a model framework system (MFS) that provides a modular structured model development platform for researchers and a user-friendly model run-time environment for practitioners.
- Development of physical models applying RU as distributed model entities to model dynamic system processes, that is, water and solute transport or erosion as means for scenario-based impact analysis of climate change on IWRM.
- Design of Web-based DIST to preserve and disseminate the knowledge obtained from the system assessment and process analysis and providing a user-friendly way for decision support to decision-makers.

## Current Knowledge

A brief review of the literature on IWRM and climate change has been dealt with elsewhere, and hence, the following discussion will focus on the current scientific potential offered by geoinformatics to address the challenges identified. Thus, the concepts, methods, and software tools that comprise the JESAT are presented and discussed in brief (Flügel, Michl, Krause, *et al.* 2009).

## Holistic Approach in Multi-scale System Assessment and Modelling

With respect to river basins, a holistic system approach can be described as the process of analysing the complex multi-process driven systems dynamics at the scale of assessment and relating the obtained insight to higher and lower system scales, hence gaining an understanding on the systems' integrated functionality. The challenge of such a complex approach is presented through two facts: (1) the interaction between natural processes and their scale-related functional significance are often not understood, and (2) socio-economic and governance activities add the human dimension to this natural complexity, making the system even more difficult to manage, with difficulty experienced in predicting the impacts of external drivers, like climate change. Figure 2 is based on numerous published hydrological catchment studies with respect to the relationship of hydrological catchment with spatial landscape heterogeneity and that of system response time with impacts, thus furthering our understanding of scale-related process interaction, and reveals the following.

1 Spatial landscape heterogeneity and the systems' response time to external impulses are increasing with scale. Increasing heterogeneity of the system component structure makes it more difficult to understand the complex process interaction between scales within the system, and therefore, the knowledge on such dynamics is decreasing with increasing system scale.
2 The distribution of hydrological catchment studies is left-skewed with the majority of the studies focusing on the micro-scale, some with focus on the meso-scale, and few studies dealing with macro-scale river basins like the Amazon, Ganges, Mekong, Mississippi or the Nile.
3 Micro- and meso-scale river systems have a low or medium degree of spatial heterogeneity and short response times to external system impacts. Hence, this can be analysed by means of measured historical time series. Macro-scale river systems are characterized by a high degree of landscape heterogeneity and have impact response times, indicating system changes that are longer than reliable hydro-meteorological data time series from the past.

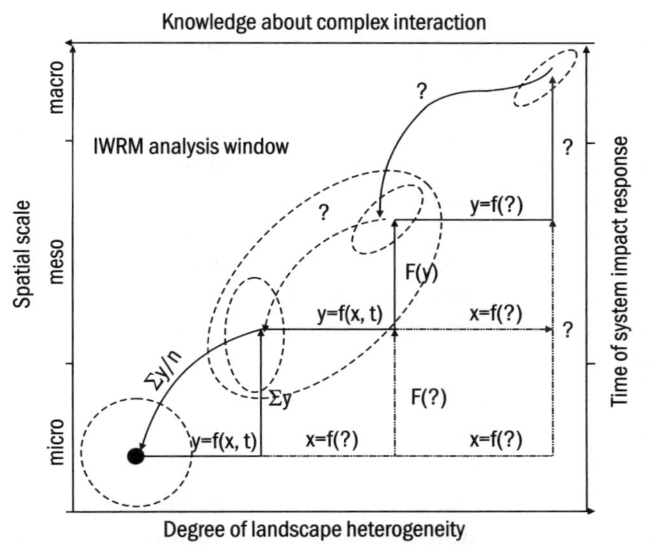


**Figure 2** Relationship between river system scale, spatial heterogeneity, response time, and process knowledge with respect to upscaling and downscaling

4 Consequently, the know-how on the complex interactive process dynamics between scales is quite well developed in the lower scale range, that is, between the micro-scale and the lower meso-scale but lacks understanding between the higher meso- and the macro-scale ("?" in Figure 2).

5 The process of upscaling and downscaling is affected by the inverse relationship between scale and knowledge of process interaction as because of (4), the transfer algorithms between scales are frequently unknown ("?" in Figure 2).

6 IWRM is typically implemented on the meso-scale and often in a transboundary domain. At present, it is possible to transfer algorithms and knowledge obtained from micro-scale catchment studies to the lower meso-scale and vice versa. However, downscaling of global macro-scale climate change dynamics to IWRM implementations on the meso-scale is still a research challenge that needs to be addressed.

From the foregoing discussion, the present constraints of climate change impact analysis for IWRM adaptation become obvious. Results from GCM and regional climate models refer to the macro-scale or the higher meso-scale (Ahrens 2003) and provide input to global water models (Alcamo and Henrichs 2002). Because the issue of downscaling to lower scales is still unresolved (Klein, Frei, Gurtz, *et al.* 2005; Kumar, Sahai, Kumar, *et al.* 2006), these models are of only limited value for the majority of IWRM applications at the micro- and meso-scale.

## Conceptual Landscape Model of Response Units

The scale-related framework from Figure 2 is also of relevance for the differentiation of process-related landscape units within a basin. Applied as model entities, they transfer basin input, that is, rainfall in corresponding basin output, that is, evapotranspiration and river run-off. The process is called regionalization and applies process knowledge into geoinformatics methods, elaborating on the landscape components within the river basin. A comprehensive overview of the different approaches developed in hydrology has been published by Kalma and Sivapalan (1995). Today, two different concepts are applied in the majority of basin models, which in different ways account for the process dynamics in the river basin.

1 Grid cells differentiate the basin into units of equal size, which are not process based.

2 RU are process-based polygons of different size, which are delineated based on the insight gained from a thorough basin hydrological system analysis.

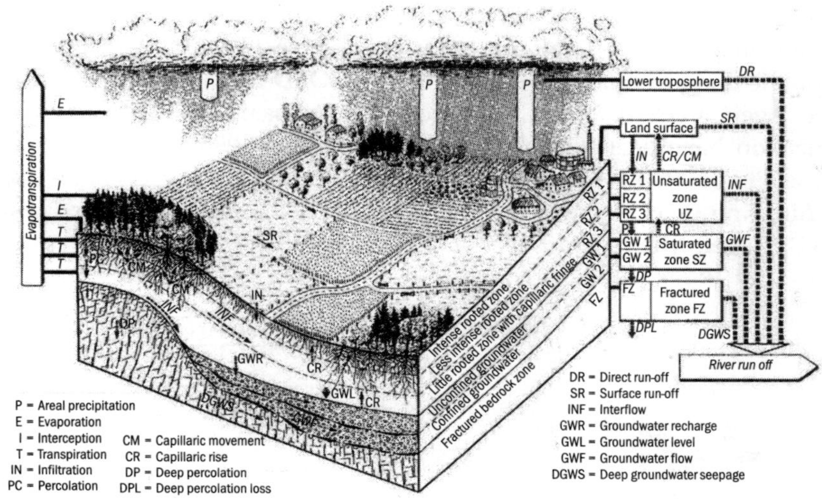


**Figure 3** Schematic presentation of the interactive hydrological process dynamics active in and between landscape components when transferring precipitation input into evapotranspiration and river run-off as corresponding system output components

The conceptual RU landscape model has been introduced by Flügel (1995, 1996) and because of its process relevance should be preferred while addressing the IWRM challenges identified from Figure 2. As shown in Figure 3, the RU landscape model is based on the fact that the hydrological process dynamics of water and solute transport active within a river basin are controlled by the vertical assembly of site-specific landscape components like topography, soil, LULC, and geology. They together constitute classes of specific process dynamics and contain spatial distributed class entities, called RU. Hence, each RU class with its RU entities represents an individual structure of interactive processes comprising storage, transfer, and output, which are unique within the RU of each class. By means of the RU landscape model, each process dynamics relevant for IWRM can be represented by respective distributed model entities, that is, erosion response unit (ERU) (Flügel and Märker 2003), hydrological response unit (HRU) (Flügel 1996; Krause 2002), and chemical hydrological response unit (CHRU) (Bende, Flügel, and Kern 1995; Fink, Krause, Kralisch, *et al.* 2007; Fink, Beisecker, Kralisch, *et al.* 2007).

## RU Delineation and Topology

The delineation of RU is done by means of GIS analysis, as shown schematically for the delineation of HRU in Figure 4. This is done by using digital data layers and spatial information obtained from remote sensing analysis. The latter can either be done by means of selected satellite images or by applying readily available LULC classification from the public domain, that is, from the GOFC-GOLD initiative of the European Space Agency (ESA). The GIS procedures necessary to carry out the RU delineation have been described in detail (Flügel 1995, 1996; Krause 2002; Krause, Bende-Michl, Bäse, *et al.* 2006; Pfennig, Kipka, Wolf, *et al.* 2009; Wolf, Pfennig, Krause, *et al.* 2009), and the tools used are software components of JESAT (Flügel, Michl, Krause, *et al.* 2009).

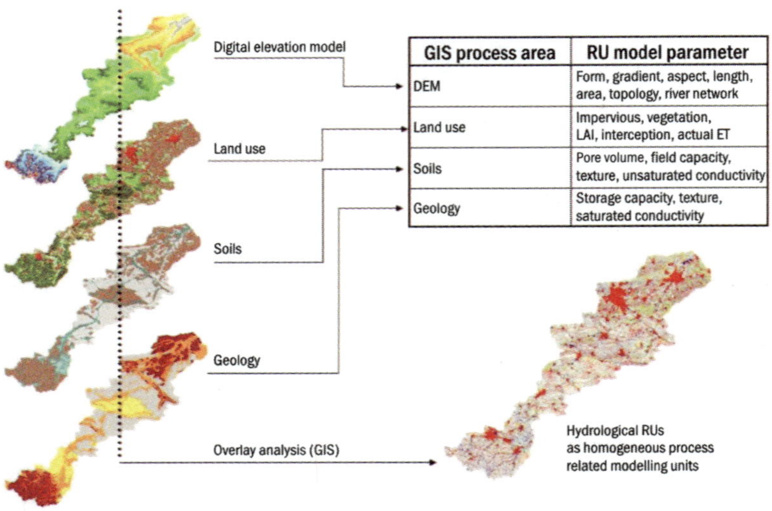


**Figure 4** Schematic presentation of hydrological response unit (HRU) delineation by means of geographic information system (GIS) overlay analysis using digital data layers

The flow paths between RU to the receiving river network are an important component for water, sediment, and solute transport. They are defined by means of a topologic model generated by one of the GIS tool components of JESAT (Wolf, Pfennig, Krause, *et al.* 2009) using a digital elevation model (DEM). The topologic flow path model is calculated from the gradient analysis of the DEM and delivers as a result the flow network between the RU towards the receiving river reaches (Figure 5). The figure also accounts for the fact that one RU can distribute its output to more than one neighbouring HRU (Pfennig, Kipka, Wolf, *et al.* 2009).

## JAMS: A Modular Model Framework Systems

Modular model framework systems (MFS) provide an object-oriented model development environment for the researcher and a user-friendly run-time environment for the practitioner and IWRM decision-maker. The motivation

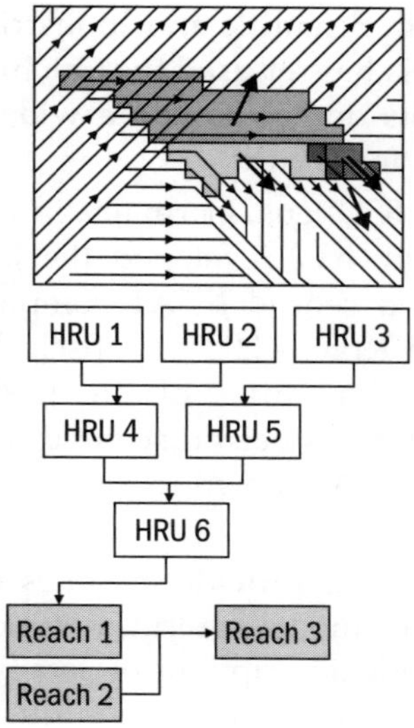


**Figure 5** Topologic model of gradient-driven flow between response units (RU) to the receiving river network

for developing MFS is derived from the fact that monolithic software models do not provide sufficient flexibility for exchanging process components, but rather encapsulate their know-how within the code, which makes it unavailable for other model applications. A modular approach overcomes this methodical shortcut and permits to assemble different process modules into an adapted model, hence accounting for individual data availability and process know-how. In the past, the following MFS have been developed.

- The Modular Modelling System (MMS) was one of the first MFS published (Leavesley, Markstrom, Brewer, *et al.* 1996) and has been successfully applied with the Precipitation–Run-off Modeling System (PRMS) model by the United States Geological Survey (USGS), Denver, Colorado, USA.
- The Object Modelling System (OMS) is a joint initiative of the United States Department of Agriculture (USDA), Fort Collins, Colorado, USA, and the Department of Geoinformatics, Hydrology and Modelling (DGHM), Friedrich-Schiller University (FSU-Jena), Jena, Germany (Kralisch, Krause, and David 2005), and is continuously enhanced for applications in Germany, the US, and internationally.
- Funded by the Framework Programme of the European Commission (EC), the HarmonIT project with the OpenMI (Moore, Tindall, and Fortune 2004) has developed a generic procedure for model interfaces.

- The Commonwealth Scientific and Industrial Research Organisation (CSIRO) in Australia has released the Catchment Management Toolkit (CMT), which in its domain offers a wide range of modelling and management tools for IWRM.
- The DGHM of the FSU-Jena, Germany, in cooperation with the OMS and the MMS team, has developed the Jena Adaptable Modelling System (JAMS) as a central JESAT component (Krause and Flügel 2005; Kralisch and Krause 2007; Kralisch, Krause, Fink, *et al.* 2007). JAMS contains the modules of the J2000 model suite, and its component structure has been discussed in detail by Kralisch, Fink, and Beckstein (2005). As shown in Figure 6, it comprises three types of libraries.
    1. *System core* libraries provide modules that describe and control the context of the model in terms of time steps and RU sequences. They also provide descriptions of data types, and organize input/output handling and visualization during run-time.
    2. *Knowledge representation* is done by modules that contain algorithms to model processes depending on the time scale and the data input available. These modules are used to design the process body of the new model under development.
    3. *Graphical user interfaces (GUI)* for model generation and parameterization allow the model developer and practitioners to adapt the model according to different defined scenarios.

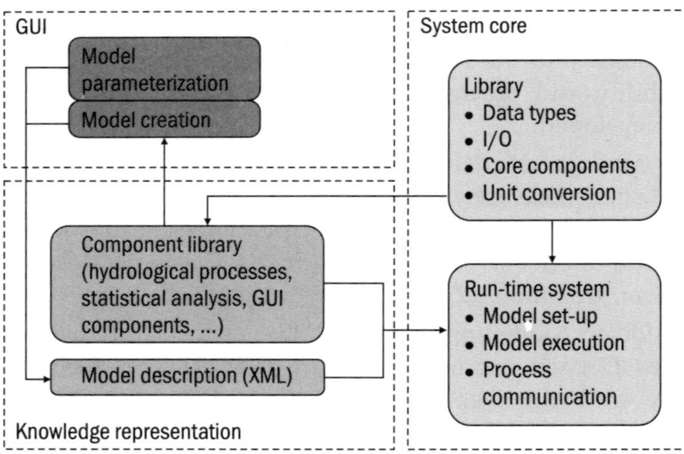


**Figure 6** Schematic design structure of the modular Jena Adaptable Modelling System (JAMS)

## J2000 Hydrological Model Suite

The concept of the RU landscape model is based on a thorough hydrological system analysis that provides process-related criteria for the RU delineation. These criteria, although related to the scale of the digital data layers used for the study, are consistent in their methodical concept so that they can be applied to higher and lower scales as well. At each scale, RU are grouped in unique process classes, which in their spatial distribution and specific arrangement represent facets of the spatial basin landscape heterogeneity. The latter controls the response time delay of the river basin when transferring basin input via internal storages to the corresponding basin output.

The J2000 hydrological model suite applies HRU as model entities for modelling the water and solute transport dynamics within the river basins. As a core modelling component of JESAT, it comprises the following three models, with each having a modular component structure, as shown in Figure 7.

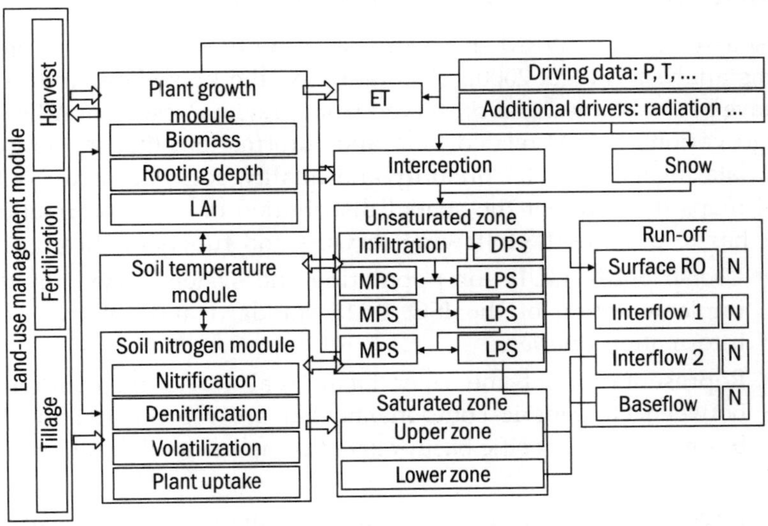


**Figure 7** Modular component structure of the J2000 (right half) and the J2000-S water and solute transport model of Jena Environment System Analysis Toolset (JESAT)

1 The *J2000* hydrological model (Krause 2002; Krause, Bende-Michl, Bäse, *et al.* 2006) comprises the atmospheric, soil related, and hydrological components of the right part of Figure 7. The model has intensively been tested in international catchment studies to analyse and quantify the hydrological dynamics of micro- till meso-scale catchments for process-based IWRM scenario analysis.

2 The *J2000-S* model was developed for process-based water and solute transport modelling (Fink, Krause, Kralisch, *et al.* 2007). It is identical in its hydrological process structure with the J2000 model but, in addition, has model components shown in the left two columns in Figure 7. They describe agricultural management, plant growth, soil temperature, and nitrate dynamics. The model has been applied successfully in river basins with intense agriculture and having the problem of seepage of nutrients, that is, nitrogen from diffuse sources generating severe water quality problems due to river and reservoir eutrophication (Fink, Beisecker, Kralisch, *et al.* 2007).

3 The *J2000-g* conceptual model is a simplified version of the J2000 model that still is process based but represents the basin hydrological dynamics in a much more conceptual way than is done in the J2000 model. J2000-g has been developed for macro-scale scenario analysis and has been applied successfully for climate impact assessments for IWRM (Krause and Hanisch 2009).

## Adaptive Integrated Data and Information System

As discussed above, IWRM should be based on the principles of integrated systems analysis (Flügel 2000), elaborating on the know-how and expertise that have been obtained and preserved from previous research. A DIST must manage various IWRM-related data and information, that is, station time series data related to water quantity and quality, socio-economic data, and digital maps that present the spatial distribution of such information, and hence, has to provide the following services and functionalities.

- User-friendly GUIs for populating the system based on the full implementation of the ISO 19115 standard and solving data and information queries.
- Representation of the river basin real world by differentiating between (1) generic components that are represented in all river basins and (2) specific basin-related components that reflect regional basin conditions.
- Import and export functionality and management of GIS digital data layers that have a geographic reference specified by a common coordinate system.
- Integration of the different kinds of data and information for the design of "what-if-scenarios" applied for the evaluation of prognostic system management alternatives in the decision-making process.
- Adaptivity of the system's data model structure to allow for extensions depending on the progress of the system's understanding and knowledge achieved by researchers and decision-makers.

The Adaptive Integrated Data Information System (AIDIS) has been designed as the JESAT data and decision support component (Flügel 2007). It provides services by using a hierarchical, multi-tier object class data model shown in Figure 8. The latter is generic in its design structure, and for each tier deploys XML-based metadata GUI to represent relationships between classes. It, therefore, can be adapted to river basins of all types of landscape and climate by enhancing the multi-tier class layer structure using new class hierarchies or modifying existing ones.

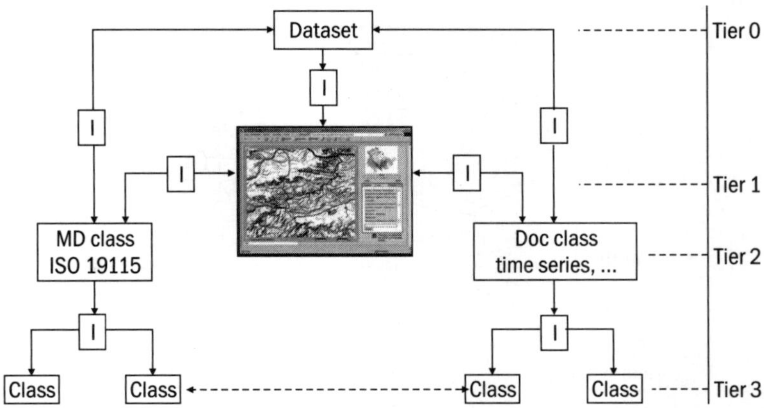


**Figure 8** Multi-tier object relational data model of the Adaptive Integrated Data Information System (AIDIS)

A pivotal role is given to the "map-class" on tier 1, which is used not only for the visualization of results but also for spatial queries that refer to metadata and information from the "MD Class" hierarchy and the "Doc Class" hierarchy, respectively. All data stored in AIDIS is described by means of the ISO 19115 metadata model, and enhancements of this model are realized by extended lookup tables (LUT). AIDIS employs the Minnesota Map Server and makes use of the GIS functionalities provided by PostGIS. GIS maps can be imported and exported in all standard formats via the Web, and thereby, AIDIS supports multi-client GIS cooperation for dissemination, that is, within a clearinghouse network (CN). If implemented in a river basin as a River Basin Information System (RBIS), it is renamed as BrahmaRBIS for the Brahmaputra river and provides interfaces to automatically supply input data to the J2000 model suite and to populate the model output into the RBIS (Kralisch, Zander, and Krause 2009).

## Jena Environment System Analysis Toolset

IWRM requires intensive and often transboundary cooperation between partners, applying the same integrated toolsets to support the decision-making processes. To address this need, JESAT components were developed (Flügel, Michl, Krause, *et al.* 2009). As shown in Figure 9, JESAT integrates components for comprehensive assessment, analysis, modelling, and resources management support of environment systems, applying a multi-scale, holistic system approach, as described by Flügel (2000). As a multifunctional software toolset, it comprises the software components described above, which can be assembled by the practitioner into a basin-specific ILMS for IWRM and sustainable land resources management.

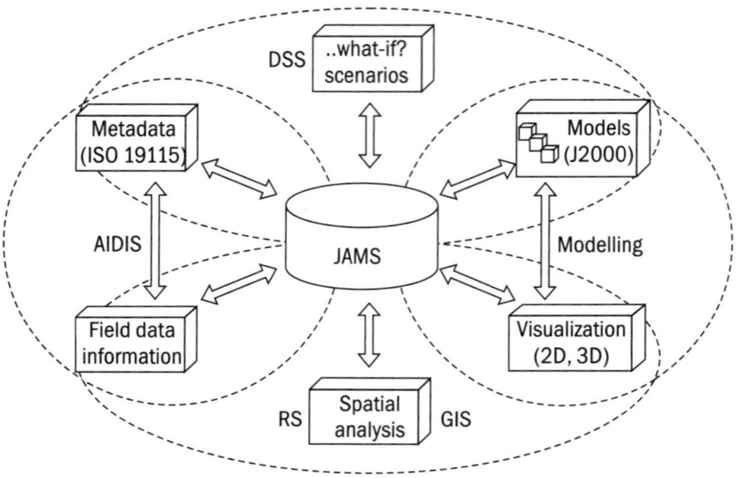


**Figure 9** Multi-component Jena Environment System Analysis Toolset (JESAT) structure to develop basin-specific Integrated Land Management System (ILMS) for sustainable integrated water resources management (IWRM)

JESAT is based on the conceptual landscape model of distributed, process-related RUs applied as model entities in the J2000 model suite within the JAMS. It considers water quantity and water quality issues in an integrated way by applying remote sensing, GIS analysis, field process data, and system modelling techniques in a consistent and scale-related way, thereby accounting for multi-scale applications. JESAT users are furthermore offered innovative techniques from remote sensing and GIS, the latter incorporated into AIDIS/RBIS for enhancing the system towards a comprehensive environment knowledge base for system management and decision support. A literature review done by the European Environmental Agency (EEA 2008a) reveals that the development of integrated toolsets like JESAT or OIKOS (Olazabal, Urzelai, García, *et al.* 2007) involves cutting edge research.

## STUDY BASINS, OBJECTIVES, METHODS, TOOLS, AND RESULTS

Geoinformatics by means of JESAT has been applied successfully in various collaborative projects related to IWRM, environment system research, and climate change impact analysis in Europe, Africa, and Asia, as discussed subsequently. The impacts of changing land use and land management on the water and solute transport, both important ecosystem processes within a river system, were assessed and analysed. As LULC changes and consequent adaptations of land and water resources management are the most likely impacts of climate change, the results obtained from these JESAT applications provided valuable tools for scenario-based climate change impact analysis.

### Mooi River Basin, Eastern Cape Province, South Africa

The location of the Mooi river (A = 307 km$^2$), a second order tributary of the Umzimvubu river (A = 19.852 km$^2$) in the Eastern Cape Province of South Africa, is shown in Figure 10. The catchment has a unique landscape structured by sandstone escarpments and valleys formed in the strata of mud stones and silt stones. Different types of wetlands have developed as unique environment features in this impressive landscape (Helmschrot and Flügel 2002).

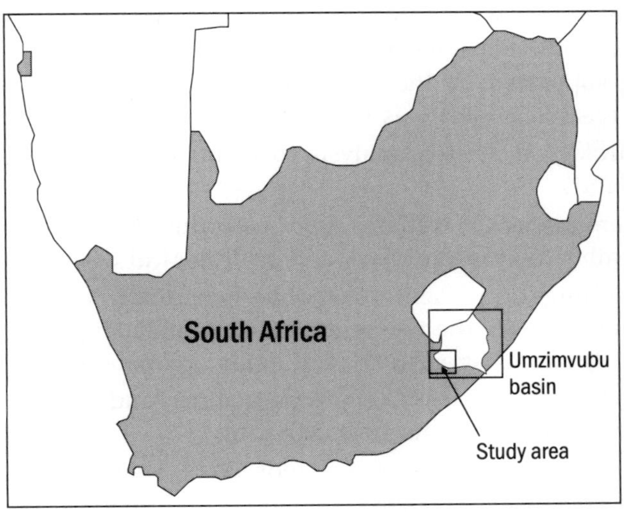


**Figure 10** Location of the study area in the Eastern Cape Province of South Africa

Since 1990, the land use in the headwater region of the Mooi river has changed significantly, as about 60 000 ha of former rangeland has

**Figure 11** Large-scale afforestation of former range land

been afforested with industrial forests comprising pines and eucalyptus (Figure 11). It was expected that this large-scale change in land use will impact the hydrological dynamics of the Mooi river system and the functioning of wetlands that are unique environmental features of the cultural landscape of the river in particular (Helmschrot 2006a, b). In collaboration with colleagues from the University of Kwazulu-Natal, South Africa, and supported by the South African Company MONDI Forest, an eight-year research project was implemented with the following objectives.

- Hydrological system analysis to identify the process dynamics of wetlands.
- Differentiation of wetland types according to landscape component assembly, vegetation types, and hydrological dynamics.
- Modelling of present discharge generation from the classified wetlands under the two scenarios of (1) unchanged grassland LULC and (2) afforestation to the full landscape potential.

  JESAT was applied in a stepwise manner, and the following results were obtained (Helmschrot 2006a, b).

- Three different types of wetlands (WL), shown in Figure 12, each with specific hydrological dynamics and adapted vegetation were defined using intense field campaigns and the hydrological system analysis (Dahlke, Helmschrot, and Flügel 2003; Dahlke, Helmschrot, and Behrens 2005).

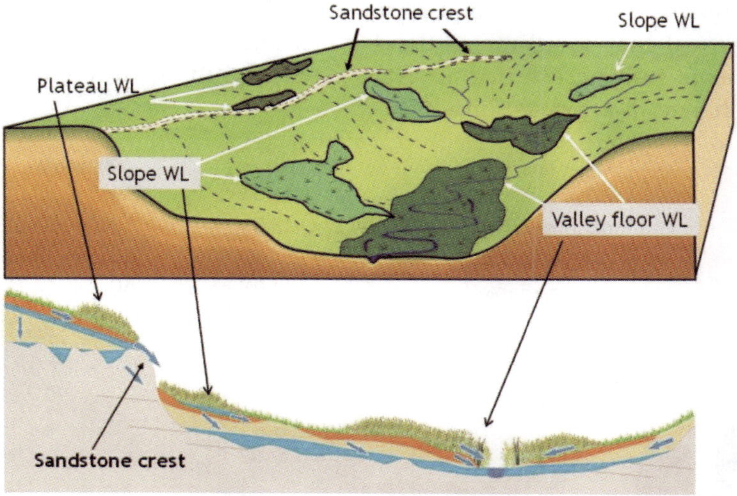


**Figure 12** Landscape-oriented distribution of wetlands

1. *Plateau WL* behind the rim of sandstone plateaus is fed by seasonal rainfall and water seeping out of groundwater from the sandy aquifer developed in the weathered zone on top of the plateaus.
2. *Slope WL* fed by rainfall and interflow, also with seasonal dynamics following the rainfall pattern and the inflow from the upslope located *plateau WL*.
3. *Valley floor WL* fed dominantly by seasonal rainfall and continuous groundwater seepage with perennial or periodic water courses meandering in the flat and broad valleys.

- A comprehensive system assessment was done by means of intensive field campaigns and landscape system analysis. The focus was on the representative micro-scale test catchment of the Weatherly Creek, and validation was done in the meso-scale Mooi river catchment.
- HRUs were delineated by means of GIS analysis using digital data layers of a terrain model, LULC classification obtained from Landsat satellite image classification, soil data from the South Africa soil survey, and geology data from a respective geological survey.
- The HRU model entities were parameterized, and the hydrological dynamics of the Weatherly Creek between 1999 and 2001 and those of the Mooi river between 1980 and 2003 were modelled using the J2000 model. The model exercises were done for the two scenarios of unchanged LULC, assuming afforestation to the full potential of the landscape.

The results obtained (Helmschrot 2006a, b) are shown in Figure 13, which reveal that the three wetland types and their run-off contributing components will be affected in ways and magnitude different from the LULC change of afforestation.

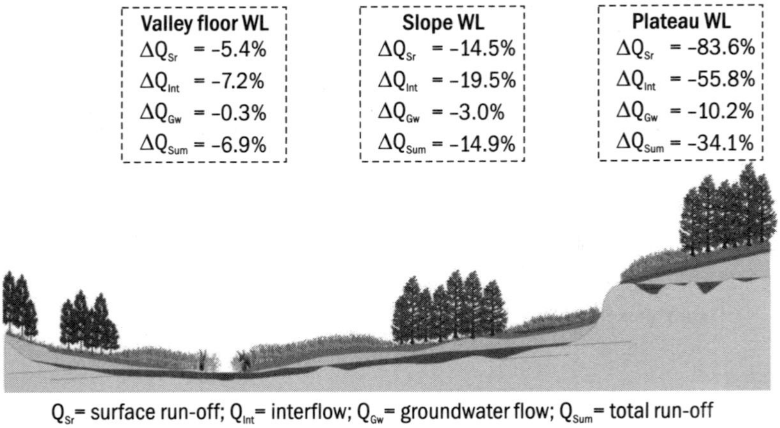


**Figure 13** Modelled change of run-off components from different wetland types due to afforestation

**Source** Adapted from Helmschrot (2006a, b)

- The total run-off contribution from small and only seasonal active *plateau WL* will be reduced by about 34% with an increase in evapotranspiration from the forests on the sandstone plateaus due to higher interception and water consumption, and thus the surface run-off contribution will almost cease.
- *Slope WL* suffers less by losing on average about 15% of its discharge contribution, as less groundwater will seep over the sandstone crest into the slope soil, and hence, less interflow and surface run-off will be generated.
- *Valley floor WL* will experience the least loss in terms of only 7% discharge contribution as they mainly get recharged by rainfall on the broad valley floor and hence depend less on the inflow from the slopes and plateaus.

## Weida River Basin, Thuringia, Germany

Weida river in East Thuringia drains an area of about 102 km$^2$ in the eastern part of Thuringia, which is a federal state of Germany (Figure 14). The basin has been described in detail by Fink (2004). Its altitude varies between 355 m and 565 m above sea level, the annual average precipitation is about 640 mm, and the temperature is about 7°C. Two-thirds of the catchment is used for intensive agriculture, and soil fertility is based on continuous nitrogen supply from manure and chemical fertilizers. Weida river flows into the Weida–Zeulenroda reservoir, which provides potable water to the population of East Thuringia. The reservoirs are impacted by a high input of nitrate from the Weida river, which receives this load by subsurface seepage, that is, interflow and groundwater flow from the adjacent agricultural lands (Fink, Beisecker, Kralisch, *et al.* 2007). During the summer season, this nitrogen input causes eutrophication, consequently giving rise to algae blooms, which result in high water purification costs to eliminate the chlorophyll content. In collaboration with the reservoir administration, an IWRM research project was implemented with the following objectives.

**Figure 14** Location of the Weida River Basin in Thuringia, Germany

- Identification and quantification of the distributed areas of nitrate input and modelled nitrogen output as a base for the design of sustainable land management options.
- Development of recommendations to optimize nitrate reducing measures like annual payments to farmers to compensate for harvest losses when using less nitrogen-based fertilizers.
- Development of land management scenarios based on tolerable nitrogen input threshold values to optimize farmer's compensation payments.

JESAT was applied in this project as described below.

- HRUs were delineated by means of digital GIS data layers of terrain, soils, LULC, and geology, and their topology was generated based on the DEM available to describe the transport of water and solutes within the HRU network towards the receiving Weida river.
- Hydrometeorological station data and fertilizer input on a field scale were collected from the German Weather Service (DWD) and the farmers, respectively.
- The J2000-S water and solute transport model from JESAT was applied to model the distributed nitrogen input and nitrogen output dynamics of the HRU within the catchment and to facilitate the flow and nitrogen output between HRU to the receiving Weida river.
- To elaborate the efficiencies of nitrate reduction measures, that is, by making compensation payments to farmers, a multidimensional sensitivity analysis was carried out by applying the HydroNet neural network tool of JAMS (Kralisch, Fink, and Beckstein 2005) to the topology of the HRU network.

The results of the J2000-S modelling for the Weida river catchment and the two reservoirs are shown in Figure 15 and can be described as follows:

- There is a significant differentiation in nitrogen output according to land use, that is, forestry versus agriculture and crop management in terms of crop rotation and fertilizer application. Forested HRUs have a very low input to Weida river, of less than 5 kg/ha*a.
- Grassland, that is, pastures and meadows, and fallow land have the second lowest nitrogen output, varying between 5 kg/ha*a and 20 kg/ha*a.
- Agriculture fields supply the bulk of the nitrogen input into the Weida river, ranging between 20 kg/ha*a and 200 kg/ha*a, depending on the crop rotation and fertilizer management. Some "hotspots" even exceed the annual output of 200 kg/ha nitrogen output if shallow soils receive high fertilization rates.

In the next step, threshold values for nitrogen inputs, which could be compensated for by the biological demineralization potential within the

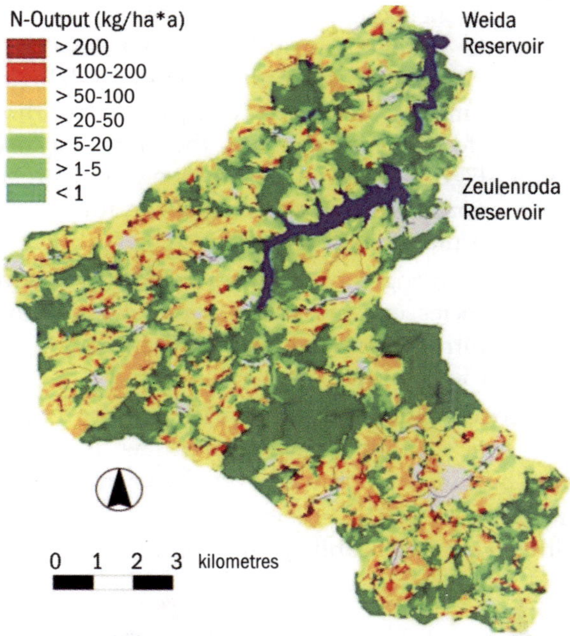


**Figure 15** Modelled distributed N-output from a hydrological response unit (HRU) network within the Weida river catchment.

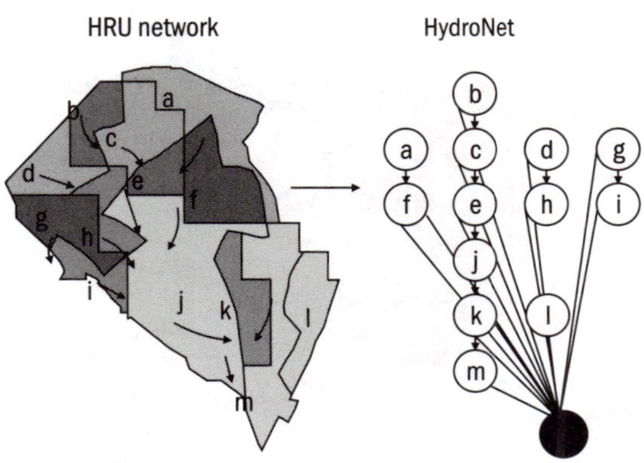


**Figure 16** Transfer of the hydrological response unit (HRU) topology into a HydroNet

**Source** Kralisch, Fink, and Beckstein (2005).

reservoir, were defined and referred to the HRU network. This was done by means of the HydroNet component of JAMS (Kralisch 2004; Kralisch, Fink, Flügel, *et al.* 2003). As shown in Figure 16, the latter is generated by transferring the topologic HRU model into an Adapted Neural Network (ANN), with linking edges defined by the vertical and lateral HRU flow paths. Each HRU neuron is associated with a polynomial activation function describing the nitrogen output of the respective HRU from scenario-based nitrogen fertilization runs of the J2000-S model (Kralisch, Krause, and David 2005). By means of a backpropagation algorithm, the nitrogen load threshold value for the reservoir was then referred to HydroNet. The results are shown in Figure 17 as the distribution of HRU-related efficiencies of farmer compensation payments to achieve the nitrogen load threshold value and can be interpreted as follows.

- Contrary to the nitrogen output distribution shown in Figure 15, the highest efficiencies of compensation payments with respect to the reduction of nitrogen input into the reservoir are distributed over all types of crop rotation.
- Higher nitrogen reduction efficiencies are not only limited to the protection zone 2, defined by German law as a buffer around the reservoirs, but are scattered all over the basin.
- High nitrogen reduction efficiencies could be achieved in shallow soils with higher permeability and higher fertilizer input.

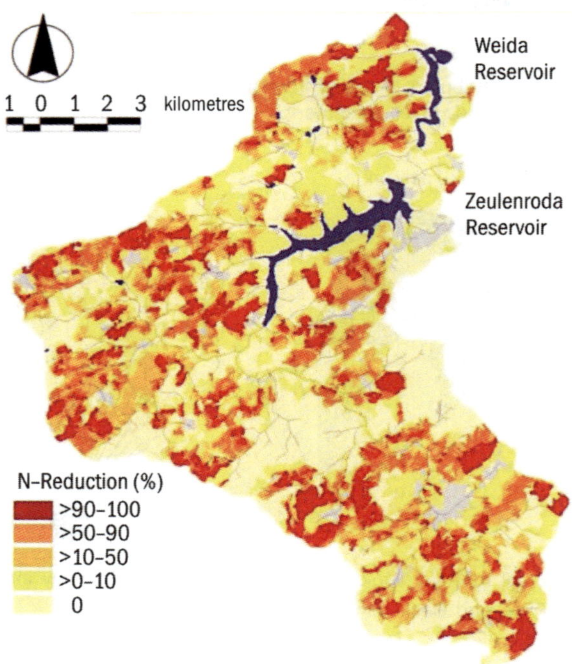


**Figure 17** Nitrogen reduction efficiencies derived from compensation payments

The HydroNet nitrogen reduction efficiency assessment provided new and valuable information to the reservoir administration on the use of limited financial resources to optimize the efficiency of this measure and has been successfully applied since then.

## Climate Change Impact Assessment for IWRM in Thuringia, Germany

Climate change scenario analyses are also of high relevance in the Free State of Thuringia, and IWRM impact assessments by means of the J2000-g model have proved their importance for sustainable water resources management (Krause and Hanisch 2009). The Free State of Thuringia covers an area of 16.172 $km^2$ and is located in the middle of Germany (Figure 14). Its landscape can be differentiated roughly into three major components.

1. *Thuringian forest* A middle mountain ridge stretching from the SW to the SE of the state, mainly covered by coniferous forest.
2. *Foreland zone* It is built from sandstones and limestones, partly forested and used for agriculture and stock breeding.
3. *Thuringian lowland* Land with intense agriculture on very fertile soils.

The annual rainfall pattern in Thuringia is characterized by a distinct windward-lee effect. The westerly winds bring moist air from the Atlantic Ocean. They are forced upward when crossing the Thuringian forest ridge at the windward side and then they cool down, hence resulting in precipitation. When moving further to the NE, they bring less and less rain to the foreland zone and the Thuringian lowland located in lee of the mountain ridge. As a result of this spatial rainfall distribution, most of the rivers in Thuringia have their headwaters in the mountain ridge. Figure 18 shows run-off generation during winter season, peaking at 450 mm in the Thuringian forest, meanwhile, in the Thuringian lowlands, this value is between 0 and 25 mm. This general rainfall and run-off distribution pattern remains unchanged in summer, but here the mountain run-off generation decreases to 100–150 mm maximum.

To study the likely impact of climate change on this run-off dynamics, a research project was initiated by the Thuringian Agency for Environment and Geology (TLUG) with the following objectives (Krause and Hanisch 2009).

- Regionalization of projected daily climate time series to available station locations.
- Delineation of HRU for Thuringia, and modelling the hydrologic water balance for Thuringia by means of the J2000-g model.
- Identification of "hotspots" for which more detailed hydrological investigations are required to implement sustainable water and land resources management, according to the EU Water Framework Directive (EU-WFD).

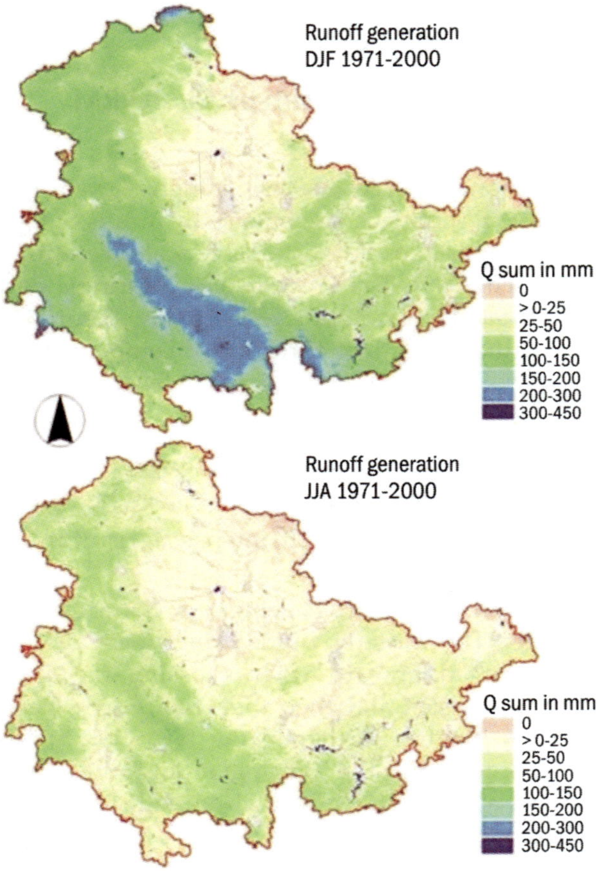


**Figure 18** Present mean DJF and JJA run-off generation in the Free State of Thuringia

JESAT was applied in this study, as described below (Krause and Hanisch 2009).

- HRUs were delineated by means of GIS analysis using digital data layers of slope, LULC, aspect, soils, and hydrogeology.
- They were applied as model entities in the JESAT model component J2000-g, a spatially distributed conceptual hydrological model.
- Hydrometeorological time series of measured station data and climate time series projections were provided by the TLUG. They have been referred to the existing climate station network to make the projection more comparable to the measured domain used to generate the results of the present scenario, as presented in Figure 18.

- The J2000-g hydrological model was successfully validated in altogether 12 test basins scattered throughout Thuringia, and a parameter set for Thuringia was derived from this exercise.

The results of the modelling were obtained for different SRES climate scenarios for which the mean monthly run-off generation (2071–2100) for the A1B scenario for the core winter and summer months is shown in Figure 19. In comparison with the present scenario given in Figure 18, they provide the following information.

- Winter precipitation will concentrate on the SE part of the Thuringian forest, and as according to the A1B scenario, temperature will increase and rainfall in winters will be more frequent.
- Winter rainfall on snow cover will increase the risk of flooding, and IWRM decision-makers will have to rethink the flood protection

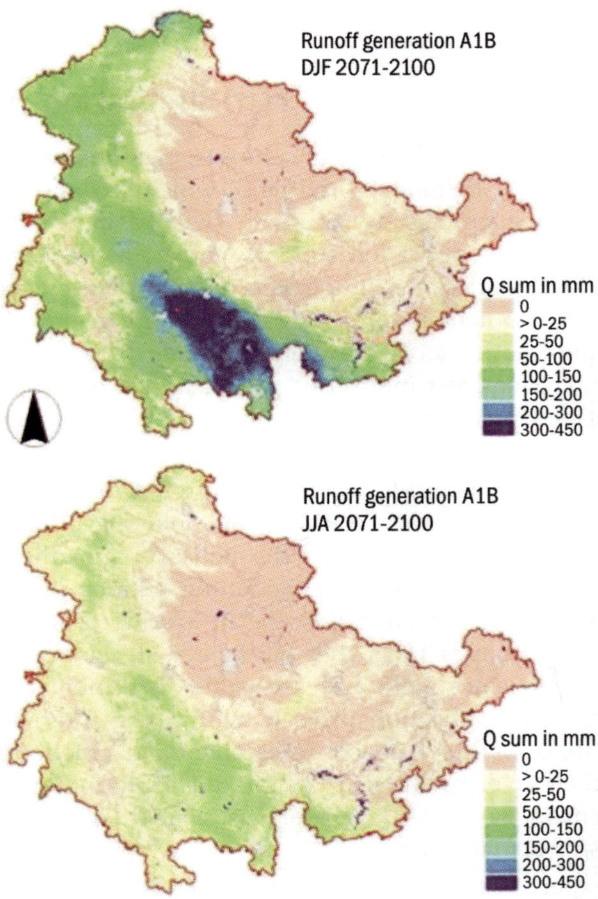


**Figure 19** Projected mean DJF and JJA run-off generation in the Free State of Thuringia

measures presently in place, that is, the operation schedule of the flood protection reservoirs.

- In the Thuringian lowland, the run-off generation is projected to reach zero, making this region winter dry, and soil water deficits are likely to develop in the upper soil zone.
- During the summer months, all regions of Thuringia will become significantly drier when compared to the present scenario, and run-off will also be reduced.
- The Thuringian lowland, however, will experience distinct dry periods, which will be a threat for agriculture in the summer months.
- Irrigation will thus become an important option for agriculture, and water managers must develop appropriate strategies to guarantee a sustainable supply from reservoirs or groundwater aquifers to account for this additional water demand.

## CONCLUSION AND FUTURE RESEARCH

Geoinformatics offers a wide range of concepts, methods, and software tools like remote sensing, GIS, models, and DIST, which account for scale-related issues of interactive process dynamics in river basins. Each of the tools used in a stand-alone application supports impact assessment of changing LULC for IWRM, that is, due to socio-economic development or climate change. Their full potential for IWRM, however, can only be realized if different geoinformatics tools are integrated into a toolset that generates synergy by exploiting their interactive potential, and makes them available to researchers and practitioners for integrated, holistic system assessments and analyses.

JESAT is one of the most developed systems in this regard and is applied successfully in many IWRM-related interdisciplinary projects on almost all continents. It is based on the holistic regionalization concepts of RUs, which are process-oriented modelling entities that represent the "real world" of the river basin landscape in a comprehensive way. JESAT comprises geoinformatics tools that have been developed and tested in numerous projects since 1995 and are under continuous enhancement, driven by ongoing application projects within the international IWRM domain. The tools of JESAT enable researchers and practitioners to assess and analyse their river basins and design and apply individual Integrated Land Management Systems (ILMS) for sustainable IWRM, accounting for likely impacts of climate change.

From the expertise gained and lessons learnt, future research regarding the application of geoinformatics for environment system assessment, analysis, IWRM, and sustainable land management should focus on the following issues.

- Scale-related process interactions need further attention, especially at meso- and macro-scale to better understand their respective integrated process dynamics. In view of the holistic system approach, the understanding of interactions between the micro- and macro-scale should be the overall objective for upscaling and downscaling exercises.
- The conceptual and methodical integration of different tools, that is, remote sensing, GIS, models, and DIST, must be improved, as these techniques are still used most often in stand-alone applications, without exploring their synergic potential. By developing holistic landscape concepts, geoinformatics can provide the theoretical background for such interdisciplinary integration.
- Integrated toolsets like the JESAT component assembly provide the means for multi-scale assessment and process integration and should be further supported in their development to make them readily available as a developing platform for researchers and a run-time environment for practitioners.
- National and international research funding programmes should relate to a higher degree to the requirements that need integrated, holistic systems research and to the potential that geoinformatics can offer to generate the required synergy for interdisciplinary cooperation.

## References

Ahrens B. 2003. **Rainfall downscaling in an Alpine watershed applying a multiresolution approach.** *Journal of Geophysical Research* **108**: D88388

Alcamo J and Henrichs T. 2002. **Critical regions: a model-based estimation of world water resources sensitive to global changes**. *Aquatic Science* **64**: 352–362

Bende U, Flügel W-A, and Kern T J. 1995. **Using GIS to delineate chemical response units (CHRU's) for hydrochemical modelling in a mesoscale catchment in Germany**. *IAHS-Publication* **226**: 133–139

Dahlke H, Helmschrot J, and Behrens T. 2005. **A GIS-based terrain analysis approach for inventory of wetland in the semi-arid headwaters of the Umzimvubu basin, South Africa**. In *Remote Sensing and GIS for Environmental Studies*, edited by S Erasmi, B Cyffka, and M Kappas. *Göttinger Geographische Abhandlungen* **113**: 78–86

Dahlke H, Helmschrot J, and Flügel W-A. 2003. **Ein integrativer Ansatz zur landschaftsbezogenen Typisierung von Feuchtgebieten im Einzugsgebiet des Umzimvubu, Ostkap-Provinz, Südafrika**. *Zentralblatt für Geologie und Paläontologie* **Teil I** (5/6): 451–466

EEA (European Environmental Agency). 1999. ***Environmental Indicators: typology and overview***. Technical Report No. 25. Copenhagen: EEA. 19 pp.

EEA (European Environmental Agency). 2005. ***Agriculture and environment in EU-15***. The IRENA Indicator Report. EEA Report No. 4. Copenhagen: EEA. 128 pp.

EEA (European Environmental Agency). 2008a. ***Modelling Environmental Change in Europe: towards a model inventory (SEIS/Forward)***. EEA Technical Report No. 11. Copenhagen: EEA. 69 pp.

EEA (European Environmental Agency). 2008b. ***Impacts of Europe's Changing Climate—2008 indicator-based assessment.*** EEA Report No. 4. Copenhagen: EEA. 246 pp.

Fink M. 2004. **Regionale Modellierung der Wasser- und Stickstoffdynamik als Entscheidungsunterstützung für die Reduktion des N-Eintrags am Beispiel des Trinkwassertalsperrensystems Weida-Zeulenroda, Thüringen.** Dissertation, Friedrich-Schiller-Universität, Jena. 190 pp.

Fink M, Krause P, Kralisch S, Bende-Michl U, Flügel W-A. 2007. **Development and application of the modelling system J2000-S for the EU Water Framework Directive**. *Advances in Geosciences* **11**: 123–130

Fink M, Beisecker R, Kralisch S, Mauden R. 2007. **Strategien zur Reduktion des diffusen Stickstoffaustrages aus landwirtschaftlich genutzten Flächen**. *Forum für Hydrologie und Wasserbewirtschaftung* **1**(20.07): 63–69

Fischer G, Shah M, and van Velthuizen H. 2002. ***Climate Change and Vulnerability***. International Institute for Applied Systems Analysis, 152 pp.

Flügel W-A. 1995. **Delineating hydrological response units (HRU's) by GIS analysis for regional hydrological modelling using PRMS/MMS in the drainage basin of the River Bröl, Germany**. *Hydrological Processes* **9**: 423–436

Flügel W-A. 1996. **Hydrological response units (HRU) as modelling entities for hydrological river basin simulation and their methodological potential for modelling complex environmental process systems. Results from the Sieg catchment**. *DIE ERDE* **127**: 42–62

Flügel W-A. 2000. **Systembezogene Entwicklung regionaler hydrologischer Modellsysteme**. *Wasser and Boden* **52**(3): 14–17

Flügel W-A. 2007. **The Adaptive Integrated Data Information System (AIDIS) for global water research**. *Water Resources Management (WARM) Journal* **21**: 199–210

Flügel W-A and Märker M. 2003. **The response units concept and its application for the assessment of hydrologically related erosion processes in semiarid catchments of Southern Africa**. *ASTM-STP 1420*: 163–177

Flügel W-A, Michl C, Krause P, Kralisch S, Fink M, Helmschrot J, Busch C, Böhm B, Bende-Michl U. 2009. **The *dewnet* network: JESAT a toolset for environment system assessment, analysis and management**. Proceedings of the 18th World IMACS/MODSIM Congress, Cairns, Australia, 13–17 July 2009

Giupponi C. 2007. **Decision support systems for implementing the European Water Framework Directive, the MULINO approach**. *Environmental Modelling and Software* **22**: 248–258

Giupponi C, Mysiak J, Fasio A, Cogen V. 2004. **MULINO, multi-sectoral, integrated and operational decision support system for sustainable use of water resources at the catchment scale**. *Mathematics and Computers in Simulation* **64**: 13–24

Giupponi C, Camera R, Fassio A, Lasut A, Mysiak J, Sgobbi A. 2006. **Network analysis, creative system modelling and decision support: the NetSyMoD approach**. The FEEM Index. Details available at <http://www.feem.it/Feem/Pub/Publications/WPapers/default.htm>

Giupponi C, Sgobbi A, Mysiak J, Camera R, Fassio A. 2008. ***NetSyMoD, An Integrated Approach for Water Resources Management. Integrated Water Management, Practical Experiences and Case Studies***, pp. 69–93. Springer

GWP (Global Water Partnership). 2009. ***A Handbook for Integrated Water Resources Management in Basins***. Sweden: Elanders. 104 pp.

GWP–TAC (Global Water Partnership–Technical Advisory Committee). 2000. **Integrated Water Resources Management**. TAC Background Paper No. 4. 67 pp.

GWP–TEC (Global Water Partnership–Technical Committee). 2004. **"...Integrated Water Resources Management (IWRM) and Water Efficiency Plans by 2005" Why, What and How?** TEC Background Paper No. 10. 45 pp.

Heathcote I W. 1998. ***Integrated Watershed Management***. New York: John Wiley & Sons. 414 pp.

Helmschrot J. 2006a. ***An Integrated, Landscape-Based Approach to Model the Formation and Hydrological Functioning of wetlands in Semiarid Headwater Catchments of the Umzimvubu River, South Africa***. Göttingen: Sierke Verlag. 314 pp.

Helmschrot J. 2006b. **Assessment of temporal and spatial effects of land use changes on wetland hydrology: a case study from South Africa**. In *Wetlands: modelling, monitoring, management*, pp. 197–204, edited by W Kotowski, E Maltby, D Miroslaw–Swiatek, T Okruszko, J Szatylowicz. The Netherlands: Taylor & Francis/A A Balkema Publisher

Helmschrot J and Flügel W-A. 2002. **Land use characterization and change detection analysis for hydrological model parameterisation of large scale afforested areas using remote sensing**. *Physics and Chemistry of the Earth* **27**: 711–718

Horlitz T. 2006. **The role of model interfaces for participation in water management**. *Water Resources Management*

IPCC (Intergovernmental Panel on Climate Change). 2000. ***Emissions Scenarios***. A Special Report of IPCC Working Group III. 27 pp.

IPCC (Intergovernmental Panel on Climate Change). 2007a. ***Climate Change 2007: the physical science basis***. Contribution of Working Group I to the Fourth Assessment Report of the IPCC. Details available at <http//www.ipcc.ch/ipccreports/ar4-wg1.htm>

IPCC (Intergovernmental Panel on Climate Change). 2007b. ***Climate Change 2007: impacts, adaptation and vulnerability***. Contribution of Working Group II to the Fourth Assessment Report of the IPCC. Details available at <http//www.ipcc.ch/ipccreports/ar4-wg2.htm>

IPCC (Intergovernmental Panel on Climate Change). 2007c. ***Climate Change 2007: mitigation of climate change***. Contribution of Working Group III to the Fourth Assessment Report of the IPCC. Details available at <http//www.ipcc.ch/ipccreports/ar4-wg3.htm>

IUCN (International Union for Conservation of Nature). 2003. ***Adaptation of Water Management to Climate Change***. Gland, Switzerland and Cambridge, UK: IUCN. ix + 53 pp.

Jain S K and Singh V P. 2003, ***Water Resources System Planning and Management***. Amsterdam: Elsevier. 858 pp.

Kalma J D and Sivapalan M. 1995. ***Scale Issues in Hydrological Modelling***. John Wiley & Sons. 489 pp.

Klein J, Frei C, Gurtz J, Lüthi D, Vidale P L, Schär C. 2005. **Hydrologic simulations in the Rhine basin driven by a regional climate model**. *Journal of Geophysical Research* **110**

Kumar K R, Sahai A K, Kumar K K, Patwardhan S K, Mishra P K, Revadekar J V, Kamaka K, Pant C B. 2006. **High-resolution climate change scenarios for India for the 21st century**. *Current Science* **90**(3): 334–345

Kralisch S. 2004. **Darstellung und Analyse hydrologischer Topologien auf der Basis künstlicher neuronaler Netze**. Dissertation, Friedrich-Schiller-Universität Jena, Institut für Geographie, Jena

Kralisch S and Krause P. 2007. **JAMS – a framework for natural resource model development and application**. In *Proceedings of the iEMSs Third Biannual Meeting "Summit on Environmental Modelling and Software"*, edited by A Voinov, A Jakeman, and A E Rizzoli. International Environmental Modelling and Software Society, Burlington, USA

Kralisch S, Fink M, and Beckstein C. 2005. ***Neural Network Based Sensitivity Analysis of Natural Resource Models***, pp. 2498–2504, edited by A Zerger and R M Argent. *Proceedings of MODSIM 2005,* December 2005

Kralisch S, Krause P, and David O. 2005. **Using the object modeling system for hydrological model development and application**. *Advances in Geosciences* **4**: 75–81

Kralisch S, Zander F, and Krause P. 2009. **Coupling the RBIS environmental information system and the JAMS modelling framework**. In *Proceedings of 18th World IMACS/and MODSIM09 International Congress on Modelling and Simulation*, pp. 902–908, edited by R Anderssen, R Braddock, and L Newham. Cairns, Australia

Kralisch S, Fink M, Flügel W-A, Beckstein C. 2003. **A neural network approach for the optimization of watershed management**. *Environmental Modelling and Software* **18**(8–9): 815–823

Kralisch S, Krause P, Fink M, Fischer C, Flügel W-A. 2007. **Component based environmental modelling using the JAMS framework**. In *MODSIM 2007 International Congress on Modelling and Simulation*, edited by D Kulasiri and L Oxley. Modelling and Simulation Society of Australia and New Zealand

Krause P. 2002. **Quantifying the impact of land use changes on the water balance of large catchments using the J2000 model**. *Physics and Chemistry of the Earth* **27**: 663–673

Krause P and Flügel W-A. 2005. **Model integration and development of modular modelling systems**. *Advances in Geosciences* **4**: 1–2

Krause P and Hanisch S. 2009. **Simulation and analysis of the impact of projected climate change on the spatially distributed water balance in Thuringia, Germany.** *Advances in Geosciences* **7**: 1–16

Krause P, Bende-Michl U, Bäse F, Fink M, Flügel W-A, Pfennig B. 2006. **Investigations in a mesoscale catchment – hydrological modelling in the Gera Catchment.** *Advances in Geosciences* **9**: 53–61

Lapin M, Damborská I, and Melo M. 2001. **Downscaling of GCM outputs for precipitation time series in Slovakia**. *Meteorologický asopis* **IV**(3): 29–40

Leavesley G H, Markstrom S L, Brewer M S, Viger R J. 1996. **The modular modeling system (MMS) – the physical process modeling component of a database-centred decision support system for water and power management**. *Water, Air and Soil Pollution*: **90**(1-2): 303–311

Moore R, Tindall I, and Fortune D. 2004. ***Update on the HARMONIT Project the OPENMI Standard for Model Linking***, edited by Liong Phoon and Babovic. 6th International Conference on Hydroinformatics. World Scientific Publishing Company

Nevo D and Yolande E C. 2007. **A Delphi study of knowledge management systems: scope and requirements.** *Information and Management* **44:** 583–597

Olazabal M, Urzelai A, García G, Herranz K, Abajo B, Feliú E, Santa Coloma O, Aspuru I. 2007. ***OIKOS: an integrated approach towards sustainable spatial planning and management***, pp. 1–15. Proceedings on International Conference on Whole Life Urban Sustainability and its Assessment, Glasgow

Pfennig B, Kipka H, Wolf M, Fink M, Krause P, Flügel W-A. 2009. **Development of an extended spatially distributed routing scheme and its impact on process oriented hydrological modelling results**. *IAHS-Publication No. 333*

Querner E P. 2002. **Analysis of basin response resulting from climate change and mitigation measures**. In *Regional Hydrology, Bridging the Gap Between Research and Practice*, pp. 77–84, edited by H A J Van Lanen and S F Demuth. IAHS Publication No. 274

Schmidli J and Frei C. 2005. **Downscaling from GCM precipitation: a benchmark for dynamical and statistical downscaling methods**. *International Journal of Climatology* **26**(3): 679–689

Wallace J S and Gregory J P. 2002. **Water resources and their use in food production systems**. *Aquatic Science* **64**: 363–375

Wilby R L, Charles S P, Zorita E, Timbal B, Whetton P, Mearns L O. 2004. **Guidelines for use of climate scenarios developed from statistical downscaling methods.** Supporting Material of the Intergovernmental Panel on Climate Change

Wolf M, Pfennig B, Krause P, Flügel W-A. 2009. **Delineation of topographic process entities using SRTM for hydrological modelling**. *Proceedings of 18*th *World IMACS/MODSIM Congress,* Cairns, Australia, 13–17 July 2009

Young P C and Romanowicz R J. 2004. **PUB and data-based mechanistic modelling, the importance of parsimonious continuous-time models**. In *Proceedings of iEMSs—Complexity and Integrated Resources Management,* Osnabrück, 14–17 June 2004

# Climate–Population–Energy: Scenarios for 2050

**Britta Stein and Christina Rullán Lemke**
*Institute of Applied Building Technology*
*Hamburg University of Technology*
*Schwarzenbergstr Hamburg, Germany*
*E-mail: b.stein@tuhh.de; christina.rullan@tuhh.de*

## INTRODUCTION

The building sector is considered to be a key sector, wherein the principles of sustainable development can be integrated. Undoubtedly, the construction, use, and demolition of buildings result in substantial social and economic benefits for society. But they may also have serious negative impacts on the environment. Energy use, coupled with associated greenhouse gas emission, is one of the biggest areas of key concern in the building sector (United Nations 2007a, p. 1), especially because 86% of the global primary energy is based on fossil fuels, with 6% based on nuclear and 8% on renewables (Energy Information Administration 2008a, p. 185).

According to the Energy Information Administration (2008b, p. 301), along with the world primary energy production, the world per capita primary energy consumption is also increasing consistently (Energy Information Administration 2009, Table E1.c). In 2005, in Organisation for Economic Co-operation and Development (OECD) countries, energy consumption per person was about 40 000–80 000 kWh, while in non-OECD Asia and Africa, it amounted to less than 10 000 kWh (Figure 1). Therefore, in 2005, about 26% of the world population (all OECD countries, non-OECD Europe, and Middle East) accounted for approximately 68% of the global primary energy consumption (Energy Information Administration 2008a, p. 103, 117).

In 2005, 16% of all the worldwide delivered energy was consumed in the residential and commercial sectors (Energy Information Administration 2008a, p. 185). Thereby, "most commercial energy use occurs in buildings or structures, accounting for services such as space heating, water heating, lighting, cooking, and cooling. Energy consumed for services not associated with buildings, such as for traffic lights and city water and sewer services, is also categorized as commercial sector energy use" (Energy Information

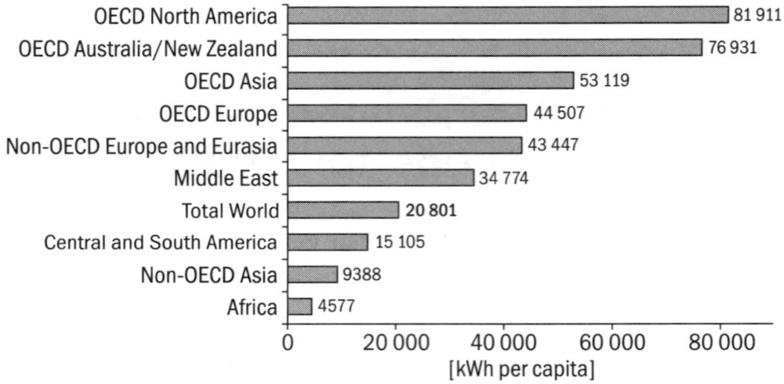


**Figure 1** World primary energy consumption per capita (2005)

**Source** Graph based on the data from Energy Information Administration (2008a), p. 103, 117

Administration 2008a, p. 12). Generally, the residential sector accounts for the major part of energy consumption in buildings—in 2005, the energy use accounted for about 11% of the worldwide delivered energy consumption (Energy Information Administration 2008a, p. 12). Nevertheless, the energy consumption in commercial buildings, such as offices and public buildings, should not be ignored.

In OECD countries, buildings are responsible for 25%–40% of total energy use (United Nations 2007a, p. 1). Data from the US Department of Energy (2008, pp. 1–3) illustrates that in 2006, buildings accounted for 39% of the entire energy consumption in the US. "In Europe, buildings account for 40%–45% of energy consumption" (United Nations 2007a, p. 1). In Germany, in 2006, 30% of the total energy was used for space heating (all sectors) (BDEW 2008, p. 10). Thus, the building sector offers the largest potential for energy efficiency in the industrialized countries.

At present, most of the energy consumption occurs during the building's operational phase, for heating and cooling purposes. Statistical data shows that the energy consumption in industrialized countries is mainly directed towards ensuring thermal comfort. In 2006, space heating and cooling accounted for 52% of the residential energy end use in the US (US Department of Energy 2008, pp. 2–5). In the same year, in the German residential sector, 74% of the energy was consumed for space heating (BDEW 2008, p. 10) (Figure 2). Therefore, the energy use in buildings is not only related to the country's level of development, but also to the building type and the climate zone where it is located (United Nations 2007a, p. v).

The 21st century will witness a shortage in fossil energy supply. In the building sector, it will be necessary to find ways to increase energy efficiency as well as substitute fossil energy. Estimations of the dimension

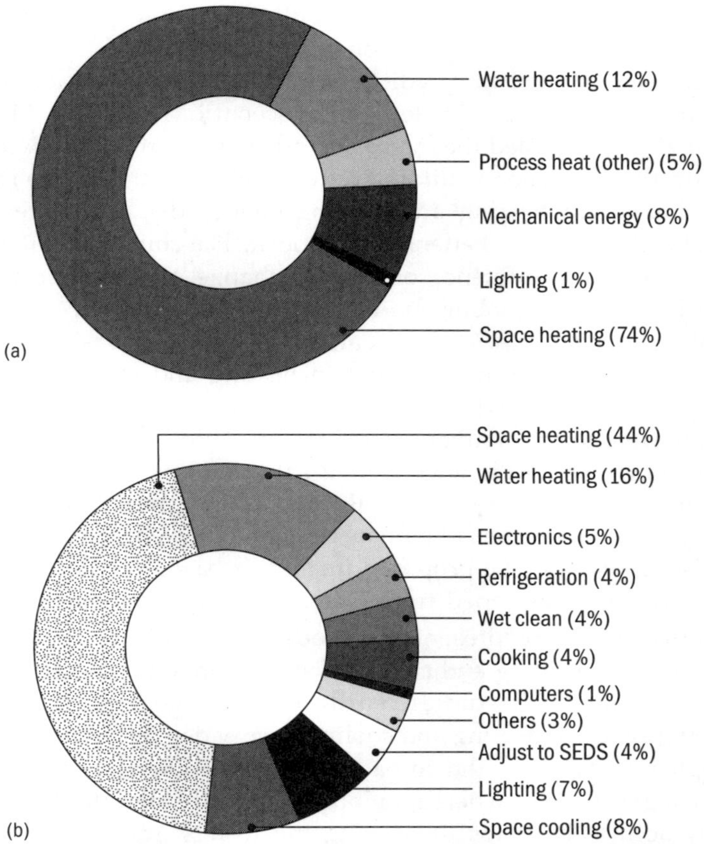


**Figure 2** Residential site energy end-use splits for Germany (a) and the US (b) (2006)

**Source** Graphs based on the data from BDEW (2008, p. 10); US Department of Energy (2008, pp. 2–5)

of future energy demand, especially for heating and cooling, will be helpful in coping with this challenge.

## CURRENT KNOWLEDGE

Studies have been conducted over the past several years on the interaction of climate change, thermal comfort, and energy use. Frank (2005) investigated the potential impacts of climate change on heating and cooling energy demand for the Zürich–Kloten location. Four climate scenarios, based on measured weather data, were used to calculate the heating and cooling energy demand of a typical building with different thermal insulation levels and internal heat gain situations. Christenson, Manz, and Gaylistras (2006) investigated the impact of global warming on Swiss building energy demand by means of the degree-days method. Holmes and Hacker (2007) addressed the challenge of designing low-energy buildings while providing thermal comfort under warmer summer conditions. The passive, or low-energy, cooling is demonstrated by predictions of the performance of some case study buildings under a climate change scenario. Aebischer, Henderson, Jacob, *et al.* (2007) regarded the increasing relevance of thermal discomfort in terms of overheating due to building retrofits and climate change in Europe. Possible changes in heating and cooling energy demand were explored for two climate variants. Pettenthaler, Gobiet, Habsburg-Lothringen, *et al.* (2007) examined the influence of climate change on the energy demand for space heating and cooling in Austria. For the period 2041–50, heating and cooling degree days were calculated on the basis of regional climate data. Thereby, future changes in heating and cooling degree-days were weighted according to altitudes and accommodation units (indirect population weighting). Wilbanks, Bhatt, and Bilello (2007) summarized what was known about the effects of climate change on energy production and use in the US. Zmeureanu and Renaud (2008) presented a method for estimating potential impacts of climate change on the energy use for heating in the existing houses. The proposed method is based on the house energy signature, which is developed from historical energy use data.

The above literature presents methods to investigate the impacts of climate change on heating and cooling energy demand at a certain location. At a global scale, such studies hardly exist. In general, the impact of population growth on heating and cooling energy demand is not considered. This chapter investigates the impacts of climate change and population growth on a global scale, whereupon both factors are allocated to the chosen grid fairly accurately. Maps illustrating the results graphically have been developed. The calculations are based on well established and commonly known data sources: climate data was calculated by the Intergovernmental Panel on Climate Change (IPCC), population data was estimated by the UN, and the data from National Aeronautics and Space Administration (NASA) was used.

## METHODOLOGY

Energy demand for heating and cooling is directly related to outside temperature. Besides, the amount of necessary energy, globally, is linked to the number of people living in certain climate zones. To estimate future changes in the share of energy demand for space conditioning, the distribution of human population and the temperature-referred climate zones was determined across the world. Following this, the generated data was correlated with each other (Figure 3).

The calculations as well as the world maps derived from them are based on a raster of 64 800 grid cells (180 lines of latitude by 360 lines of longitude).

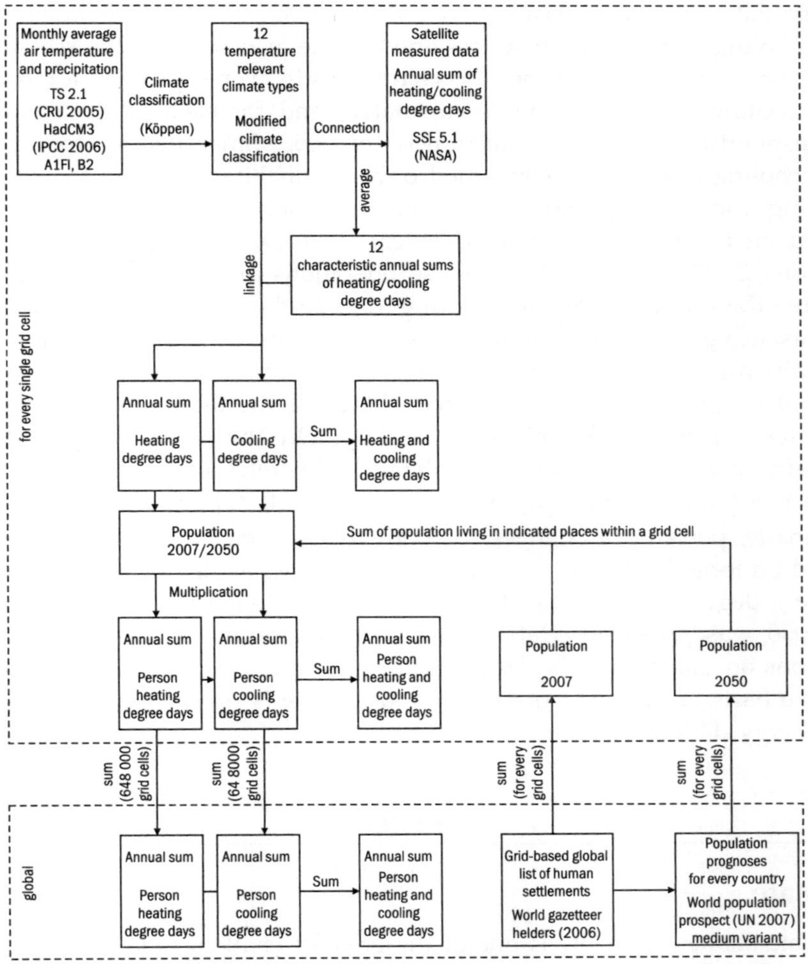


**Figure 3** Paradigm of the study

First, each grid cell was assigned to either a land surface or a sea surface (land–sea mask). The grid cells that belong to landmass were connected to further data; in addition to this, new data was generated. To show future changes, calculations were done using measured data for the present scenario on the one hand and data for two IPCC scenarios (A1FI and B2) for 2050 on the other.

For 2007 and 2050, the total number of people and a climatic type were assigned to all cells containing land surface for the scenarios A1FI and B2. Climate types were determined in accordance with the Köppen–Geiger climate classification based on the data derived by the IPCC (IPCC Data Distribution Centre n.d.) and the CRU (2005). Furthermore, on the basis of the data that was measured by NASA (2006), annual totals for heating and cooling degree-days were assigned to the grid cells for the scenario in 2007. For each climatic type, the characteristic average values for heating and cooling degree-days were calculated, and for each of the scenarios, corresponding data was determined for 2050. Since energy for heating and cooling purposes is consumed only in inhabited regions, the sums of heating and cooling degree-days were multiplied by the population data based on the World Gazetteer (Helders 2006) and UN prognoses (United Nations 2007b). The results of these operations are called person heating degree-days and/or person cooling degree-days. That way, grid cells representing extreme climate zones, for example, polar regions, may contain high totals of degree-days but lack population. Because of the lack of people, no heating is needed; therefore, the total amount of person heating degree-days is reduced to zero. On the contrary, a grid cell containing a high number of inhabitants illustrates high energy demand, even when the value of degree-days is comparatively low.

Finally, percentage changes for both future scenarios were determined based on today's situation, on the basis of global person heating and person cooling degree-days. Using the current data, the energy demand resulting in 2050 was projected. Data sources and methods are described in the sections on *Climate* and *Population*. The method to calculate energy demand on the basis of data on population and climate is presented in the section on *Energy* (Figure 3).

## RESULTS

### Climate

Climate change probably is one of the greatest challenges facing the earth. It threatens the basic elements of sustenance of life for people across the world—access to water, food, healthcare facilities as well as use of land and other environment resources (Stern 2006, p. 56).

Apart from an increase in global-scale surface temperatures, changes in other variables are expected, such as ocean climate change, sea level rise, atmospheric circulation changes as well as changes in precipitation and cryosphere (IPCC 2007a). Freshwater availability is projected to become a major problem, which might affect more than 1 billion people in Asia (IPCC 2007b, p. 8). The most significant change, however, will be the rise in the globally averaged surface temperature, which is projected to increase by 1.4°C to 5.8°C over the period 1990–2100. According to IPCC (2001, p. 69, 72) "higher maximum temperatures and more hot days over nearly all land areas" as well as "higher minimum temperatures, fewer cold days and frost days over nearly all land areas" are considered to be very likely during the 21st century.

This development will also affect the heating and cooling energy demands of buildings. The heating (or cooling) requirements for a given structure at a specific location are considered to be directly proportional to the number of heating degree-days at that location. These indices are usually derived from daily or even hourly temperature observations. Since the data obtained from the IPCC scenarios used for this study only gives monthly average values, which is furthermore only available for a low resolution raster of 7008 grid cells (73 lines of latitude by 96 lines of longitude), heating and cooling degree-days for the 64 800 grid cells, determined by NASA (2006) (for definitions of NASA heating and cooling degree-days, see section on *Energy*), were used. To be able to transfer heating and cooling degree-days to the year 2050, 12 temperature-relevant climate types were defined according to the Köppen climate classification. Afterwards, each climate type was linked with average sums of heating and cooling degree-days.

It was found that in terms of the totals of heating and cooling degree-days—so as to say the thermal fingerprint—the affiliation to a certain climate type is of much more importance than, for example, the affiliation to a certain latitude belt. In this study, it was possible to determine a characteristic sum of heating and cooling degree-days for each of the climate types. Therefore, the calculations of existing and future demands for heating and cooling are linked to a system of climate classification.

### *Climate Classification*

The effective climate classification was developed by Wladimir Köppen at the beginning of the 20th century. Originally, climate zone boundaries have been selected considering vegetation distribution. Today, Köppen system of classification is one of the most widely used methods for classifying the world's climates. It recognizes five main zones and several types and subtypes. It is based on the annual and monthly averages of temperature and precipitation, which are considered to be the most important factors in determining an area's climate. According to the original Köppen system, each climate type is represented by a two- to three-letter symbol: the climate zones are designated by a capital letter (A–E). The zones A, C, D, and E differ in temperature; zone B is differentiated by the annual precipitation in combination with temperature (aridity threshold). A second letter indicates the precipitation pattern and the third, the degree of summer heat (Heyer 1972, pp. 169–171). As for examining coherences between climate and energy use, the precipitation pattern is of least interest. We characterized 12 temperature-relevant climate types, referring to the climate groups and degrees of summer heat Köppen identified.

To classify climate types according to the Köppen system, temperature and precipitation data from the Hadley Centre climate model (HadCM3) (IPCC Data Distribution Centre n.d.) was used. The HadCM3 simulations offer the widest range for the predicted changes in temperature. Furthermore, their resolution is fairly high (7008 grid cells). Apart from this, population data used in this article is in accord (more or less) with that assumed by HadCM3. First, the classification of climate types was done for the period 1961–90. Furthermore, data by CRU (2005) for the same period of time was used for comparison. In doing so, a close analogy was discovered.

To classify climate types for 2050, temperature and precipitation data from the HadCM3 was used. In 2001, the IPCC published a set of scenarios to forecast future climate conditions. The scenarios A1FI (biggest change in temperature) and B2 (minimum change in temperature), chosen for our research, are based on some socio-economic assumptions, as discussed subsequently (IPCC 2001, p. 18).

The A1FI scenario assumes a mid-century peak in global population, very rapid economic growth, rapid introduction of new and more efficient technologies as well as fossil-intensive energy production and consumption. Compared with the other IPCC scenarios, calculations of the A1FI scenario project the highest temperature for the end of this century. Data from HadCM3 predicts an average temperature rise of 3.33°C for the period 1980 (measured average 1961–90) to 2050 (calculated average 2040–69).

The B2 scenario describes a world with continuously increasing global population, intermediate levels of economic development, and less rapid and more diverse technological changes compared to A1FI. Compared with the other IPCC scenarios, the projected temperatures for the end of this century are lower or equal to the ones projected by the A1 scenario family, but higher than those in the B1 scenario. For the period 1980–2050, HadCM3 predicts an average temperature rise of 2.25°C within the B2 scenario.

According to our calculations, hot climate types are going to increase (Figure 4, Table 1). In 1980, the cumulative area for climate types A and Bh was 52 million km$^2$, and therefore, accounted for 35% of the global landmass. According to B2 scenario in 2050, hot climate types account for 59 million km$^2$ of area, which equates to 40% of landmass. On the contrary, cold climate types like EF, ET, Dd, and Bk are going to decrease. In 1980, total area under these climate types was 39 million km$^2$ (27%), while in the B2 scenario, the area was 29 million km$^2$ (20%).

Thereby, the increase in hot climate types occurs irregularly. Despite higher temperatures, climate type A (hot humid) is only increasing slightly, whereas climate type Bh (hot arid) is growing at an above average

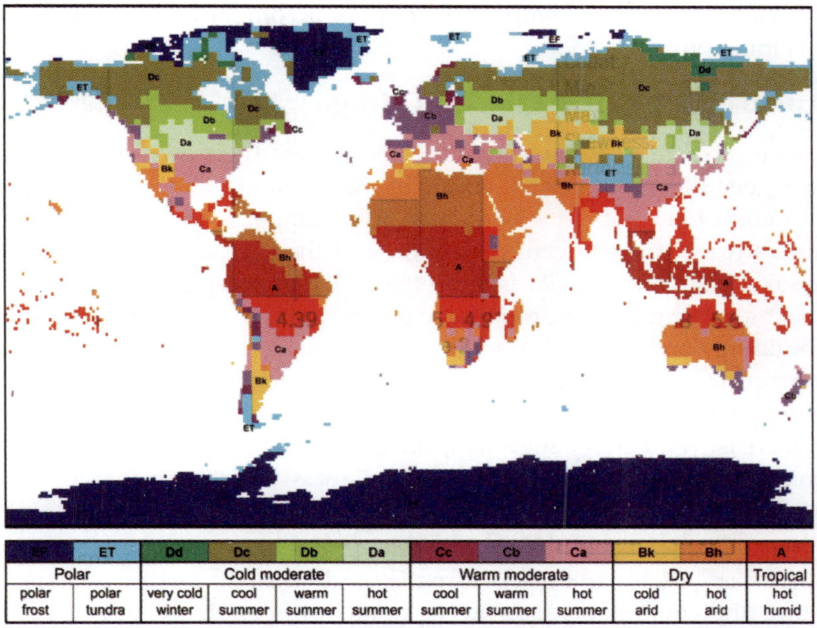


**Figure 4** Twelve temperature relevant climate types (2050), B2 scenario
**Source** Map based on the data from the IPCC Data Distribution Centre (n.d.)

**Table 1** Global allocations of the 12 temperature-relevant climate types

| Type | EF | ET | Dd | Dc | Db | Da | Cc | Cb | Ca | Bk | Bh | A |
|---|---|---|---|---|---|---|---|---|---|---|---|---|
| **Subtype** | **Polar** | | **Cold moderate** | | | | **Hot moderate** | | | **Dry** | | **Trop.** |
| **Share of landmass** | | | | | | | | | | | | |
| 1980 (m km²) | 15.86 | 9.47 | 3.78 | 18.88 | 9.93 | 3.30 | 0.44 | 7.13 | 16.96 | 10.15 | 24.41 | 27.65 |
| 2050 B2 (m km²) | 15.49 | 3.88 | 1.10 | 21.33 | 7.95 | 8.31 | 0.51 | 4.60 | 16.54 | 8.97 | 29.61 | 29.68 |
| Difference in % | –2.4 | –59.0 | –71.1 | +13.0 | –19.9 | +152.2 | +17.4 | –35.5 | –2.5 | –11.7 | +21.3 | +7.3 |
| **Percentage allocation** | | | | | | | | | | | | |
| 1980 (%) | 10.72 | 6.40 | 2.56 | 12.76 | 6.71 | 2.23 | 0.30 | 4.82 | 11.46 | 6.86 | 16.50 | 18.69 |
| 2050 B2 (%) | 10.47 | 2.63 | 0.74 | 14.41 | 5.37 | 5.62 | 0.35 | 3.11 | 11.18 | 6.06 | 20.01 | 20.06 |
| Difference in % | –0.3 | –3.8 | –1.8 | +1.7 | –1.3 | +3.4 | +0.1 | –1.7 | –0.3 | –0.8 | +3.5 | +1.4 |

rate, because the rise in temperatures is not accompanied by a similar rise in precipitation.

## Excursus: Cities and Climate Change

Because of their built environment, cities create their own microclimate, which generally enhances the effects of global climate change. The impacts of different land uses in urban areas, for example, lead to an increase in temperature at the city centre compared to the surrounding countryside, known as the "urban heat island effect". Urban designs and forms that neglect local climate conditions lose on the cooling effects of green areas, as buildings and paved surfaces heat up considerably. Thus, the average temperature of towns or cities becomes 2°C to 6°C more than that of the surrounding countryside. As urbanization and the standard of living increase, climate change will, on the one hand, enhance the energy demand for air conditioning in urban areas and, on the other hand, contribute to the urban heat island effect through heat pollution (United Nations Population Fund 2007, p. 59).

Apart from micro- and macroclimatic conditions, the energy demand for heating and cooling depends on the number of people living in a certain climate zone. In this backdrop, big changes are expected to happen within the next 50 years.

### *Population*

The United Nations (2007b) expects the world population to grow rapidly between 2007 and 2050. Over this period, the world's population is predicted to grow by 38%, from 6.7 billion to 9.2 billion. Generally speaking, the more economically developed a country is today, the slower its population will grow. In some developed regions, for example, in Europe and Japan, population is even expected to decrease (Figure 5). Therefore, the growth will be absorbed mostly by developing countries. While the population of the more developed regions is projected to remain largely unchanged, 86% of the world population is expected to live in less developed regions by 2050, including 19% in the least developed countries (Table 2) (United Nations 2007b, p. 1).

In 2007, almost 40% of the world's population lived in China and India (United Nations 2007b, p. 2), which in 2005 also accounted for the largest number of urban dwellers in the world (United Nations 2006, p. 9). By 2050, Bangladesh, the Democratic Republic of Congo, and Ethiopia, three least developed countries, will be among the 10 most populous countries in the world. Besides, the world population increase is largely witnessed in a few countries, which are generally the most populous. Eight countries

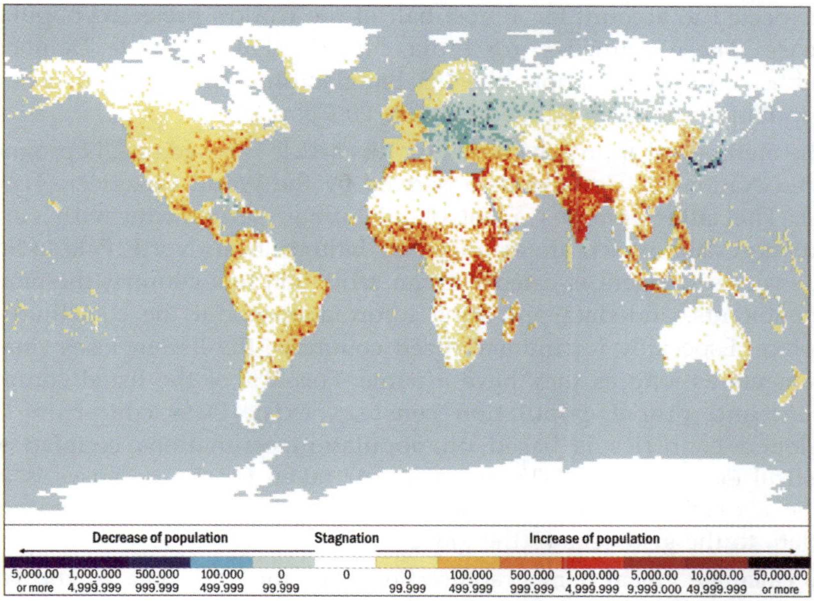


**Figure 5** Population change (2007–50), UN medium variant

**Source** Map based on the data from Helders (2006); UN (2007b)

**Table 2** Population of the world, major development groups and major areas, 1950, 1975, 2007 and 2050 medium variant

| Major area | Population (million) | | | |
|---|---|---|---|---|
| | **1950** | **1975** | **2007** | **2050 Medium variant** |
| World | 2 535 (100%) | 4 076 (100%) | 6 671 (100%) | 9 191 (100%) |
| More developed regions | 814 (32%) | 1 048 (26%) | 1 223 (18%) | 1 245 (14%) |
| Less developed regions | 1 722 (68%) | 3 028 (74%) | 5 448 (82%) | 7 946 (86%) |
| Least developed countries | 200 (8%) | 358 (9%) | 804 (12%) | 1 742 (19%) |
| Less without least developed countries | 1 521 (60%) | 2 670 (65%) | 4 644 (70%) | 6 204 (67%) |
| Africa | 224 (9%) | 416 (10%) | 965 (14%) | 1 998 (22%) |
| Asia | 1 411 (56%) | 2 394 (59%) | 4 030 (60%) | 5 266 (57%) |
| Europe | 548 (21%) | 676 (17%) | 731 (11%) | 664 (7%) |
| Latin America and the Caribbean | 168 (7%) | 325 (8%) | 572 (9%) | 769 (8%) |
| Northern America | 172 (7%) | 243 (6%) | 339 (5%) | 445 (5%) |
| Oceania | 13 (< 1%) | 21 (< 1%) | 34 (1%) | 49 (1%) |

**Source** United Nations 2007b, p. 1

are expected to account for almost half of the world's projected population increase. These countries are India, Nigeria, Pakistan, the Democratic Republic of Congo, Ethiopia, the US, Bangladesh, and China (in the order of population increase) (United Nations 2007b, p. 2).

The methodical approach chosen for this article is based on UN prognoses (United Nations 2007b) and data derived by the World Gazetteer (Helders 2006). The latter contains a list of all human settlements with current population data, which are arranged by latitude from North Pole to South Pole. Every settlement is listed with an affiliation to a country, the number of inhabitants, and data providing information about the coordinates of the place. Especially for industrialized countries, the list encloses smallest settlements as long as they have a name. For 40% of the listed countries, current and proper population censuses exist. Data concerning less developed countries is based on population estimations coupled with statistical extrapolations. The average year taken for the censuses is 2000. As the list can be sorted according to coordinates, it is possible to assign the data to the grid cells of the globe.

To calculate global demographic allocations for 2050, data by the World Gazetteer was connected to UN prognoses. This article relates to the UN medium variant, which assumes medium fertility, normal mortality, and normal international migration (United Nations 2007b, p. 33). Thus, for every country, a number predicting inhabitants for 2050 is available, and of course, these figures can be related to the current numbers. A development factor was derived for every country, with which it was possible to calculate the future inhabitants for every settlement according to the affiliation to a country in 2050. Afterwards, the numbers of inhabitants of all settlements affiliated to a certain grid cell were summed up. This way it was ensured that many grid cells include several countries that are connected to different prognoses. To discern demographic changes geographically, the data for all grid cells with similar latitudes was summed up. The results show that the dimension of population growth varies at different latitudes of the globe (Figure 6).

By 2050, population living between 50°N and 70°N (2007: 374 million, 2050: 341 million people) is predicted to have declined by 9%, while during the same period, population residing between 20°N and 20°S (2007: 2039 million, 2050: 3550 million people) is expected to have increased by 74%. Except for latitudes between 30°N and 20°S, all regions are characterized by below average population growth. The figures mentioned above reveal a unilateral population growth in equatorial regions, between the Tropic of Capricorn and Tropic of Cancer.

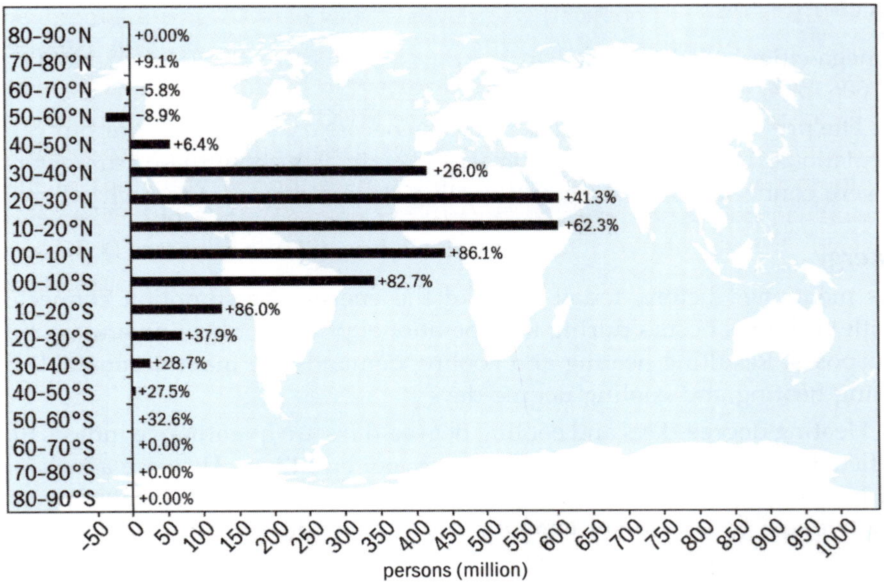


**Figure 6** Population change (2007–50) at different latitudes, UN medium variant

**Source** Map based on the data from Helders (2006); UN (2007b)

## Excursus: Urbanization

While so far, majority of the world population has lived primarily in rural areas, it is the global population growth, which is now becoming an urban phenomenon. Today, approximately half of the world's population lives in cities. By 2030, roughly 60% of the population is likely to reside in the urban centres. In comparison, the world's rural population is expected to decrease (United Nations 2006, p. 5, 9, 14). Thus, nearly the entire future population growth is expected to be in towns and cities.

Furthermore, the projected population growth will be primarily accounted for by the growth in the urban areas of the less developed regions (United Nations 2006, p. 14). By 2030, almost half of the world's population will live in the cities of developing countries (United Nations 2006, p. 5). Therefore, what happens in the cities of the less developed world in coming years will not only affect global economic growth, but also environmental sustainability, including the energy issues. Evidence suggests that the number of mega-cities with 10 million or more inhabitants has increased from 2 in 1950 to 20 in 2005. It is projected that two additional mega-cities will have emerged by 2015. By then, 17 of the 22

*Box Contd...*

*Box Contd...*

mega-cities will be located in the developing countries (United Nations 2006, p. 4, pp. 19–20).

The previous discussion described the necessary parameters for our calculations. The following section will explain as to how these parameters were connected to heating and cooling degree-days and to each other.

### Energy

As mentioned before, today, most of the energy comsumption connected with buildings occurs during the operational phase, for heating and cooling purposes. Resulting heating and cooling demands are made comparable by using heating and cooling degree-days.

Heating degree-days and cooling degree-days are quantitative indices that reflect the demand for energy to heat or cool buildings. They are a measure of how cold or hot a location was over a period of time (typically a year), relative to a balance point temperature. Heating degree-days are calculated by adding up the differences between each day's daily mean temperature and the target inside temperature till the daily mean temperature drops below the balance point temperature. Similarly, a cooling degree-day is the difference between that day's average temperature and the target inside temperature, if the daily average is greater than the balance point temperature.

The use of target inside temperatures as well as the balance point temperatures varies from country to country. Therefore, results may differ.

As already mentioned, the data derived by NASA (2006) was used for the calculations that follow. This data was measured from 1983 to 1993. Heating and cooling degree-days calculated by NASA from these measures relate to a target inside temperature and a balance point temperature of 18°C, a value that is commonly used in Europe and the US.

The determined average heating and cooling degree-days for each of the climate types (measured average from 1951 to 2000) (Kottek, Grieser, Beck, *et al.* 2006) were linked to the projected climate types in 2050 (IPCC B2 and A1FI scenarios). Figure 9 shows a global map with the total sums of the average heating and cooling degree-days in 2050 for every grid of the globe, based on the B2 scenario.

Furthermore, for every grid of the globe, the number of people was multiplied with the specific heating and cooling degree-days according to the climate type in order to compute population weighted data. As a newly defined term of energy equivalent, the result is called person heating or person cooling degree-days. Person degree-days (PersonDD) are calculated by adding person heating degree-days (PersonHDD) and person cooling degree-days (PersonCDD). In order to compare the current and future energy demands, person heating degree-days and person cooling degree-days were calculated for every grid of the globe for 2007 and 2050 (Table 3). Furthermore, the results were compared with each other (Figures 7 and 8).

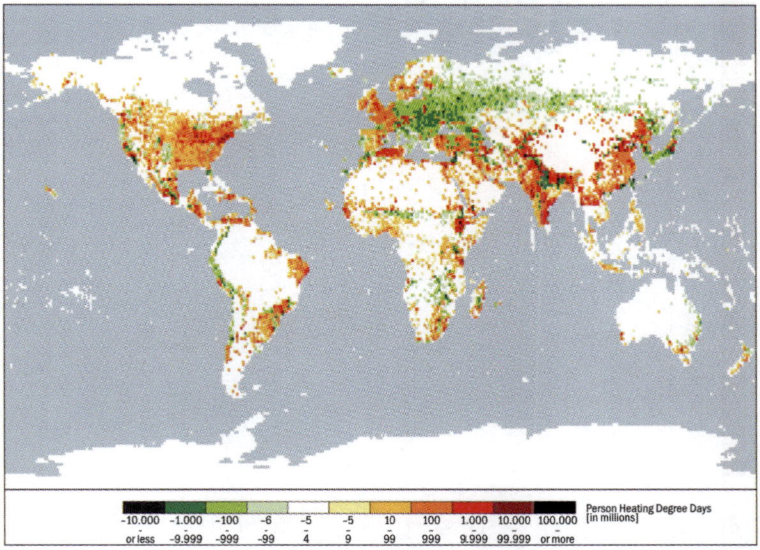


**Figure 7** Change in person heating degree-days (2007–50), B2 scenario

**Source** Map based on the data from IPCC Data Distribution Centre (n.d.); NASA (2006); Helders (2006); UN (2007b)

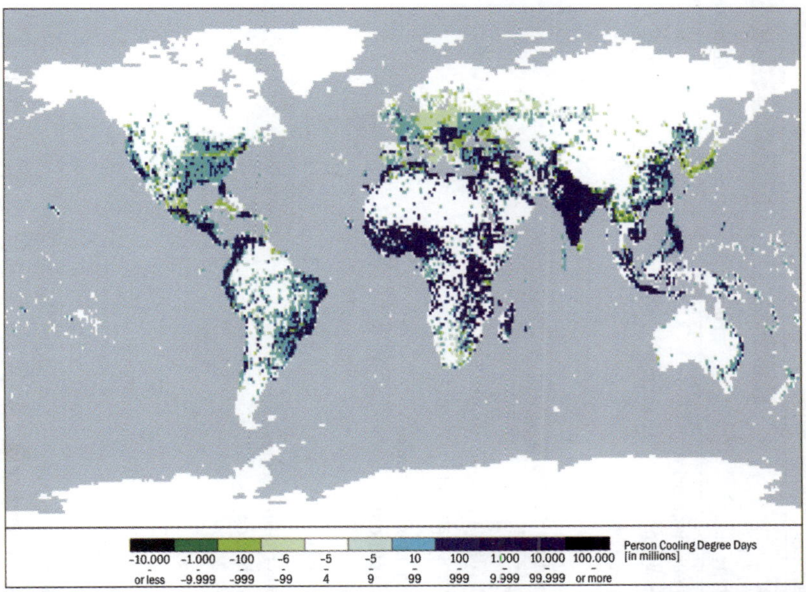


**Figure 8** Change in person cooling degree-days (2007–50), B2 scenario

**Source** Map based on the data from the IPCC Data Distribution Centre (n.d.); NASA (2006); Helders (2006); UN (2007b)

**Table 3** Average heating degree-days (HDD), cooling degree-days (CDD), population, PersonHDD, and PersonCDD related to the 12 temperature climate zones

| Type | EF | ET | Dd | Dc | Db | Da | Cc | Cb | Ca | Bk | Bh | A |
|---|---|---|---|---|---|---|---|---|---|---|---|---|
| Subtype | Polar | | Cold moderate | | | | Hot moderate | | | Dry | | Trop. |
| **Thermal fingerprint** | | | | | | | | | | | | |
| Average HDD | 19 410 | 10 800 | 11 996 | 8645 | 5663 | 4656 | 4346 | 2617 | 1409 | 3760 | 518 | 19 |
| Average CDD | 0 | 3 | 24 | 24 | 130 | 365 | 8 | 230 | 857 | 446 | 2199 | 2400 |
| HDD + CDD | 19 410 | 10 803 | 12 020 | 8670 | 5793 | 5021 | 4354 | 2847 | 2260 | 4206 | 2717 | 2419 |
| **Population (million)** | | | | | | | | | | | | |
| 2007 | 0.001 | 6 | 0.5 | 49 | 359 | 385 | 6 | 674 | 2021 | 249 | 779 | 2053 |
| 2050 (B2) | 0.02 | 7 | 0.05 | 70 | 210 | 748 | 11 | 632 | 2065 | 209 | 2182 | 3092 |
| Difference absolute | 0.02 | +1 | -0.5 | +21 | −149 | +363 | +5 | −42 | +44 | −40 | +1.403 | +1.309 |
| **Person heating degree days (billion)** | | | | | | | | | | | | |
| 2007 | 0.03 | 67 | 5 | 422 | 2030 | 1791 | 25 | 1763 | 2847 | 936 | 404 | 39 |
| 2050 (B2) | 0.3 | 72 | 0.5 | 602 | 1192 | 3483 | 46 | 1653 | 2910 | 785 | 1130 | 59 |
| **Person cooling degree days (billion)** | | | | | | | | | | | | |
| 2007 | 0.00 | 0.02 | 0.01 | 1 | 47 | 140 | 0.05 | 155 | 1732 | 111 | 1714 | 4926 |
| 2050 (B2) | 0.00 | 0.02 | 0.001 | 2 | 27 | 273 | 0.09 | 145 | 1770 | 93 | 4798 | 7422 |
| **Person degree days (billion)** | | | | | | | | | | | | |
| 2007 | 0.03 | 67 | 5 | 424 | 2077 | 1931 | 25 | 1917 | 4579 | 1047 | 2117 | 4965 |
| 2050 (B2) | 0.3 | 72 | 0.5 | 603 | 1219 | 3757 | 46 | 1799 | 4679 | 878 | 5928 | 7481 |
| Difference absolute | +0.3 | +5 | −5 | +181 | −858 | +1825 | +21 | −120 | +101 | +169 | +3811 | +2516 |
| Difference (%) | +909 | +7 | −90 | +42 | −41 | +95 | +88 | −6 | +2 | −16 | +180 | +51 |

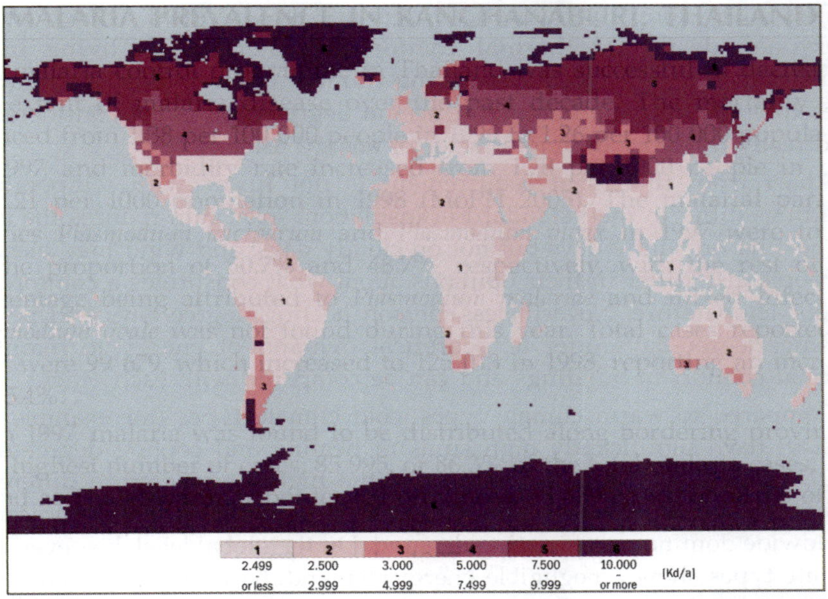


**Figure 9** Total sums of heating and cooling degree-days per climate type (2050) (B2 scenario)

**Source** Map based on the data from the IPCC Data Distribution Centre (n.d.); NASA (2006)

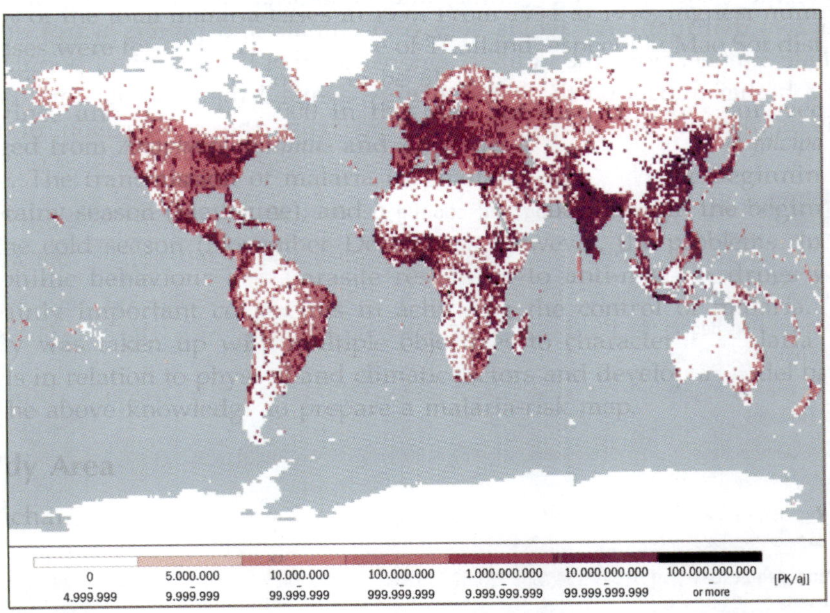


**Figure 10** Person degree-days (PersonHDD + PersonCDD) (2050) (B2 scenario)

**Source** Map based on the data from the IPCC Data Distribution Centre (n.d.); NASA (2006); Helders (2006); UN (2007b)

PersonDD show a theoretical demand in an ideal world, where everyone's desire for thermal comfort, whether one is living in an industrialized country or a developing country, is given equal importance. Therefore, the absolute number of person degree-days is not considered to be directly transformed into a factor of energy use.

Apart from that, heat requirement is not linear with temperature. Buildings differ in their standard of insulation; solar gain reduces the need for heating on sunny days but not on cloudy days, and wind may increase it. People also differ in their opinions about what constitutes a comfortable indoor temperature. Nevertheless, person heating degree-days and person cooling degree-days offer a theoretical demand of energy to maintain thermal comfort in buildings and can be compared with each other.

Compared to warm climate types, cold climate types (for example, ET, EF, Dd) have significantly high sums of heating and cooling degree-days (HDD + CDD, see Table 3) because of the higher temperature difference with respect to the balance point. At the first sight, this may indicate a worldwide dominance of heating demand. On the global level, however, cold climate types show a negligible energy demand, because these regions are less densely populated (for example, Siberia). While Figure 9, showing the total sums of heating and cooling degree-days per climate type, highlights the dominance of heating demand, Figure 10 reveals a higher demand for cooling, which is caused by densely populated regions in warm areas, like India, the eastern coast of China, and equatorial regions of Africa. This indicates that mankind avoids very cold climate and prefers warm or hot climate. Furthermore, the figures demonstrate that the growth of population has a higher impact on the global energy demand than climate conditions. This is proved by our results derived from the comparison of two

**Table 4** Comparison of heating and cooling demand (B2 and A1FI scenarios)

| | 2007 | 2050 B2 | 2050 A1F1 | Diff. B2 | Diff A1FI |
|---|---|---|---|---|---|
| Population (million) | 6 671 | 9 191 | 9 191 | 2 520 (+38%) | 2 520 (+38%) |
| Person heating degree days (billion) | 10 329 (54%) | 11 932 (45%) | 10 819 (42%) | 1 603 (+16%) | 489 (+5%) |
| Person cooling degree days (billion) | 8 826 (46%) | 1 4530 (55%) | 15 175 (58%) | 5 704 (+65%) | 6 349 (+72%) |
| Person degree days (billion) | 19 155 (100%) | 26 462 (100%) | 29 994 (100%) | 7 307 (+38%) | 6 838 (+36%) |
| Cooling degree days per capita | 1 341 (46%) | 1 575 (55%) | 1 645 (59%) | 234 (+17%) | 304 (+23%) |
| Heating degree days per capita | 1 570 (54%) | 1 293 (45%) | 1 173 (42%) | −277 (−18%) | −397 (−25%) |
| CDD per capita + HDD per capita | 2 911 (100%) | 2 868 (100%) | 2 818 (100%) | −43 (−2%) | −94 (−3%) |

different scenarios: B2 (main variant, moderate climate change) and A1FI (fossil intense, extreme climate change) (Table 4).

## DISCUSSION

As Figure 11 shows, the total number of person degree-days increases from 2007 to 2050 by about 38% in B2 scenario and 36% in A1 scenario. Both the scenarios show a strong correlation with population growth (38%). On the contrary, the heating and cooling energy demand per capita is declining until 2050 (B2: –2%; A1FI: –3%). This is due to the shrinking population in cold regions (Europe, Russia) and population growth in warm regions (India, Africa) (Figure 5).

Figure 11 shows a significant increase of person cooling degree-days (B2: +65%; A1FI: +72%), while the increase in person heating degree-days is very negligible (+16% in B2 and +5% in A1). The reason for this imbalance is the interaction of demographic and climate changes—the unilateral population growth in warm regions and the global warming.

While in 2007, heating demand dominated, with the demand being 54%, it is displaced by cooling demand in 2050 (55% in B2 scenario and 58% in A1 scenario).

In 2007, global primary energy consumption (commercially traded fuels) accounted for 464 705 PJ (BP 2008, pp. 40–41). The further estimations are

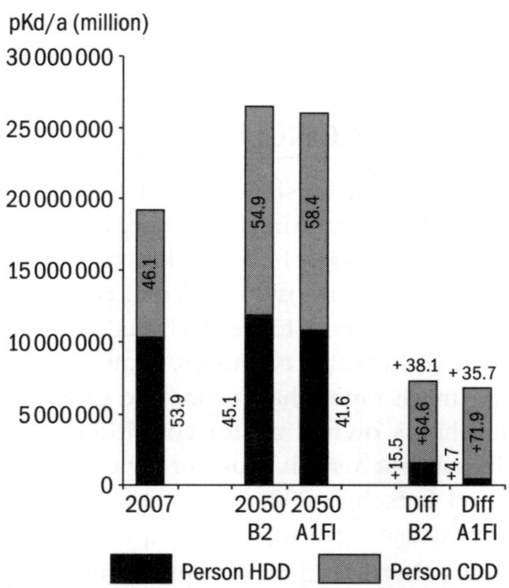


**Figure 11** Person heating degree-days and person cooling degree-days (2007, 2050, B2 scenario, and A1FI scenario)

based on the assumption that the energy used for space conditioning lies between 15% und 45%. This range was chosen because in 2006, 16% of the globally delivered energy was consumed in the two predominant sectors for heating and cooling purposes: residential and commercial (Energy Information Administration 2008a, p. 185). Besides, in Europe, the share of energy used in buildings accounted for 45% (United Nations 2007a, p. 1). As expected, with growing standards of living, the value of energy consumption in developing and newly industrializing countries is going to be equivalent to the values of energy consumption in industrialized countries; an equivalent rise in the share of space conditioning is also assumed. Based on the described assumptions, the theoretically needed energy demand for heating and cooling was extrapolated (Table 5).

It is, however, to be considered that the figures related to the world primary energy consumption comprise commercially traded fuels only, and therefore, do not reflect the real amount of energy consumption, let alone demand. In addition to that, the extrapolation assumes constant efficiency and a constant share of people who are able to afford energy prices.

As human beings living in cold regions need heating energy to survive, most of the energy demand for heating might be met as calculated. Since today, 1.6 billion people, living in warm climate, do not even have access to electricity (International Energy Agency 2002, p. 32), the use of cooling energy in 2007 is certainly overestimated by the absolute figures. On the contrary, the percentage growth of cooling energy use in 2050 is probably underestimated because of rising standards of living in some populous developing countries. Therefore, the absolute figures shown in this article are only meant as a rough estimate.

## CONCLUSION

Today, architecture and urban design very often do not relate to climate conditions. Therefore, a lot of buildings are energy-inefficient. By 2050, some developing countries will have reached a higher standard of living, and their inhabitants are expected to desire more thermal comfort. Already today, an air-conditioned house is deemed to be a status symbol. Apart from that, recent history shows that in warm countries, growth in the ownership of air conditioners may be more rapid than economic growth. In 1990, less than 1% of Chinese households owned an air conditioner; this figure was 62% by 2003! This might suggest a similar use of air conditioners in many other countries (McNeil and Letschert 2008, p. 1). Thus, the growing demand for cooling energy might cause significant consequences.

Our research shows that hot climate types will have increased by 2050. In addition to that, population growth is likely to increase in the very same regions. This is especially true for non-OECD Asia, which is characterized

**Table 5** Projection of theoretical global energy demand

| | 2007 Total | 2007 Heating | 2007 Cooling | 2050 B2 Total | 2050 B2 Heating | 2050 B2 Cooling | 2050 A1F1 Total | 2050 A1F1 Heating | 2050 A1F1 Cooling |
|---|---|---|---|---|---|---|---|---|---|
| World primary energy consumption (PJ)[a] | 46 4705 | | | | | | | | |
| Increase in person heating degree days | | | | +38% | | | +36% | | |
| Share used for heating and cooling | | 54% | 46% | | 45% | 55% | | 42% | 58% |
| Share used for heating and cooling 15% (PJ) | 69 706 | 37 641 | 32 065 | 96 194 | 43 287 | 52 907 | 94 800 | 39 816 | 54 984 |
| Share used for heating and cooling 30% (PJ) | 139 412 | 75 282 | 64 129 | 192 388 | 86 575 | 105 813 | 189 600 | 79 632 | 109 968 |
| Share used for heating and cooling 45% (PJ) | 209 117 | 112 923 | 96 194 | 288 582 | 129 862 | 158 720 | 284 400 | 119 448 | 164 952 |

**Data source** BP (2008, pp. 40–41)

[a]Primary energy comprises commercially traded fuels only. Excluded, therefore, are fuels such as wood, peat, and animal waste which, though important in many countries, are unreliably documented in terms of consumption statistics. Also excluded are wind energy, geothermal, and solar power generation.

*The primary energy values for both nuclear and hydroelectric power generation have been derived by calculating the equivalent amount of fossil fuel required to generate the same volume of electricity in a thermal power station, assuming a conversion efficiency of 38% (the average for OECD thermal power generation).

by the second lowest energy use per capita. In some regions, increase in cooling energy demand due to global warming will probably be outweighed by reductions in the need for heating energy. Especially, the rising demand for electricity as a consequence of the growing use of air conditioning might still cause an increase in primary energy consumption even when the overall demand for delivered energy is reduced. Furthermore, our results demonstrate that the growth of population has a higher impact on the global energy demand than climate change.

From a technical point of view, the production of cooling energy is mired in more problems than the supply of heating energy. Conventional cooling technologies use electricity-driven vapour-compression refrigerating machines and chlorofluorocarbons (CFCs) with high ozone depletion potential (ODP). After cooling the air below the dew point to reduce humidity, it has to be heated up again. For that reason, a lot more energy is needed for cooling than for heating. The main problem of cooling, however, is the dependency on electricity and the use of advanced technologies, as for example, heat pumps, as well as the low effectiveness of the cooling process.

"Globally, the energy efficiency of converting primary energy to useful energy is estimated at 37% at present. Moreover, considering the work capacity (exergy) of primary energy compared with exergy of useful energy, according to the Second Law of thermodynamics, the efficiency of today's energy systems in industrialised countries is less than 15%, and in developing countries even less" (Jochem, Spreng, Semadeni 2000). If highly inhabited developing nations follow the same unsustainable path as industrialized countries, the consequences will be significant, especially because resources of fossil energy are finite, and today's technologies are not yet adequate for the use of renewable energies.

Our projections are not statements on what will happen, but what might happen given the specific assumptions and methodologies used. Trends depicted in the analysis are indicative of tendencies in the real world rather than representations of specific real-world outcomes. Assumptions concerning future demographic and climate changes are necessarily uncertain. Nevertheless, in our opinion, the analysis identifies a great demand on research in solar cooling systems, because in general, the demand for cooling energy is geographically and temporarily correlated with solar energy supply.

Engineers in industrialized countries should focus more on developing energy-efficient cooling systems, for example, the desiccant evaporative cooling (DEC) system, wherein humidity is removed by a hygroscopic desiccant wheel, which can be regenerated by solar-heated hot air (Casas 2006). The correlation of cooling energy demand and solar insolation also focuses on solar building envelopes and solar urban planning. To find solutions for cities and agglomerations with problematic microclimates, high energy demand and shading problems will be especially encountered; but considering urban growth and a related augmentation in energy demand, these will be very important. In our opinion, this will be one of the major challenges for architects and urban planners in the 21st century.

## References

Aebischer B, Henderson G, Jacob M, Catenazzi G. 2007. **Impact of climate change on thermal comfort, heating and cooling energy demand in Europe**. Details available at <http://www.cepe.ethz.ch/publications/Aebischer_5_110.pdf>, last accessed on 12 February 2009

BDEW (Bundesverband der Energie und Wasserwirtschaft e.V.). 2008. **Energie-Info. Endenergieverbrauch in Deutschland 2006**. Details available at <http://www.ag-energiebilanzen.de/cms/verwaltung/files.php?path=../../daten/1204730661_91.0.74.253.pdf&name=Endenergieverbrauch_2006.pdf&mime=application/pdf>, last accessed on 07 November 2008

BP (British Petroleum). 2008. **Statistical review of world energy**. Details available at <http://www.deutschebp.de/liveassets/bp_internet/germany/STAGING/home_assets/assets/deutsche_bp/broschueren/statistical_review_of_world_energy_full_review_2008.pdf>, last accessed on 07 November 2008

Casas W. 2006. ***Untersuchung und Optimierung sorptionsgestützter Klimatisierungsprozesse***. Göttingen: Cuvillier Verlag

Christenson M, Manz H, and Gyalistras D. 2006. **Climate warming impact on degree-days and building energy demand in Switzerland**. *Energy Conversion and Management* **47**: 671–686

CRU (Climatic Research Unit). 2005. **High-resolution gridded datasets**. Details available at <http://www.cru.uea.ac.uk/cru/data/hrg.htm>, last accessed on 07 November 2007

Energy Information Administration. 2008a. ***International Energy Outlook 2008***. Washington, DC: Office of Integrated Analysis and Forecasting. Details available at <http://www.eia.doe.gov/oiaf/ieo/pdf/0484(2008).pdf>, last accessed on 09 February 2009

Energy Information Administration. 2008b. ***Annual Energy Review 2007***. Washington, DC: Office of Energy Markets and End Use, US Department of Energy. Details available at <http://www.eia.doe.gov/emeu/aer/pdf/aer.pdf>, last accessed on 09 February 2009

Energy Information Administration. 2009. **International Energy Annual 2006**. Details available at <http://www.eia.doe.gov/iea/>, last accessed on 09 February 2009

Frank Th. 2005. **Climate change impacts on building heating and cooling energy demand in Switzerland**. *Energy and Buildings* **37**: 1175–1185

Helders S. 2006. **World Gazetteer**. Details available at <http://www.world-gazetteer.com>, last accessed on 20 July 2007

Heyer E. 1972. ***Witterung und Klima. Eine allgemeine Klimatologie***, 7th edn. Leipzig: Teubner Verlagsgesellschaft

Holmes M and Hacker J. 2007. **Climate change, thermal comfort and energy: meeting the design challenges of the 21st century**. *Energy and Buildings* **39**: 802–814

International Energy Agency. 2002. **World Energy Outlook 2002**. Paris. Details available at <http://www.iea.org/textbase/nppdf/free/2000/weo2002.pdf>, last accessed on 31 July 2007

IPCC (Intergovernmental Panel on Climate Change). 2001. ***Climate Change 2001: the scientific basis***. Cambridge: Cambridge University Press. Details available at <http://www.grida.no/climate/ipcc_tar/wg1/index.htm>, last accessed on 31 July 2007

IPCC (Intergovernmental Panel on Climate Change). 2007a. **Understanding and attributing climate change**. Working Group I Contribution to the IPCC Fourth Assessment Report, Chapter 9. Details available at <http://ipcc-wg1.ucar.edu/wg1/Report/AR4WG1_Pub_Ch09.pdf>, last accessed on 30 July 2007

IPCC (Intergovernmental Panel on Climate Change). 2007b. **Climate change 2007: impacts, adaptation and vulnerability**. Summary for Policymakers. Working Group II Contribution to the Intergovernmental Panel on Climate Change Fourth Assessment Report. Details available at <http://www.ipcc.ch/SPM13apr07.pdf>, last accessed on 25 July 2007

IPCC (Intergovernmental Panel on Climate Change) Data Distribution Centre (n.d.). **HadCM3 climate scenario data**. Details available at <http://www.ipcc-data.org/sres/hadcm3_download.html>, last accessed on 31 July 2007

Jochem E, Spreng D, and Semadeni M. 2000. **Steps towards a 2000 Watt society: challenges for the technological development in Switzerland – analysis of technological and organisational long-term potentials**. Details available at <http://www.cepe.ch/research/projects/2000_watt_society/2000_watt_society.htm>, last accessed on 31 July 2007

Kottek M, Grieser C, Beck, Rudolf B, Rubel F. 2006. **World map of the Köppen-Geiger climate classification updated**. Details available at <http://koeppen-geiger.vu-wien.ac.at>, last accessed on 31 July 2007

McNeil M and Letschert V. 2008. **Future air conditioning energy consumption in developing countries and what can be done about it: the potential of efficiency in the residential sector**. Details available at <http://repositories.cdlib.org/cgi/viewcontent.cgi?article=6379&context=lbnl>, last accessed on 12 February 2002

NASA (National Aeronautics and Space Administration). 2006. **SSE – Surface Meteorology and Solar Energy (release 5.1) NASA's Science Mission Directorate**. Earth–Sun System Division, Applied Sciences Program. Details available at <http://eosweb.larc.nasa.gov/sse/>, last accessed on 27 July 2007

Pettenthaler F. Gobiet A, Habsburg-Lothringen C, Steinacker R, Töglhofer C, Türk A. 2007. **Auswirkungen des Klimawandels auf Heiz- imd Kühlenergiebearf in Österreich**. Details available at <http://www.austro-clim.at/fileadmin/user_upload/reports/StCl06F.pdf>, last accessed on 11 February 2009

Stern N. 2006. **Stern Review on the economics of climate change**. Details available at <http://www.hm-treasury.gov.uk/independent_reviews/ stern_review_economics_climate_change/stern_review_report.cfm>, last accessed on 30 July 2007

United Nations. 2006. ***World Urbanization Prospects. The 2005 Revision***. New York: Department of Economic and Social Affairs, Population Division, United Nations. Details available at <http://www.un.org/esa/population/publications/WUP2005/2005WUPHighlights_Final_Report.pdf>, last accessed on 25 July 2007

United Nations. 2007a. ***Buildings and Climate Change: status, challenges and opportunities.*** United Nations Environment Programme. Details available at <http://www.unep.fr/pc/sbc/documents/Buildings_and_climate_change.pdf>, last accessed on 25 July 2007

United Nations. 2007b. ***World Population Prospects: the 2006 revision***. Highlights. New York: Department of Economic and Social Affairs, Population Division, United Nations. Details available at <http://www.un.org/esa/population/publications/wpp2006/wpp2006_highlights.pdf>, last accessed on 17 July 2007

United Nations Population Fund. 2007. ***State of World Population 2007. Unleashing the Potential Urban Growth***. New York. Details available at <http://www.unfpa.org/swp/2007/presskit/pdf/sowp2007_eng.pdf>, last accessed on 26 July 2007

US Department of Energy. 2008. ***2008 Buildings Energy Data Book***. Buildings Technologies Program and Office of Planning, Budget and Analysis, Energy Efficiency and Renewable Energy. Details available at <http://buildingsdatabook.eren.doe.gov/docs%5CDataBooks%5C2008_BEDB_Updated.pdf>, last accessed on 09 February 2009

Wilbanks T J, Bhatt V, and Bilello D E. 2007. **Effects of climate change on energy production and use in the United States**. US Climate Change Science Program. Synthesis and assessment product 4.5. Details available at <http://www.climatescience.gov/Library/sap/sap4-5/sap4-5-draft3.pdf>, last accessed on 12 February 2009

Zmeureanu R and Renaud G. 2008. **Estimation of potential impact of climate change on the heating energy use of existing houses**. *Energy Policy* **36**: 303–310

# Geoprocessing for Soft Mapping of Sparse and Inaccurate Evapotranspiration Data

**Nazzareno Diodato***, **Michele Ceccarelli**†, **and Gianni Bellocchi**#

**Met European Research Observatory – GEWEX/CEOP Network, World Climate Research Programme, Benevento, Italy*

*†Research Center on Software Technologies-RCOST, University of Sannio, Benevento, Italy*

*#Grassland Ecosystem Research Unit, French National Institute of Agricultural Research, Clermont-Ferrand, France*

*E-mail: nazdiod@tin.it; ceccarelli@unisannio.it; giannibellochi@yahoo.com*

## INTRODUCTION

Evapotranspiration monitoring has important implications for global and regional climate modelling, as well as for acquiring knowledge on the hydrological cycle and assessing environmental stress that affects forest and agricultural ecosystems (Cristóbal, Poyatos, Ninyerola, *et al.* 2009). With global warming, practical decision-making in spatially distributed water conservation management is increasingly becoming an objective of government policy internationally (Jakeman and Letcher 2003). However, evapotranspiration phenomena are characterized by multiple ecological processes that operate at more than one spatial scale (after Fortin, Dale, and ver Hoef 2002), making the actual evapotranspiration (AET) prediction in unsampled location difficult. Also AET measurements require in-depth research, monitoring, field data collection, and programming conservation oriented land management, with high cost involved in acquiring this data (Shevenell 1999; Dalezios, Loukas, and Bampzelis 2002), especially estimating data derived from remote sensing methods (Kite and Drooger 2000). Before these procedures are implemented, especially in the absence of large primary data set, it would be beneficial to acquire and evaluate the uncertainty associated with detailed information collected at a coarse scale towards a finer scale (after Philips and Marks 1996). In response to similar situations, various multivariate geostatistical procedures (for example, cokriging) were developed (Journel and Huijbregts 1978) and applied while predicting environmental and ecological data (Webster and Oliver 2001; Caeiro, Goovaerts, Painho, *et al.* 2003; Saito, Mckenna, and Goovaerts 2005), including climate variables (Hartkamp, De Beurs, Stein, *et al.* 1999; Deirasme, Humbert, Drogue, *et al.* 2000; Goovaerts 2000; Monestiez, Courault, Allard, *et al.* 2001; Diodato 2005; Diodato and Ceccarelli 2005). A lot of data was specifically concerned with evapotranspiration, such as that in Martínez-Cob (1996), Hashmi and Garcia (1998), Antonić, Križan, Marki, *et al.* (2001), Dalezios, Loukas, and Bampzelis (2002), Li, Tarboton, and McKee (2003), and Mardikis, Kalivas, and Kollias (2005). A common drawback of multivariate geostatistical techniques is that inference demands a high number of primary and co-located auxiliary data with appropriate spatial scales in order to have detailed specification of the spatial covariance structure (Rodriguez-Iturbe, Marani, D'dorico, *et al.* 1998).

Alternatively, we suggest, after Abbaspour, Schulin, van Genuchten, *et al.* (1998), the transformation of the initial AET–sparse sampled data ($AET_{SSD}$) into a larger primary one to generate a preliminary large grid-data ($AET_{PLGD}$), combining radial basis function and topographical information with a topographically weighted radial basis function (TWRBF). Later on, $AET_{PLGD}$ was assessed with non-parametric geostatistical approach to final climate probability mapping. For this purpose, a GIS–geostatistical integrated approach (for example, indicator kriging [IK]—Journel 1983) in water protection studies, with the acronym GED (geoprocessing in evapotranspiration data), is presented in this article. Our objective here is to demonstrate a method of incorporating uncertainty associated with more topographical attributes into a model accounting for AET, and to illustrate how the model performs at a basin scale. The methods and tools we use are well known in literature. The novelty of the approach lies in how different tools, in this case a deterministics AET model, a stochastic (geostatistical) indicator, and GIS framework, are put together and applied to solve a specific problem in AET estimate at a local scale. GED is applied to a test site cropland of the Tammaro mountainous sub-regional basin (South Italy), where IK was utilized to mitigate the evapotranspiration spatial uncertainty estimates connected with the preliminary large grid data.

## RESEARCH DESIGN FOR EVAPOTRANSPIRATION MODELLING AND SPATIAL ANALYSIS

Evapotranspiration, an important component of terrestrial hydrological cycle, is the rate at which the water (in the form of vapour) moves from the land surface, through evaporation from the soil and transpiration from vegetation, towards the troposphere. There are different possibilities for making such an estimation using regionalization processes. When there is no observation data on the primary variable, radial basis function–model could be improved by relying on auxiliary variable that has correlation with the estimate variable (for example, topographical information). Here, we show that GED may be used to transfer sampled-scarce data into the (AET)-large primary data to create input for geostatistical diagnostic model. GED is based on a two-step process (Figure 1). First step combines radial basis function with topographical information to create a preliminary (AET)-large grid data ($AET_{PLGD}$), and the second step comprises input preliminary $AET_{PLGD}$ in geostatistical flow for (AET)-spatial uncertainty assessment using probability maps.

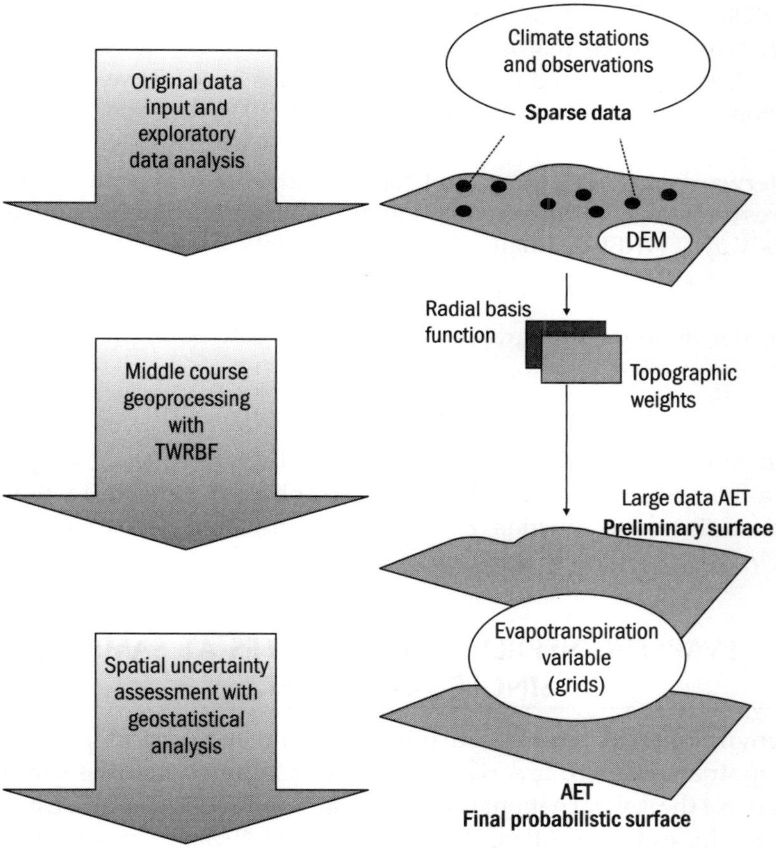


**Figure 1** Flow chart of general geoprocessing in evapotranspiration data (GED) procedure DEM – digital elevation map; AET – actual evapotranspiration; TWRBF – topographically weighted radial basis function

## GENERATED LARGE PRIMARY DATA ($Z(x_i)$ = $AET_{PLGD}$)

In the first step, TWRBF procedure was used to generate input primary large grid data for geostatistical analysis. Let us consider a set of grid locations $x_1$, ..., $x_N$, where we want to estimate our variable, and let us consider, respectively, aspect and slope of the terrain at point $x_i$: $\theta(x_i)$ and $\rho(x_i)$. TWRBF procedure combines a preliminary continuous surface $Z^*$ with the topographic weight at the grid-point, where $Z$ should be estimated. The surface $Z^*$ is created by a radial basis function of the evapotranspiration variable at sample stations. The motive for using radial basis function approximation lies in its robustness and operational simplicity, predicting better results than the inverse weighting distance (Price, McKenney, Nalder, *et al.* 2000). It is widely used in many areas such as machine learning and artificial intelligence (Girosi, Jones, and Poggio 1995), attracting a lot of interest for solving spatial interpolation problems in recent years (Hickernell and Hon 1999; Gutmann 2001; Lin and Chen 2004; Larsson and Fornberg 2005).

Afterwards, in order to account for the influence of the different terrain features on the evapotranspiration, a topographic weight has been introduced. In this way, TWRBF is finally defined in the following form.

$$Z(x_i) = Z^*(x_i) \,.\, W(x_i) \qquad (1)$$

where the adopted topographic weight $W(.)$ can be expressed as (Golf 1981)

$$W(x_i) = 1 + \rho(x_i) \cdot (1.605 \cdot 10^{-2} \cdot \sin(\theta(x_i) - 90) - 2.5 \cdot 10^{-4}) \qquad (2)$$

The main objective in this chapter is, therefore, the estimation of uncertainty associated with the evapotranspiration variable in the model (Equation 2) as the function $Z$, which we considered in the sequel of the paper $AETP_{PLGD}$–data at (hillyslope) grid-point $x_i$.

## EVAPOTRANSPIRATION ESTIMATES AT SAMPLED SINGLE POINT ($AET_{SSD}$)

The environmental variables driving the climate physical process, such as evapotranspiration, are readily available through routine monitoring networks of the several national and regional climatic data centres. Established methods (for example, Allen, Pereira, Raes, *et al.* 1998; Xu and Li 2003) exist to calculate point-specific evapotranspiration from point meteorological data. Mean annual AET from a catchment is determined largely by precipitation and references evapotranspiration (Zhang, Hickel, Dawes *et al.* 2004). In this way, AET-datum was estimated using the relationship given by Zhang, Yu, Liu, *et al.* (2001), which was derived from long-term hydrologic water budget data from over 250 watersheds across the world.

$$AET_{SSD}(s_\alpha) = \left\{1 + \left(w\,\frac{\overline{ET}_o}{P}\right) \cdot \left[1 + \left(w\,\frac{\overline{ET}_o}{P}\right) + \left(\frac{P}{\overline{ET}_o}\right)\right]^{-1}\right\} \cdot P \qquad (3)$$

where, $AET_{SSD}(s_\alpha)$ is the expected annual climatological mean water loss by evapotranspiration (mm/year) at location $s_\alpha$; $w$ is the plant-available water coefficient derived from tables (Sun, McNulty, Steve, *et al.* 2005) and reflecting the relative differences of water use for transpiration; $P$ is the average annual precipitation amount (mm/year); and $\overline{ET}_o$ is the average annual reference evapotranspiration amount ($\overline{ET}_o$, mm/year). A Hargreaves-based method (Hargreaves, Hargreaves, and Riley 1985) was used for the computation of reference evapotranspiration, in the form revised by Diodato and Bellocchi (2007).

$$ET_0 = \frac{R_{so}}{\lambda} \cdot (T + 17.8) \cdot \left(\sqrt{T_{\max} - \mathrm{ref}(T_{\min})}\right) - f(w) \cdot d_{\mathrm{wet}} \cdot f(\mathrm{clim}) \qquad (4)$$

where $R_{so}$ is the mean monthly extraterrestrial solar radiation (MJ/m$^2$/day) obtained from tables in Allen, Pereira, Raes, *et al.* (1998); $\lambda = 2.501 - (2.361 \cdot 10^{-3})$ $T$ is latent heat of vaporization (MJ/kg) according to Harrison (1963); $T$ is average monthly air temperature (°C); $T_{\max}$ is mean monthly maximum air temperature (°C); $d_{\mathrm{wet}}$ is a rainfall frequency term (number of rainy days in a month); ref($T_{\min}$) is a reference minimum air temperature; and $f$(clim) and $f(w)$ are semi-empirical functions modulating the rate at which the energetic forcing involved in the evapotranspirative processes is approached as air temperatures and rainy days change. A complete description of the functions ref($T_{\min}$), $f(w)$, and $f$(clim) was given by Diodato and Bellocchi (2007) and is not reproduced here. The annual reference evapotranspiration amount was computed as

$$\overline{ET}_o = \sum_{m=1}^{12} ET_o \cdot d_m \qquad (5)$$

where $m$ = 1 (January) to 12 (December), and $d$ is the number of days in a month.

## SPATIAL UNCERTAINTY ASSESSMENT IN LARGE PRIMARY GRID DATA–(AET$_{LPGD}$)

In the second step, the $Z(x_i)$ variable was used in order to invoke an uncertainty analysis within geostatistical tools. It is well known that uncertainty exists in every phase of the geographic data, arising from sampling, modelling, and interpolation, which must be quantified to allow an evaluation of the risk involved in any decision (Buttafuoco, Castrignanò, and Stelluti 2000; Wilson, Mitasova, and Wright 2000). Non-parametric geostatistics allows to assess such uncertainty through the determination of a conditional cumulative distribution function (ccdf) of the unknown attribute value. Geostatistics is not only used to describe spatial uncertainty, but can also be used to understand or explore the underlying processes responsible for the variation of a wide range of environmental data (Trangmar, Yost, and Uehara 1985; Oliver 1987; Davidson and Watson 1995). We used a non-parametric geostatistical technique, called IK (Journel 1983; Isaaks and Srivastava 1989). When applying IK at a range of cut-off values (multiple indicator kriging, MIK) to water resource estimation, it is necessary to settle on a finite number of thresholds that adequately represent the input data distribution shape. In this article, a range $AET_{PLGD}(x_i)$ for cut-off value $z_k$ was utilized. Consider the values of the random variable $Z$, at a set of locations $x_i$, $z(x_i)$, over the basin area. Indicating with $F$ the ccdf of the variable $Z$, the following expression is derived.

$$F(x_o; z_k|(n)) = \text{Prob}\{Z(x_o) < z_k|(n)\} \tag{6}$$

where the notation $(n)$ expresses conditioning to the $n$ data $\{z(x_i); i = 1, 2 \ldots n\}$, retained in the neighbourhood of $x_o$. In this approach, the probabiliy distribution is regarded as the conditional expectation of the indicator random variable $I(x_o; z_k)$, given the information $(n)$ data as

$$F(x_o; z_k|(n)) = E\{I(x_o; z_k)|(n)\} \tag{7}$$

Then, the pattern of variability of the range of cut-off values $z_k$ can be characterized by semivariograms defined on an indicator coding of one-zero of that threshold value

$$I(x_o; z_k) = \begin{cases} 1 \rightarrow z(x_o) > z_k \\ 0 \rightarrow z(x_o) \leq z_k \end{cases} \tag{8}$$

The indicator transform is consistent with a soft description of evapotranspiration, used in this study to mitigate the uncertainty of the $AET_{PLGD}(x_n)$ estimates. The least square estimate of the indicator $I(x_o; z_k)$ is also the least square estimate of its conditional expectation. Thus, the ccdf can be estimated by kriging the indicator $I(x_o; z_k)$ using transformation of data. The ordinary indicator kriging (oIK) estimator is a linear combination of the $n$ indicator variable in the neighbourhood $x_o$.

$$[\text{Prob } (Z(x_o) > z_k|(n)_{\text{oIK}} = \sum_{j=1}^{n} \lambda_{j,k} (x_o; z_k) \cdot i(x_j; z_k) \tag{9}$$

where $\lambda_j$ is the weight factor calculated by solving the kriging simultaneous equation system (Goovaerts 1997).

## CASE STUDY: DATA SOURCES AND AREA DESCRIPTIONS

The demonstration site (Tammaro basin) is a rural mountainous region in the southern Italy (Figure 2a), the primary mainstay of which is agriculture. The Tammaro river basin flows down into the great basin of the Volturno river, occupying its surface in a ratio of about 12% (675 $km^2$ on 5600 $km^2$). Altitudes of its land range from 128 masl to 1429 masl, with average spatial slope of about 3%; only 19% of the area has mean greater slope, in the range of 13%–15%. However, many hillslopes reach a slope of 30%. The natural vegetation of the zones of the Italian Apenninic inland is similar to the typical vegetation of Mediterranean mountainous and sub-mountainous sectors. The combination of wet winters and dry summers produces a termophyla vegetation, such as *Quercus pubescens* Willd., *Colutea arborescens* L., *Acer campestre* L. and, for high-mounainous sector, *Fagus sylvatica* L., *Acer obtusatum* Willd., and *Ilex aquifolium* L. The secondary formation of agricultural shrublands, originating from environmental degradation, comprised *Robinia pseudoacacia* L., *Pinus pinea* L., and a variety of resinous plants. Major crops in the area include grain, forage, olives, grapes, and potatoes. The rain gauge stations indicate a moderate rainfall regime: annual precipitation is about 900 mm in the southern part of the basin, ranging from 1000 mm to 1400 mm in the remaining part, while exceeding 1400 mm on the north-western watershed line.

### Topoclimatic Data

The period from 1955 to 1994 has been selected for spatial analysis, with the availability of quality data on temperature and rainfall estimation at 31 point stations (Figure 2). Average monthly temperature and rainfall data was obtained from former National Hydrographic and Marine Service (SIMN 1954–97) and former Central Office of Agricultural Ecology (UCEA 1994–2003) data sets. Average, long-term data values were desired, which did not emphasize on possible extreme annual variability, which might occur if stations with (for example) 40 years of records were used.

Slope and aspect grids were constructed for about 100 × 100 m digital elevation model (DEM). The slope and orientation of each cell are computed from elevation gradients between adjacent cells and assigned to an aspect bin on 8-point compass. DEM originates from SRTM 90 m digital elevation data (csi@cgiar.org).

### Distribution of the Data

The first step in the environmental impact assessment within a geostatistical study is seeking a preliminary knowledge on the phenomenon features and data distribution. Data analysis and interpretation cannot be completely

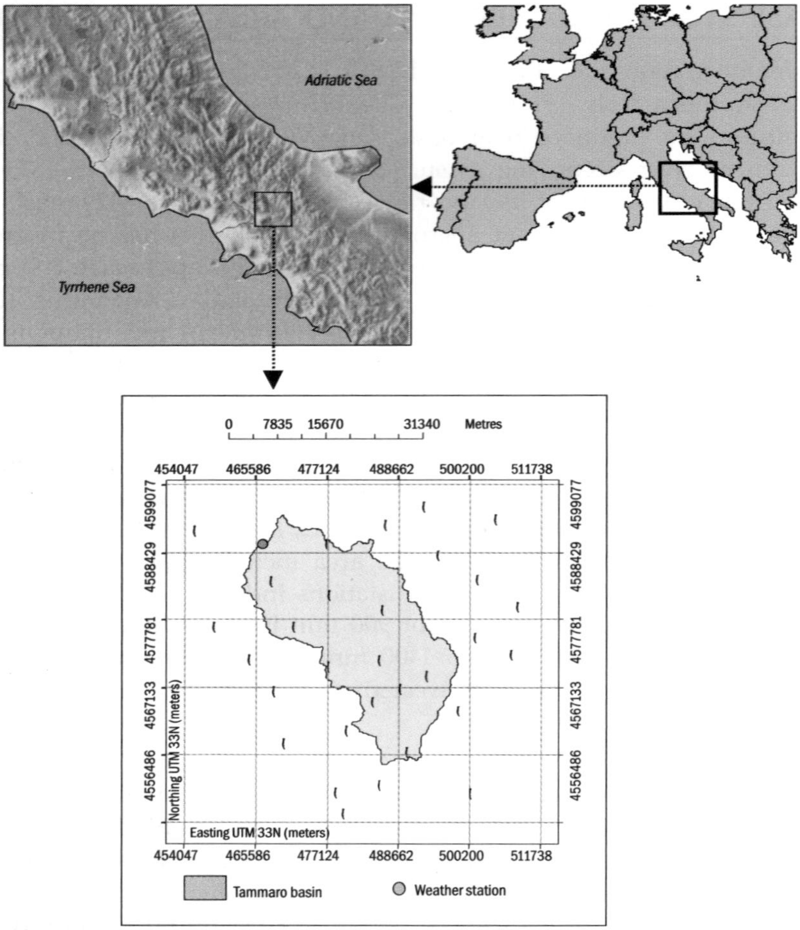


**Figure 2** Geographic location of the area under study (above) and Tammaro basin (below) (the points indicate the 31 thermo-pluviometrical stations used in this work to sample data)

automated, particularly when making crucial modelling choices (Kitanidis 1997). The sampling histograms of AET, as its descriptive statistics, for initial data set and geoprocessing data, are presented in Figure 3a, b, respectively. The distribution is far from normality for initial data and Gaussian for geoprocessing data.

## Spatial Structural Geostatistical Modelling

A model of spatial structure was fitted using the procedure developed by Johnston, Ver Hoef, Krivoruchko, *et al.* (2001). This procedure begins by assuming an isotropic model, computing the empirical and modelling

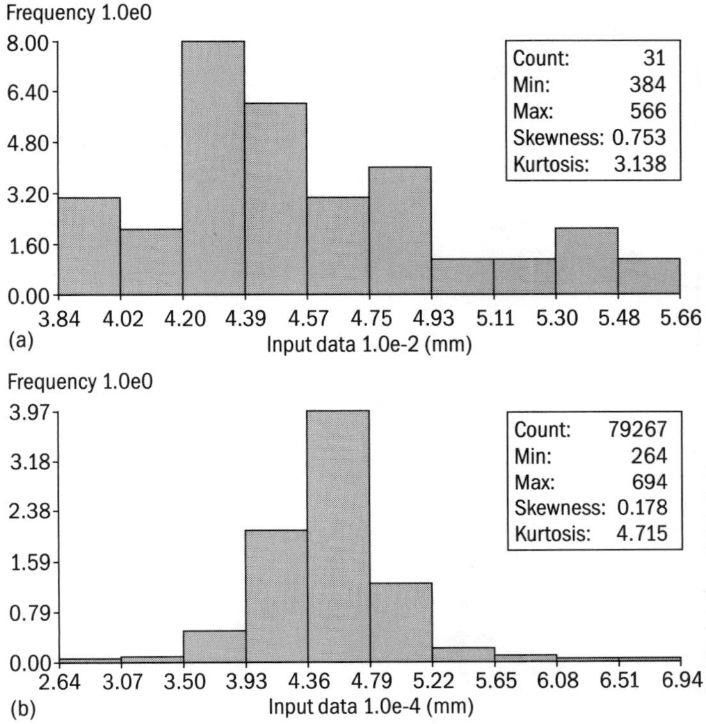


**Figure 3** Histograms and descriptive statistics for sampled AETSPD-data (a), and for generated AETLPGD–large data within topographically weighted radial basis function (TWRBF) procedure (b)

semivariance on the scaled data with respect to the standard deviation. In this way, assuming an isotropic model, any parameter, such as number of lag (assumed equal 7), lag size h (assumed equal 100 meters), range a, nugget $C_o$, sill $C_1$ and model type, is calibrated interactively (Table 1).

**Table 1** Parameters of indicator semivariograms computed for large preliminary grid data $AET_{LPGD}$ $(x_i)$-data threshold values

| AET threshold values (mm/year) | Nugget effect | Structure | | Model |
|---|---|---|---|---|
| | | Partial sill | Range (m) | |
| 350 | 0.00158 | 0.00357 | 491 | Hole effect |
| 400 | 0.01500 | 0.03645 | 669 | Hole effect |
| 450 | 0.03500 | 0.09963 | 696 | Hole effect |
| 500 | 0.03500 | 0.09963 | 696 | Hole effect |
| 550 | 0.00300 | 0.00400 | 696 | Hole effect |

AET – actual evapotranspiration

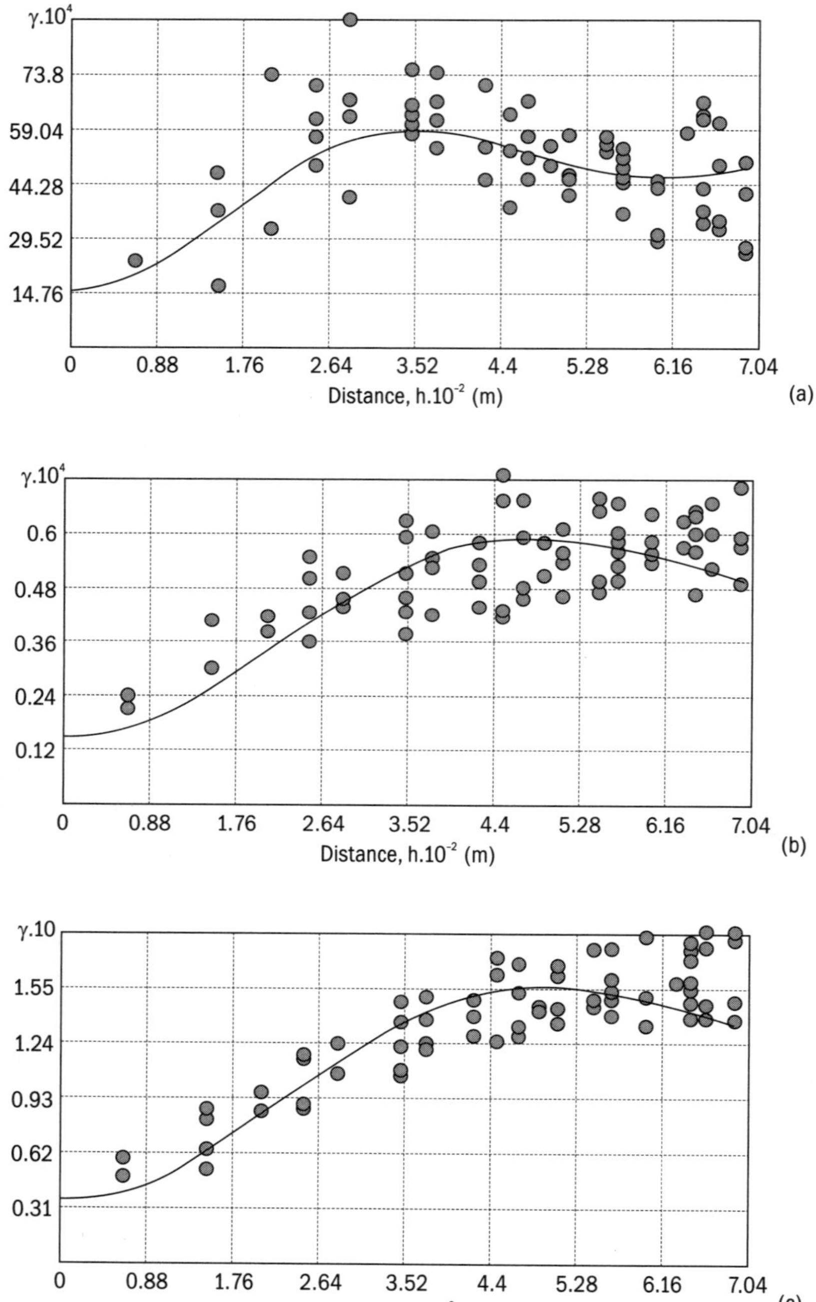


**Figure 4** Experimental indicator semivariograms (dots) and model of regionalization (solid curve) for a range of cut-off values AETPLGD $(x_i)$-large data = $z_k$: 350 mm/year (a), 400 mm/year (b), 450 mm/year (c), 500 mm/year (d), 550 mm/year (e)

*(Fig. 4 Contd..)*

*(Fig. 4 Contd..)*

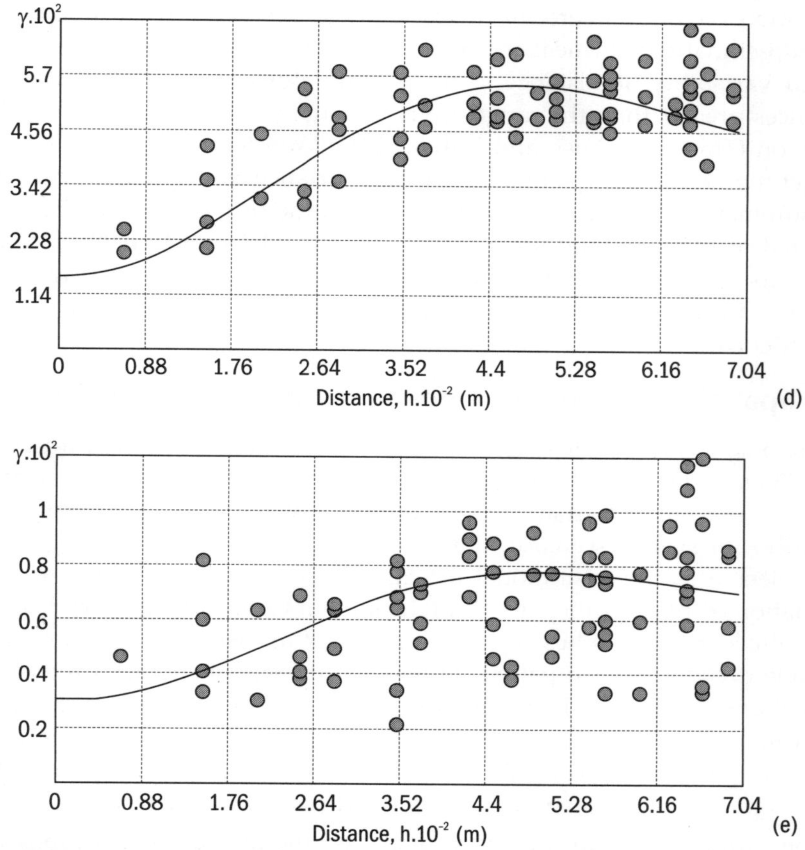


**Figure 4** Experimental indicator semivariograms (dots) and model of regionalization (solid curve) for a range of cut-off values AETPLGD ($x_i$)-large data = $z_k$: 350 mm/year (a), 400 mm/year (b), 450 mm/year (c), 500 mm/year (d), 550 mm/year (e)

Figure 4 shows the experimental indicator semivariograms computed from the 79 267 data of $AET_{PLGD}(x_i)$ predicted for a range of cut-off values, with Hole effect permissible models fitted. Semivariogram values increase with the separation distance, reflecting the assumption that AET data nearby tends to be more similar than data that is farther apart. The indicator semivariogram reaches from 491–670 m, for $z_k$ = 350–400 mm/year, to about 696 m, for $z_k$ = 450 mm/year, 500 mm/year, and 550 mm/year, before dipping and fluctuating around a sill value. It can be seen that the hole effect reduces with the increase in the threshold values, and vice versa, leading to a fragmented and more complex evapotranspiration pattern at lower values. This was also confirmed from the increase in spatial dependence when model moves to higher evapotranspiration threshold values.

Ideally, the value of the semivariograms should be zero when the separation vector $h$ is zero; in the case study, this is not true; first, because in pedo-climatic data measurement, error does exist, and second, because spatial variability at distance <350 m is unknown. Samples separated by distances greater than the range are not spatially related, implying random variation (Trangmar, Yost, and Uehara 1985; Webster and Oliver 1990). The nugget represents unexplained or random variance, which may be caused by measurement error and/or variability within the property, which cannot be detected at the sampling scale (Burrough and McDonnell 1998). However, the ratio of nugget to total sill for all the indicator semivariograms of recharge ranged from 26% to 43%, indicating strong to moderate spatial dependence.

## AET Spatial Variation and Uncertainty in its Assessment

Figure 5 shows indicator-kriged probability maps based on the threshold of AET 350 m × 350 m grid. The maps indicate that the phenomenon accounted by IK is not smooth (that is, AET indicator values change strongly with distance). In such a scenario, the non-linear variogram with nugget effect was selected as the base model for calculations, and the variogram model estimations differ significantly from the known value even at short distances. Hole effect model implies that the variations in evapotranspiration reflect possible cyclic landform pattern (after the periodic phenomena referred by Journel and Huijbregts 1978), due to continuous alternation between reliefs and valleys over a wide range of scales. Since the process determining the landform pattern also controls the ecosystem functioning, self-similarity in ecosystem functional attributes is expected (Lathrop and Peterson 1992).

The influence of exposure is strong even though it is superimposed on varying aspect and land cover. Surface on the N-slope resulted in a reduction in evapotranspiration (Figure 5b), whereas S-slope evapotranspiration clearly reveals high values of evapotranspiration, especially evident in Figure 5d, where a terrain map was overlaid. Evapotranspiration on the E and W slopes was similar, fluctuating primarily with precipitation received at sub-regional scale. This finding suggests that increasing variability in topographic properties, producing a mosaic pattern of source and sink area, may present an effective difficulty in the management of evapotranspiration control.

The most frequent annual values for AET are between 410 mm/year and 470 mm/year (Figure 5f). High probability of AET, exceeding the median recharge (450 mm/year), occurs generally around the River Tammaro (Figure 5c).

Because over large parts of the basin, AET values are two times higher than the precipitation values, during the period May–September, there is a considerable moisture deficit in the soils. In this way, the coverage of the Tammaro basin for a probability range exceeding a set of recharge threshold

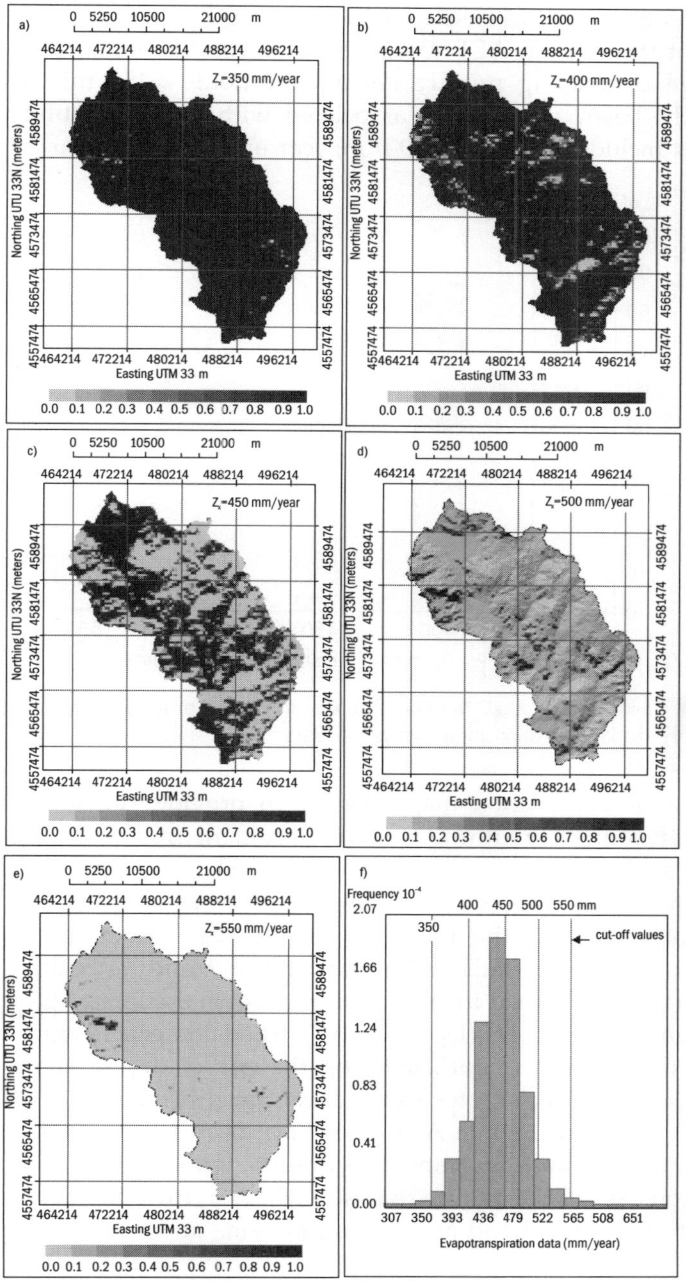


**Figure 5** Kriged probability map of evapotranspiration for a range of cut-off values AETLPGD ($x_i$)-large data equal to $z_k$: 350 mm/year (a), 400 mm/year (b), 450 mm/year (c), 500 mm/year (d), 550 mm/year (e), and respective input-data histogram (f)

values was observed, as dispayed in Figure 6. The histogram in Figure 6 shows that the median AET value (450 mm/year) is about 50%, with land percentage decreasing rapidly above thresholds >450 mm/year. Large parts of the basin are, instead, associated with high probability to match thresholds included between 350 mm/year and 400 mm/year.

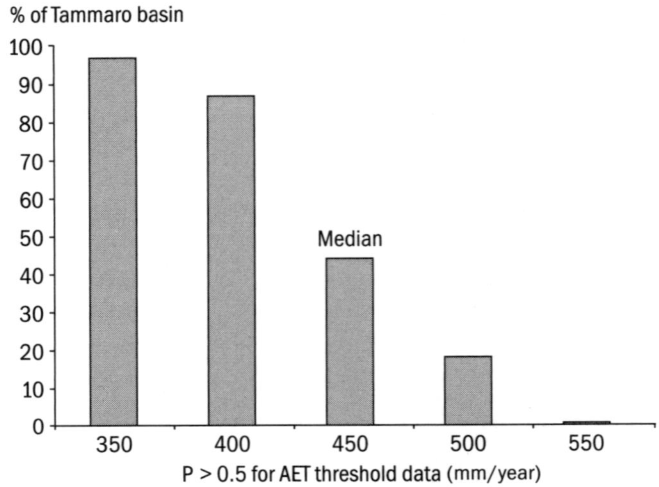


**Figure 6** Coverage of the Tammaro basin for high probability ($p > 0.5$) of an evapotranspiration threshold range data

The spatial distribution maps (Figure 5) provide valuable information on water resources planning and management in the catchment, since spatial distribution of annual values of AET is an important driving force in the hydrological cycle. During dry seasons, in particular, potential evapotranspiration is the main driver of AET. Combining the spatial distribution maps of AET with the spatial trend provides important background and physical interpolation for regional climate change studies. As an example, we report (Figure 7) AET trends on coarse maps developed for September (critical month marking the end of dry season) since 1950. High evapotranspiration rates were concentrated before 1970–80 decade (in 1950, 1955, and 1965), while AET rates have been lowered after 1975. This was due to an increase in evapotranspiration, while precipitation remained roughly constant in the Tammaro basin, which in turn had decreased water availability for natural and agricultural systems and human needs.

## CONCLUSION

Geoprocessing may be necessary before programming conservation oriented land management measures are undertaken. In this article, a

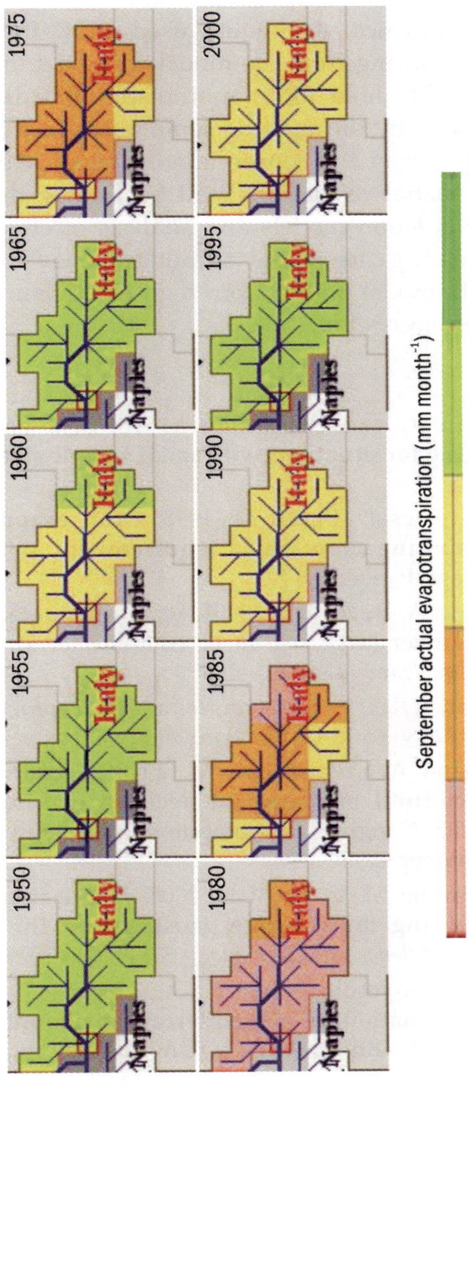


**Figure 7** September monthly time evolution of AET over Calore River Basin including at north the Tammaro sub-basin

**Source** From Global Rapid Indicator Mapping System for Water Cycle and Water Resource Assessment, Global-RIMS, <http://rbis.sr.unh.edu>

topographically weighted radial basis function has been used to create evapotranspiration data input for a spatial uncertainty assessment in the absence of large climate data set in water protection studies. GED procedure, which was developed in this work, is a particularly promising GIS–geostatistical integrated approach to mitigate the spatial uncertainty connected with large deficiencies in the primary data set. In particular, the non-parametric geostatistical techniques used allow the mitigation of the spatial uncertainty resulting from the model while defining probability of exceedance of given threshold for evapotranspiration data. This work could be further developed by formulating some guidelines for sustainable agricultural water management and conservation. Strategies to assess results obtained on different spatial resolutions will be further developed with techniques based on simulation models and analysis of error propagation. This will involve comparing the model output from low resolution, low accuracy input data with model output from high resolution, high accuracy input data at various disaggregation levels.

## References


Abbaspour K C, Schulin R, van Genuchten M Th, Schläppi E. 1998. **An alternative to cokriging for situations with small sample sizes**. *Mathematical Geology* **30:** 259–274

Allen R G, Pereira L S, Raes D, Smith M. 1998. **Crop evapotranspiration. Guidelines for computing crop water requirements,** pp. 161–194. FAO Irrigation and Drainage Paper 56.

Antonić O, Križan J, Marki A, Bukovec D. 2001. **Spatio-temporal interpolation of climatic variables over large region of complex terrain using neural networks**. *Ecological Modelling* **138:** 255–263

Burrough P A and McDonnell R A. 1998. ***Principles of Geographical Information Systems,*** pp. 98–161. New York: Oxford University Press

Buttafuoco G, Castrignanò A, and Stelluti M. 2000. **Accounting for local uncertainty in agricultural management decision making,** pp. 510–517. 7th ICCTA International Congress for Computer Technology in Agriculture, Florence, 15–18 November 1998

Caeiro S, Goovaerts P, Painho M, Costa M H. 2003. **Delineation of estuarine management areas using multivariate geostatistics: the case of Sado Estuary**. *Environmental Science and Technology* **37:** 4052–4059

Cristóbal J, Poyatos R, Ninyerola M, Pons X, Llorens P. 2009. **Actual evapotranspiration estimation in a Mediterranean mountain region by means of Landsat-5 TM and TERRA/AQUA MODIS imagery and sap flow measurements in *Pinus sylvestris* forest stands**. *Geophysical Research Abstracts* **11:** EGU2009-11496-3

Dalezios N R, Loukas A, and Bampzelis D. 2002. **Spatial variability of reference evapotranspiration in Greece**. *Physics and Chemistry of the Earth* **27:** 1031–1038

Davidson D A and Watson A I. 1995. **Spatial variability in soil moisture as predicted from airborne thematic mapper (ATM) data**. *Earth Surface Processes and Landforms* **20:** 219–230

Deirasme J, Humbert J, Drogue G, Freslon N. 2000. **Geostatistical interpolation of rainfall in mountainous areas.** In *Proceedings of Meeting geoENV III – Geostatistics for Environmental Application,* pp. 57–66. Avignone: Kluwer Accademic Publisher

Diodato N. 2005. **The influence of topographic co-variables on the spatial variability of precipitation over little regions of complex terrain.** *International Journal of Climatology* **25:** 351–363

Diodato N and Bellocchi G. 2007. **Modeling reference evapotranspiration over complex terrains from minimum climatological data**. *Water Resources Research* **43**

Diodato N and Ceccarelli M. 2005. **Processes using multivariate geostatistics for mapping interpolation of climatolgical precipitation mean in the Sannio Mountains (Southern Italy)**. *Earth Surface Processes and Landforms* **30:** 259–268

Fortin M-J, Dale M R T, and ver Hoef J. 2002. **Spatial analysis in ecology.** In *Encyclopedia of Environmetrics,* pp. 1–8, edited by A H El-Shaarawi and W W Piegorsch. Chichester: John Wiley & Sons, Ltd

Girosi F, Jones M, and Poggio T. 1995. **Regularization theory and neural networks architectures.** *Neural Computation* **7:** 219–269

Golf W. 1981. **Ermittlung der Wasserressourcen im Mittelgebirge**. *Wasserwittschaft Wassertechnik* **31:** 93–95

Goovaerts P. 1997. ***Geostatistics for Natural Resources Evaluation***. New York: Oxford University Press. 483 pp.

Goovaerts P. 2000. **Geostatistical approaches for incorporating elevation into the spatial interpolation of rainfall.** *Journal of Hydrology* **228:** 113–129

Gutmann H-M. 2001. **On the semi-norm of radial basis function interpolants.** *Journal of Approximation Theory* **111:** 315–328

Hargreaves G L, Hargreaves G H, and Riley J P. 1985. **Agricultural benefits for Senegal River Basin**. *Journal of Irrigation and Drainage Engineering ASCE* **111:** 113–124

Harrison L P. 1963. **Fundamentals concepts and definitions relating to humidity.** In *Humidity and Moisture,* vol. 3, edited by A Wexler. New York: Reinhold

Hartkamp A D, De Beurs K, Stein A, White J W. 1999. ***Interpolation Techniques for Climate Variables***. NRG-GIS Series 99-01. Mexico, DF: CIMMYT

Hashmi M A and Garcia L A. 1998. **Spatial and temporal errors in estimating regional evapotranspiration**. *Journal of Irrigation and Drainage Engineering* **124**(2): 108–114

Hickernell F J and Hon Y C. 1999. **Radial basis function approximations as smoothing splines**. *Applied Mathematics and Computation* **102:** 1–24

Isaaks E H and Srivastava R M. 1989. ***An Introduction to Applied Geostatistic***. New York: Oxford University Press. 561 pp.

Jakeman A J and Letcher R A. 2003. **Integrated assessment and modelling: features, principles and examples for catchment management**. *Environmental Modelling and Software* **18:** 491–501

Johnston K, Ver Hoef J M, Krivoruchko K, Lucas N. 2001. ***Using ArcGis Geostatistical Analyst***. ESRI. 300 pp.

Journel A G. 1983. **Non-parametric estimation of spatial distributions**. *Mathematical Geology* **15:** 445–468

Journel A G and Huijbregts C J. 1978. ***Mining Geostatistics***. New York: Academic Press. 600 pp.

Kite G W and Drooger P. 2000. **Comparing evapotranspiration estimates from satellites, hydrological models and field data**. *Journal of Hydrology* **229:** 3–18

Kitanidis P K. 1997. ***Introduction to Geostatistics: applications to hydrogeology***. Cambridge University Press. 246 pp.

Larsson E and Fornberg B. 2005. **Theoretical and computational aspect of multivariate interpolation with increasingly flat radial basis functions**. *Computers and Mathematics with Apllications* **49:** 103–130

Lathrop R G and Peterson D L. 1992. **Identifying structural self-similarity in mountainous landscapes**. *Landscape Ecology* **6:** 233–238

Li S, Tarboton D, and McKee M. 2003. ***Development of an ArcMap Toolbar for Regional Evapotranspiration Modeling***. Report presented at 23rd ESRI Internation User Conference, 7–11 July 2003, California, 12 pp.

Lin G-F and Chen L-H. 2004. **A spatial interpolation method based on radial basis function networks incorporating a semivariogram model**. *Journal of Hydrology* **288:** 288–298

Mardikis M G, Kalivas D P, and Kollias V J. 2005. **Comparison of interpolation methods for the prediction of reference evapotranspiration – an application in Greece**. *Water Resources Management* **19:** 251–278

Martínez-Cob A. 1996. **Multivariate geostatistical analysis of evapotranspiration and precipitation in mountainous terrain**. *Journal of Hydrology* **174:** 19–35

Monestiez P, Courault D, Allard D, Ruget F. 2001. **Spatial interpolation of air temperature using environmental context: application to a crop model**. *Environmental and Ecological Statistics* **8:** 297–309

Oliver M A. 1987. **Geostatistics and its application to soil science**. *Soil Use and Management* **3:** 8–20

Philips D L and Marks D G. 1996. **Spatial uncertainty analysis: propagation of interpolation errors in spatially distributed models**. *Ecological Modelling* **91:** 213–229

Price D T, McKenney D W, Nalder I A, Hutchinson M.F, Kesteven J L. 2000. **A comparison of two statistical methods for spatial interpolation of Canadian monthly mean climate data**. *Agriculture and Forest Meteorology* **101:** 81–94

Rodriguez-Iturbe I, Marani M, D'dorico P, Rinaldo A. 1998. **On space-time scaling of cumulated rainfall fields**. *Water Resources Research* **34:** 3461–3469

Saito H, Mckenna S, and Goovaerts P. 2005. **Accounting for geophysical information in geostatistical characterization of unexploded ordnance (UXO) sites**. *Environmental and Ecological Statistics* **12:** 7–25

Shevenell L. 1999. **Regional potential evapotranspiration in arid climates based on temperature, topography and calculated solar radiation**. *Hydrological Processes* **13:** 577–596

SIMN (Servizio Idrografico and Mareografico Nazionale Italiano). 1954–97. ***Annali Idrologici***. Rome: SIMN (in Italian)

Sun G, McNulty, Steve G, Lu J, Amatya Devendra M, Liang Y, Kolka R K. 2005. **Regional annual water yield from forest lands and its response to potential deforestation across the southeastern United States**. *Journal of Hydrology* **308:** 258–268

Trangmar B B, Yost R S, and Uehara G. 1985. **Application of geostatistics to spatial studies of soil properties**. *Advances in Agronomy* **38:** 45–94

UCEA (Ufficio Centrale di Ecolgia Agraria). 1994–2002. ***Bollettino agrometeorologico nazionale***. Roma: Ministero Risorse Agricole e Forestali

Webster R and Oliver M A. 1990. ***Statistical Methods in Soil and Land Resource Survey***. Oxford: Oxford University Press

Webster R and Oliver M A. 2001. ***Geostatistics for Environmental Scientist***. Chichetser: John Wiley & Sons, Ltd

Wilson J P, Mitasova H, and Wright D J. 2000. ***Water Resource Applications of Geographic Information Systems***. *URISA Journal* **12:** 61–79

Xu Z X and Li J Y. 2003. **A distributed approach for estimating catchment evapotranspiration: comparison of the combination equation and the complementary relationship approaches**. *Hydrological Processes* **17:** 1509–1523

Zhang Y, Yu Q, Liu C, Jiang J, Zhang X. 2001. **Estimation of winter wheat evapotranspiration under water stress with two semiempirical approaches**. *Agronomy Journal* **96:** 159–168

Zhang L, Hickel K, Dawes W R, Chiew F H S, Western A W, Briggs P R. 2004. **A rational function approach for estimating mean annual evapotranspiration**. *Water Resources Research* **40**

# Impact of Climate Variability on Human Health

**Nitin Kumar Tripathi[*], Nakarin Chaikaew[†], and Phisan Jeefoo[*]**
[*]*Asian institute of Technology, Bangkok, Thailand*
[†]*Naresuan University, Phayao, Thailand*
*E-mail: nitinkt@ait.asia*

## INTRODUCTION

Climate has been changing due to human activities, but the process of climate change has intensified in the last two decades and is causing serious concern due to its impacts on various aspects of human life. The effect has been profound on the vegetation pattern worldwide and is endangering the agriculture. Disasters are becoming frequent due to rising temperature and unpredictable rainfall worldwide. These factors are becoming highly variable and are causing impacts on human health. Several papers have indicated that climate variability has direct bearing on the diseases caused by mosquitoes, and respiratory and water-borne diseases.

Climate change is one of the greatest challenges of our time and will affect, in profoundly adverse ways, some of the most fundamental determinants of health: food, air, and water (WHO 2008). It has begun to negatively affect human health, with larger burdens projected in the future as weather patterns continue to change. The climate change related health consequences of diarrhoeal diseases, malnutrition, and malaria are projected to pose large risks to future populations. Such risks may vary with scales and geographic regions. In this context, it is worth assessing the importance of geospatial tools for the diversity of health issues. Research community and planners might be interested in the prediction of a disease prevalence on a global scale, which might vary in the climate change scenarios (Martens 1998).

Geographic information system (GIS) has emerged as a major tool in all the areas of health research, ranging from spatial variation of diseases to access to health services, including identifying hotspots of illness, and effective planning (Gatrell and Senior 1999). The research in this area has demonstrated utility in geographic visualization, explanatory data analysis, and modelling health outcomes in relation to environmental covariates. Some of these aspects are illustrated by Gatrell and Loytonen (1998). Further, vector-based GIS is being used in exploring blood flows in microvascular networks (Roth and Kiani 1999) as well as identifying problems at more conventional geographical scales. The methods and techniques of spatial data analysis, coupled with GIS, can thus highlight a wide variety of health problems, from the microscopic to the global scale. Such assessments are urgently needed to predict climate change scenarios.

Some of the global initiatives target researchers, public health professionals, policy-makers, and other users for them to better understand the geographic relationship that affects health outcomes, risks, disease transmission, and access to health care and other facilities. The GIS and Public Health of CDC National Center for Health Statistics provides information on the larger public health community. GIS is increasingly being used to address neighbourhood, local, state, national, and international public health issues. Similarly ESRI GIS for health and human services highlights geographic context of human health problems and services. The Agency for Toxic Substances and Disease Registry (ATSDR) has been in the forefront in the use of GIS for public health applications, and has been instrumental in popularizing this technology. The Atlas of United States Mortality (1997) was probably the first to reveal all leading causes of death, by race and sex, for small US geographic areas referred to as health service areas.

This chapter presents two case studies representing different provinces of Thailand. These studies highlight the effect of climate variability on human health. One of the studies shows the impact of climate variability on diarrhoea, while the other study focuses on malaria and its relationship with climate parameters. GIS is extensively used to integrate various parameters and identify risk areas of these diseases in respective provinces. The outcomes are expected to help the Ministry of Public Health officers in adopting strategies to prevent outbreak in future and save thousands of lives.

## CLIMATIC VARIABILITY AND DIARRHOEA INCIDENCES IN CHIANG MAI, THAILAND

Diarrhoea is one of the leading causes of morbidity and mortality in the developing world. The World Health Organization (WHO) estimates that there are approximately 4 billion cases and 2.2 million deaths annually, generally among children under the age of five. These deaths represent approximately 15% of all child deaths under the age of five in Asia, Africa, and South America (WHO 2000). Diarrhoea is highly correlated with nutritional status. It is an important area of focus due not only to its high worldwide prevalence and health costs, but also because it can be significantly reduced by appropriate interventions and treatment.

Diarrhoea can manifest itself in many ways because it has many possible causes. Most cases of diarrhoea result due to bacteria, viruses or parasites. In developing countries, most infections of diarrhoea take the faecal–oral route. The infections result from environmental causes such as poor sanitation, decreased access to clean water, and a poor understanding of transmission and treatment of disease. These conditions are witnessed most commonly in the developing world, though the disease affects both rural and urban populations (Richard and Martin 2000).

In Thailand, diarrhoea has been a major public health problem for many years now. The Bureau of Epidemiology (2004), Ministry of Public Health, estimated nearly 1 million cases every year (1 020 377, 1 055 393, 966 760, 1 161 877, and 1 142 581 cases in 2001, 2002, 2003, 2004, and 2005, respectively, with corresponding deaths being 176, 160, 124, 93, and 77, respectively). In 2006, the diarrhoea incidence was reported to be 1 245 022 cases and 9 deaths, with the highest incidences occurring in Chiang Mai, Chiang Rai, Khon Kaen, and Roi Et provinces, all in the northern and north-eastern regions of Thailand. Diarrhoea is most commonly caused by gastrointestinal infections (bacterial, viral, and parasitic organisms). The infection is spread through contaminated food or drinking water, or from person to person as a result of poor hygiene (WHO 2005).

Chiang Mai province is the endemic area of diarrhoea in the north of Thailand. The epidemic pattern of diarrhoea in this area has fluctuated every year from 2001 to 2006. The highest number of cases was recorded in 2004, particularly in Doi Tao, Samoeng, and Hot districts, with incidence rates of 6345, 4905, and 4493 per 100 000 people, respectively. The incidence of diarrhoea marks high variability at district, sub-district, and village levels. Socio-demographic, environmental and sanitation, and climate factors are thought to be associated with the outbreak of diarrhoea. The burden of diarrhoea is well described in this area, as there is a mandatory reporting of the cases, which is recorded at the Chiang Mai Provincial Public Health Office (CMPHO). All confirmed diarrhoea cases are entered into a computerized public health surveillance system of the CMPHO.

Spatial statistical methods and GIS are considered to be the important tools in epidemiological studies, particularly for identifying the disease outbreaks in Thailand. The space–time analysis detects diarrhoea outbreaks when available signs (for example, number of cases) for a particular area are being monitored. The daily notification data of Chiang Mai was analysed for the period 2001–06 to determine local clustering of cases (the village outbreak identification) by using space–time permutation scan statistic (Kuldorff, Zhang, Hartman, *et al.* 2004).

## Study Area

The province of Chiang Mai was taken as the study area as it represents climate variability in Thailand. Other parts of Thailand are more or less of uniform climate. It is the biggest province in northern Thailand, with a geographical area ranging between the latitude 17°25′ N to 20°15′ N and longitude 98°00′ E to 99°58′ E (UTM: 390000 E–560000 E, 1900000 N–2230000 N) (Figure 1), thus covering an overall area of 22 061.17 km$^2$. The region is mostly covered with forested mountain and has an approximate elevation of 310 m above mean sea level. Chiang Mai province is divided administratively into 24 districts (*amphoe*), 204 sub-districts (*tambon*), and 2070 villages (*mooban*).

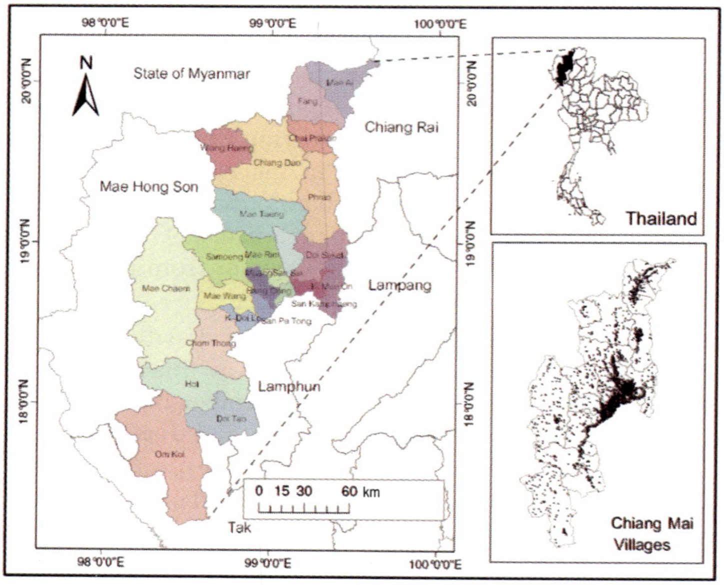


**Figure 1** Chiang Mai, Thailand

Chiang Mai has a population of about 1.6 million. The high densities occur in the vicinity of Chiang Mai metropolitan area (namely, Muang, Mae Rim, San Sai, Hang Dong, San Kamphaeng, Saraphi, and San Pa Tong districts) (Figure 2b).

Chiang Mai's climate is subtropical, but is cooler and less humid than other parts of Thailand. It can be clearly divided into three distinct seasons: hot, rainy, and cold. The hot season starts from early March and lasts till the end of May, with the temperature ranging between 19°C and 36°C. During the daytime, the average temperature is somewhere around the 35°C mark. The extreme high temperature usually occurs in April. The rainy season

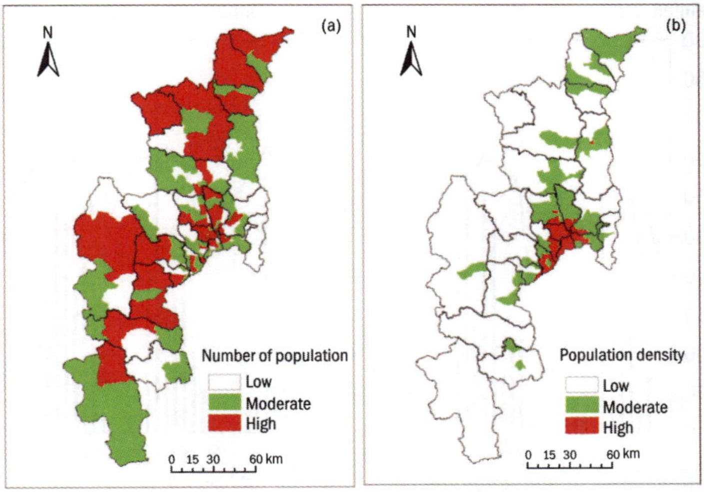


**Figure 2** (a) Population, (b) population density

begins in early June and lasts till the end of October, with the heaviest rainfall usually recorded in the months of August/September. The cold season lasts from late October till the end of February. In this period, the temperature ranges between 14°C and 30°C, and nights are much colder. The coldest months are December/January (Figure 3).

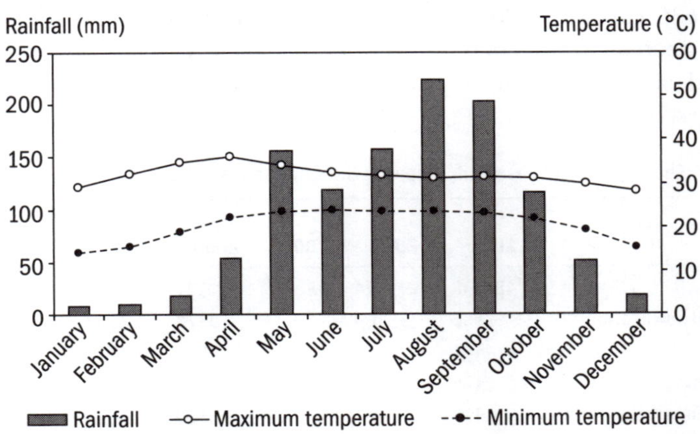


**Figure 3** Temperature and rainfall patterns for the 30-year period (1971–2000)

Northern Meteorological Center (NMC) reported the average monthly rainfall of Chiang Mai province, which was obtained from the meteorological station. The average rainfall recorded is 1304 mm in six years (2001–06). During the same period, the average temperature and humidity values were recorded as 26°C and 75%, respectively (Figure 4).

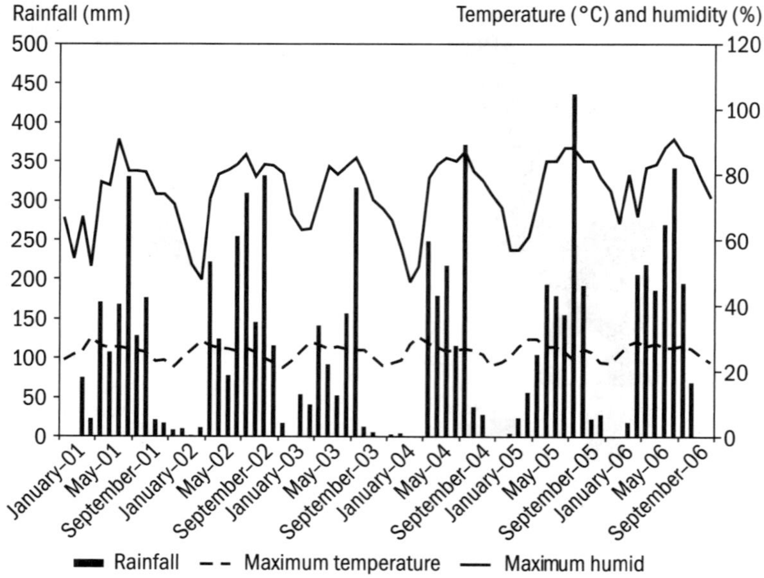


**Figure 4** Weather data from the meteorological station during 2001–2006

Table 1 illustrates the annual incidence cases of eight endemic diseases in Chiang Mai, between 2001 and 2006. During this period, the number of diarrhoea cases reported was higher than other diseases, which increased from 18 636 cases in 2001 to 35 550 cases in 2006. While the number of patients decreased in 2006 for many diseases, dengue fever (DF) cases continued showing an increasing trend, which is reflected in the recent years as well.

**Table 1** The incidence cases of diseases in Chiang Mai, 2001–06

| Disease | 2001 | 2002 | 2003 | 2004 | 2005 | 2006 | Mean 6 years |
|---|---|---|---|---|---|---|---|
| Diarrhoea | 18 636 | 26 624 | 33 493 | 39 171 | 39 641 | 35 550 | 32 186 |
| Fever of unknown origin | 3860 | 5189 | 5587 | 5378 | 4332 | 4164 | 4752 |
| Pneumonia | 4441 | 4366 | 4758 | 4089 | 4773 | 3731 | 4360 |
| Food poisoning | 2475 | 3305 | 3457 | 3631 | 3174 | 2901 | 3157 |
| Dengue haemorrhagic fever | 919 | 427 | 1108 | 500 | 351 | 393 | 616 |
| Malaria | 716 | 379 | 670 | 877 | 525 | 300 | 578 |
| Influenza | 249 | 551 | 723 | 516 | 404 | 150 | 432 |
| Dengue fever | 244 | 162 | 487 | 256 | 352 | 402 | 317 |

**Source** Chiang Mai Provincial Public Health Office, 2001–06

## Data Acquisition and Processing

Relevant data for Chiang Mai province was classified into six different geographic data sets (Table 2). The daily data at village level on diarrhoea patients less than five years of age was obtained from the CMPHO. The spatial data related to the village location was collected from the Geo-Informatics and Space Technology Centre (Northern Region) (GISTC), Thailand. All this data was incorporated into the system GIS. The data for age group less than five years was adopted, as this age group was found to be the most vulnerable group.

**Table 2** Summary of the data sets

| Age | Spatial resolution | Locations | Year | Total cases (n) |
|---|---|---|---|---|
| < 5 years | Villages | 1532 | 2001 | 18 979 |
| < 5 years | Villages | 1897 | 2002 | 26 904 |
| < 5 years | Villages | 1941 | 2003 | 29 743 |
| < 5 years | Villages | 1923 | 2004 | 38 087 |
| < 5 years | Villages | 1983 | 2005 | 38 481 |
| < 5 years | Villages | 2003 | 2006 | 32 863 |

**Source** Chiang Mai Provincial Public Health Office

Space–time permutation scan statistics (Kulldorff 1997) is used to identify diarrhoea outbreaks by considering both spatial and temporal aspects of the indicator. This is done by modifying the scanning window so that instead of circles across space, cylinders are tested, with time being represented as the height of the cylinder. The spatial statistic is implemented in the SaTScan software (Kulldorff, Fang, and Walsh 2003). A space–time retrospective analysis is used to include all confirmed cases during the week. For analysis, the upper limit on the geographical size of the outbreak is taken to be a circle with 5.2 km radius (mean disperse distance of villages), and the maximum temporal length is set at 7 days. To ensure sufficient statistical power, the number of Monte Carlo replications is set at 999, and outbreaks with statistical significance of $p \leq 0.001$ are reported.

## Seasonal Analysis

The climate pattern of Chiang Mai can be categorized into three seasons. The hot season starts from early March and lasts till the end of May. The rainy season begins in early June and lasts till the end of October, and the cold season spans from late October to the end of February. The epidemic patterns of diseases were observed for all the three seasons.

The distribution of foodborne and waterborne diseases during the six years (2001–06) is shown in Figure 5. The diarrhoea incidence cases were found to increase from the end of the winter season to rainy season (February–October). Every year, the highest number of cases was reported during the rainy season. During the period 2001–06, a high correlation (Pearson r = 0.82) between the monthly cases for diarrhoea and food poisoning was found, which was statistically significant at 0.01 level. Thus, it is confirmed that with the increase in the monthly cases of diarrhoea during the hot and rainy seasons, the monthly cases of food poisoning also increase.

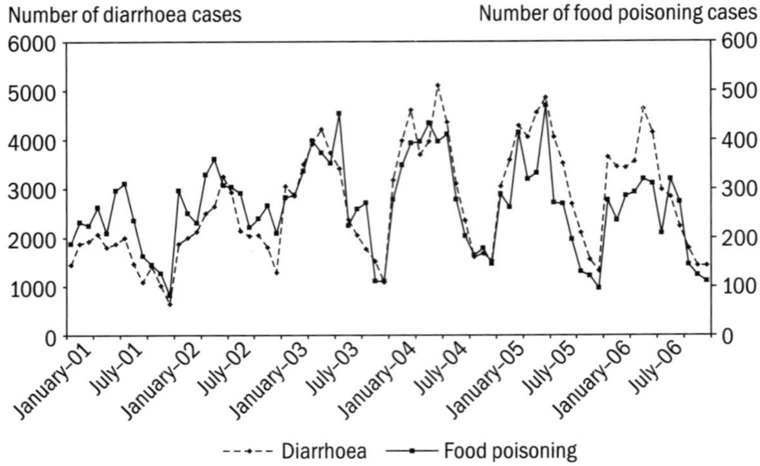


**Figure 5** Monthly cases of foodborne and waterborne diseases, 2001–06

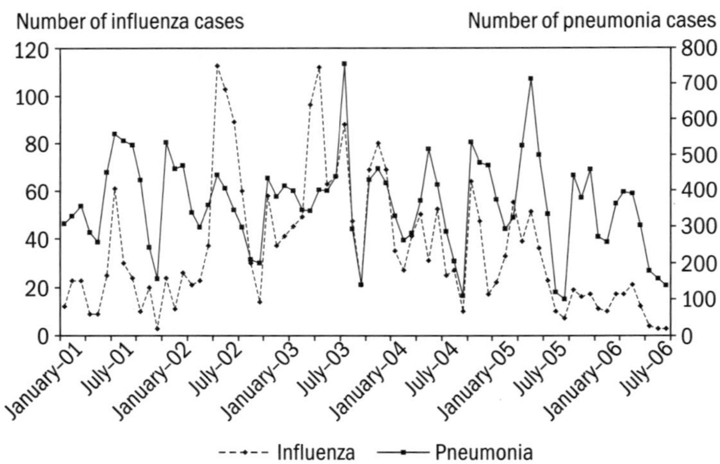


**Figure 6** Monthly cases of respiratory diseases, 2001–06

Figure 6 shows the distribution of respiratory diseases during 2001–06. The influenza incidence cases were highest in the rainy season (June–October) during the period 2001–03, while during 2003–06, these cases showed an increase from the last month of the winter season to the beginning of the hot season (February–March) and the rainy season. Similarly, the incidence cases of pneumonia increased during both hot and rainy seasons. The study revealed that the correlation between the monthly cases of influenza and pneumonia during 2001–06 was relatively low (Pearson r = 0.49) at 0.01 level of significance.

The seasonal patterns of vector-borne diseases dengue fever (DF), dengue haemorrhagic fever (DHF), and malaria are represented in Figure 7. The incidence cases for all the diseases were highest during the rainy season every year. In the study period, there was a high correlation between monthly cases of DF and DHF (Pearson r = 0.85), and a low correlation of malaria with DF (Pearson r = 0.42) and DHF (Pearson r = 0.46) at 0.01 level of significance.

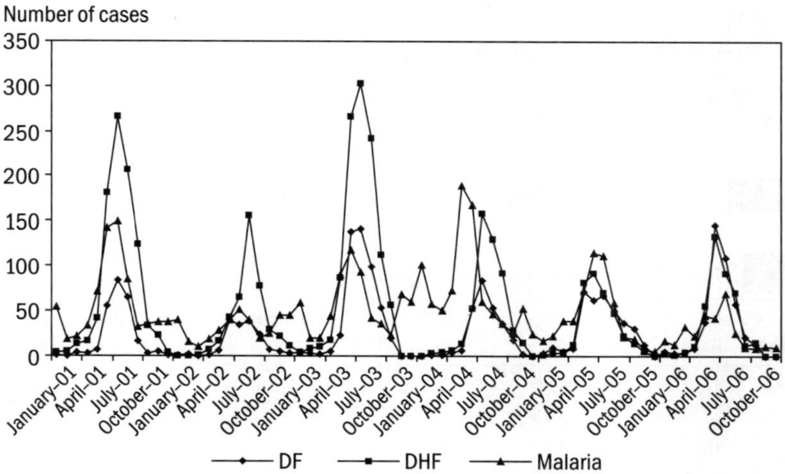


**Figure 7** Monthly cases of vector-borne diseases, 2001–06

The distribution of the fever of unknown origin during 2001–06 is shown in Figure 8. The incidence cases were highest during the rainy season (May–October) every year. The highest number of cases was reported during the rainy season of 2002. The monthly cases of the fever of unknown origin were highly correlated with DF, DHF, and influenza, with the Pearson correlation being 0.78, 0.75 and 0.67, respectively.

## Diarrhoea Outbreak Detection

For the space–time retrospective analysis, top 10 outbreaks from 45 detections with $p \leq 0.001$ are listed in Table 3 and depicted in Figures 9–11.

In the northern part of Chiang Mai (zone A), most of the outbreak signals were concentrated within the urban areas of Fang and Mae Ai districts at the beginning of the rainy season (May–June) for 2001, 2003, and 2004 (Figures 9 and 12). The highest outbreak in this zone (A1) was reported during 30 April to 6 May 2001; there had been 26 observed cases, of which 0.52 case was expected theoretically (relative risk = 49.97).

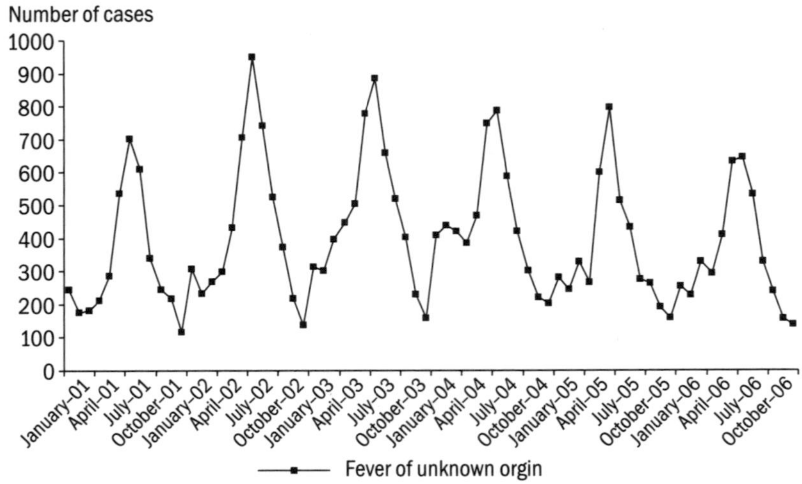


**Figure 8** Monthly cases of fever of unknown origin, 2001–06

**Table 3** Top 10 outbreaks from 1 January 2001 to 31 December 2006

| Location | Date | Villages in signal | Observed cases | Expected cases | Relative risk |
|---|---|---|---|---|---|
| C1 | 5–11 February 2001 | 1 | 32 | 0.42 | 76.45 |
| A1 | 30 April to 6 May 2001 | 1 | 26 | 0.52 | 49.97 |
| B1 | 16–22 January 2006 | 23 | 35 | 2.18 | 16.07 |
| A2 | 19–25 February 2001 | 1 | 17 | 0.43 | 39.80 |
| C2 | 1–7 January 2001 | 1 | 11 | 0.07 | 163.50 |
| C3 | 1–7 January 2001 | 3 | 11 | 0.15 | 74.32 |
| A3 | 7–13 January 2002 | 3 | 11 | 0.19 | 57.30 |
| B2 | 23–29 January 2006 | 2 | 24 | 2.48 | 9.70 |
| B3 | 14–20 March 2005 | 1 | 15 | 0.70 | 21.30 |
| C4 | 4–10 December 2006 | 3 | 6 | 0.01 | 439.63 |

In the middle zone (zone B), the strongest signal (B1) was reported during 16–22 January 2006 (the end of the winter season), covering 23 villages around the highland in the west (Mae Cheam district). The signal had 35 observed cases, of which 2.18 cases were expected (relative risk = 16.07). On the other hand, the outbreaks occurred at the end of the rainy season (August–October); these were concentrated within the urban area of San Kamphaeng district (B11) in 2002 and San Pa Tong district (B9) in 2003 (Figures 10 and 12).

The strongest signal among the list (C1) was reported during 5–11 February 2001, which was located in the southern zone (zone C) of Chiang Mai (Hot district). This signal reported 32 cases, which were observed

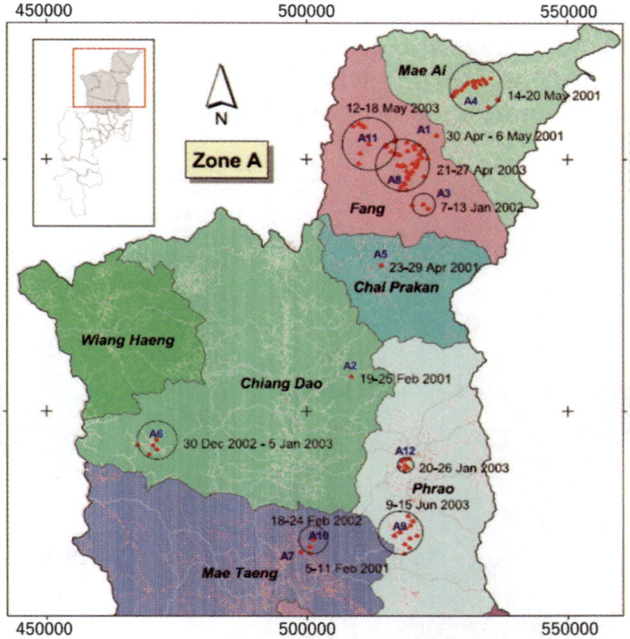


**Figure 9** Diarrhoea outbreaks in zone A

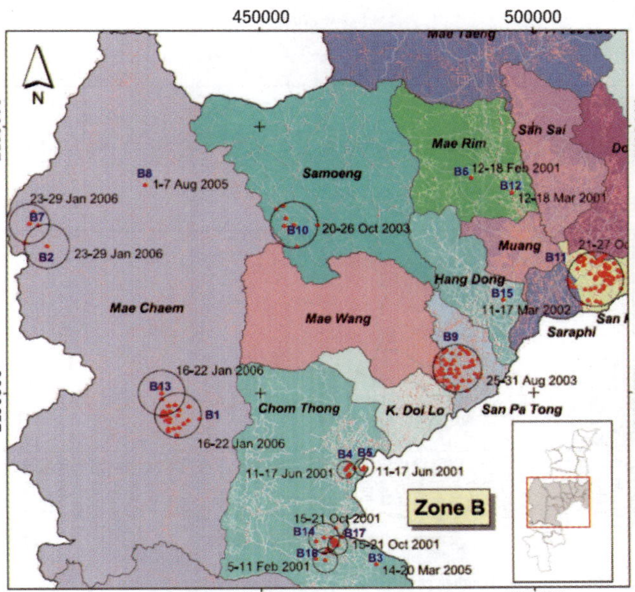


**Figure 10** Diarrhoea outbreaks in zone B

during the seven days when 0.42 case was expected (relative risk = 76.45). Most of the signals were distributed around the highland in Om Koi district, prevalent from the end of the winter season (January–February) to the beginning of the summer season (March–April) (Figures 11 and 12).

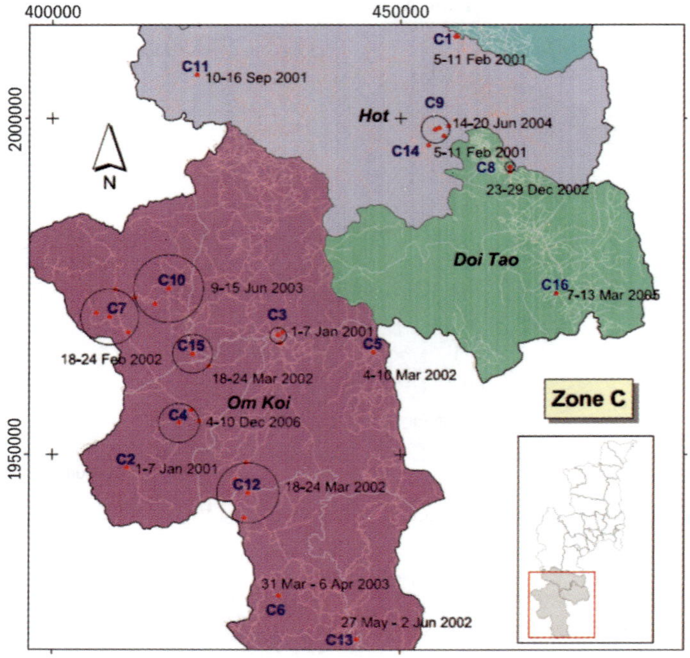


**Figure 11** Diarrhoea outbreaks in zone C

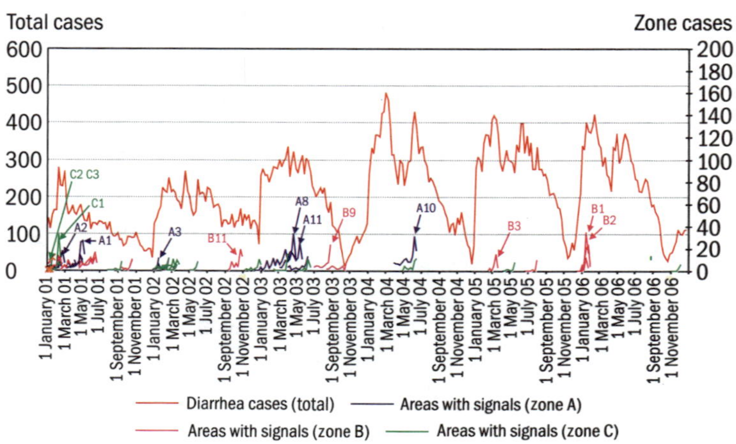


**Figure 12** The weekly temporal pattern of emergency diarrhoea cases in Chiang Mai province, 1 January 2001 to 31 December 2006

## MALARIA PREVALENCE IN KANCHANABURI, THAILAND

The malaria control programme in Thailand was successful in decreasing the extent of malaria disease over the past decade. The mortality rate reduced from 1.38 per 100 000 people in 1996 to 1.26 per 100 000 population in 1997, and morbidity rate increased from 1.78 per 1000 people in 1997 to 2.21 per 1000 population in 1998 (MoPH 2005). The malarial parasite species *Plasmodium falciparum* and *Plasmodium vivax* in 1997 were found in the proportion of 50.7% and 48.7%, respectively, with the rest of the percentage being attributed to *Plasmodium malariae* and mixed infection. *Plasmodium ovale* was not found during this year. Total cases reported in 1997 were 99 679, which increased to 125 013 in 1998, reporting an increase of 25.4%.

In 1997, malaria was found to be distributed along bordering provinces. The highest number of cases, 85 995, or 86.3% of the total malaria cases, was found at Thai–Myanmar border (10 provinces, 58 439 cases, or 58.6% of the total malaria cases). From 1993 to 1997, foreign cases were found to range between 51 090 and 66 622, and in 1997, annual parasite index (API) rate decreased by 14.8%. The provinces of Tak, Mae Hong Sorn, Kanchanaburi, Yala, Trad, Suratthani, Ranong, Krabi, and Chanthaburi had the highest local transmitted cases in 1998. The above-mentioned provinces also had the highest incidence rate, reporting 82 120 cases, or 65.69% of the total malaria cases in Thailand, recording an increase from 75 357 cases, representing 7.77% of the total malaria cases in 1997. From 1994 to 1998, highest numbers of cases were found in Tak province of Thailand, especially Mae Sot district. In 1998, incidences were found to be 61.03 per 1000 population in the Tak province and 53.85 per 1000 in the Mae Sot district. The main vectors ranged from *Anopheles minimus* and *Anopheles dirus* to infected *P. falciparum* type. The transmission of malaria in Thailand occurs in the beginning of the rainy season (May–June), and the second peak occurs in the beginning of the cold season (November–December). However, the problems due to exophillic behaviour and parasite resistance to anti-malaria drugs were certainly important constraints in achieving the control of malaria. The study was taken up with multiple objectives to characterize malaria risk levels in relation to physical and climatic factors and develop a model based on the above knowledge to prepare a malaria-risk map.

### Study Area

Kanchanaburi province is located in the western Thailand, with geographical location being 14°2′ north and 99°32′ east (Figure 13). It covers an area of 19 486 km$^2$, most of which comprises forested mountains. Northern boundary is connected with Tak, Uthai Thani, and Chainat provinces, while the southern part is allied with Ratchaburi and Nakhon Pathom provinces.

The eastern boundary is connected with Suphanburi province, and the western part is aligned with the Union of Myanmar (Burma).

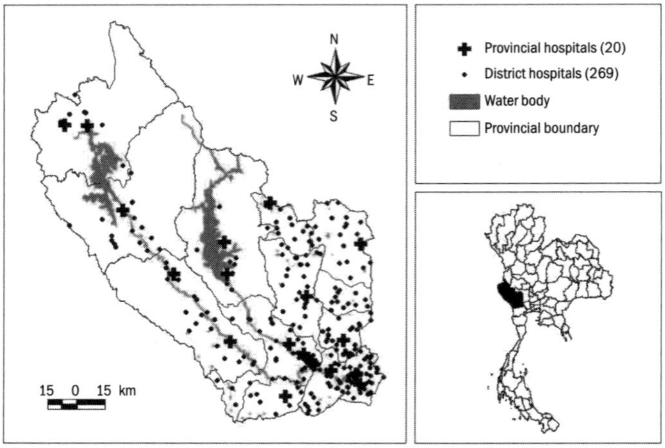


**Figure 13** Study area

## Seasonal Characteristics of Malaria

The climate pattern of Thailand can be categorized into three seasons. The cold season is from October to February, hot season from February to May, and rainy season ranges from May to October (Keola and Tokunaga 2001). Observations on the epidemiological characteristics of diseases were divided into these three seasons. Data during 1999–2005, provided by the Provincial Health Care Department, Kanchanaburi, was adopted. The distribution of malaria disease during seven years (1999–2005) is shown in Figure 14. The Figure shows that malaria incidence cases were highest during the rainy season (May–September). The same trend of high incidence is observed every year during the rainy season.

## Climate Parameters Influencing Malaria Incidence

The cause of malaria incidence is directly related to rainfall. Table 4 demonstrates that during the rainy season, rainfall and humidity values are high along with the number of malaria cases.

Table 4 shows that the rainfall and humidity values are the highest in September compared to other months, and malaria incidence was reported to be the highest in June compared to other months. Analysis indicated that malaria cases rise during the beginning of the rainy season and decrease as the rainy season intensifies.

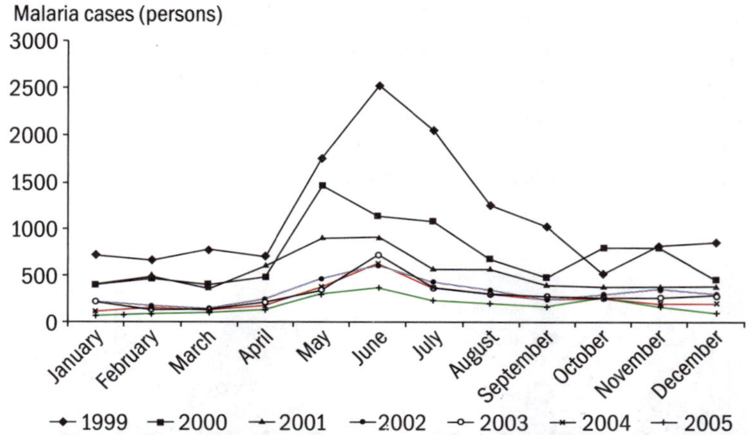


**Figure 14** The distribution of malaria disease during the period 1999–2005 The malaria distribution in the entire province, reporting highest incidences in the rainy season, has similar characteristic for each year from 1999 to 2005.

**Table 4** Summary of monthly average of rainfall, temperature, humidity, and malaria cases during seven years, 1999–2005

| Description | Jan | Feb | Mar | Apr | May | Jun | July | Aug | Sep | Oct | Nov | Dec |
|---|---|---|---|---|---|---|---|---|---|---|---|---|
| Rainfall (mm.) | 5.3 | 12.4 | 41.5 | 87.3 | 134.7 | 102.1 | 132.6 | 105.4 | 139.4 | 127.9 | 24.8 | 4.4 |
| Temperature(°C) | 25.9 | 27.4 | 29.3 | 30.4 | 28.7 | 28.0 | 27.9 | 27.6 | 27.5 | 27.2 | 26.1 | 24.8 |
| Humidity (%) | 69.3 | 65.2 | 63.6 | 66.6 | 77.5 | 78.7 | 79.1 | 79.4 | 81.6 | 81.4 | 75.6 | 70.3 |
| Malaria cases | 313 | 314 | 292 | 371 | 805 | 995 | 733 | 527 | 416 | 401 | 436 | 375 |

**Source** Report of Meteorological Department of Thailand in seven years (1999–2005). Vector Borne Disease Control Center 4.1, 2004. Annual Epidemiological Surveillance Report 2004: Kanchanaburi province. Vector Borne Disease Control Center 4.1, Kanchanaburi, Thailand

## Physical Parameters Influencing Malaria Incidence

Physical environmental factors were analysed in order to identify the area and percentage of the land cover types within 1, 2, and 5 km buffer on malaria risk area, considering the locations of each village (768 villages) (Figure 15). It can be concluded that malaria incidence case was reported to be the highest in the agricultural area, comprising approximately 6204.4 km² (61% of total area), and generally, the types of agriculture area are sugar cane area, transplanted paddy field, cassava area, corn area, and so on. The rest of the area comprises forest area of about 11 697.0 km² (32% of the total area), urban and built-up area, and waterbody. Elevation data was derived using spatial analysis tools and information value approach so as to create elevation area over 13 districts. The results show that the malaria-risk maps are classified into three categories, including low, moderate, and high, depending on the elevation levels. The highest malaria risk area was covered under the elevation area between 100 m and 300 m, followed by elevation area between 300 m and 500 m, between 500 m and 700 m, and elevation area between 700 m and 900 m, and 900 m and 1100 m.

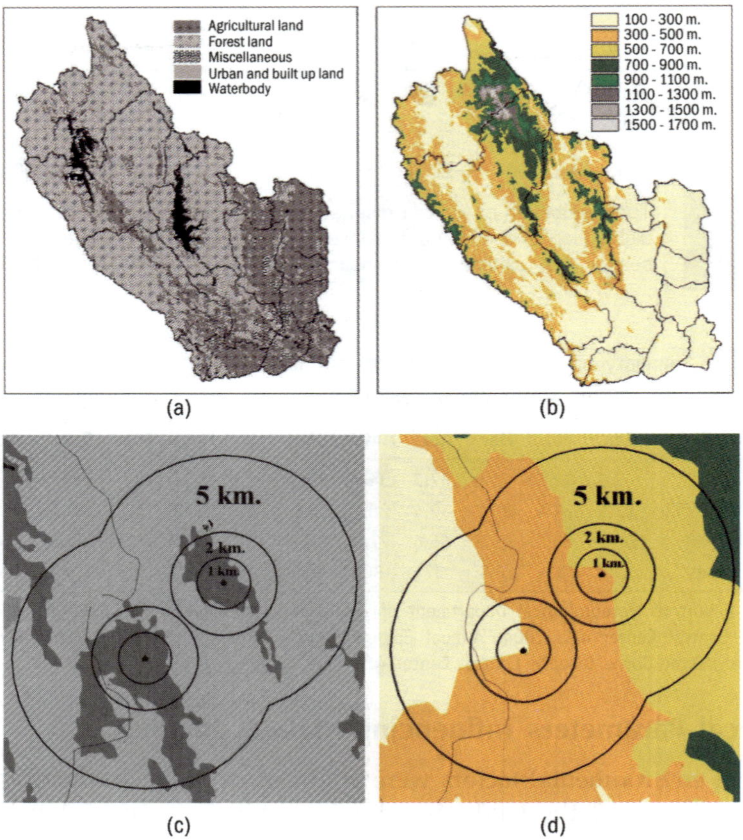


**Figure 15** (a) and (c) are the land use types and buffering of 1, 2, and 5 km, (b) and (d) are the elevation classification and buffering of 1, 2, and 5 km.

## MODEL DEVELOPMENT—INFORMATION VALUE APPROACH

In this study, the parameter values required to calculate information value are obtained by crossing malaria map with a certain parameter map. Map crossing results in a cross-table, showing the number of pixels per class occupied by malaria and total number of pixels in each class. The remaining values, necessary to calculate the information value, are obtained from these values, which actually can be represented in terms of information value process; therefore

|  |  |  |  |
|---|---|---|---|
|  | i-value | = log (classden/mapden) | (1) |
| where | classden | = nmclass/nclass |  |
|  | mapden | = sum (nmclass)/sum (nclass) = nmmap/nmap |  |
| in which | nmclass | = number of pixels with malaria in a class |  |
|  | nclass | = number of pixels in the class |  |
|  | nmmap | = total number of malaria pixels in the map (thematic layer) |  |
|  | nmap | = total number of pixels in the map |  |

This study used the observation data from Landsat satellite for March and December 2003. The statistical model is developed based on the information value derived for each factor by using information value method given by Yin and Yan (1988), as provided in Prakash (1998).

## PROBABILITY INTERPRETATION FOR MALARIA APPEARANCE

Malaria incidence could be confirmed about the accuracy and the probability of appearance with the help of factor prediction. This study applied the relation testing between the number of pixel with malaria cases in class (nmclass) and information value (i-value), to be considered with "correlation statistic". In this chapter, evidence is classified into two main factors: climatic factor (rainfall classes) and physical environmental factor (land use classes).

### Rainfall Classes

Rainfall is the main climatic factor determining malaria occurrence. The total area under average annual rainfall for the entire province is about 19 396.5 km². Results of the average annual rainfall (1999–2005) classification in the study area are shown in Table 5.

**Table 5** Information value for average annual rainfall, 1999–2005

| Rainfall | nmclass | nclass | nmmap | nmap | classden | mapden | i-value | Area (km²) |
|---|---|---|---|---|---|---|---|---|
| 400–900 | 442 203 | 3 283 068 | 1 058 628 | 7 758 600 | 0.1347 | 0.1364 | −0.0056 | 8207.8 |
| 900–1500 | 591 412 | 3 001 097 | 1 058 628 | 7 758 600 | 0.1971 | 0.1364 | 0.1597 | 7502.6 |
| 1500–2000 | 25 013 | 1 474 435 | 1 058 628 | 7 758 600 | 0.0170 | 0.1364 | −0.9054 | 3686.1 |
| Sum | 1 058 628 | 7 758 600 |  |  |  |  |  | 19 396.5 |

The class having average annual rainfall (1999–2005) of 900–1500 mm has the highest possibility of having malaria occurrence, with 591 412 pixel image, that is, an area of about 7502.6 km², which is 38.68% of the total average annual rainfall formation area. The second class is 400–900 mm, with an area of about 8207.8 km² (42.32%) and malaria occurrence image having 442 203 pixels. The last class is 1500–2000 mm, with 25 013 pixels and an area of about 3686.1 km² (19%).

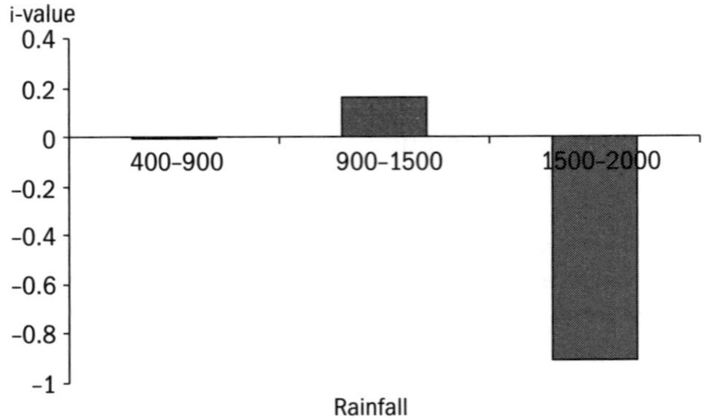


Figure 16 Information value of average annual rainfall, 1999–2005

As is clear from Figure 16, positive values were reported for average annual rainfall ranging from 900 m to 1500 mm, but negative values have been reported for average annual rainfall in the range 400–900 mm and more than 1500 mm.

## Physical Environmental Factors

### *Land Use*

Land use plays an important role in malaria occurrence. Table 6 presents the results of spatial statistical relationship of various land use classes and i-value. The highest i-value for malaria occurrence is found in urban and built-up land with an area of about 447.6 km² (76 298 pixels, 2% of the total land use class). The second class is the agriculture area having an area of about 6204.4 km² (670 521 pixels, 32% of the total land use class), and the third class is the forest area having an area of about 11 697.0 km² (44 005 pixels, 61% of the total land use class).

Figure 17 presents a positive relationship among agriculture, forest, and urban and built-up areas. The negative values were shown in the miscellaneous and water areas only.

**Table 6** Information value for physical environmental factor (land use classes)

| Land use | nmclass | nclass | nmmap | nmap | classden | mapden | i-value | Area (km²) |
|---|---|---|---|---|---|---|---|---|
| Agricultural land | 670 521 | 2 485 523 | 1 059 133 | 7 762 239 | 0.26977 | 0.1364 | 0.2960 | 6204.4 |
| Forestland | 44 005 | 174 846 | 1 059 133 | 7 762 239 | 0.05397 | 0.1364 | 0.2659 | 11 697.0 |
| Miscellaneous | 252 280 | 4 674 851 | 1 059 133 | 7 762 239 | 0.25168 | 0.1364 | −0.4028 | 437.3 |
| Urban and built-up land | 76 298 | 179 125 | 1 059 133 | 7 762 239 | 0.42595 | 0.1364 | 0.4944 | 447.6 |
| Waterbody | 16 029 | 247 894 | 1 059 133 | 7 762 239 | 0.06466 | 0.1364 | −0.3243 | 619.7 |
| Sum | 1 059 133 | 7 762 239 | | | | | | 19 406.0 |

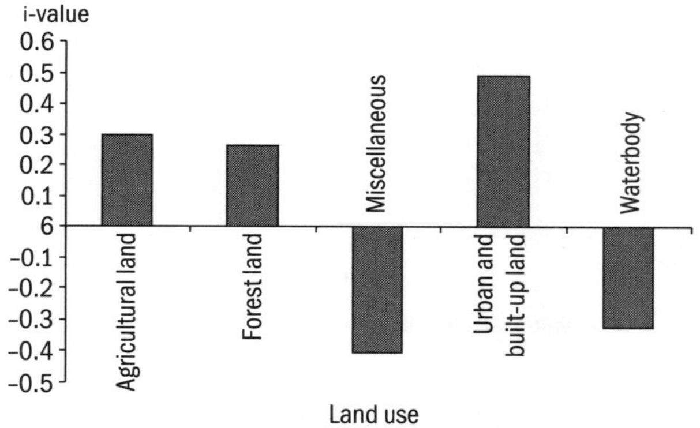


**Figure 17** Information value of land use

## *Elevation*

Elevation class is the one of the main factors in the occurrence of malaria. For this study, five elevation classes have been used, as shown in Table 7. Total elevation area is about 19 073.70 km² (100%, 1 795 089 pixels). It is observed that an elevation between 100 m and 300 m has the highest information value of malaria incidence occurrences, which is about 10 757.99 km² (94.13%, 1 689 757 pixels). The second class, having an elevation between 300 m and 500 m, has an area of about 3289.71 km² (4.15%, 74 481 pixels), and the last class, having an elevation between 900 m and 1100 m, has an area of about 510.95 km² (0.04%, 723 pixels).

In Figure 18, a positive value of elevation is witnessed in elevation class 100–300 m only. A statistical approach has been used to find out the accuracy in terms of the correlation between information value of the occurrence of malaria and the number of pixels of satellite image that have malaria

**Table 7** Information value for physical environmental factor (elevation classes)

| Elevation (m) | nmclass | nclass | nmmap | nmap | classden | mapden | i-value | Area (km²) |
|---|---|---|---|---|---|---|---|---|
| 100–300 | 1 689 757 | 4 303 307 | 1 795 089 | 7 629 498 | 0.3927 | 0.2353 | 0.2224 | 10757.99 |
| 300–500 | 74 481 | 1 315 907 | 1 795 089 | 7 629 498 | 0.0566 | 0.2353 | -0.6188 | 3289.71 |
| 500–700 | 19 892 | 1 102 893 | 1 795 089 | 7 629 498 | 0.0180 | 0.2353 | -1.1154 | 2757.36 |
| 700–900 | 10 236 | 703 004 | 1 795 089 | 7 629 498 | 0.0146 | 0.2353 | -1.2084 | 1757.69 |
| 900–1100 | 723 | 204 387 | 1 795 089 | 7 629 498 | 0.0035 | 0.2353 | -1.8229 | 510.95 |
| Sum | 1 795 089 | 7 629 498 | | | | | | 19 073.7 |

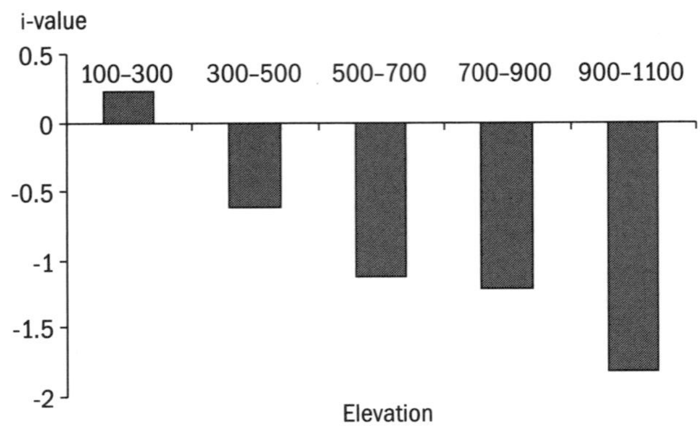


**Figure 18** Information value of elevation (m)

occurrences, and then classify these incidences to be the circumstances of malaria risk area. Table 8 presents the classification theme for malaria risk area in terms of rainfall, land use, and elevation.

**Table 8** Accuracy for malaria risk area

| Risk | Rainfall | | Land use | | Elevation | |
|---|---|---|---|---|---|---|
| | T | R | T | R | T | R |
| Low | – | – | – | – | 2 010 284 | 30 851 |
| Moderate | 1 474 435 | 25 013 | 4 922 745 | 268 309 | 1 315 907 | 74 481 |
| High | 6 284 165 | 1 033 165 | 2 839 494 | 790 824 | 4 303 307 | 1 689 757 |

T-Total number of pixels; R-Pixels having malaria risk

## Malaria Incidence Classification Area

The malaria risk area has been classified into three classes, namely, low, moderate, and high. The three factors that appear in all these classes are rainfall, land use, and elevation (Table 9).

**Table 9** Image type and malaria class area

| Risk (Information Value Class) | | | |
|---|---|---|---|
| **Type** | **Low (<–1.1)** | **Moderate (–1.1 to –0.3)** | **High (–0.3 to 0.5)** |
| Rainfall | – | 3686.10 (19%) | 15 710.4 (81%) |
| Land use | – | 1056.95 (5.45%) | 18 349.1 (94.55%) |
| Elevation | 5026.00 (26.35%) | 3289.71 (17.25%) | 10 757.99 (56.4%) |

Table 10 gives the location of malaria incidence.

**Table 10** Summary of malaria incidence

| | **Location and Situation** | | |
|---|---|---|---|
| **Type** | **Low** | **Moderate** | **High** |
| Rainfall | – | 1500–2000 mm | 400–1500 mm |
| Elevation | 500–1100 m | 300–500 m | 100–300 m |
| Land use | | • Miscellaneous (wetland, bare land, and so on)<br>• Waterbody (river, reservoir) | • Agricultural area<br>• Forest area<br>• Urban and built-up area |
| Number of villages affected | 3 (0.57%) | 13 (2.48%) | 508 (96.95%) |

It can be concluded that climatic and physical environmental factors are very important, and malaria occurrence depends on changes in physical environment such as elevation, waterbody, forest, and agriculture, and climate changes such as rainfall, humidity, and temperature. Malaria risk zone map is presented in Figure 19.

Table 10 shows the summary of malaria risk zone and malaria-affected villages. It can be concluded that the highest malaria risk zone covered 508 villages (96.95%), moderate malaria risk zone was composed of 13 villages (2.48%), and low malaria risk zone comprised 3 villages (0.57%). In conclusion, the total malaria-affected villages were 524.

## CONCLUSION

The analysis of disease patterns in terms of seasonal and population patterns reveals that the outbreak of foodborne and waterborne diseases (diarrhoea

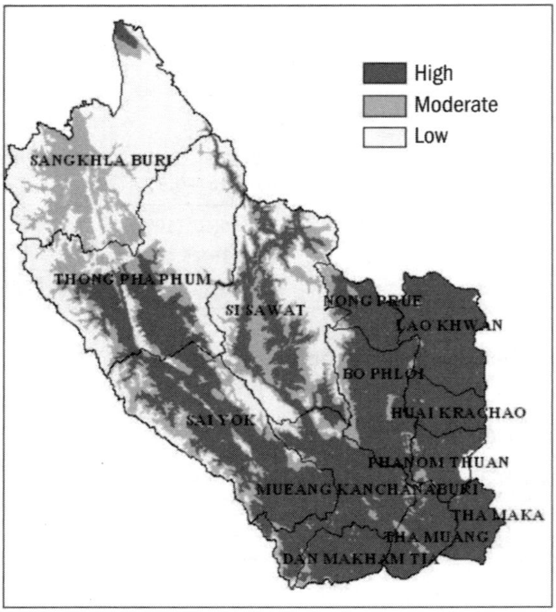


**Figure 19** Malaria risk zone map

and food poisoning) occurred during the period of hot and rainy seasons and was mainly manifested in the children under the age of five years. This confirmed that the foodborne and waterborne diseases could be epidemic in many climate change scenarios, and these diseases were reported to be high in children (FAO 2004).

Using space–time permutation scan statistics in GIS, we explored the outbreaks of diseases. The first part of the study shows that statistical methods and tools can be useful for the surveillance for public health officials. It also demonstrates that using existing health data, space–time analysis and GIS can provide opportunity to specify the health burden within the infected areas. To implement specific and geographically appropriate risk reduction programmes for public health officers, the use of such space–time analysis and tools should become an integral component in the epidemiologic description and risk assessment. In the second part of the study, malaria risk zonation maps were developed by statistically establishing the relationship of various physical environmental and climatic factors by using remotely sensed data and GIS. To identify the statistical correlations between cases and climatic factors, the information value method was applied. The risk maps obtained are able to support public health officers in space and time so as to control and predict malaria spread over extensive areas. Moreover, the risk maps can be beneficial for warning public and generating awareness.

The study approach is to apply GIS and remote sensing for health planning. The results retrieved from the model can be applied along with other epidemiological information so as to predict the spread of the current epidemic and in order to indicate where the resources might be best mobilized. Also, the approaches could be applied for the mapping of high-risk areas of vector-borne disease and distribution is dependent on climatic and environmental factors. Therefore, the use of GIS in these applications is very important in decision support system in order to enhance the awareness and warn people. Public health officers can employ the model contributed from this study to control the spread of malaria through the factors mentioned above and predicted scenario of rainfall. Not only is the model applicable in health issues, but such models also can be used in different predicted climate change patterns for better socio-physical disaster management, health care, and so on.

## References

Bureau of Epidemiology. 2004. **Malaria**. Department of Disease Control, Ministry of Public Health, Thailand. Details available at <http://epid.moph.go.th/dssur/vbd/malaria.htm>

Keola S and Tokunaga M. 2001. **Provincial risk maps for highest tendency ranking epidemiological surveillance diseases in Ayutthaya province, Thailand**. Centre for Remote Imagine, Sensing and Processing (CRISP). Proceedings of 22nd Asian Conference on Remote Sensing

Kulldorff M. 1997. **A spatial scan statistic**. *Communications in Statistics: theory and methods* **26**: 1481–1496

Kulldorff M, Fang Z, and Walsh S J. 2003. **A tree-based scan statistic for database disease surveillance**. *Biometrics* **59**(2): 323–331

Kulldorff M, Zhang Z, Hartman J, Heffernan R, Huang L, Mostashari F. 2004. **Benchmark data and power calculations for evaluating disease outbreak detection methods**. *Morbidity and Mortality Weekly Report* **53**:144–151

MoPH (Ministry of Public Health). 2005. **Malaria Control Program in Thailand.** August 2005. Details available at <http://eng.moph.go.th/SpecificHealth/malaria/malaria.htm>

Prakash C. 1998. **Landslide hazard zonation in Doon valley using remote sensing and GIS**. Masters Thesis. Kanpur, Indian Institute of Technology, India

Richard S and Martin B. 2000. ***Nutrition and Health in Developing Countries.*** Totowa, NJ: Humana Press

Yin K L and Yan T Z. 1988. **Statistical prediction model for slope instability of metamorphosed rocks**. In *Proceedings of Fifth International Symposium on Landslides, Lausanne*, pp. 1269–1272, vol. 2, edited by C Bonnard. Rottterdam: A.A. Balkema

# Geo-information-based Approach for Monitoring Climate-induced Land Degradation in Nigeria

**Ayobami T Salami*** **and Kayode A Adepoju**†

*Institute of Ecology and Environmental Studies, Obafemi Awolowo University, Ile-Ife, Nigeria

†African Regional Centre for Space Science and Technology Education English, Obafemi Awolowo University Campus, Ile-Ife, Nigeria

E-mail: ayobasalami@yahoo.com; asalami@oauife.edu.ng

## INTRODUCTION

Land degradation due to climate change or anthropogenic factors has been recognized as one of the major environmental problems plaguing the world today. Land degradation results in reduced biological productivity and is, therefore, a major source of poverty in developing countries, where most of the population ekes a living out of land. UNCED (1992), for instance, estimated that over 250 million people are directly affected by desertification, while about one billion people in over 100 countries are at risk. In order to combat land degradation, there is a need to develop spatial monitoring systems that could provide a framework for informed policy formulation at global, national, and local levels. The choice of appropriate indicator(s) of land degradation is vital in addressing this problem. Legg, Kormawa, Maziya-Dixon, *et al.* (2005) noted that the main challenge in identifying indicators is to select those that are sufficiently representative and, at the same time, easy to understand and measure on a routine basis. An indicator may be a quantitative or a qualitative factor or a variable that provides a simple and reliable basis for assessing achievement, change or performance. A key land quality indicator generally agreed upon is the net primary productivity (NPP), since vegetation cover is considered perhaps the single most important biophysical indicator for land degradation and desertification. NPP measures overall land productivity and ecosystem health and also provides some indication of land degradation and soil productivity (FAO 2000). NPP is defined as one of the primary processes describing the vegetation activity in terms of mass and energy exchanges between the earth's surface and atmosphere. Turner, Ritts, and Cohen (2006) noted that it is the rate of net production of organic matter (the total green biomass produced by all the plants) by the terrestrial vegetation and represents the efficiency with which the carbon dioxide is absorbed by it.

The two principal sources of data on land degradation are the global estimates of desertification by Dregne and Chou (1994) and estimates of land degradation by the International Soil Reference and Information Centre (Oldeman, Hakkeling, and Sombroek 1991). The major difference between the two approaches is the status of vegetation. There is also a difference in the terminology used to express the severity of land degradation. Dregne and Chou (1994) used the terms like slight, moderate, severe, and very severe to designate the severity of degradation, while Oldeman, Hakkeling, and Sombroek (1991) used the terms like light, moderate, strong, and extreme, and these terms may not be comparable.

Early efforts on the assessment of land degradation were mainly based on expert opinions. For instance, Oldeman, Hakkeling, and Sombroek (1991), on the basis of expert judgment, attempted to differentiate natural degradation from human-induced degradation. However, some recent fundamental advances in methods and technologies, especially, space technology, have tremendous applications for land degradation assessment. Biophysical indicators of land degradation can now be derived using remote sensing, and this technique has been recognized as a viable method for assessing land degradation and desertification by Food and Agricultural Organization (FAO 2000). Ustin, Jacquemoud, Palacios-Oreuta, *et al.* (2005), for example, present a review of spectral characteristics of plants and soils that are detectable using optical sensors and methods to identify and quantify properties that have potential for monitoring arid ecosystem processes. Also, changes in NPP may be measured by remote sensing of vegetation indices to identify hot spots as well as bright spots, while its deviation from the local norm may be taken as a measure of land degradation or improvement (Bai and Dent 2006). Scientists have shown that satellite data can be used in the development of efficiency models to estimate NPP at global, regional or national scales (Running, Heinsch, Zhao, *et al.* 2004; Singh, Meirelles, and Costa 2006). These models are based on the concept of light-use efficiency, and the scientists use the strong linear relationship between Normalized Differential Vegetation Index (NDVI) and the fraction of photosynthetically active radiation (PAR) absorbed by the plant (fPAR) to set the upper limit for unstressed NPP. NDVI (the greenness indicator) measured by satellite-borne instruments may, therefore, be used as a proxy for NPP.

In Nigeria, conventional methods of land degradation assessment and monitoring are more often localized to specific regions and exist in a disaggregated format, often inaccurate, time consuming, and difficult to obtain in near-real time. In addition to this, many dynamic geophysical processes occurring in remote and ecologically sensitive locations are difficult to access and survey through traditional means. In contrast, remote sensing offers alternative techniques, often more efficient and cost-effective, for data capture and management in a very timely manner, providing crucial inputs for carrying out operationally viable strategies. This study was, therefore, based on satellite measurements of the NDVI, or greenness index, for the period 1982–2006 as a proxy for NPP. This approach demonstrates a great promise for improving the existing quality and quantity of information on degradation trends in Nigeria as a basis for early warning system or disaster management. The study serves as the first step towards the development of a national framework for future assessment and monitoring of land degradation in Nigeria.

## STUDY AREA

Nigeria covers an area of 923 768 km$^2$ and is located within longitudes 3°E and 14°E of the Greenwich Meridian and latitudes 4°N and 14°N of the Equator. About 98.6% of the total area is land, while the rest is water. It is the most populous African country with a total population of 140 million (in 2006). About 70 million ha of the land in the country is considered to be suitable for arable cultivation, but only about 50% of this area is currently under cultivation. The country is characterized by a strong climatological gradient north to south, with definitive dry and wet seasons. It is this gradient that defines Nigeria's ecological zones.

In terms of administrative arrangement, Nigeria comprises 36 states and a Federal Capital Territory (FCT), which is grouped into six geo-political zones (Figure 1).

Human activities have exacerbated the ecological problems in Nigeria, while the institutional capacity for combating the problems remains weak. For instance, land-use change due to rapid urbanization, deforestation, desertification, agricultural practices, and other reasons affects the physical and biological properties of the land surface in the country. The population of the country is expected to exceed 200 million in 2100, based on the current growth rate. To feed the burgeoning population, agriculture and livestock grazing have expanded correspondingly, often into marginal areas that are not ecologically suitable for such activities. Since the 1900s, Nigeria has seen the disruption of farming systems, degradation of natural resources, and a decline in the production of many staples against the backdrop of mounting pressure on natural resources from its growing population, now increasing at about 2.5% annually (World Bank 2005).

Nigeria enjoys a warm tropical climate condition with relatively high temperatures throughout the year, marked by the dry and wet seasons. The

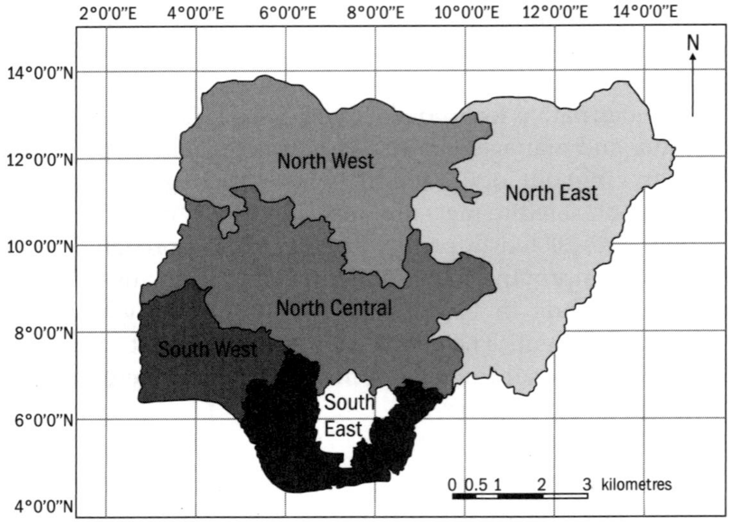


**Figure 1** Geo-political zones in Nigeria

climate of the country is influenced by the interactions of the two major air masses: the relatively warm and moist tropical maritime air mass (mT), which originates from the Atlantic Ocean and is associated with South West Trade winds, and the relatively cool, dry, and stable continental air mass (cT), which originates from Sahara Desert and is associated with the dry, cool, and dusty North East Trade winds. The boundary surface area between the two air masses is known as Inter-tropical Discontinuity (ITD), or the Inter-tropical Convergence Zone (ITCZ). The ITD migrates north and south of the country, bringing rainfall or causing dryness in different areas of the country at different times of the year. Similarly, annual rainfall ranges from over 2500 mm along the coast in the south to less than 400 mm in parts of the extreme north.

## DATA COLLECTION

This study utilized NDVI data from January 1982 to December 2006, generated by the Global Inventory Modelling and Mapping Studies (GIMMS) group from measurements made by the advanced, very high resolution radiometer (AVHRR) on board US National Oceanic and Atmospheric Administration (NOAA) satellites. The fortnightly images at 8 km spatial resolution were corrected for calibration, view geometry, volcanic aerosols, and other effects not related to the vegetation cover (Tucker, Pinzon, and Brown 2005). This data is compatible with the data from other sensors such as MODIS, SPOT, and Landsat ETM+ (Tucker, Pinzon, and Brown 2005; Brown, Pinzon, Didan, *et al.* 2006). Rainfall data for 36 stations was collected from the Nigeria Meteorological Agency for the period between 1982 and 2006 and was used for the analysis. The data was supplemented by the Global Precipitation Climatology Centre (GPCC) full re-analysis product (Schneider, Fuchs, Meyer-Christoffer, *et al.* 2008) to generate rainfall values matching the GIMMS NDVI data.

The global soil and terrain (SOTER) database comprises harmonized spatial and soil attributes data, incorporating the 90-m-resolution Shuttle Radar Topography Mission (SRTM) digital elevation model (Engelen, Batjes, Dijkshoorn, *et al.* 2005). For this study, a consistent data set of key soil attributes was prepared at scale 1:1M (Engelen, Dijkshoorn, and Huting 2008). With respect to land cover data, global land cover data (GLC 2000; Figure 2) was generalized for Nigeria for preliminary comparison with NDVI trends. The CIESIN Global Rural–Urban Mapping, which provides data for population and urban extent, gridded at 30 arc-second resolution, was used for this study (CIESIN 2004).

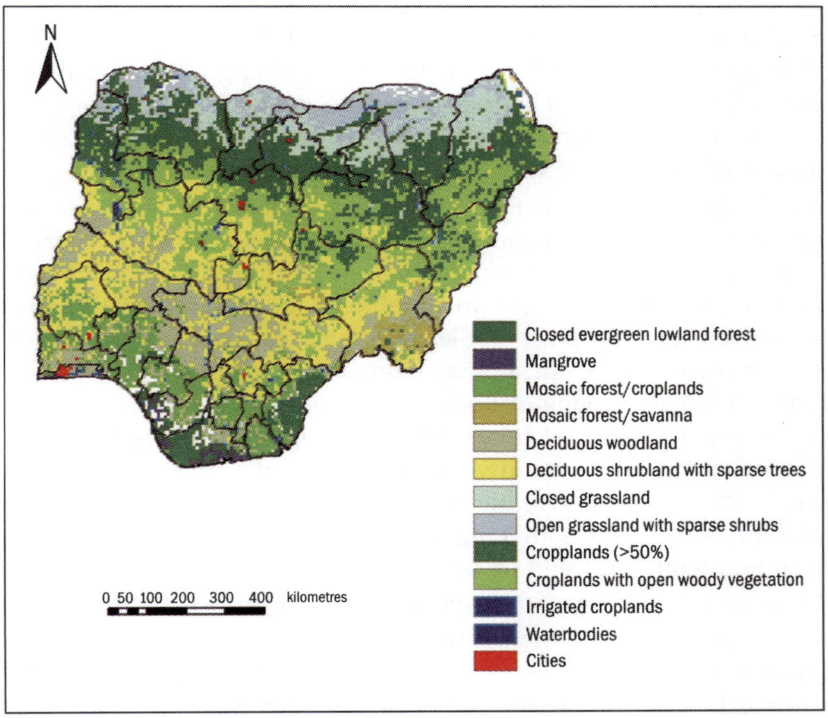


**Figure 2** Map of Nigeria showing main land cover types
**Source** GLC (2000)

## DATA ANALYSIS

Areas of land degradation and improvement were identified by a sequence of analyses of the remotely sensed data. Simple NDVI indicators (NDVI minimum, maximum, maximum–minimum, mean, sum, standard deviation [STD], and coefficient of variation [CoV]) were computed for the calendar years. The annual sum NDVI, the annual aggregate of greenness, was chosen as the standard proxy for annual biomass productivity. To distinguish between declining productivity caused by land degradation and declining productivity caused by other factors, false alarms were eliminated. Rainfall variability and irrigation were accounted for by identifying areas where there is a positive relationship between NDVI and rainfall, that is, where rainfall determines productivity. For those areas where rainfall determines productivity, rain-use efficiency (RUE) was considered; for the areas where NDVI declined but RUE increased, the declining productivity was attributed to declining rainfall, and those areas were masked (urban areas were also masked). For the remaining areas with a positive relationship between NDVI and rainfall but declining RUE, and also for all areas where there is a negative relationship between NDVI and rainfall, that is, where rainfall does not determine productivity, NDVI trend was calculated as RUE-adjusted NDVI. Land degradation was indicated by a negative trend in RUE-adjusted NDVI. The indices of land degradation and improvement were compared with land cover, land use, population density (2006), and mean household income, which was used as an index of poverty.

ArcGIS Spatial Analyst, ERDAS IMAGINE, and ArcView were used to calculate NDVI minimum, maximum, maximum–minimum, mean, sum, STD, and CoV, as well as climate variables. The fortnightly NDVI data was geo-referenced and averaged to monthly data; annual NDVI indicators were derived for each pixel, while their temporal trends were determined by linear regression at an annual interval and mapped to depict spatial changes. A negative slope of linear regression indicates a decline in green biomass and a positive slope, indicating an increase (except for STD and CoV, which indicate trends in variability). Monthly grids of rainfall for the period 1982–2006 were geo-referenced and re-sampled to the same spatial resolution as the NDVI (8 km), using neighbourhood statistics. Spatial pattern and temporal trend of rainfall and RUE (the ratio of annual NDVI and annual rainfall) for each pixel were determined by regression.

The trend analysis assumes that the data is spatially and temporally independent. This was tested by examining autocorrelation coefficients following Livezy and Chen (1983). When the absolute values for the autocorrelation coefficients of lag-1 to lag-3, calculated for a time series consisting of n observations, are not larger than the typical critical value corresponding to 5% significance level, that is, 1.96/n, the observations in this time series can be accepted as being independent from each other. t-test was used to arrange the slope values in classes showing strong or weak positive or negative trends.

$$t = b/\text{se}(b) \quad (1)$$

where $b$ is the calculated slope of the regression line between the observation values and time, and se($b$) represents the standard error of $b$. The class boundaries were defined for 95% confidence level; trends were labelled *high* if the t-values of the slope exceeded the 0.025 p-value of either tail of the distribution; lesser t-values were labelled *low*. In addition, SPSS and MS Excel were employed to analyse trends, correlations, and significance of the non-gridded variables.

Minimum NDVI is defined as the lowest value that occurs in any one year (annual), which is usually at the end of the dry season. Variation in minimum NDVI may serve as a baseline for other parameters. Maximum or peak NDVI represents the maximum green biomass. The large spatial variations reflect the diverse landscapes and climate. Maximum–minimum NDVI describes the difference between annual maximum and minimum. NDVI reflects annual biomass productivity for areas with one, well-defined growing season but may not be meaningful for areas with bimodal rainfall. Sum NDVI is defined as the sum of the fortnightly NDVI values for the year, most nearly aggregate annual biomass productivity. NDVI STD was calculated as the root mean square deviation of the NDVI time series values (annual) from their arithmetic mean. It is a measure of statistical dispersion, assessing the spread of NDVI values. CoV can be used to compare the amount of variation in different sets of sample data. NDVI CoV images were generated by computing for each pixel the STD of the set of individual NDVI values and dividing this by the mean (M) of these values. This represents the dispersion of NDVI values relative to the mean value. The long-term trends of the indicators of biological productivity may be taken as the indicators of land degradation (where the trend is declining) or land improvement (where the trend is increasing). A positive change in the value of a pixel-level CoV over time relates to increased dispersion of values, not increasing NDVI; similarly, a negative CoV dispersion means decreasing dispersion of NDVI around mean values, not decreasing NDVI.

## RESULTS AND DISCUSSION

### Spatial Pattern of Biomass, Rainfall, and Rain-use Efficiency

Biomass productivity fluctuates according to rainfall cycles. Countrywide, greenness increased over the period 1982–2006 (Figure 3).

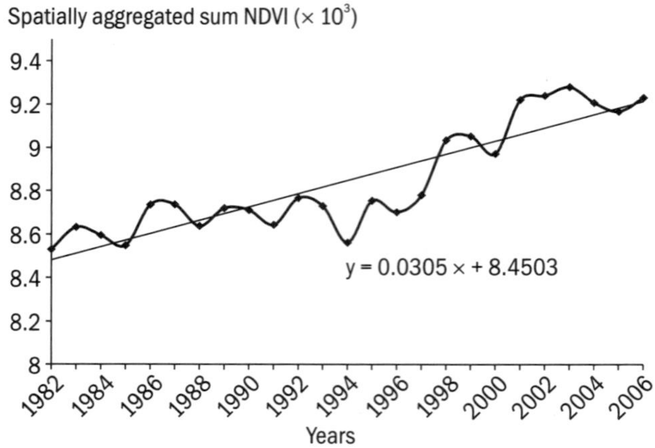


**Figure 3** Spatially aggregated annual sum NDVI (1982–2006, $p < 0.05$)

Biomass fluctuates according to rainfall, stage of growth, and changes in land use, as well as land quality. In Nigeria, mean annual biomass productivity (represented by sum NDVI in Figure 4) essentially follows rainfall, which has fluctuated significantly, both cyclically (Figure 5) and spatially (Figure 6). During the period 1982–2006, rainfall increased over nearly the entire country, and so did overall biomass production; but the correlation between spatially aggregated annual sum NDVI and annual rainfall is only moderate. It seems there are other factors at play.

Figure 7 shows mean annual RUE and its trend over the period 1982–2006: RUE is generally higher in the drylands than the humid areas, which generate significant run-off. Over the period, RUE decreased over 22% of the country and increased over 78% of the country. Confidence levels were assessed by the t-test.

Allowance may be made for the effects of fluctuations in rainfall on biomass productivity by considering RUE, that is, production per unit of rainfall. RUE may fluctuate significantly in the short term, and often there is a sharp decline in RUE in a wet year; and we may assume that the vegetation, whether cultivated or semi-natural, cannot make use of the additional rain. However, where rainfall is the main limiting factor on biomass productivity, we assume that the long-term trend of RUE is a good indicator of land degradation or improvement (Houérou 1989; Snyman 1998; O'Connor, Haines, and Snyman 2001). RUE also accommodates the effects of local variations in slope, soil, and vegetation (Justice, Dugdale, Townshend, *et al.* 1991). In North China and Kenya, Bai, Dent, and Schaepman (2005) and Bai, Dent, Olsson, *et al.* (2008) demonstrated that values for RUE calculated from NDVI, which are easy to obtain, were comparable with those calculated from

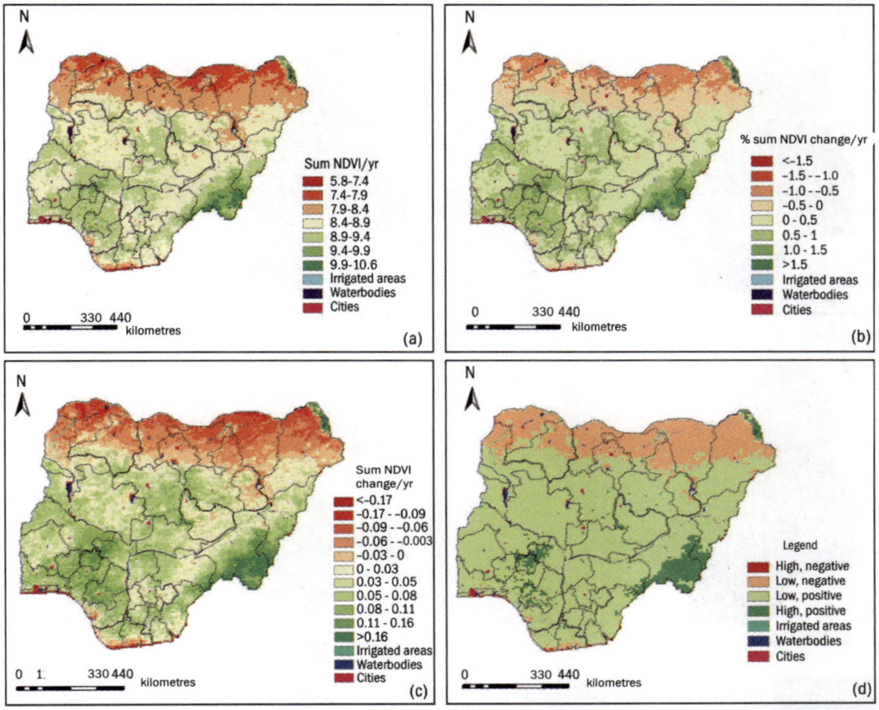


**Figure 4** Annual sum Normalized Differential Vegetation Index 1982–2006: (a) mean, and trends in (b) percentage, (c) absolute change, and (d) confidence level

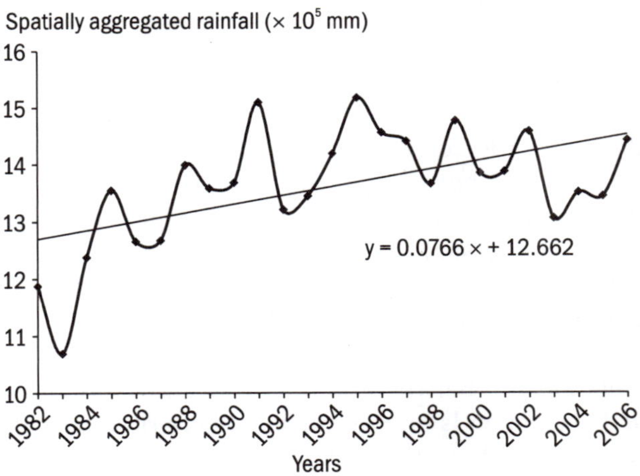


**Figure 5** Spatially aggregated annual sum rainfall (1982–2006, $p < 0.05$)

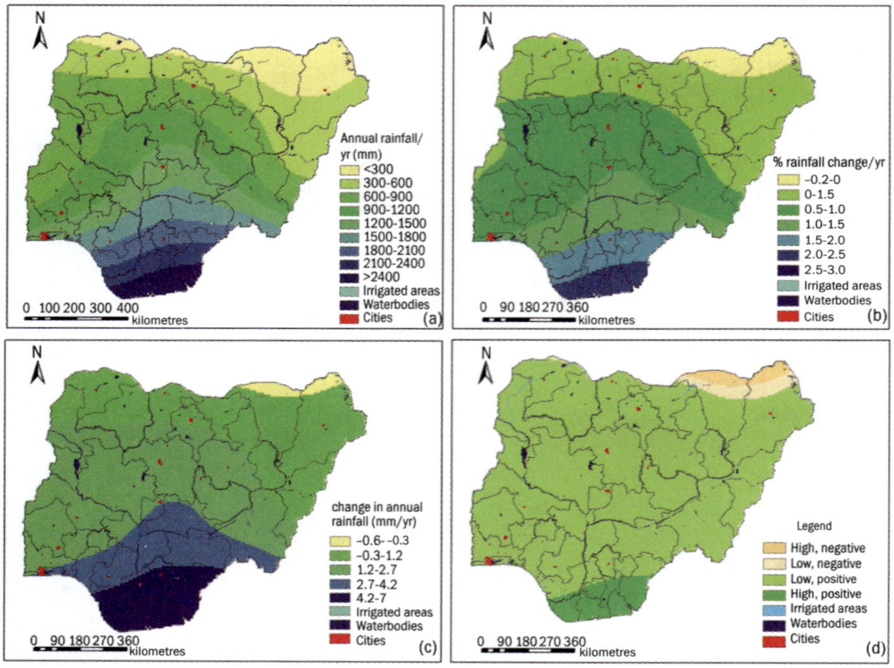


**Figure 6** Annual rainfall 1982–2006: multi-year (a) mean, and trends in (b) percentage change, (c) absolute change, and (d) confidence levels

field measurements of NPP, which are not easy to obtain. For this study, RUE was calculated as the ratio of annual sum NDVI and station-observed annual rainfall.

## Land Degradation

Land degradation in the context of this study means a loss of NDVI, but a decrease in NDVI is not necessarily land degradation. To distinguish between decline in productivity caused by land degradation and decline due to other factors, it is necessary to eliminate false alarms arising from climatic variability and changes in land use and management. Rainfall variability was accounted for using both RUE and RESTREND (Wessels, Prince, and Malherbe 2007). RUE was considered by first identifying pixels where there is a positive relationship between productivity and rainfall. For those areas where productivity depends on rainfall *and* where productivity declined but RUE increased, we attribute the decline of productivity to drought. Those areas were masked (and urban areas were also masked). NDVI trends are presented for the remaining parts of the country as RUE-adjusted NDVI.

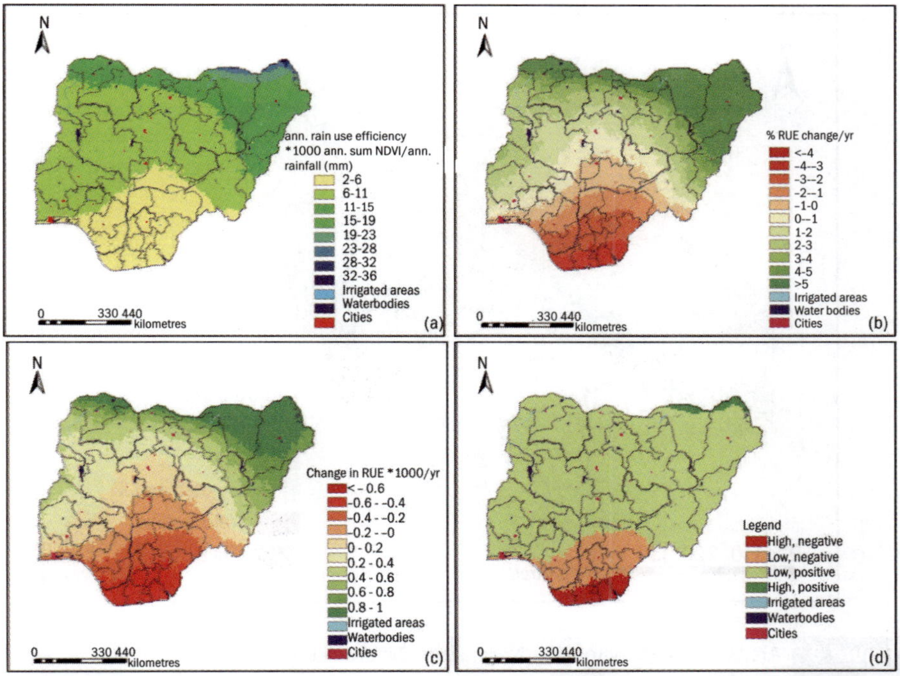


**Figure 7** Rain use efficiency 1982–2006: (a) mean, and trends in (b) percentage change, (c) absolute change, and (d) confidence levels

Figure 7b depicts the negative trend of RUE-adjusted NDVI over 1982–2006. About 18% of the country suffered declining RUE-adjusted NDVI, generally in the south of the country and especially in the south-east and south-south. As with rainfall variability, land use change may also generate false alarms about land degradation. Conversion of forest or grassland to cropland will usually result in an immediate reduction in NDVI (and NPP) but may well be profitable and sustainable, depending on management. Likewise, bush encroachment into cropland or rangeland will result in an increase in NDVI, but this would not be considered land improvement by the land users (Justice, Dugdale, Townshend *et al.* 1991). Lack of consistent time series data for land use and management precludes a generalized analysis of land use change. In this study, land improvement was identified by a combination of a positive trend in sum NDVI for those areas where there is no correlation between rainfall and NDVI, and for areas where NDVI is correlated with rainfall, and a positive trend is observed in RUE. These areas account for about 45% of the country. Most of the improving areas are found in the south-western, north-central, and part of north-eastern zones around Lake Chad. Figure 8 shows the gain in NPP in those areas.

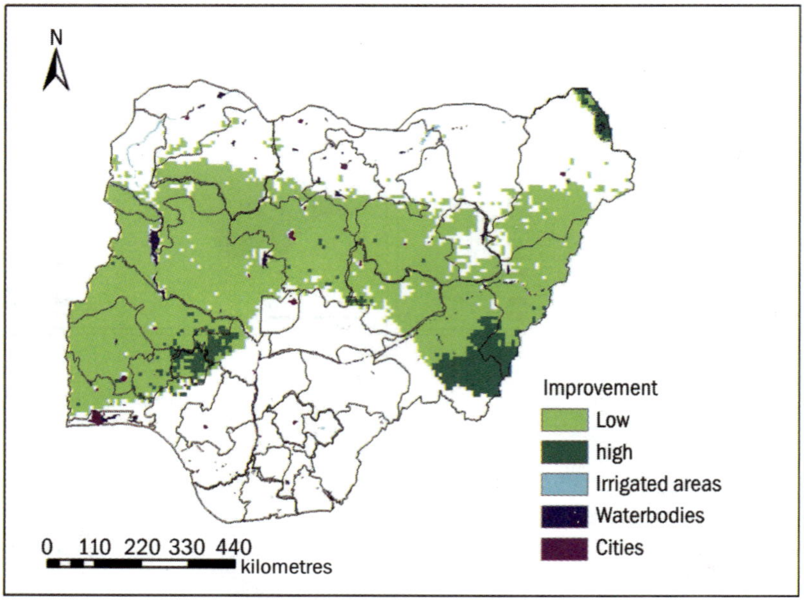


**Figure 8** Areas showing increasing Normalized Differential Vegetation Index and rain-use efficiency

Comparing the combined index of land degradation and improvement with land cover (Table 1), it was noted that 27% of the degrading area is cropland and a further 7.5% a mosaic of cropland, scrub, and grassland; in all, more than 10% of arable land is degrading. Areas of severe land degradation are located in the North-East and along the coast in the South-South (Figure 9).

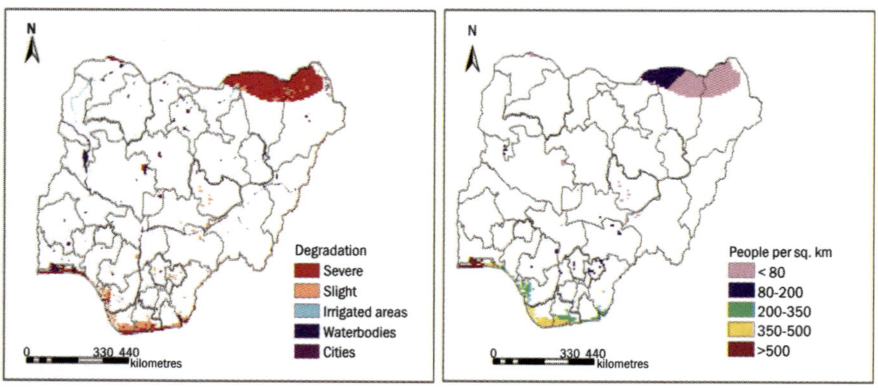


**Figure 9** (a) Areas with severe land degradation, (b) population density of areas of severe land degradation

**Table 1** Degrading and improving land by land cover

| Code | Land cover | Total pixels (TP)[a] | Degrading pixels (DP)[b] | DP/TP (%) | DP/TDP[c] (%) | Improving pixels (IP) | IP/TP (%) | IP/TIP[d] (%) |
|---|---|---|---|---|---|---|---|---|
| 1 | Closed evergreen, lowland forest | 29 380 | 14 | 0.04 | 1.64 | 0 | 0 | 0 |
| 2 | Degraded evergreen, lowland | 10 273 | 16 | 0.16 | 1.87 | 2 | 1.95 | 0.03 |
| 6 | Mangrove | 6760 | 20 | 0.3 | 2.34 | 6 | 0.09 | 0.09 |
| 7 | Mosaic forest/ croplands | 72 182 | 8 | 0.01 | 0.94 | 120 | 0.17 | 1.85 |
| 8 | Mosaic forest/ savanna | 11 850 | 9 | 0.08 | 1.05 | 216 | 1.82 | 3.33 |
| 10 | Deciduous woodland | 137 265 | 8 | 0.006 | 9.36 | 817 | 0.60 | 12.62 |
| 11 | Deciduous shrub with sparse trees | 160 340 | 7 | 0.004 | 0.8 | 814 | 0.51 | 12.56 |
| 13 | Closed grassland | 60 771 | 14 | 0.02 | 1.64 | 174 | 0.29 | 2.69 |
| 14 | Open grassland with sparse shrub | 59 772 | 14 | 0.02 | 1.64 | 147 | 0.25 | 2.27 |
| 15 | Open grassland | 3139 | 58 | 1.85 | 6.78 | 173 | 5.51 | 2.67 |
| 17 | Swamp bushland and grassland | 3596 | 21 | 0.58 | 2.46 | 279 | 7.76 | 4.31 |
| 18 | Croplands (>50%) | 182 232 | 35 | 0.02 | 4.09 | 1003 | 0.56 | 15.5 |
| 19 | Croplands with open woody vegetation | 188 780 | 10 | 0.005 | 1.17 | 1980 | 1.04 | 30.5 |
| 20 | Irrigated croplands | 1742 | 4 | 0.22 | 4.68 | 28 | 1.61 | 0.43 |
| 26 | Waterbodies | 7994 | 5 | 0.06 | 0.58 | 2 | 0.03 | 0.03 |
| 27 | Cities | 2425 | 7 | 0.29 | 8.19 | 1 | 0.04 | 0.02 |

[a]Pixel size 8 × 8 km; [b]urban extents are excluded; [c]TDP – total degrading pixels; [d]TIP – total improving pixels

Based on the 2006 Census, it was estimated from this study that about 43 million people live within states affected by land degradation, which is about 34% of the country's total landmass. In most of the affected areas, population density is greater than the national average. Although, the extent and severity of desertification in Nigeria have not been fully established and neither has the rate of its progression properly documented, there is a general consensus that desertification is by far the most pressing environmental problem in Nigeria. According to the UNCCD (1999), between 50% and 75% of Adamawa, Bauch, Borno, Gombe, Jigawa, Kano, Katsina, Kebbi, Sokoto, Yobe, and Zamfara states in Nigeria are affected by desertification. These states with a population of about 50 million people account for about 43% of the country's total land, and exploitation of these marginal lands has aggravated desertification and drought. The pressure of the migrating human and livestock populations from these areas is absorbed by pressure point buffer states such as the Federal Capital Territory, Plateau, Taraba, Niger, Kwara, and Kaduna states. It is reported that these buffer states have about 10%–15% of their land area threatened by desertification (UNCCD 1999).

## CONCLUSION

Land degradation and improvement have been assessed in this study by remotely sensed indicators of biomass productivity over a period of 25 years, based on NDVI, which may be translated in terms of NPP. It was shown that RUE could be used for early warning of land degradation, or as a herald of improvement. Instances where NPP is rising but RUE declining, some process of land degradation might be under way, which is not yet reflected in declining NPP. The methodology adopted in this study presents a different picture from previous assessments of land degradation (Oldeman, Hakkeling, and Sombroek 1991). It should, however, be noted that remote sensing provides only indicators of trends in biomass productivity. The various kinds of land degradation and improvement are not distinguished, and the interpretation is preliminary until validated by extensive fieldwork. This is not straightforward: an 8-km pixel cannot be checked by a windscreen survey, and a 25-year trend cannot be checked by a snapshot. A rigorous procedure must be followed. The coarse resolution of the GIMMS data is a limitation: an 8-km pixel integrates the signal from a wider surrounding area. Many symptoms of even severe degradation, such as gullies, rarely extend over such a large area; degradation must be severe indeed to be seen against the signal of surrounding unaffected areas. Nevertheless, this study provides a first step in adopting remote sensing methodology for developing early warning system for land management in a developing country like Nigeria.

### References

Bai Z G and Dent D L. 2006. ***Global Assessment of Land Degradation and Improvement: pilot study in Kenya***. Report 2006/01. Wageningen: ISRIC–World Soil Information

Bai Z G, Dent D L, and Schaepman M E. 2005. ***Quantitative Global Assessment of Land Degradation and Improvement: pilot study in North China***. ISRIC Report 2005/06. Wageningen

Bai Z G, Dent D L, Olsson L, Schaepman M E. 2008. ***Global Assessment of Land Degradation and Improvement 1. Identification by Remote Sensing***. ISRIC Report 2008/1, Wageningen

Brown M E, Pinzon J E, Didan K, Morisette J T, Tucker C J. 2006. **Evaluation of the consistency of long-term NDVI time series derived from AVHRR, SPOT-Vegetation, SeaWIFS, MODIS and Landsat ETM+.** *IEEE Transactions on Geoscience and Remote Sensing* **44**: 1787–1793

CIESIN (Center for International Earth Science Information Network). 2004. ***Global Rural–Urban Mapping Project: urban/rural extents.*** Center for International Earth Science Information Network, Columbia University, Palisades, NY. Details available at <http://sedac.ciesin.columbia.edu/gpw>

Dregne H E and Chou N T. 1994. **Global desertification dimensions and costs.** In *Degradation and Restoration of Arid Lands,* edited by H E Dregne. Lubbock: Texas Technical University

Engelen V W P van, Dijkshoorn J A, and Huting J R M. 2008. ***Global Assessment of Land Degradation and Improvement 2. Soil and Landform Properties for SOTER Units.*** ISRIC Report 2008/02. Wageningen

Engelen V W P van, Batjes N H, Dijkshoorn K, Huting J. 2005. ***Harmonized Global Soil Resources Database.*** Report 2005/06. Wageningen: ISRIC – World Soil Information

FAO (Food and Agricultural Organization). 2000. ***Food Insecurity: when people must live with hunger and fear starvation. The State of Food Insecurity in the World.*** Rome: FAO

GLC (Global Land Cover). 2000. Project. Details available at <http://glcf.umiacs.umd.edu>

Houérou H N Le. 1989. ***The Grazing Land Ecosystems of the African Sahel.*** Berlin: Springer-Verlag

Justice C O, Dugdale G, Townshend J R G, Narracott A S, Kumar M. 1991. **Synergism between NOAA AVHRR and Meteosat data for studying vegetation development in semi-arid West Africa.** *International Journal of Remote Sensing* **12**: 1349–1368

Legg C, Kormawa P, Maziya-Dixon P, Okechukwu R, Ofodile S, Alabi T. 2005. **Report on mapping livelihoods and nutrition in Nigeria using data from the national rural livelihoods survey and the national food consumption and nutrition survey.** Details available at <gisweb.ciat.cgiar.org/povertymapping/download/Nigeria.pdf>

Livezy R E and Chen W Y. 1983. **Statistical field significance and its determination by Monte Carlo techniques.** *Monthly Weather Review* **111**: 46–59

O'Connor T G, Haines L M, and Snyman H A. 2001. **Influence of precipitation and species composition on phytomass of semi-arid African grassland.** *Journal of Ecology* **89**: 850–860

Oldeman L R, Hakkeling R T A, and Sombroek W G. 1991. ***World Map of the Status of Human-induced Soil Degradation: an explanatory note.*** Second revised edn. Wageningen: International Soil Reference and Information Centre (ISRIC)

Running S W, Heinsch F A, Zhao M, Reeves M, Hashimoto H. 2004. **A continuous satellite-derived measure of global terrestrial production.** *Bioscience*: **54**: 547–560

Schneider U, Fuchs T, Meyer-Christoffer A, Rudolf B. 2008. ***Global Precipitation Analysis Products of the GPCC, Full Data Re-analysis Product Version 4***. Deutscher Wetterdienst, Offenbach: Global Precipitation Climatology Centre. Details available at <ftp://ftp-anon.dwd.de/pub/data/gpcc/html/fulldata_download.html>

Singh D, Meirelles M S P, and Costa G A. 2006. **Environmental degradation analysis using NOAA/AVHRR data.** *Advances in Space Research* **37**: 720–727

Snyman H A. 1998. **Dynamics and sustainable utilization of rangeland ecosystems in arid and semi-arid climates of southern Africa.** *Journal of Arid Environments* **39**: 645–666

Tucker C J, Pinzon J E, and Brown M E. 2005. **An extended AVHRR 8-km NDVI dataset compatible with MODIS and SPOT vegetation NDVI data.** *International Journal of Remote Sensing* **26**: 4485–4498

Turner D P, Ritts W D, and Cohen W B. 2006. **Evaluation of MODIS NPP and GPP products across multiple biomes.** *Remote Sensing of Environment* **102**: 282–292

UNCCD (United Nations Convention to Combat Desertification). 1999. National Report on the Implementation of the United Nations Convention to Combat Desertification in those Countries Experiencing Serious Drought and/or Desertification, Particularly in Africa for Submission at the Third Session of the Conference of The Parties, Recife, Brazil

UNCED (United Nations Conference on Environment and Development). 1992. **Managing fragile ecosystems: combating desertification and drought Agenda 21, Chapter 12**. UNCED. Details available at <http://wwwunorg/esa/sustdev/documents/agenda21>

Ustin S L, Jacquemoud S, Palacios-Oreuta A, Li L, Whiting, M L. 2005. **Remote sensing based assessment of biophysical indicators for land degradation and desertification.** Proceedings of the 1st International Conference on Remote Sensing and Geoinformation Processing in the Assessment and Monitoring of Land Degradation and Desertification, Trier, Germany, 7–9 September 2005

Wessels K J, Prince S D, and Malherbe J. 2007. **Can human-induced land degradation be distinguished from the effects of rainfall variability? A case study in South Africa.** *Journal of Arid Environments* **68**: 271–297

World Bank. 2007. ***The Little Green Data Book.*** Washington, DC: World Bank

# Assessing Land Surface Temperature over Bangladesh using MODIS Satellite Images—An Indicator of Climate Change

**Akm Saiful Islam**

*Institute of Water and Flood Management*

*Bangladesh University of Engineering and Technology, Dhaka, Bangladesh*

*E-mail: akmsaifulislam@iwfm.buet.ac.bd*

## INTRODUCTION

Recent studies have shown that climate has been changing due to anthropogenic factors. According to the *Fourth Assessment Report* of the Intergovernmental Panel on Climate Change (IPCC), global surface temperature is changing globally, and the average global surface temperature has increased by about 0.74 ± 0.18°C during the last 100 years, ending in 2005 (IPCC 2007). The last decade has witnessed a significant and consistent increase in global temperature. Therefore, there is a need to know the trends in land surface temperature (LST) over the landmasses. This especially assumes significance for a developing country like Bangladesh, where the rate of urbanization is very high, and its unique geographic location makes it vulnerable to climate variations. However, due to limitation of resources, it is not viable to collect temperature data for every kilometre of the country by using ground-based instruments. Remote sensing can be an effective tool for estimating the changes in temperature with a dense spatial resolution.

LST is computed using airborne or space-borne thermal infrared remote sensing data, which provides a synoptic and uniform means for studying land surface. Satellite-measured LST has been utilized in various heat balance, climate modelling, and global change studies, since it is determined by the effective radiating temperature of the earth's surface, which controls surface heat and water exchange with the atmosphere (Chen, Zhao, Li, *et al.* 2006). Higher vegetation index values typically indicate a larger fraction of vegetation in a pixel. The LST is also determined by the amount of vegetation, which is characterized by the latent heat flux from the surface to the atmosphere through evapotranspiration. Lower LSTs are usually found in areas with high vegetation index. The non-linear relationship between Normalized Difference Vegetation Index (NDVI) and LST is well known, and predominantly bare soil locations experience a wider variation in surface radiant temperature compared to densely vegetated locations (Carlson, Gillies, and Perry 1994; Gillies and Carlson 1995; Owen, Carlson, and Gillies 1998; Price 1990). Satellite thermal infrared (TIR) sensors measure the top of the atmosphere (TOA) radiances, from which brightness temperatures (also known as blackbody temperatures) can be derived using Plank's law (Dash, Gottsche, Olesen, *et al.* 2002). The TOA radiances are the mixed results of (1) emitted radiance from the earth's surface, (2) upwelling radiance from the atmosphere, and (3) downwelling radiance from the sky. The atmospheric effects, including absorption, upward emission, and downward irradiance, reflected from the surface (Franca and Cracknell 1994) are corrected before the land surface brightness temperatures are obtained. These brightness temperatures are also corrected with spectral emissivity values prior to the computation of LST to account for the roughness properties of the land surface, the amount and nature of vegetation cover, and the thermal properties and moisture content of the soil (Friedl 2002; Weng, Lu, and Schuring 2004; Yuan and Bauer 2007).

Satellite-derived LST measurements have been conducted primarily using National Oceanic and Atmospheric Administration (NOAA), advanced very high resolution radiometer (AVHRR) data (Li and Becker 1993; Gallo and Owen 1998; Streutker 2002, 2003), and moderate resolution imaging spectroradiometer (MODIS) (Dozier and Wan 1994; Neteler 2010) for regional-scale urban temperature mapping. Landsat thematic mapper (TM), enhanced thematic mapper plus (ETM+) (Weng, Lu, and Schuring 2004; Yuan and Bauer 2007), and advanced spaceborne thermal emission and reflection radiometer (ASTER) TIR data (Gillespie, Rokugawa, Matsunaga, *et al.* 1998) have also been utilized for local-scale studies of urban heat island (UHI) (Chen, Wang, and Li 2002; Weng 2001).

In this study, changes in LST over Bangladesh have been analysed using MODIS land surface images (MODIS 2009). Data from MODIS satellites is freely available on the website, with an interval of 24 hours. Spatial resolutions of these images are moderate (250 m for visible band and 1000 m for thermal band) as compared to other satellites and are geometrically corrected. Finally, MODIS images are compared with data from ground-based measurement stations located across the country.

## STUDY AREA

In this study, changes in temperature over Bangladesh have been analysed using remote sensing data. Bangladesh is located between latitudes

20°N and 27°N and longitudes 88°E and 93°E. The country is located in a floodplain delta of three major river basins: the Ganga, the Brahmaputra, and the Meghna (GBM). A map of the study area is shown in Figure 1. There are six distinct seasons in Bangladesh: summer (April–May), rainy (June–July), autumn (August–September), late autumn (October–November), winter (December–January), and spring (February–March). The highest temperature occurs in summer season, while the lowest temperature occurs in winter season. Most of the rainfall, which is about 80% of the total rainfall, is received during the monsoon period (June–September).

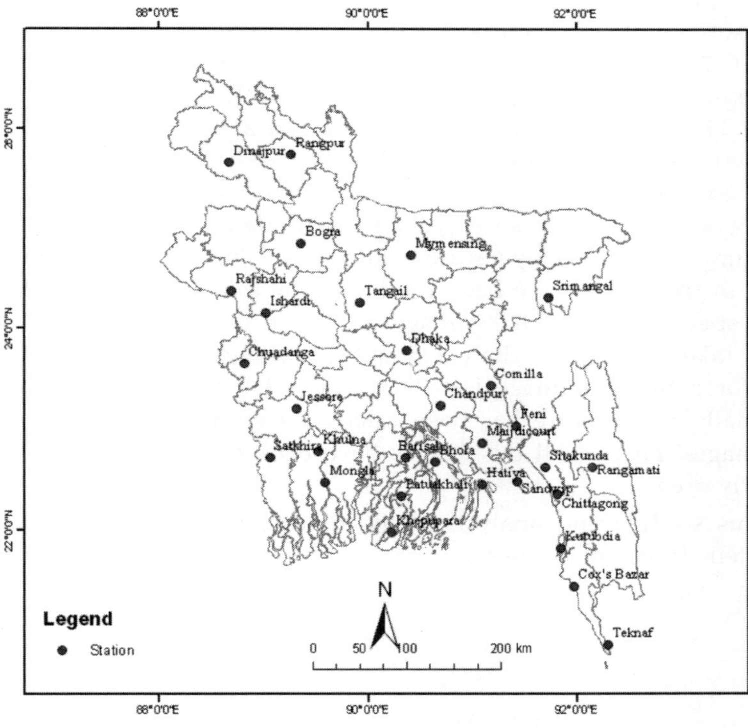


**Figure 1** Study area

## DATA USED

In this study, both satellite images and data from meteorological stations have been used. Satellite images used for the study are MODIS/TERRA images. An average of eight days' temperature data, labelled as "Land Surface Temperature/Emissivity 8-Day L3 Global 1 km" has been used. Images have been downloaded from January 2000 to December 2007 for every eight-day interval. Daily mean temperature has been calculated from the average of day and night images. However, in situ station data has been collected from Bangladesh Meteorological Department (BMD). Maximum and minimum daily temperature data for the last 60 years (1948–2007), collected in 34 stations of BMD located all over Bangladesh, has been used in this study. Figure 1 shows the location of the BMD stations across Bangladesh. These stations are mainly located at the city centres of urban districts.

## METHODOLOGY

Integrated Land and Water Information System (ILWIS 3.6) open source geo-spatial software has been used for processing images and analysing spatial data (ITC 2007). All TERRA/MODIS images downloaded in raw format have been geo-referenced using the software MODIS Re-projection Tools (MRT). Files of TERRA/MODIS images are presented as Geo-TIF format. Geo-TIF formatted images are geo-referenced to WGS-84 Coordinate system. Geo-referenced images, afterwards, have been masked for the study area for carrying out further analysis for each event for determining daily maximum, minimum, and mean temperature values. The values of temperature MODIS indices in the image have been multiplied by a factor as per MODIS data format specification. MODIS image is stored as positive integer because integer takes less space than a double precision number. Multiplication factor for a thermal image is 0.01 to convert it into degree Kelvin (°K). Mean daily temperature has been calculated by averaging day and night time images. Finally, all the images have been processed and masked for the study area.

In this study, trend analysis of the time series thermal data has been conducted. Trend in a time series data can be found by using the following formula.

$$b = \frac{\Sigma(x_i - \overline{x})(y_i - \overline{y})}{\Sigma(x_i - \overline{x})^2} \tag{1}$$

where, $x_i$ is an independent variable, $\overline{x}$ is the average of the independent variable, $y_i$ is the dependable variable, and $\overline{y}$ is the average of dependable variables. One way of testing the significance of trends in temperature is calculating the coefficient of determination, $R^2$, of the trend. Values of $R^2$ vary between 0 and 1, which can be found by using the following formula.

$$R^2 = \frac{[\Sigma(x_i - \overline{x})(y_i - \overline{y})]^2}{\Sigma(x_i - \overline{x})^2 \, \Sigma\,(y_i - \overline{y})^2} \tag{2}$$

Highest correlation of the data set can be found at 1, and it gradually reduces towards 0. Value less than 0.5 has been considered as a less significant correlation.

## Daily Maximum, Minimum, and Average Temperatures over Bangladesh

Using MODIS images, monthly mean values of daily maximum, minimum, and average temperatures across Bangladesh have been calculated for eight years (2000–07). These values have been compared with the BMD station data. Figure 2 shows the plot of mean monthly values of maximum, minimum, and average temperature across Bangladesh using both MODIS images and station data. Due to excessive clouds, it is not possible to correctly calculate temperature from MODIS images during June, July, and August. The mean temperatures from MODIS images are found to be much lower than those obtained from station data. It is also found that during winter (December to February), maximum temperature from MODIS images is lower than that from stations. However, during winter, minimum temperature from MODIS images is higher than that from the stations. Mean temperature data from these two sources shows a good agreement.

Historic data for daily maximum, minimum, and mean temperatures from 34 meteorological stations has been analysed for the last 60 years (1948–2007). Figure 3 shows the trends and $R^2$ values of daily maximum, minimum, and mean temperatures across Bangladesh. Maximum temperature shows a positive trend of 0.0062°C per year. However, the analysis of daily minimum temperature for the past 60 years shows a positive trend of 0.0154°C per year. Daily minimum temperature has increased at a much faster rate than the maximum temperature across Bangladesh. The mean trend in the daily mean temperature has been found to be 0.0103°C per year (or 1.03°C per 100 years) for the country.

Monthly average trend in maximum, minimum, and mean temperatures has been calculated by using station data and MODIS images. A comparison of trends and $R^2$ values of these two data sets has been presented in Table 1. Mean temperature change for the past eight years in the country is found to be 0.035°C per year from the data obtained from stations and 0.105°C per year from MODIS images. Trend in temperature data from MODIS images shows much higher value than that obtained from the stations. The highest trend in mean monthly temperature is found to be 0.194°C per year from station and 0.332°C from MODIS images obtained during May.

Maximum, minimum, and mean annual temperatures from station data and MODIS images have been plotted in Figure 4. In all the three cases, temperature based on station data showed a higher value than that computed from MODIS images. Unavailability of the satellite images during June, July, and August, due to excessive clouds, has lowered the mean annual values of temperature compared to the data obtained from the stations.

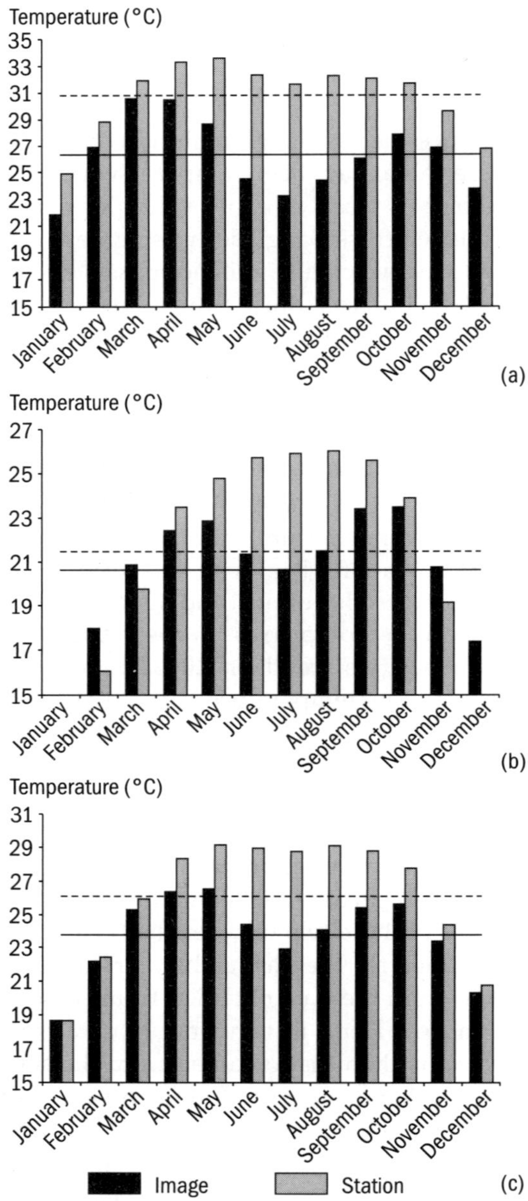


**Figure 2** Comparison of mean monthly value of the daily (a) maximum, (b) minimum, and (c) average temperatures across Bangladesh, calculated from MODIS images and ground-based meteorological station data

## TEMPERATURE TRENDS, BY STATION

Temperature data from each station has also been compared with the corresponding values from MODIS images. Trends and $R^2$ values for

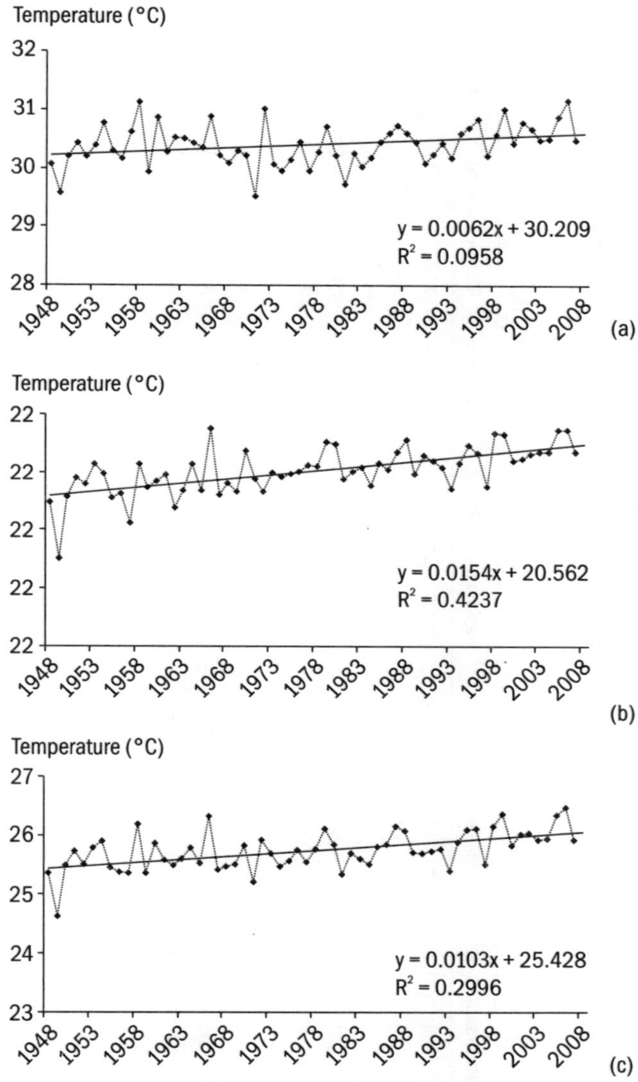


**Figure 3** Trends in daily (a) maximum, (b) minimum, and (c) mean temperatures based on the data for the past 60 years (1948–2007), obtained from all the meteorological stations

maximum, minimum, and mean temperatures collected from 34 stations across Bangladesh have been presented in Table 2. Similarly, trends and $R^2$ values for maximum, minimum, and mean temperatures from the corresponding locations in the MODIS images for the past eight years (2000–07) are also presented in Table 2. It was found that the average trends in daily maximum and minimum temperatures from MODIS images are higher than those obtained from station data. The highest daily maximum

**Table 1** Trend in daily maximum, minimum, and mean temperatures for each month and $R^2$ value by using ground-based station data and MODIS images

| Month | Ground-based station data | | | | | | MODIS image | | | | | |
|---|---|---|---|---|---|---|---|---|---|---|---|---|
| | Maximum | $R^2$ | Minimum | $R^2$ | Mean | $R^2$ | Maximum | $R^2$ | Minimum | $R^2$ | Mean | $R^2$ |
| January | 0.042 | 0.01 | 0.031 | 0.00 | 0.039 | 0.01 | 0.129 | 0.06 | 0.046 | 0.02 | −0.039 | 0.02 |
| February | −0.048 | 0.01 | 0.326 | 0.40 | 0.140 | 0.10 | −0.107 | 0.06 | 0.214 | 0.43 | 0.061 | 0.03 |
| March | −0.061 | 0.02 | 0.039 | 0.01 | −0.011 | 0.00 | 0.243 | 0.08 | 0.221 | 0.52 | 0.225 | 0.26 |
| April | −0.031 | 0.01 | 0.017 | 0.01 | −0.006 | 0.00 | −0.171 | 0.03 | −0.082 | 0.04 | −0.061 | 0.01 |
| May | 0.234 | 0.40 | 0.154 | 0.39 | 0.194 | 0.41 | 0.493 | 0.77 | 0.279 | 0.54 | 0.332 | 0.52 |
| June | 0.257 | 0.47 | 0.108 | 0.34 | 0.188 | 0.44 | 0.396 | 0.27 | 0.014 | 0.00 | 0.264 | 0.20 |
| July | −0.079 | 0.20 | −0.016 | 0.03 | −0.049 | 0.15 | −0.125 | 0.18 | −0.096 | 0.06 | −0.043 | 0.02 |
| August | −0.044 | 0.07 | −0.003 | 0.00 | −0.025 | 0.05 | 0.182 | 0.17 | 0.136 | 0.07 | 0.196 | 0.14 |
| September | −0.085 | 0.12 | 0.037 | 0.19 | −0.025 | 0.03 | 0.143 | 0.11 | 0.325 | 0.31 | 0.261 | 0.28 |
| October | −0.031 | 0.02 | −0.033 | 0.02 | −0.031 | 0.03 | −0.057 | 0.04 | −0.075 | 0.08 | −0.011 | 0.00 |
| November | −0.010 | 0.02 | −0.037 | 0.01 | −0.015 | 0.00 | 0.114 | 0.47 | 0.029 | 0.01 | 0.068 | 0.15 |
| December | 0.037 | 0.04 | −0.002 | 0.00 | 0.018 | 0.01 | 0.007 | 0.00 | −0.029 | 0.03 | 0.007 | 0.00 |
| Average | 0.015 | | 0.052 | | 0.035 | | 0.104 | | 0.082 | | 0.105 | |

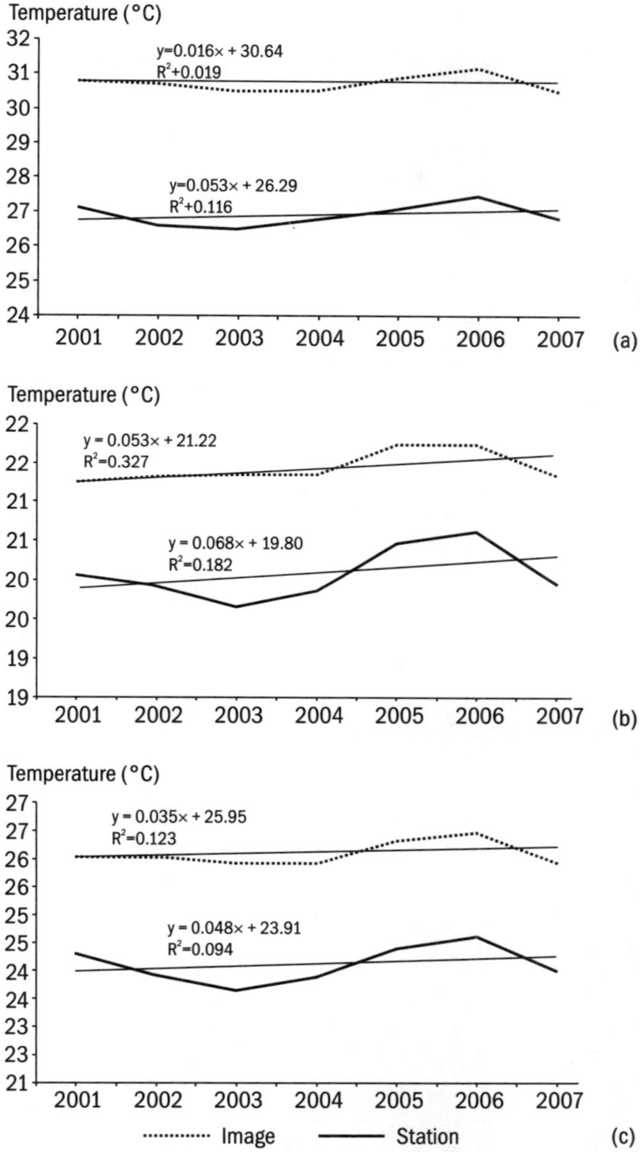


**Figure 4** Comparison of the trends in daily (a) maximum, (b) minimum, and (c) average temperatures across Bangladesh by using MODIS image and ground-based station data

temperature was found to be 0.129°C in Ishurdi, from station data, and 0.214°C in Rajshahi, from MODIS images. However, the highest trend in daily minimum temperature was found to be 0.13°C in Kutubdia, from station data, and 0.193°C in Dhaka, from MODIS images. Dhaka city has

**Table 2** Trend in daily maximum, minimum, and mean temperature and $R^2$ values by using ground-based station data and MODIS images for the past eight years (2000–07)

| | Ground-based station data | | | | | | MODIS image | | | | | |
|---|---|---|---|---|---|---|---|---|---|---|---|---|
| | Maximum | | Minimum | | Mean | | Maximum | | Minimum | | Mean | |
| Station | Trend | $R^2$ | Trend | $R^2$ | Trend | $R^2$ | Trend | $R^2$ | Trend | $R^2$ | Trend | $R^2$ |
| Barisal | 0.029 | 0.04 | 0.117 | 0.45 | 0.073 | 0.30 | −0.033 | 0.03 | −0.038 | 0.02 | −0.037 | 0.03 |
| Bhola | 0.041 | 0.09 | 0.065 | 0.33 | 0.014 | 0.01 | −0.023 | 0.01 | −0.001 | 0.00 | −0.052 | 0.05 |
| Bogra | 0.086 | 0.23 | 0.023 | 0.11 | 0.055 | 0.19 | 0.173 | 0.26 | 0.065 | 0.25 | 0.035 | 0.08 |
| Chandpur | 0.079 | 0.30 | 0.079 | 0.30 | 0.079 | 0.40 | 0.075 | 0.13 | 0.121 | 0.60 | 0.066 | 0.18 |
| Chittagong | 0.053 | 0.15 | 0.085 | 0.31 | 0.069 | 0.25 | −0.047 | 0.02 | 0.062 | 0.14 | −0.044 | 0.07 |
| Chuadanga | 0.040 | 0.05 | −0.018 | 0.03 | 0.011 | 0.01 | −0.133 | 0.22 | 0.002 | 0.00 | −0.099 | 0.20 |
| Comilla | −0.019 | 0.02 | 0.061 | 0.30 | 0.021 | 0.03 | 0.179 | 0.27 | 0.147 | 0.81 | 0.113 | 0.72 |
| Cox's Bazar | −0.080 | 0.50 | 0.063 | 0.20 | −0.008 | 0.01 | −0.018 | 0.01 | 0.022 | 0.01 | −0.015 | 0.01 |
| Dhaka | 0.100 | 0.36 | 0.120 | 0.47 | 0.110 | 0.45 | 0.129 | 0.18 | 0.193 | 0.57 | 0.126 | 0.26 |
| Dinajpur | −0.055 | 0.08 | 0.049 | 0.26 | −0.003 | 0.00 | 0.154 | 0.29 | 0.115 | 0.10 | 0.123 | 0.16 |
| Faridpur | 0.038 | 0.07 | — | — | — | — | −0.001 | 0.00 | 0.035 | 0.04 | −0.026 | 0.03 |
| Feni | −0.061 | 0.19 | 0.050 | 0.27 | 0.007 | 0.00 | 0.030 | 0.02 | 0.029 | 0.03 | −0.049 | 0.11 |
| Hatiya | 0.043 | 0.11 | 0.078 | 0.39 | 0.061 | 0.24 | −0.193 | 0.48 | 0.149 | 0.49 | −0.030 | 0.04 |
| Ishardi | 0.129 | 0.55 | 0.013 | 0.02 | 0.072 | 0.41 | 0.034 | 0.01 | 0.125 | 0.23 | 0.053 | 0.06 |
| Jessore | 0.099 | 0.50 | 0.118 | 0.63 | 0.108 | 0.78 | 0.083 | 0.22 | 0.115 | 0.14 | 0.047 | 0.03 |
| Khepupara | 0.040 | 0.12 | 0.015 | 0.02 | 0.027 | 0.07 | −0.013 | 0.01 | 0.012 | 0.01 | −0.012 | 0.00 |
| Khulna | −0.058 | 0.19 | 0.062 | 0.33 | 0.002 | 0.00 | −0.002 | 0.00 | 0.089 | 0.20 | 0.056 | 0.08 |
| Kutubdia | 0.027 | 0.06 | 0.130 | 0.60 | 0.085 | 0.46 | −0.047 | 0.10 | 0.011 | 0.00 | −0.025 | 0.03 |
| Madaripur | −0.016 | 0.03 | — | — | — | — | 0.047 | 0.16 | 0.061 | 0.18 | 0.021 | 0.02 |
| Maijdicourt | 0.073 | 0.19 | 0.012 | 0.01 | 0.050 | 0.09 | 0.096 | 0.24 | 0.043 | 0.09 | 0.014 | 0.01 |

| Station | Ground-based station data | | | | | | MODIS image | | | | | |
|---|---|---|---|---|---|---|---|---|---|---|---|---|
| | Maximum | | Minimum | | Mean | | Maximum | | Minimum | | Mean | |
| | Trend | $R^2$ | Trend | $R^2$ | Trend | $R^2$ | Trend | $R^2$ | Trend | $R^2$ | Trend | $R^2$ |
| Mongla | −0.002 | 0.00 | 0.009 | 0.01 | 0.003 | 0.00 | 0.099 | 0.54 | 0.117 | 0.18 | 0.121 | 0.28 |
| Myensing | 0.003 | 0.00 | 0.023 | 0.13 | 0.013 | 0.02 | 0.065 | 0.09 | 0.161 | 0.51 | 0.067 | 0.14 |
| Patuakhali | 0.007 | 0.00 | 0.052 | 0.27 | 0.029 | 0.07 | 0.191 | 0.40 | 0.111 | 0.21 | 0.096 | 0.16 |
| Rajshahi | 0.052 | 0.15 | 0.031 | 0.21 | 0.042 | 0.20 | 0.214 | 0.44 | 0.017 | 0.01 | 0.042 | 0.04 |
| Rangaati | −0.071 | 0.19 | 0.014 | 0.01 | −0.029 | 0.06 | 0.009 | 0.01 | −0.120 | 0.47 | −0.120 | 0.38 |
| Rangpur | −0.001 | 0.00 | 0.049 | 0.53 | 0.025 | 0.06 | 0.209 | 0.34 | 0.136 | 0.24 | 0.122 | 0.19 |
| Sandwip | −0.015 | 0.01 | 0.060 | 0.20 | 0.023 | 0.03 | 0.040 | 0.17 | 0.082 | 0.40 | 0.009 | 0.01 |
| Satkhira | −0.004 | 0.00 | 0.078 | 0.54 | 0.037 | 0.17 | 0.106 | 0.21 | 0.122 | 0.29 | 0.075 | 0.13 |
| Sayedpur | −0.013 | 0.01 | — | — | — | — | 0.064 | 0.08 | 0.153 | 0.19 | 0.076 | 0.08 |
| Sitakunda | 0.007 | 0.00 | 0.073 | 0.33 | 0.040 | 0.10 | 0.029 | 0.02 | 0.045 | 0.05 | −0.010 | 0.00 |
| Sriangal | — | — | 0.071 | 0.33 | 0.028 | 0.04 | 0.108 | 0.36 | 0.039 | 0.03 | 0.061 | 0.07 |
| Sylhet | −0.016 | 0.01 | 0.041 | 0.15 | — | — | 0.007 | 0.00 | 0.093 | 0.09 | 0.067 | 0.06 |
| Tangail | 0.065 | 0.18 | 0.042 | 0.23 | 0.054 | 0.21 | 0.121 | 0.20 | −0.032 | 0.05 | −0.003 | 0.00 |
| Teknaf | −0.032 | 0.08 | 0.026 | 0.05 | −0.003 | 0.00 | 0.077 | 0.06 | 0.036 | 0.02 | 0.009 | 0.00 |

been considered as one of the megacities of the world. Due to its high population density, it has become a UHI in recent years. MODIS images have also shown a high positive trend in minimum temperature in Dhaka city.

In the southern parts of the country, majority of the stations, such as Barisal, Patuakhali, Bhola, Kehpurpara, Chittagong, Rangamati, and Kutubdia, have shown negative trends in mean daily temperature, obtained from MODIS images. However, trends in daily minimum temperature, obtained from MODIS images, have shown positive values. Most of the stations in central and north of the country have shown positive trends in daily maximum, minimum, and mean temperatures, using both stations' data and MODIS images.

### Spatial Distribution of Mean Monthly and Annual Temperatures and their Trends

Spatial distribution of mean temperature for each month for the past eight years (2000–07) has been constructed for Bangladesh using MODIS images (Figure 5). Lowest temperature is found during January, whereas highest temperature occurred during May. However, due to excessive clouds in the sky, it is not possible to find cloud-free images during June, July, and August. Spatial distribution of trends in mean monthly temperature during 2000–07 has been shown in Figure 6. The highest and lowest values of trends in mean temperature were found in May and April, respectively. As mentioned before, it is not possible to construct spatial distribution map during June, July, and August because of the monsoon clouds.

In Figure 7, using MODIS images, spatial distribution of (a) maximum, (b) minimum, and (c) mean temperatures, their trends, and $R^2$ values have been shown. The north and the north-east parts of the country have lower temperature, whereas the middle, south, and south-east parts of the country have temperature higher than other parts of the country. Spatial distribution of daily minimum temperature is uniform and higher than the mean daily temperature across the country. Similarly, good correlation ($R^2$) has been found for daily minimum temperature compared to daily maximum temperature.

## CONCLUSION

The study has investigated trends in temperature across Bangladesh, using MODIS images available for the past eight years (2000–07). The results have been compared with the temperature data from ground-based stations located all over Bangladesh. It can be concluded that the eight-day average thermal data from TERRA/MODIS can be successfully used to determine trends in temperature across Bangladesh. MODIS images can be a viable

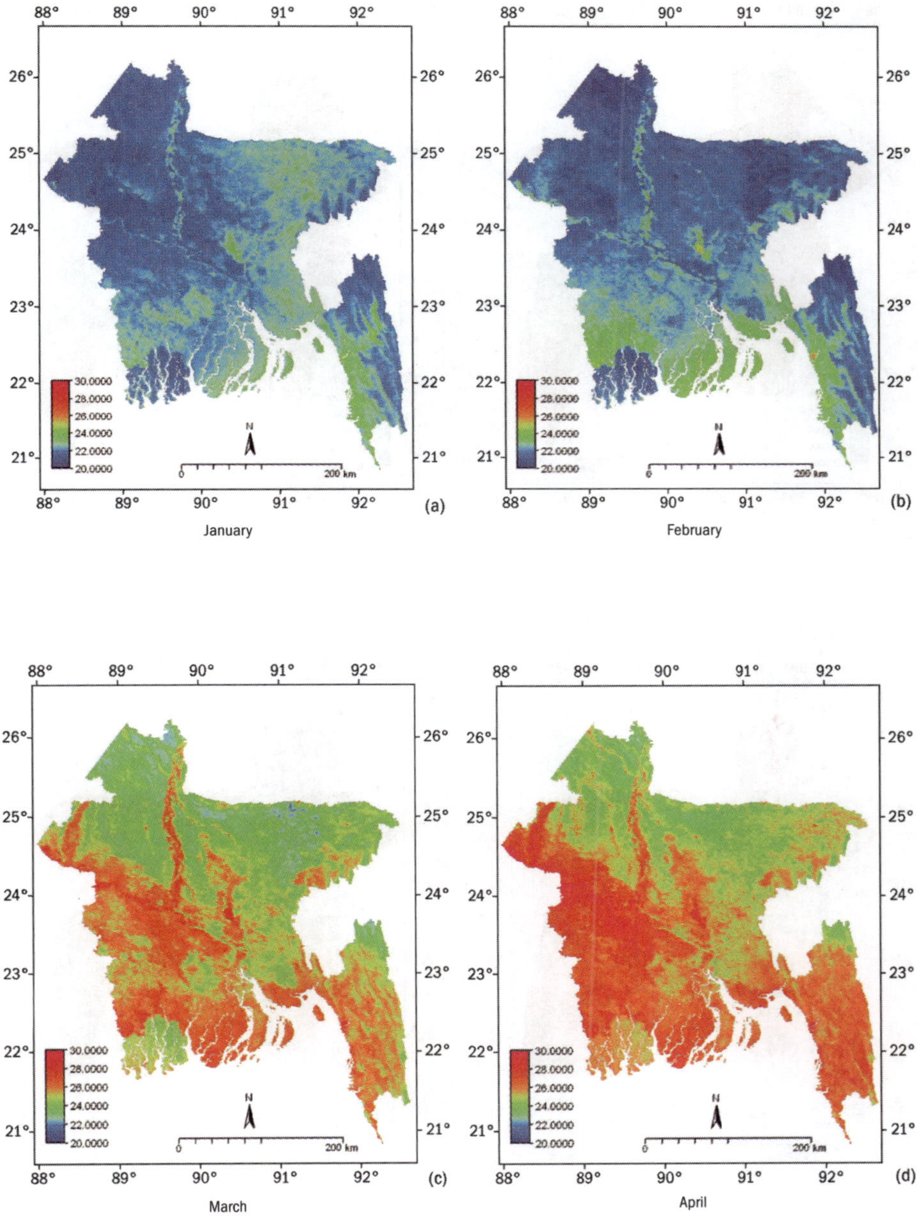


**Figure 5** Spatial distribution of mean monthly temperature across Bangladesh for the past eight years (2000–07)

*(Fig. 5 Contd..)*

*(Fig. 5 Contd..)*

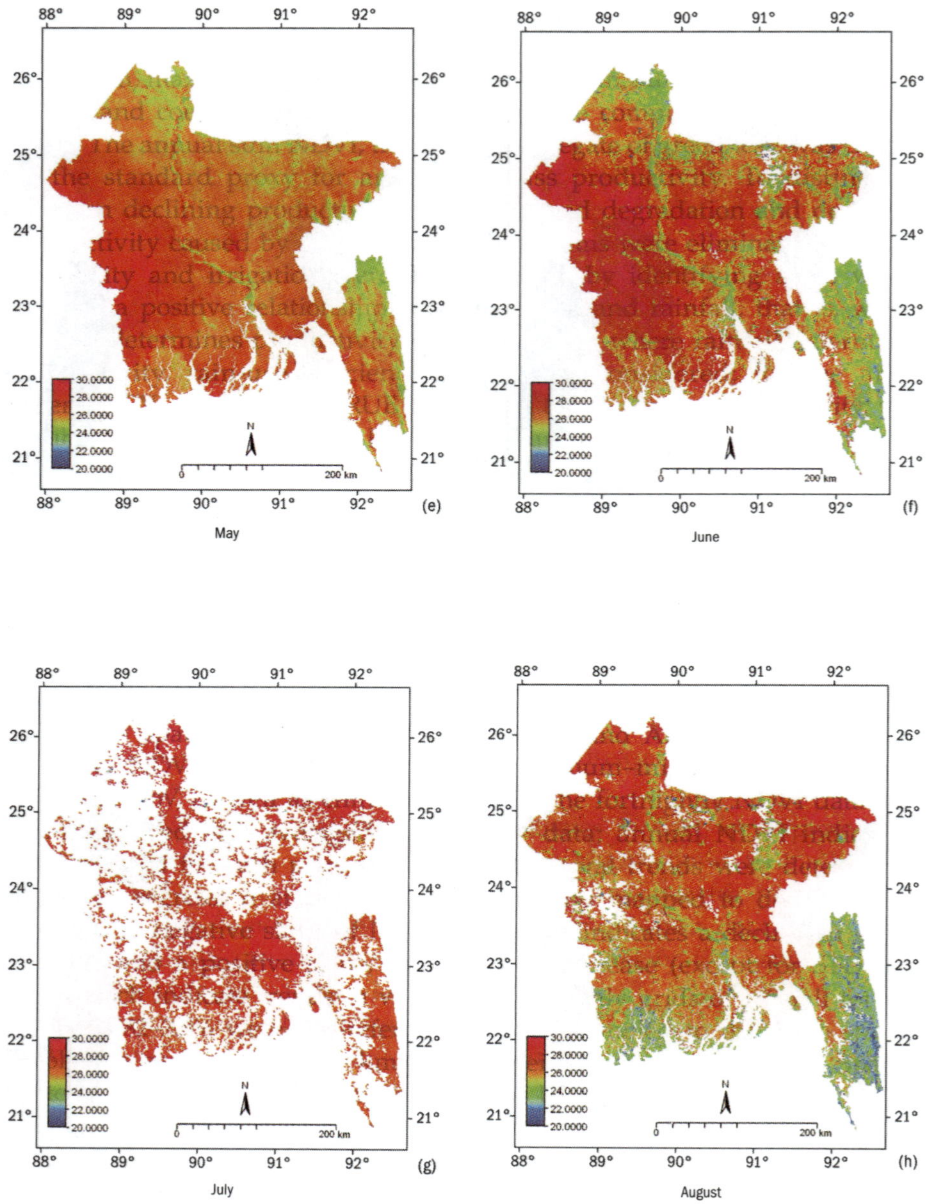


**Figure 5** Spatial distribution of mean monthly temperature across Bangladesh for the past eight years (2000–07)

*(Fig. 5 Contd..)*

*(Fig. 5 Contd..)*

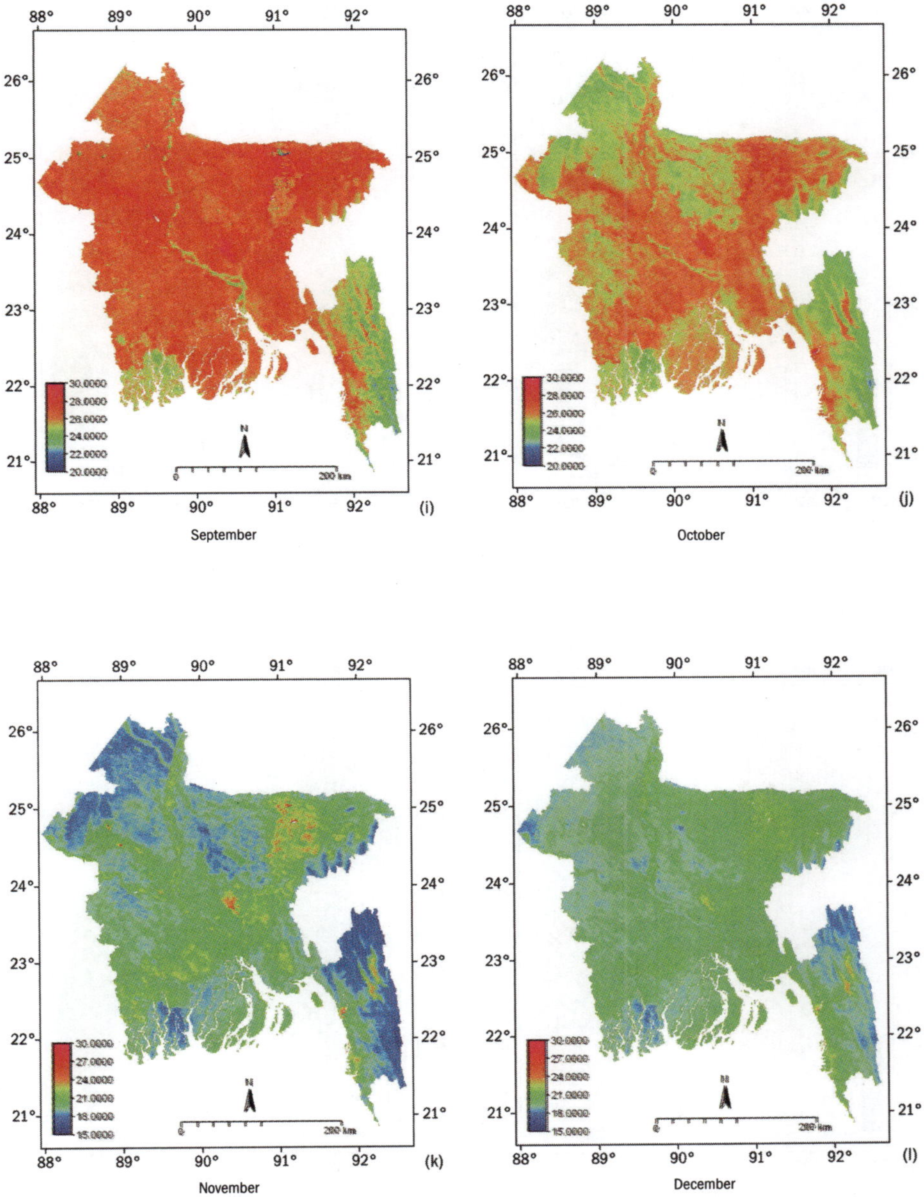


**Figure 5** Spatial distribution of mean monthly temperature across Bangladesh for the past eight years (2000–07)

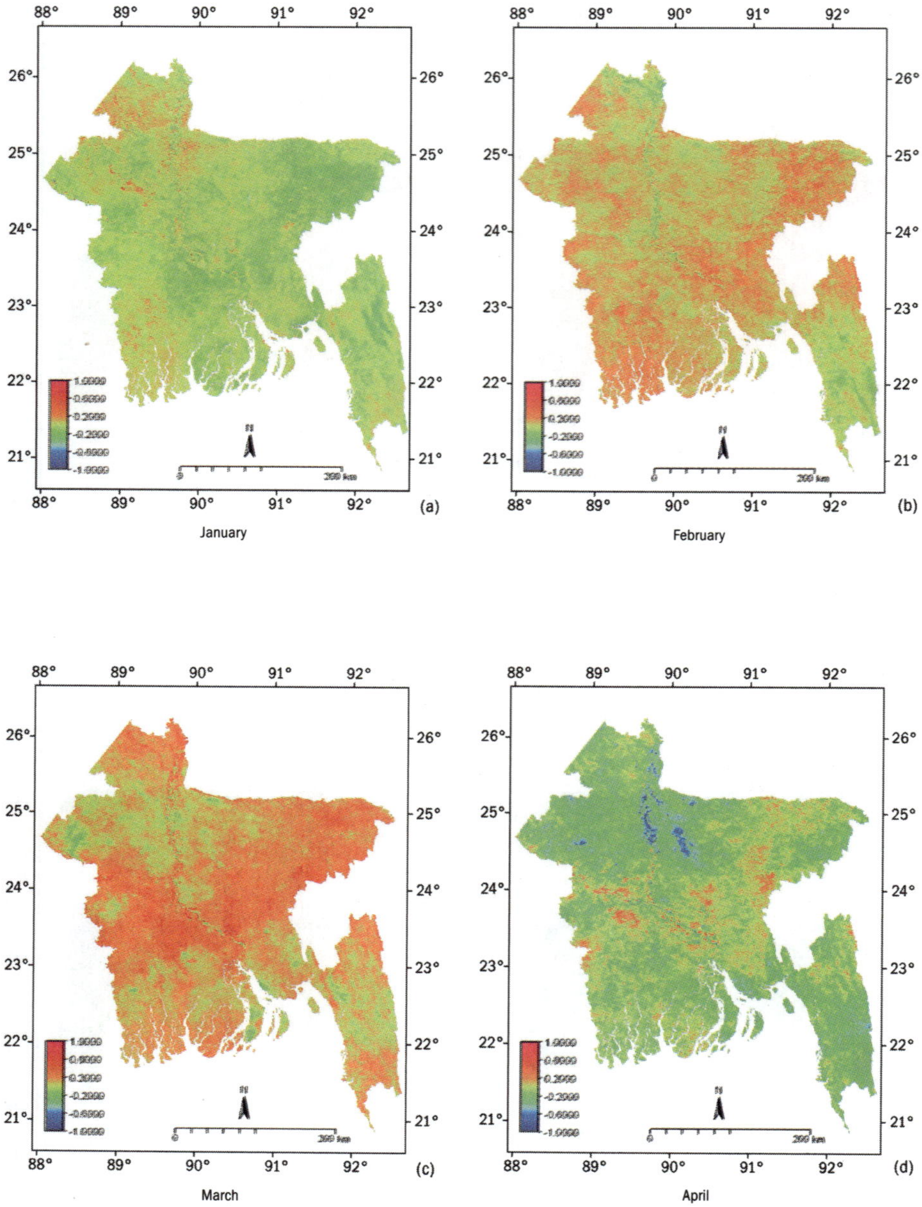


**Figure 6** Spatial distribution of trends of mean monthly temperature across Bangladesh for the past eight years (2000–07)

*(Fig. 6 Contd..)*

*(Fig. 6 Contd..)*

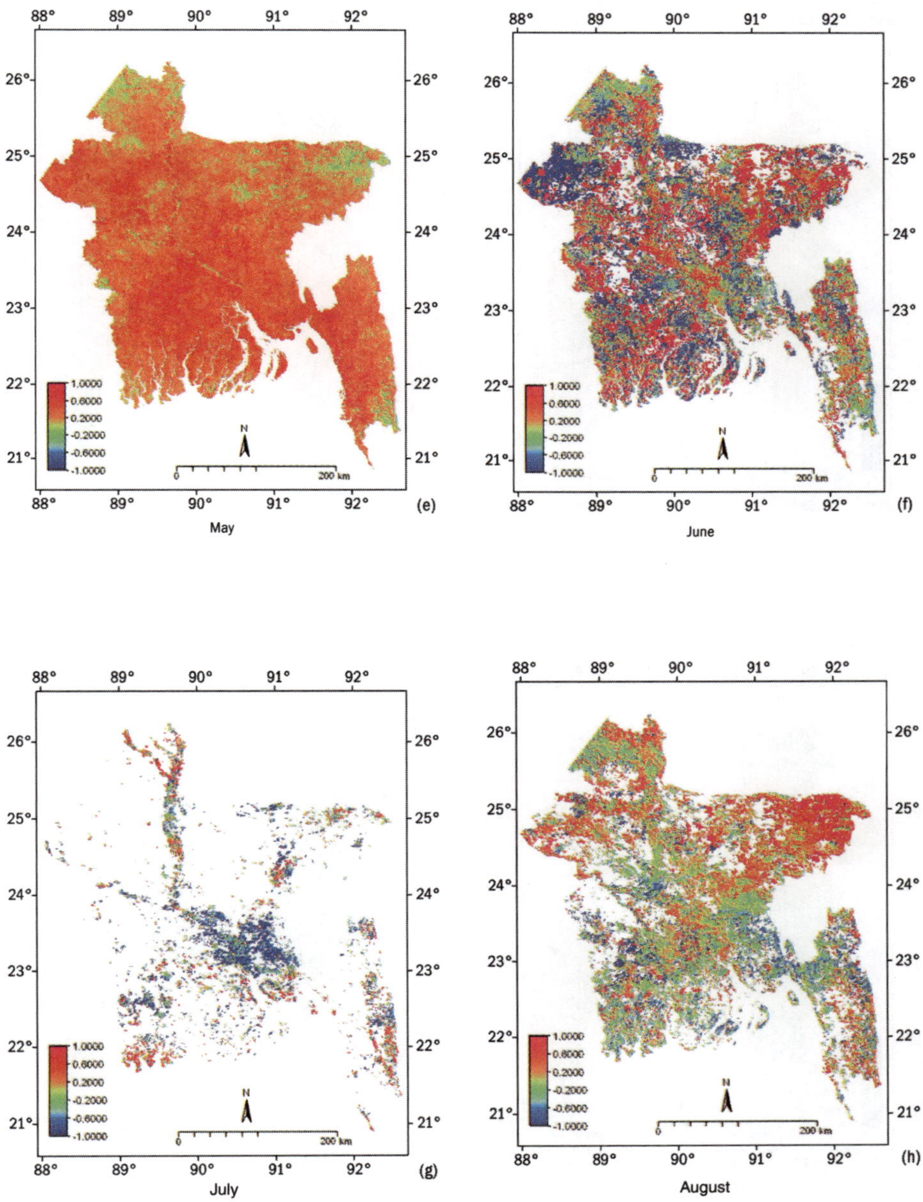


**Figure 6** Spatial distribution of trends of mean monthly temperature across Bangladesh for the past eight years (2000–07)

*(Fig. 6 Contd..)*

*(Fig. 6 Contd..)*

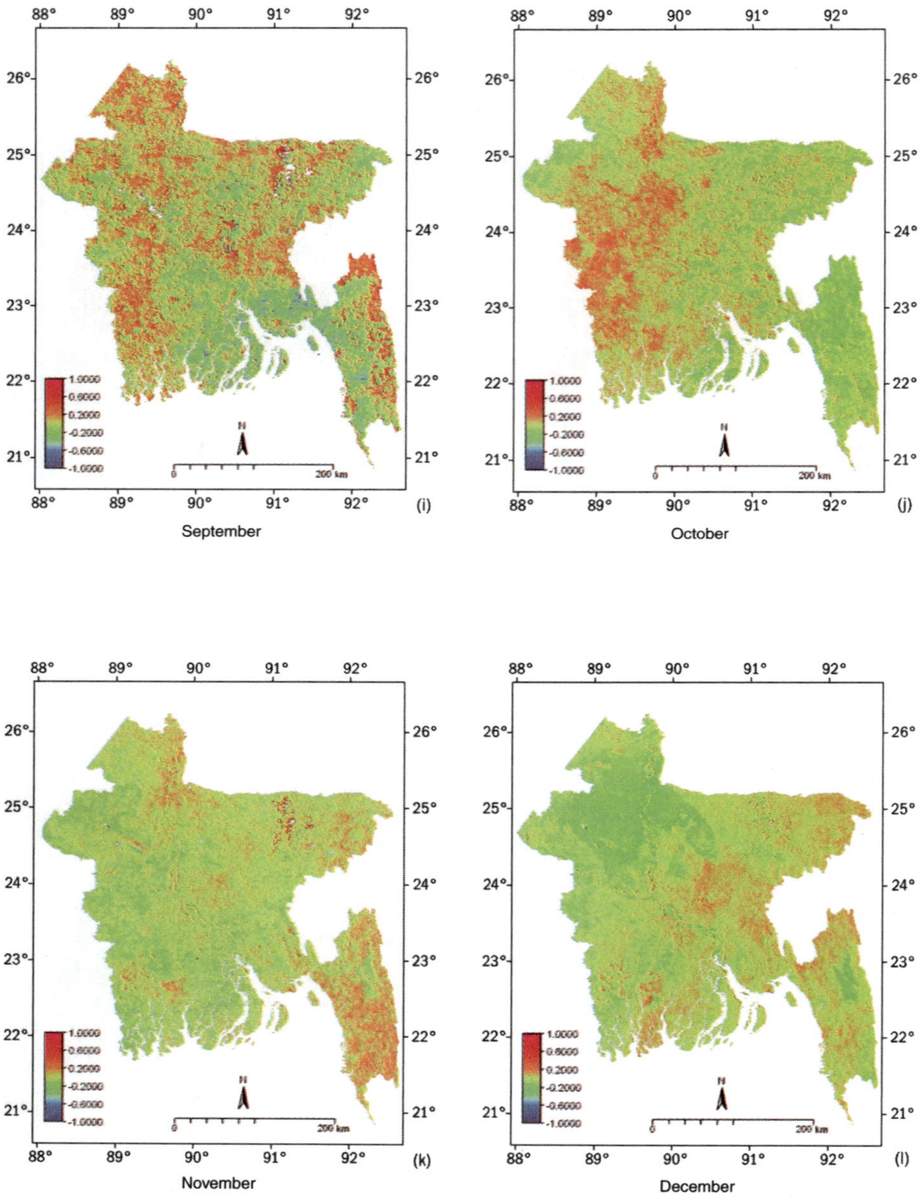


**Figure 6** Spatial distribution of trends of mean monthly temperature across Bangladesh for the past eight years (2000–07)

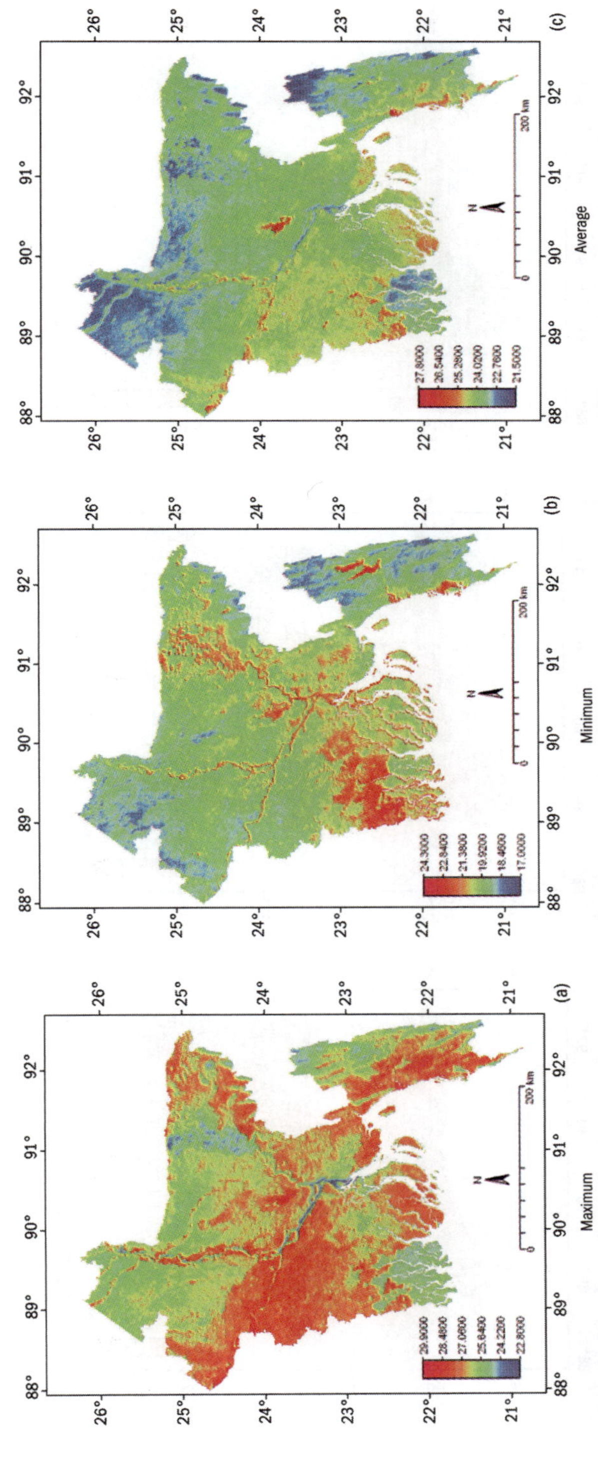


**Figure 7** Spatial distribution of (a) maximum, (b) minimum, and (c) mean temperatures, their trends, and $R^2$ values from MODIS image

*(Fig. 7 Contd..)*

*(Fig. 7 Contd..)*

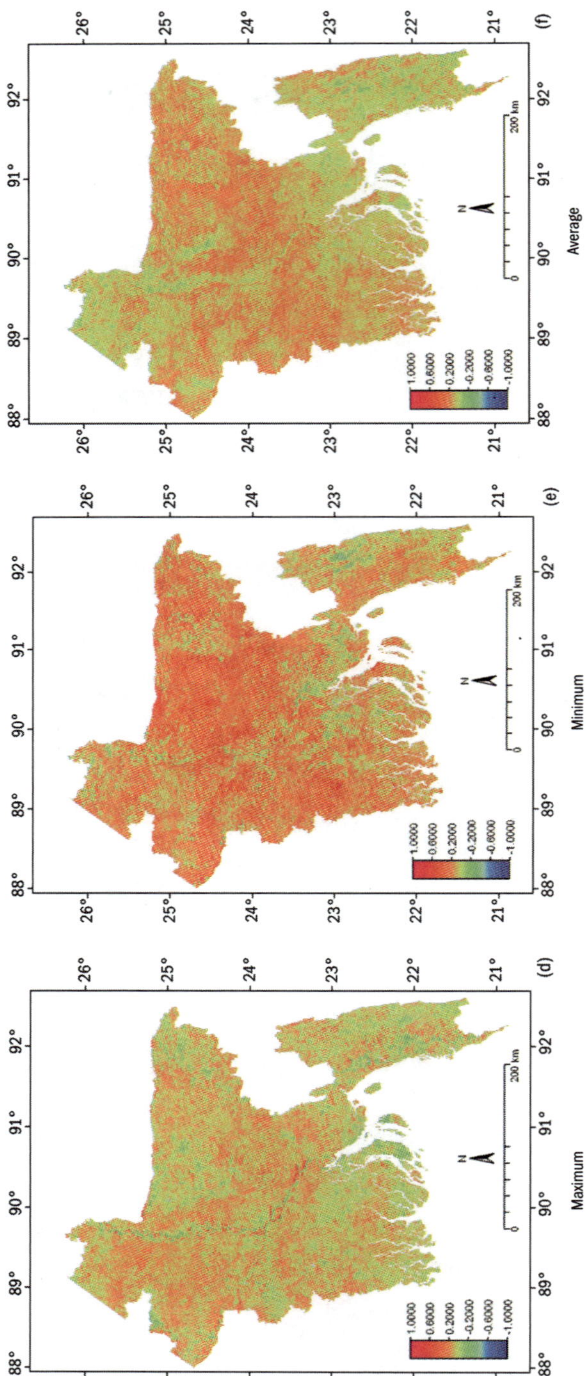


**Figure 7** Spatial distribution of (a) maximum, (b) minimum, and (c) mean temperatures, their trends, and $R^2$ values from MODIS image

*(Fig. 7 Contd..)*

*(Fig. 7 Contd..)*

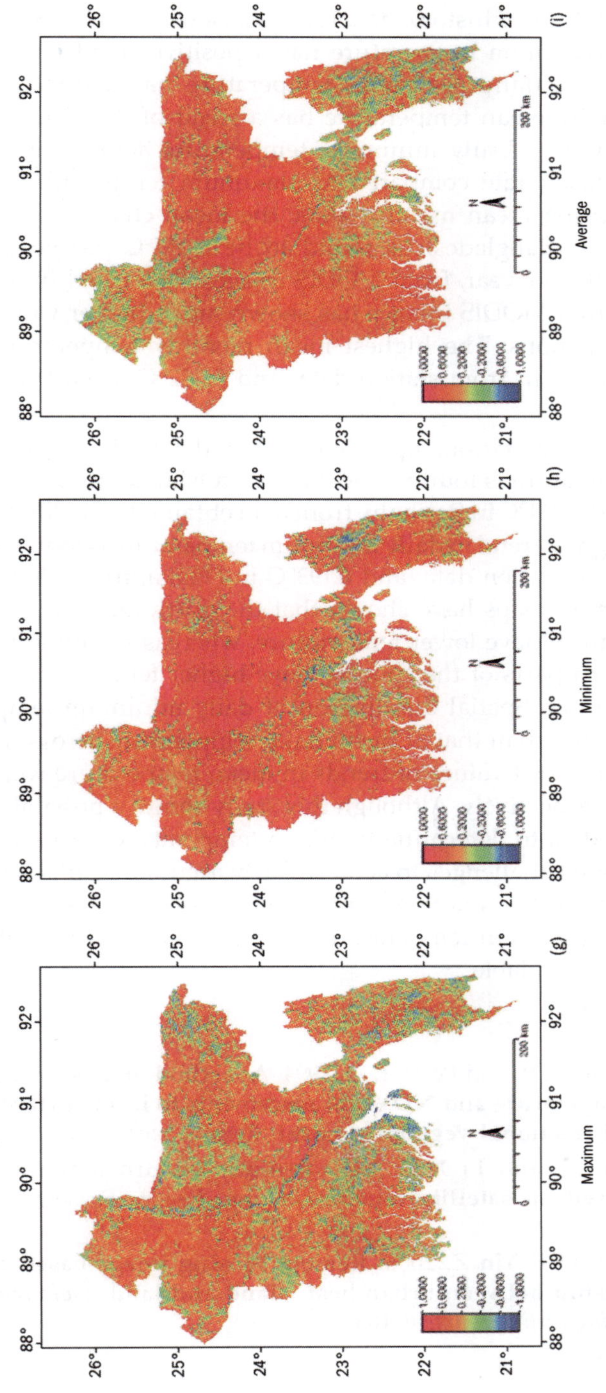


**Figure 7** Spatial distribution of (a) maximum, (b) minimum, and (c) mean temperatures, their trends, and $R^2$ values from MODIS image

alternative to meet the need of fine resolution ground-based measurement instruments. Analysis of historic data for the past 60 years (1948–2007) shows that the daily maximum temperature has a positive trend of 0.0062°C per year. However, the daily minimum temperature has a trend of 0.0154°C per year, and daily mean temperature has a trend of 0.0103°C per year (or 1.03°C per 100 years). Daily minimum temperature across Bangladesh has increased at a faster rate compared to maximum temperature during the past 60 years. Using mean monthly data, the mean change during the last eight years across Bangladesh is found to be 0.035°C per year, from stations, and 0.105°C per year, from MODIS images. The trend in temperature data obtained from MODIS images has shown much higher value than that obtained from stations. The highest mean monthly temperature is found to be 0.194°C per year, from station data, and 0.332°C, from MODIS image, during May.

Data analysis, by station, has shown that the highest value in daily maximum temperature is found to be 0.129°C in Ishurdi, from data obtained from station, and 0.214°C in Rajshahi, from data obtained from MODIS images. However, the highest trend in daily minimum temperature is found to be 0.13°C in Kutubdia, from station data, and 0.193°C in Dhaka, from MODIS images. Spatial distribution maps have shown that the north and the north-eastern parts of the country have lower temperature, whereas the middle, southern, and south-eastern parts of the country have higher temperature than other parts of the country. Spatial distribution of daily minimum temperature is uniform and higher than that of mean daily temperature across the country. The highest and lowest values of trends in mean temperature were found in May and April, respectively. Although this study shows a possibility of using remote sensing data to determine trends in temperature across Bangladesh, there are still many challenges to achieve fairly accurate results. The possible future directions of the study could be the use of high resolution images. The accuracy of trends in temperature can be increased if the satellite data of many years is available.

## References

Carlson T N, Gillies R R, and Perry E M. 1994. **A method to make use of thermal infrared temperature and NDVI measurements to infer surface soil water content and fractional vegetation cover**. *Remote Sensing Reviews* **9**:161–173

Chen Y H, Wang J, and Li X B. 2002. **A study on urban thermal field in summer based on satellite remote sensing**. *Remote Sensing for Land and Resources* **4:** 55–59

Chen X, Zhao H, Li P, Yin Z. 2006. **Remote sensing image-based analysis of the relationship between urban heat island and land use/cover changes**. *Remote Sensing of Environment* **104**: 133–146

Dash P, Gottsche F-M, Olesen F-S, Fischer H. 2002. **Land surface temperature and emissivity estimation from passive sensor data: theory and practice—current trends**. *International Journal of Remote Sensing* **23**(13): 2563–2594

Dozier J and Wan Z. 1994. **Development of practical multiband algorithms for estimating land-surface temperature from EOS/MODIS data**. *Advances in Space Research* **13**(3): 81–90

Franca G B and Cracknell A P. 1994. **Retrieval of land and sea surface temperature using NOAA-11 AVHRR data in north-eastern Brazil**. *International Journal of Remote Sensing* **15**: 1695–1712

Friedl M A. 2002. **Forward and inverse modeling of land surface energy balance using surface temperature measurements**. *Remote Sensing of Environment* **79**: 344–354

Gallo K P and Owen T W. 1998. **Assessment of urban heat islands: a multi sensor perspective for the Dalas Ft Worth, USA region.** *Geocarto International* **13**: 35–41

Gillespie A R, Rokugawa S, Matsunaga T, Cothern J S, Hook S J, Kahle A. B. 1998. **A temperature and emissivity separation algorithm for advanced spaceborne thermal emission and reflection radiometer (ASTER) images**. *IEEE Transactions on Geoscience and Remote Sensing* **36**: 1113– 1126

Gillies R R and Carlson T N. 1995. **Thermal remote sensing of surface soil water content with partial vegetation cover for incorporation into climate models**. *Journal of Applied Meteorology* **34**: 745–756

IPCC (Intergovernmental Panel on Climate Change). 2007. ***Climate Change 2007: the physical science basis.*** IPCC Working Group I Fourth Assessment Report. Summary for Policymakers. Geneva: IPCC

ITC (International Institute for Geo-information Science and Earth Observation). 2007. **Integrated Land and Water Information System (ILWIS)**. 2009. Details available at <http://www.itc.nl/ilwis>, last accessed on 11 November 2009

Li Z-L and Becker F. 1993. **Feasibility of land surface temperature and emissivity determination from AVHRR data**. *Remote Sensing of Environment* **43**: 67–85

MODIS. 2009. **MODerate resolution Imaging Spectroradiometer**. Details available at <http://modis.gsfc.nasa.gov/>, last accessed on 11 November 2009

Neteler M. 2010. **Estimating daily land surface temperatures in mountainous environments by reconstructed MODIS LST data**. *Remote Sensing* **2**(1): 333–351

Owen T W, Carlson T N, and Gillies R R. 1998. **An assessment of satellite remotely-sensed land cover parameters in quantitatively describing the climatic effect of urbanization**. *International Journal of Remote Sensing* **19**: 1663–1681

Price J. C. 1990. **Using spatial context in satellite data to infer regional scale evapotranspiration**. *IEEE Transactions on Geoscience and Remote Sensing* **28**: 940–948

Streutker D R. 2002. **A remote sensing study of the urban heat island of Houston, Texas.** *International Journal of Remote Sensing* **23**: 2595–2608

Streutker D R. 2003. **Satellite-measured growth of the urban heat island of Houston, Texas.** *Remote Sensing of Environment* **85**: 282–289

Weng Q. 2001. **A remote sensing and GIS evaluation of urban expansion and its impact on surface temperature in Zhujiang Delta, China**. *International Journal of Remote Sensing* **22**(10): 1999–2014

Weng Q Q, Lu D, and Schuring J. 2004. **Estimation of land surface temperature–vegetation abundance relationship for urban heat island studies**. *Remote Sensing of Environment* **89** : 467–483.

Yuan F and Bauer M E. 2007. **Comparison of impervious surface area and normalized difference vegetation index as indicators of surface urban heat island effects in Landsat imagery**. *Remote Sensing of Environment* **106**: 375–386

# High Resolution Climate Change Scenarios for Morocco for the 21st Century


**Abdelaziz Babqiqi[*, †], Mohamed Messouli[†], and Atika Kasmi[#]**

[*]*Ministry of Environment, Rabat, Morocco*

[†]*University Cadi Ayyad, Marrakech, Morocco*

[#]*Moroccan Meteorological Services, Casablanca, Morocco*

*E-mail: abdelaziz.babqiqi@gmail.com; messouli@gmail.com; kasmiatika@gmail.com*


## INTRODUCTION

Moroccan climate is affected by different factors at large scale as well as at small scale. It is regulated by the position of two centres of action: Azores anticyclone and Saharan depression. Mountain channels, latitudinal extension, geographic diversity, and two oceanic openings on Atlantic Ocean and Mediterranean Sea, all of them play a crucial role in determining the different types of climate in Morocco. Precipitation is most intense in mountainous area and in the west of the Atlas Mountains. All these characterize a temperate to semi-arid climate in the north and semi-arid to arid climate in the south of Morocco.

Further, winters in Morocco are influenced by the North Atlantic Oscillation (NAO) phenomenon. NAO is related to the position, trajectories, and pressure values of two centres of action: Azores anticyclone and Iceland depression. When NAO is in positive phase, a pressure gradient between Iceland low pressure and Azores high pressure cells generates a strong westerly wind across the mid-latitude of North Atlantic, thus resulting in a wetter and warmer winter towards the northwest of Africa. Conversely, under the negative phase of NAO, pressure gradient is weaker, thus the winter is drier than normal.

A study of the De-Martonne Aridity Index revealed that the climate in the north of Morocco is semi-humid to semi-arid, and that in the south is arid to desertical. In the same context, a comparison of this index during two time periods, namely, 1961–70 and 1971–2000, as illustrated in Figure 1, showed northwards migration of arid climate.

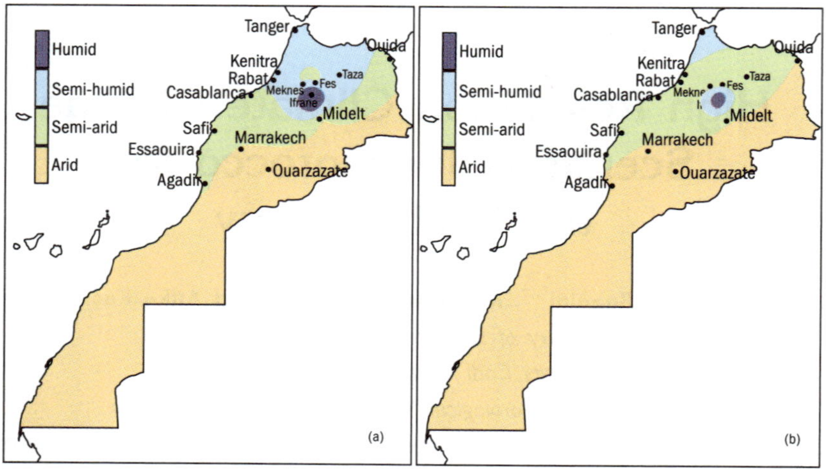


**Figure 1** De-Martonne Aridity Index over Morocco during (a) 1961–70 and (b) 1971–2000

The signs of climate change are already manifesting themselves in Morocco through changes in temperatures and precipitation. A study conducted by Moroccan Meteorological Services over the last 45 years (1961–2005) shows a significant negative trend in rainfall and positive trend in temperatures. The *Fourth Assessment Report* of Intergovernmental Panel on Climate Change (IPCC) predicts a decrease in rainfall over Morocco, which could decrease by 20% by the end of the century. The temperature is expected to increase by 2.5°C to 5.5°C, according to the IPCC scenarios, by the end of the century. Further, global climate models used until now by the IPCC have low resolutions and do not include a very precise information on topography, vegetation, and soil. However, assessment of climate change impact on socio-economic sectors, such as agriculture, water resources, and health, requires climate change scenarios at high resolution with detailed information at the local level.

The objective of this study is to use the Statistical Downscaling Model (SDSM) (Figure 2) and climate change scenarios from low resolution models (General Circulation Models, or GCMs) to generate climate change scenarios at high resolution and at daily time step. In this work, we have used daily observations from 20 meteorological stations and daily National Center for Environmental Prediction (NCEP), USA, re-analysis at large scale for the present climate (1961–2000) in Morocco. For future climate, we have used climate change scenarios at low resolution from the English model HadCM3 for the period 1961–2099. During this study, we have developed future climate change scenarios in Morocco at high resolution using daily time step during the period 1961–2099 and for four parameters: mean temperature, maximum temperature, minimum temperature, and precipitation.

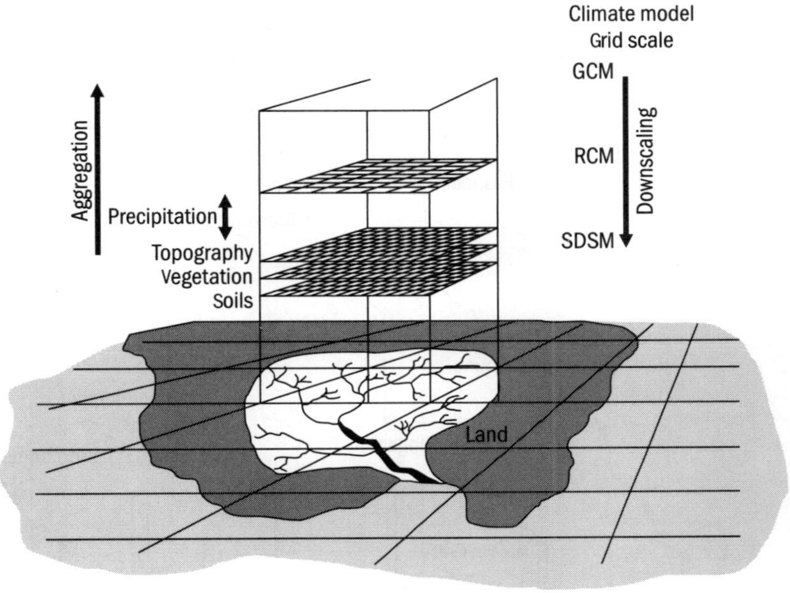


**Figure 2** Principle of the downscaling technique and the reverse technique (aggregation)

GCM – Global Climate Model; RCM – Regional Climate Model; SDSM – Statistical Downscaling Model

This article is presented in three parts, the *first* part details the data used in this work, the *second* describes the methodology and tools, and finally, in the *third* section, results have been discussed. It is very important to note that the availability of meteorological data set is the fundamental requirement, and the selection of an appropriate algorithm and techniques is the analytical necessity for such type of studies.

## DESCRIPTION OF DATA USED

### Observed Data

Daily observed data for 40 years (1961–2000) for mean temperatures, maximum temperatures, minimum temperatures, and precipitation is investigated in this study to develop high resolution climate change scenarios. These observations have been compiled from 16 Moroccan meteorological stations and four stations from Mauritania in order to have more observations over the Sahara region. These 20 stations were chosen to represent different agro-ecological zones. Figure 3 summarizes distribution of 20 observation stations chosen in this study.

**Figure 3** Observation stations chosen in this study

## NCEP Re-analysis at Low Resolution

The data was provided by NCEP centre. This new atmospheric re-analysis was produced by using historical data (all observations) and data assimilation system, with resolution of 2.5° × 2.5° and is available for every 6 hours since 1948.

In this study, we were interested only in the daily data between 1961 and 2000 after interpolation to HadCM3 model grid described in the following paragraph.

## Climate Change Scenarios of HadCM3 at Low Resolution

These climate change scenarios describe possible future climate, relying on the functioning of climate, evolution of population, economic activity, and greenhouse gas effect. In this study, we have used two climate change scenarios $A_2$ and $B_2$ (*Fourth Assessment Report* of IPCC), produced by the English model HadCM3 and used by the IPCC. These scenarios were used over the period 1961–2099 with a resolution of 2.5° × 3.75°. Figure 4 shows the position of grid points of climate model HadCM3 over the north of Morocco.

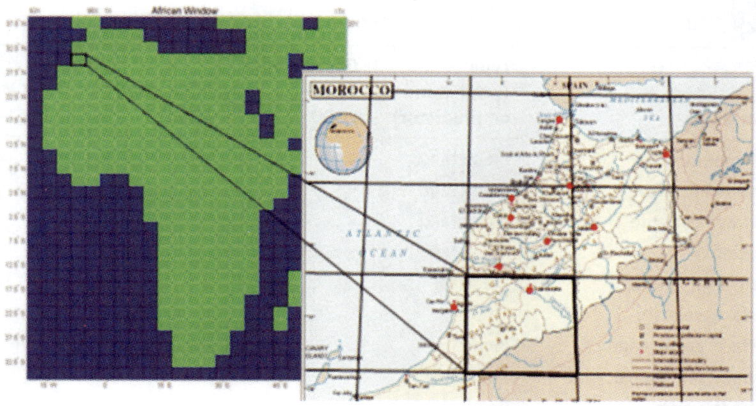


**Figure 4** Grid points of climate model HadCM3 over the north of Morocco

## TOOLS AND METHODOLOGY

SDSM is a model based on the statistical techniques of downscaling, developed by the environmental agency of England (Environment Agency of England and Wales) (Wilby, Dawson, and Barrow 2002). This model is recommended by the World Bank and the Canadian Climate Impacts Scenarios (CCIS), international sponsors of climate change projects.

The downscaling technique uses empirical relationships between observed variables called predictands (mean temperature, minimum temperature, maximum temperature or precipitations) (Table 1) and large-scale variables called predictors (data from NCEP or $HadCM_3$) (Table 1). This allows constructing climate change scenarios at high resolution with daily time step. These scenarios were developed for four parameters over the period 1961–2099: mean temperature, maximum temperature, minimum temperature, and precipitations.

Scenarios at high resolution over the period 1961–2000 (present climate) are created only for scientific reasons. We need to have a reference (1961–2000) for the same conditions of creation of scenarios for the period 2001–99 (future climate).

The major advantage of SDSM is that it allows taking into account local factors such as topography, vegetation, and soil through local observations. It also produces rapidly the daily meteorological parameters (mean temperature, maximum temperature, minimum temperature, and precipitation) for the entire period of 1961–2099. This facilitates interpretation of trends of interannual variability of these parameters. However, this is only possible in areas where observations are available consistently and reliably. Figure 5 shows appropriate operations for the creation of high resolution climate change scenarios. These operations are summarized in five main steps.

**Table 1** All predictors and predictands used in the Statistical Downscaling Model

| Daily variables | Type (predictand or predictor) | Observations 1961–2000 | NCEP (1961–2000) | HadCM3 (1961–2009) |
|---|---|---|---|---|
| Mean temperature | Predictand | × | | |
| Maximum temperature | Predictand | × | | |
| Minimum temperature | Predictand | × | | |
| Precipitations | Predictand | × | | |
| Mean sea level pressure | Predictor | | × | × |
| Near surface velocity | Predictor | | × | × |
| Near surface zonal wind | Predictor | | × | × |
| Near surface meridional wind | Predictor | | × | × |
| Near surface vorticity | Predictor | | × | × |
| Near surface wind direction | Predictor | | × | × |
| Near surface divergence | Predictor | | × | × |
| Zonal velocity component at 500 hpa | Predictor | | × | × |
| Meridional velocity component at 500 hpa | Predictor | | × | × |
| Vorticity at 500 hpa | Predictor | | × | × |
| Geopotentiel height 500 hpa | Predictor | | × | × |
| Wind direction at 500 hpa | Predictor | | × | × |
| Divergence at 500 hpa | Predictor | | × | × |
| Relative humidity at 500 hpa | Predictor | | × | × |
| Near surface specific humidity | Predictor | | × | × |
| Velocity at 850 hpa | Predictor | | × | × |
| Zonal velocity at 850 hpa | Predictor | | × | × |
| Meridional velocity at 850 hpa | Predictor | | × | × |
| Vorticity at 850 hpa | Predictor | | × | × |
| Geopotential height at 850 hpa | Predictor | | × | × |
| Wind direction at 850 hpa | Predictor | | × | × |
| Divergence at 850 hpa | Predictor | | × | × |
| Relative humidity at 850 hpa | Predictor | | × | × |

NCEP – National Center for Environmental Prediction

## Data Monitoring and Processing

The first step in the SDSM technique is the verification of observed data before model calibration. It is also sometimes necessary to make some changes in predictands (observations) and/or predictors (data from large-scale NCEP or HadCM3), for example, logarithmic, inverse or binomial transformation.

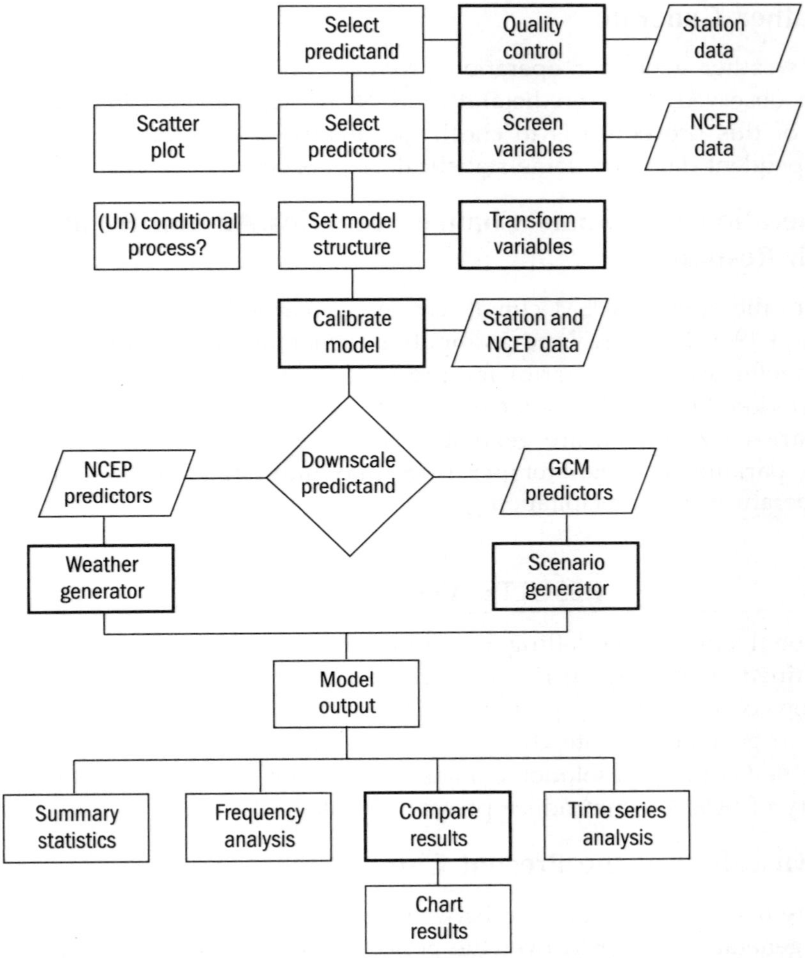


**Figure 5** Different operations for producing high resolution climate change scenarios

## Screening of Predictors

Identifying empirical relationships between predictors (such as mean sea level pressure) and predictands (such as observed precipitation) is very important for statistical downscaling methods. The main purpose of the screening operation is to assist the user in making an optimal selection of appropriate predictors.

## Model Calibration

The model calibration operation comprises computing all the coefficients of multiple regression equations through an optimal algorithm.

## Weather Generator

The weather generator operation produces an artificial daily time series from observed data (predictand) and NCEP re-analysis (predictors). The aim of this procedure is to check the quality of calibrated models using independent data and some statistical diagnostics.

## Generation of Climate Change Scenarios A2 and B2 at High Resolution

After calibration of SDSM, climate change scenarios A2 and B2 from $HadCM_3$ (period 1961–2099) are used as inputs to generate daily series (20 series by parameter and by scenario) for present climate and future climate over the period 1961–2099 and for each meteorological station. Climate change scenarios A2 and B2 are generated separately for each scenario and for each parameter: mean temperature, maximum temperature, minimum temperature, and precipitation.

# RESULTS AND DISCUSSION

Regional climate modelling in mountainous, semi-arid, and arid areas remained extremely difficult until now. In this study, 20 observation stations covering a large part of spatial variability of Morocco have provided data to generate climate change scenarios at local scale. Before discussing the results of high resolution climate change scenarios, we will illustrate the ability of SDSM to reproduce present climate.

## Downscaling of the Present Climate

Figure 6 shows a comparison between daily average temperature observed and generated by SDSM over the period 1 January 1999 to 1 January 2000 at Beni-Mellal station. This station was chosen because it represented an agricultural and semi-arid area. SDSM reproduced the variability of daily average temperature observed in this area (Figure 6). This pattern was observed on interannual scale also, wherein SDSM reproduced correctly (with a very small error) the seasonal mean temperature (Figure 7) between 1961 and 2000. In general, SDSM effectively reproduces daily seasonal temperature, despite the uncertainty caused by the lack of observation during this period from this station.

Figure 8 presents a comparison between monthly precipitation observed and reproduced by SDSM over the period 1973–98. There is a general agreement between both types of data, except for the period January and February when there is an underestimation by SDSM. This underestimation may be related to the uncertainty introduced by the lack of data, which characterizes some periods of observation. In fact, Beni-Mellal was chosen

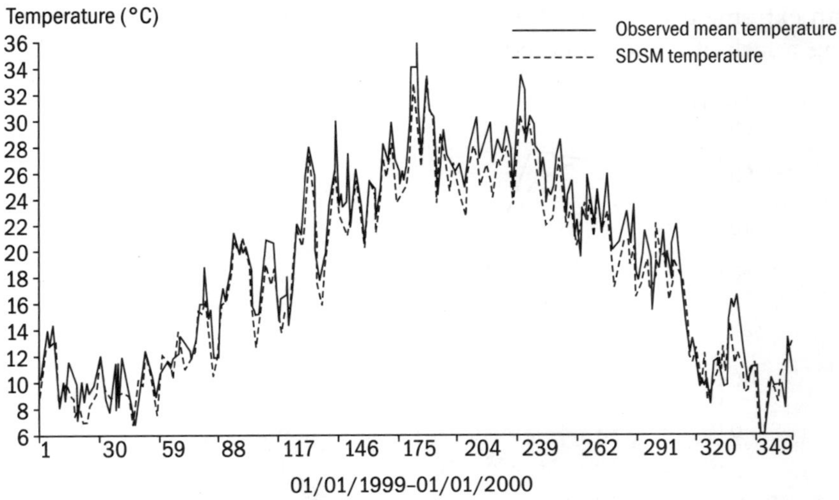


**Figure 6** Comparison between daily observed mean temperature and that generated by over the period 1 January 1999 to 1 January 2000 for the meteorological station at Beni-Mellal

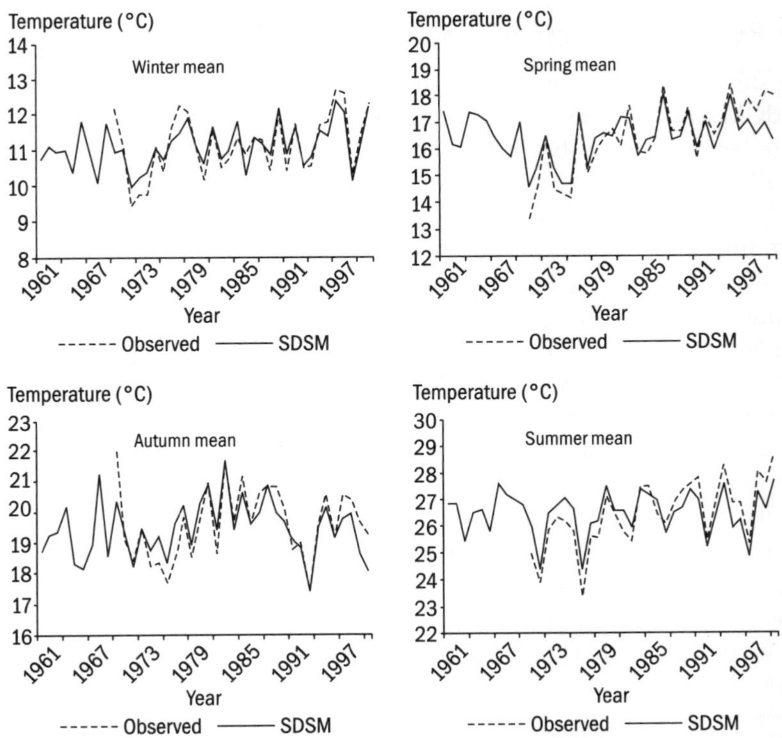


**Figure 7** Comparison between seasonal mean temperature observed and generated by SDSM over the period 1961–2000 at meteorological station Beni-Mellal

as an extreme example to illustrate the performance of SDSM despite there being a lack of some observed data during the study period.

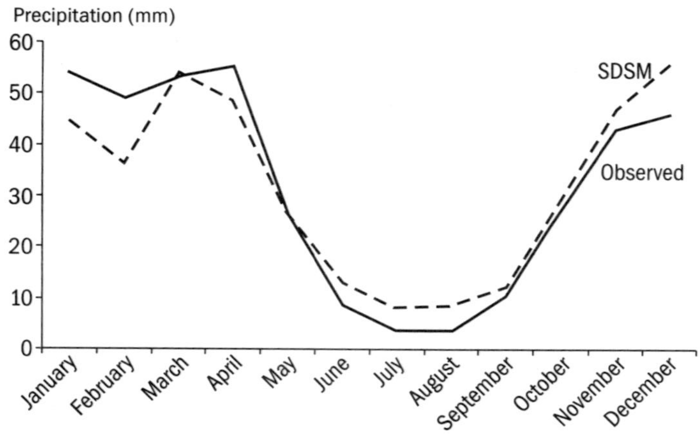


**Figure 8** Comparison between monthly precipitation observed and reproduced by SDSM during the period 1973–98.

## High Resolution Climate Change Scenarios

Using downscaling technique of the SDSM and both A2 and B2 scenarios of $HadCM_3$ model at low resolution, for the period 1961–2099, future climate change scenarios at high resolution have been generated. Statistical anomalies were calculated for the years 2020 (2011–40), 2050 (2041–70), and 2080 (2071–99) compared to the reference period 1961–90 for mean temperature, maximum temperature, minimum temperature, and precipitation.

Figure 9 presents seasonal and annual anomalies of mean, maximum, and minimum temperature of Beni-Mellal for three future horizons: 2020, 2050, and 2080. We found that the downscaling with SDSM provides a seasonal global warming for three temperatures, for both A2 and B2 scenarios and for the three future horizons: 2020, 2050, and 2080 (Figure 9). This warming is more pronounced under IPCC-A2 scenario than under IPCC-B2 scenario. This finding is justified because A2 scenario is more pessimistic than B2 scenario.

In order to get an idea on estimated annual warming in Morocco using SDSM, we present in Figure 10 a spatialization of future scenarios obtained for mean, maximum, and minimum temperatures at 20 observation stations. It can be observed that the annual warming for three horizons—2020, 2050, and 2080—is characterized by an east–west gradient with larger amplitudes in the east than in the west. This finding is valid for three parameters: mean temperature (Figure 10: a1, a2, and a3), maximum temperature

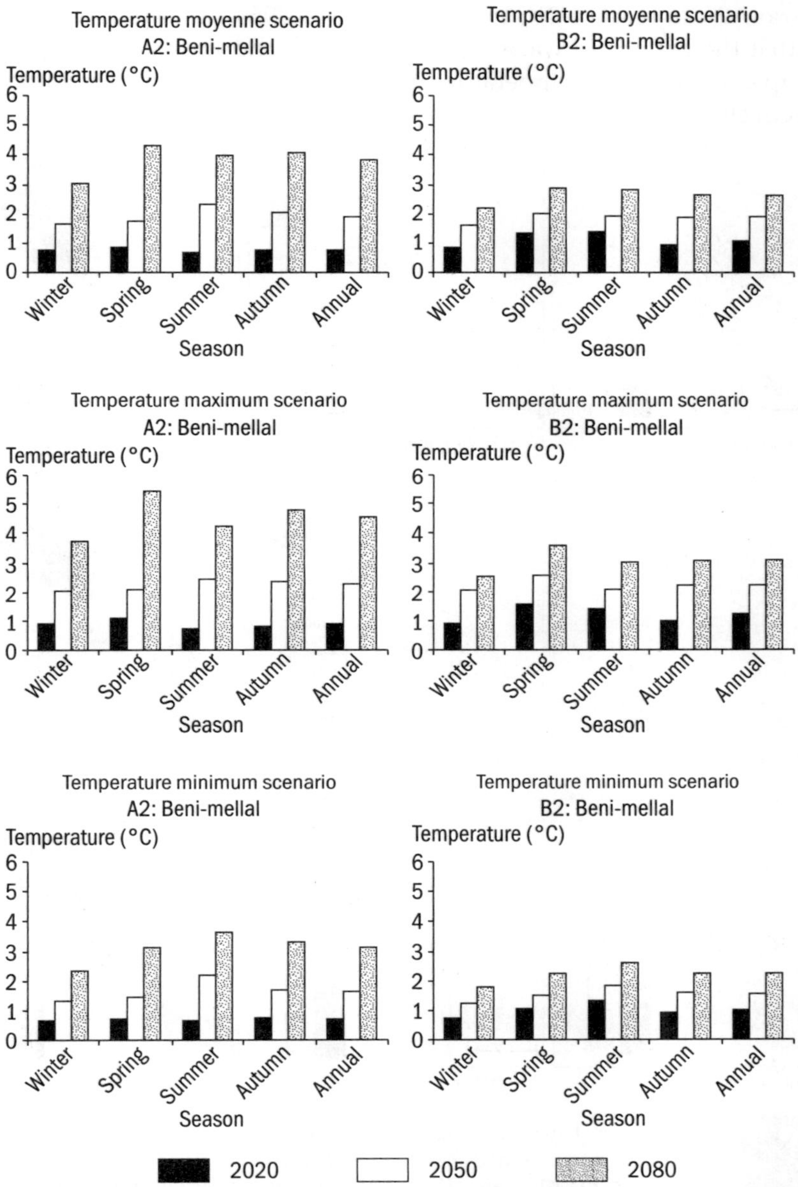


**Figure 9** Seasonal and annual anomalies (°C) of mean temperature, maximum temperature, and minimum temperature in the agricultural area Beni-Mellal, for both A2 and B2 scenarios and for three future horizons: 2020 (2011–40), 2050 (2041–70), and 2080 (2071–99)

(Figure 10: b1, b2, and b3), and minimum temperature (Figure 10: c1, c2, and c3). With respect to amplitude, increase in annual mean temperature with scenario A2 can reach 1–2°C by 2020, 3–4°C by 2050, and 4–5°C by 2080. We note that the pattern of warming in maximum and minimum temperature is comparable to mean temperature but with great and moderate amplitude, respectively.

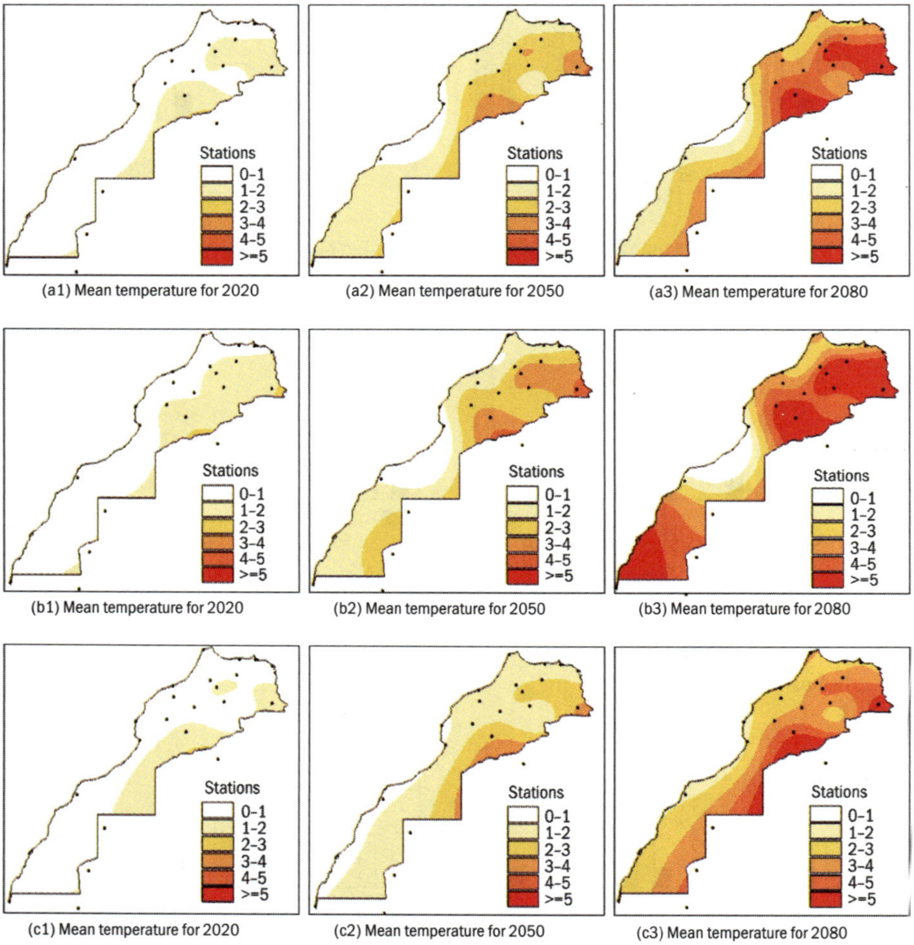


**Figure 10** Spatialization of results of scenario A2: mean temperature anomaly, maximum temperature anomaly, and minimum temperature anomaly (°C) for three future horizons—2020 (2011–40), 2050 (2041–70), and 2080 (2071–99) compared to the reference period 1961–90

Figure 11 presents the percentage change in seasonal and annual total precipitations for the three future horizons: 2020, 2050, and 2080. Change shown in this figure is for both A2 and B2 scenarios at three stations: Meknes representing an agricultural area, Ouarzazate representing a dry

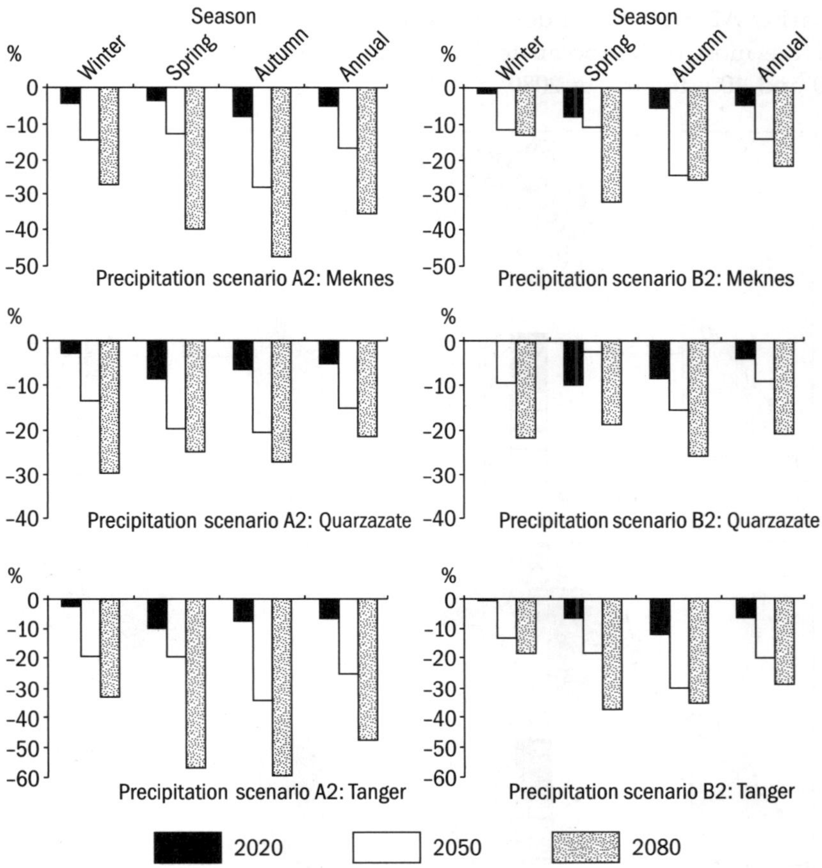


**Figure 11** Percentage change in seasonal precipitation (winter, spring, and autumn) and annual precipitation for the three future horizons, namely, 2020, 2050, and 2080, and for both A2 and B2 scenarios at the three meteorological stations: Meknes, Ouarzazate, and Tanger

area, and Tanger representing a semi-humid area. The downscaling with SDSM provides global future drying for three areas (Meknes, Ouarzazate, and Tanger) and for three seasons: winter, spring, and autumn. Even a naturally dry area as Ouarzazate may be further drained. The percentage of estimated annual drainage is more significant with A2 than B2, particularly in 2050 and at the end of century (2080), when reduced rainfall constructed by SDSM can exceed 40%, as in the case of Tanger (Figure 11).

We show in Figure 12 the percentage change in precipitation at 20 stations and for both horizons 2050 and 2080. It is observed that SDSM constructs large drought in the north and the west of the country under the two scenarios A2 and B2. Under this technique, the decrease in precipitation over the most of Moroccan territory may be to the tune of 5% to 40% in 2050 and 40% to 60% in 2080.

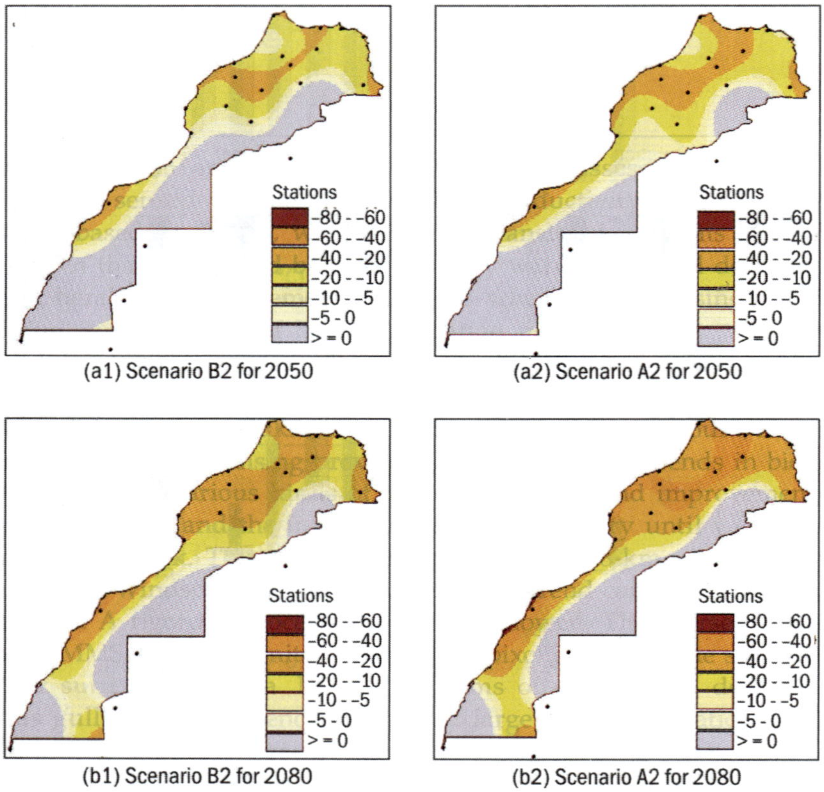


**Figure 12** Spatialization of percentage change of precipitation for 20 stations for both horizons 2050 (2041–70) and 2080 (2071–99) compared to the reference period 1961–90

## CONCLUSION

In this study, we used SDSM to generate the future climate change scenarios at high resolution for mean temperature, maximum temperature, minimum temperature, and precipitation in Morocco. In this context, we used three types of data. Local observations of 20 meteorological stations, NCEP re-analysis, and climate change scenarios at low-resolution from HadCM3 model.

The comparison of the present climate (1961–2000) based on observations and that based on data generated by SDSM gave a match concordance between both data sets. This demonstrates the ability of SDSM to reproduce scenarios at high-resolution with high performance for four parameters (mean temperature, maximum temperature, minimum temperature, and precipitation).

On the other hand, future climate change scenarios at high resolution obtained in this work remain very similar to those shown in the IPCC *Fourth Assessment Report*. In fact, calculation of future climate anomalies (2011–40, 2041–70, and 2071–99) compared to present climate (1961–2000) for the four parameters studied (mean temperature, maximum temperature, minimum temperature, and precipitation) shows an increase in temperature and a decrease in precipitation. This increase in mean temperature in some areas of Morocco may be to the tune of 1–2°C, 3–4°C, and 4–5°C for 2011–40, 2041–70, and 2071–99, respectively. The decrease in precipitation, according to SDSM, may be of the order of 5%–10% for 2011–40, 10%–40% for 2041–70, and 40%–60% by the end of the century (2071–99).

Climate change scenarios generated at high resolution in this work provide a reference database as input for climate change impact model on socio-economic sectors such as agriculture, water resources, and health. Recently, scenarios obtained in this work are being used by the World Bank and the Food and Agricultural Organization (FAO) to conduct a study on the impact of climate change on agriculture in Morocco. This study gives future projections for 50 crops under both A2 and B2 scenarios.

## Reference

Wilby R L, Dawson C W, and Barrow E M. 2002. **SDSM – a decision support tool for the assessment of regional climate change impacts**. *Environmental and Modelling Software* **17**: 145–157

## Bibliography

Bennani A, Buret J, and Senhaji F. 2001. **Communication Nationale Initiale a la Convention Cadres des Nations Unies sur les changements climatiques**, Maroc.

DMN (Direction de la Météorologie Nationale). 2006. **Etude des indices de changements climatiques sur le Maroc: températures et précipitations**. DMN "INFOMET", Casablanca.

DMN (Direction de la Météorologie Nationale). 2007. **Les changements climatiques au Maroc: observations et projections**, Maroc-Météo, Casablanca.

GIEC (Groupe Intergouvernemental d'Experts sur l'Evolution du Climat). 2007. le quatrième rapport du groupe de travail I

Gommes R, Kanamaru H, El hairech T, Rosillon D, Babqiqi A, Balaghi R, Doukkali R, Driouech F, El ouali A, Jlibene M, Khannoufi A, Wilby R. 2008. ***World Bank–Morocco Study on the Impact of Climate Change on the Agricultural Sector, FAO component: impacts on crop yields.*** FAO and Agricultural Ministry of Morocco, Maroc

Kundzewicz Z W, Mata L J, Arnell N W, Döll P, Kabat P, Jiménez B, Miller K A, Oki T, Sen Z, Shiklomanov I A. 2007. **Freshwater resources and their management**. In *Climate Change 2007: impacts, adaptation and vulnerability*. Contribution of Working Group II to the Fourth Assessment Report of the Intergovernmental Panel on Climate Change, edited by Parry *et al.* Cambridge: Cambridge University Press

Osborn T J, Briffa K R, Tetts S F B, Jones P D, Trigo R M. 1999. **Evaluation of the North Atlantic Oscillation as simulated by a coupled climate model.** *Climate Dynamics* **15**: 685–702

Rousset N and Arrus R. 2006. **L'agriculture du Maghreb au défi du changement climatique: quelles stratégies d'adaptation face à la raréfaction des ressources hydriques.** Communication à WATMED 3,3e conférence internationale sur les Ressources en Eau dans le Bassin Méditerranéen, Tripoli.

# GIScience Tools for Climatic Change Related Natural Hazards and Modelling

**Biswajeet Pradhan**
*Institute of Advanced Technology, University Putra Malaysia, UPM, Serdang, Malaysia*
*E-mail: Biswajeet.Pradhan@mailbox.tu-dresden.de*

## INTRODUCTION

Today, the world is experiencing severe global warming and uncertain climate changes. Climatic change and the environmental degradation, in particular, are the results of a long-term evolution. These phenomena consist of components that can be situated in the past and the present, and can be extrapolated towards the future. The process leading to climatic changes strongly differs with the regional geography, and more precisely, with the human and physical climate of each region. The combined effect of a high-energy environment and human land use is that a disproportionate number of natural disasters are causing serious damage to mankind.

With the advent of the earth-observing sensors in orbit in association with geographic information technology (GIT), climatologists, environmentalists, and earth scientists have access to data and technology that can be used to assess climatic changes and dynamics of natural hazards. The impact of natural disasters, such as flooding, forest fires, storms, and earthquakes, can be lessened through effective hazard management and mitigation. Hazard is defined as an incident that occurs in a sudden manner, and which is complex in nature and causes loss of lives and damage to property, and adversely impacts natural environment and local activities. Such an incident needs management, which involves effective coordination between many agencies for extensive resources, equipment, skills, and manpower, which possibly demands a complex action and would require a long time. While the vulnerability to natural hazards varies from region to region, a large part of the country is exposed to such natural hazards, which often turn into disasters, causing significant disruption in the socio-economic life of communities, leading to loss of life and property (BMTPC 2001). It is necessary to develop and enhance prediction models, monitor facilities, and draft regulations and contingency plans vis-à-vis natural or man-made disasters. The term "risk" is often confused with "hazard". Risk is the probability or chance that hazard poses. Hazard is an inescapable part of life (Smith 1996). Hazard signifies potential, while disaster is the actual event (Drabek 1997). Disaster is the result of a hazard impacting a community (Blanchard 1999). Disaster is a source of danger the evaluation of which encompasses three elements: risk of personal harm; risk of property; risk of environmental damage and the acceptability of the degree of risk (Kovach 1995; Smith 1996). Natural or environmental disaster includes both natural and man-made dimensions, such as lithosphere disasters (landslide, subsidence, earthquake), atmospheric disasters (rain, lighting, temperature), hydrosphere disasters (flooding, coastal erosion), biologic disasters (forest fires and wildfire), and technological disasters (oil spills, transport accidents, and failures of constructions), which cause substantial damage to property, pollution, and injury to or death of people. The risk is the probability or chance that the hazard poses. Consequently, it can be reduced by primarily developing a suitable risk management plan. Risk management is important in ensuring community and environment safety, providing better information to take decisions, enabling better asset and monetary management, and improving the perception of community for risks. Geographic information system (GIS) provides tools for spatial analysis and for modelling the changing climate and surface processes and their factors at different scales. Consequently, geosciences and environmental investigations are increasingly utilizing satellite imagery, aerial photos, digital elevation models (DEM), spatial analysis, and modelling techniques to study the spatial and temporal scale dependency processes and resources (Price and Heywood 1994).

The purpose of this chapter is to demonstrate the complexity of relationship between the climatic changes and their direct impacts on natural hazards. The effective utilization of GIT to study these natural disasters and to address technical and some application issues in Malaysia has been discussed. Further, this chapter will discuss the different types of prevailing natural hazards (flood, forest fire, and landslide) to reduce the impacts through measures that can be taken prior to their occurrence, which includes adopting spatial information technologies to support increased coordination among multiple programmes of risk management by examining the application of these technologies to the task of identifying, analysing, assessing, treating, and monitoring.

## REMOTE SENSING IN NATURAL HAZARDS

Remote sensing covers all techniques related to the analysis and use of data from satellites, such as METEOSAT, NOAA-AVHRR, Landsat, SPOT, ERS-

SAR, and from aerial photographs. Remote sensing data has been extensively used to map and monitor the earth's resources and environment.

The early applications of satellite remote sensing included meteorological applications using TIROS/NOAA since 1960s. The use of remote sensing for land resource and environmental studies started in the early 1970s with the launch of Landsat series of satellites. Table 1 lists the various remote sensing sensors and their spatial resolution for major satellite systems. Most of these satellites operated using the optical wavelength of the electromagnetic spectrum. The major drawback of using optical imagery is the cloud cover problem inherent in tropical areas. In the early 1990s, the use of microwave (radar) satellite data gained prominence, which could overcome the cloud cover problem. At present, both optical and microwave data compliment each other for various applications.

**Table 1** Spatial resolution of some major remote sensing satellites

| Launch | Satellite (country) | Sensor | Spatial resolution |
|---|---|---|---|
| March 1985 | Landsat-5 (US) | MX | 30 m, 79, 120 m |
| January 1990 | SPOT-2 (France) | MX | 20 m |
| | | PAN | 10 m |
| August 1991 | IRS-1B (India) | MX | 36.5 m, 72 m |
| October 1994 | IRS-P2 (India) | MX | 36.25 m |
| November 1995 | RADARSAT-1 (Canada) | SAR | 8 m |
| December 1995 | IRS-1C (India) | MX | 23.5 m, 70 m, 188 m |
| | | PAN | 5.8 m |
| August 1997 | Orb View-2 (US) | MX | 1100 m |
| September 1997 | IRS-1D (India) | MX | 23.5 m, 70 m, 188 m |
| | | PAN | 5.8 m |
| March 1998 | SPOT-4 (France) | MX | 20 m, 1000 m |
| | | PAN | 10 m |
| April 1999 | Landsat-7 (US) | MX | 30 m |
| | | | 60 m |
| | | PAN | 15 m |
| December 1999 | Terra (US) | MX | 15 m, 30 m, 90 m, 240 m, 250 m, 500 m |
| December 1999 | KOMPSAT (Korea) | PAN | 6.6 m |
| October 2001 | QuickBird (US) | PAN | 0.61 m |
| | | MX | 2.5 m |
| October 2001 | TES (India) | PAN | 1 m |
| March 2002 | ENVISAT (ESA) | SAR | 10 m |
| May 2002 | SPOT-5 (France) | MSI Veg | 10 m, 1000 m |
| | | PAN | 2.5 m |

MX – multispectral imager; PAN – panchromatic; SAR – synthetic aperture radar

## HISTORY OF NATURAL HAZARDS IN MALAYSIA

Malaysia is a tropical country endowed with breathtaking natural beauty, exotic cultures, and natural resources. Peninsular Malaysia and the two states of Sabah and Sarawak on the island of Borneo are part of the country, which is located between 2° N and 7° N of the Equator. Peninsular Malaysia is separated from the states of Sabah and Sarawak by the South China Sea. To the north of the Peninsular Malaysia is Thailand, while its southern neighbour is Singapore. Sabah and Sarawak are bounded by Indonesia, while Sarawak also shares a border with Brunei. Recently, the country has experienced frequent natural and man-made disasters, with the most recent ones occurring in 2000, 2001, 2002, 2004, 2006, 2007, 2008, and 2009. These disasters have adversely affected settlements and structures that support transportation, natural resources, and tourism. They have caused considerable damage to highways, waterways, properties, and livestock. Natural disasters in tropical terrain are natural degradation processes. However, the recent severe man-made pollution and climate changes have direct impacts on the disasters. Figure 1 shows the climate changes and warming rates in Malaysia for the last 40 years (Tangang 2007).

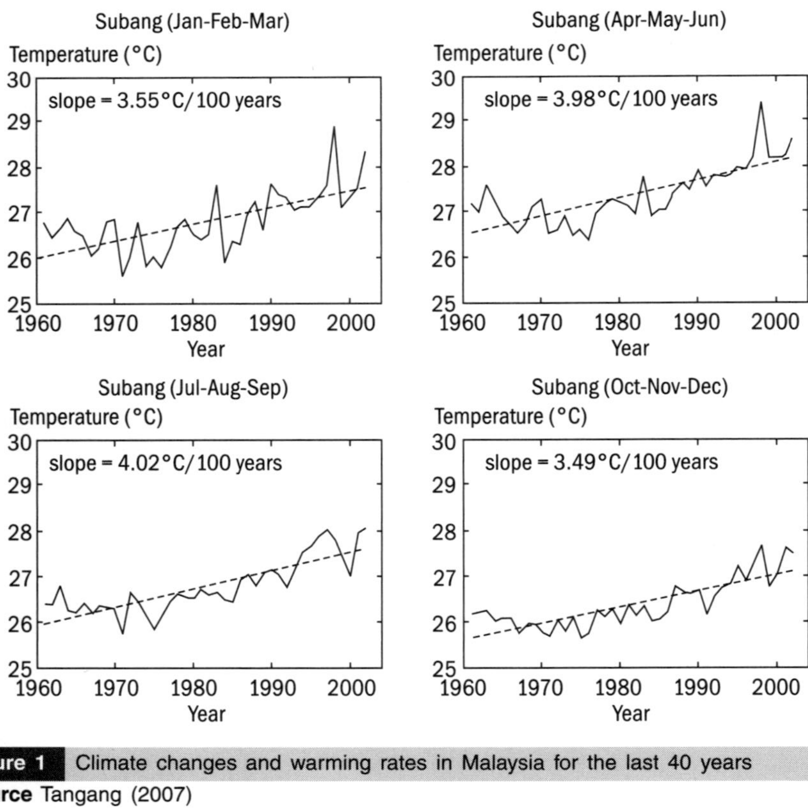


**Figure 1** Climate changes and warming rates in Malaysia for the last 40 years

**Source** Tangang (2007)

## METHODOLOGICAL MODELLING APPROACH USING GIT

In the discussion on natural hazard assessment, Aleotti and Chowdhury (1999) distinguished between qualitative methods in which the hazard is determined directly in the field by combining different index maps and quantitative methods that use statistical analysis or deterministic methods that use probabilistic approaches.

The most basic approach to hazard assessment and modelling relies on field analysis (Wohl and Oguchi 2004). Such assessment usually evolves from the detailed maps of the geomorphic process and history of hazard occurrence. Advantages of field analysis are that it can be applied successfully at any scale and adapted to specific local requirements. The field work is also likely to be time- and labour-intensive and may be dangerous in all cases of natural hazards. The use of different index maps to assess and model hazards is mostly based on the selection and mapping of factors believed to affect hazard occurrences. Historically, many statistical models are used for hazard mapping and risk analysis. Here, we demonstrate a few of those models.

### Frequency Ratio Model

Frequency ratio model computes the spatial relationships between the hazard location and each hazard-related factor. It is a statistical model, wherein the frequency ratio, a ratio between the occurrence and absence of hazard in each cell, was calculated for each factor's type or range, which had been identified as significant with respect to causing hazards. An area ratio for each factor's type or range to the total area can be calculated. Finally, frequency ratios for each factor's type or range can be calculated by dividing the hazard occurrence ratio by the area ratio. If the ratio is greater than 1, the relationship between hazard and the factors is higher, and if the ratio is less than 1, the relationship between hazard and each factor's type or range is lower. Pradhan and Lee (2007) have applied this model for landslide susceptibility and hazard mapping. The ratios were used for calculating the Landslide Susceptibility Index and mapping. The ratios of each factor's type or range were summed to calculate the Landslide Hazard Index (LHI), as shown in Equation (1).

$$\mathrm{LHI} = \Sigma \mathrm{Fr} \tag{1}$$

where Fr is the frequency ratio of each factors' type or range.

### Logistic Regression Model

Logistic regression allows one to form a multivariate regression relation between a dependent variable and several independent variables. Logistic regression, which is one of the multivariate analysis models, is useful for predicting the presence or absence of a characteristic or outcome based on values of a set of predictor variables. The advantage of logistic regression is that through the addition of an appropriate link function to the usual linear regression model, the variables may be either continuous or discrete, or any combination of both types, and they do not necessarily have normal distributions. In the case of multi-regression analysis, the factors must be numerical, and in the case of a similar statistical model, discriminant analysis, the variables must have a normal distribution. In the present situation, the dependent variable is a binary variable representing the presence or absence of flood. Where the dependent variable is binary, the logistic link function is applicable (Atkinson and Massari 1998). For hazard modelling analysis, the dependent variable must be input as either 0 or 1, so the model applies well to susceptibility analysis. Logistic regression coefficients can be used to estimate ratios for each of the independent variables in the model.

Quantitatively, the relationship between the occurrence of a hazard and its dependency on several variables can be expressed as

$$p = 1/(1 + e^{-z}) \tag{2}$$

where $p$ is the probability of an event occurring. In the present situation, the value $p$ is the estimated probability of hazard areas. The probability varies from 0 to 1 on an *S*-shaped curve and z is the linear combination. It follows that logistic regression involves fitting an equation of the following form to the data

$$z = b_0 + b_1x_1 + b_2x_2 + \ldots + b_nx_n \tag{3}$$

where $b_0$ is the intercept of the model, the $b_i$ ($i = 0, 1, 2, \ldots, n$) are the slope coefficients of the logistic regression model, and the $x_i$ ($i = 0, 1, 2, \ldots, n$) are the independent variables. The linear model formed is then a logistic regression of the presence or absence of hazard areas (present conditions) on the independent variables (pre-failure conditions).

## Fuzzy Logic Model

The fuzzy set theory introduced by Zadeh (1965) is one of the tools used to handle the complex problems. Therefore, the fuzzy set theory has been commonly used for many scientific studies in different disciplines. The idea of fuzzy logic is to consider the spatial objects on a map as members of a set. In the classical set theory, an object is a member of a set if it has a membership value of 1, or is not a member if it has a membership value of 0. In the fuzzy set theory, membership can take any value between 0 and 1, reflecting the degree of certainty of membership. The fuzzy set theory employs the idea of a membership function that expresses the degree of membership with respect to some attribute of interest.

Generally, with maps, the attribute of interest is measured over discrete intervals, and the membership function can be expressed as a table relating map classes to membership values. Fuzzy logic is attractive because it is straightforward to understand and implement. It can be used with data from any measurement scale and the weighting of evidence is controlled entirely by the expert. The fuzzy logic method allows for more flexible combinations of weighted maps, and could be readily implemented with a GIS modelling language. This is different from data-driven approaches such as weights of evidence or logistic regression, which use the locations of known objects, such as known hazards, to estimate weights or coefficients. The idea of using fuzzy logic in natural hazard mapping is to consider the spatial objects on a map as members of a set. For example, the spatial objects could be areas on an evidence map and the set could be defined as "areas hazardous to landslide". Fuzzy membership values must lie in the range (0, 1), but there are no practical constraints on the choice of the fuzzy membership values. Values are chosen to reflect the degree of membership of a set, based on subjective judgment. Given two or more maps with fuzzy membership functions for the same set, a variety of operators can be employed to combine the membership values.

Zimmerman (1996) discussed a variety of combination rules. Bonham-Carter (1994) discussed five operators, namely, the fuzzy and, fuzzy or, fuzzy algebraic product, fuzzy algebraic sum, and fuzzy gamma operator. This study uses the five fuzzy operators for combining the fuzzy membership functions. The fuzzy and is equivalent to a Boolean AND (logical intersection) operation on classical set values of Equation (4). It is defined as

$$\mu\text{combination} = \text{MIN}\ (\mu A,\ \mu B,\ \mu C,\ \ldots) \tag{4}$$

where μcombination is the calculated fuzzy membership function, $\mu A$ is the membership value for map $A$ at a particular location, $\mu B$ is the value for map $B$, and so on.

The fuzzy or is like the Boolean OR (logical union) in that the output membership values are controlled by the maximum values of any of the input maps. The fuzzy or is defined as

$$\mu\text{combination} = \text{MAX}\ (\mu A,\ \mu B,\ \mu C,\ \ldots) \tag{5}$$

The fuzzy algebraic product is defined as

$$\mu\text{combination} = \prod_{i=1}^{n} \mu i, \tag{6}$$

where $\mu i$ is the fuzzy membership function for the $i$th map, and $i = 1, 2, \ldots, n$ maps are to be combined.

The fuzzy algebraic sum is complementary to the fuzzy algebraic product, being defined as

$$\mu\text{combination} = 1 - \prod_{i=1}^{n} (1 - \mu i) \tag{7}$$

The gamma operation is defined in terms of the fuzzy algebraic product and the fuzzy algebraic sum by

$$\mu\text{combination} = (\text{Fuzzy algebraic sum})\lambda \times (\text{Fuzzy algebraic product})1 - \lambda \quad (8)$$

where λ is a parameter chosen in the range (0,1), and the fuzzy algebraic sum and fuzzy algebraic product are calculated using Equations (7) and (8), respectively. In the fuzzy gamma operation, when λ is 1, the combination is the same as the fuzzy algebraic sum, and when λ is 0, the combination equals the fuzzy algebraic product. Judicious choice of λ produces output values that ensure a flexible compromise between the "increase" tendencies of the fuzzy algebraic sum and the "decrease" effects of the fuzzy algebraic product.

## Artificial Neural Network Model

An artificial neural network (ANN) is a "computational mechanism able to acquire, represent, and compute a mapping from one multivariate space of information to another, given a set of data representing that mapping" (Paola and Schowengerdt 1995). An ANN is composed of a set of nodes and a number of interconnected processing elements. ANNs use learning algorithms to model knowledge and save this knowledge in weighted connections, mimicking the function of a human brain. Neural networks allow the target classes to be defined in relation to their distribution in the corresponding domain of each data source (Zhou 1999), and therefore, integration of remote sensing data or GIS data is convenient. The most popular ANN model used in the prediction and regression tasks is Multi-layer Perceptron (MLP) with Feed-forward Back-error Propagation (BP) type of learning algorithm, or simply MLP-BP. The back-propagation learning algorithm (MLP) has multi-layered neural network. This is a supervised, pattern recognition model that needs to be trained using a data set for which both the input values (x) for a set of predictors and the correct output values (y) are known for a set of examples. A three-layered interconnected neural network (Figure 2), consisting of input layer, hidden layer, and an output layer, was used for landslide susceptibility analysis by Pradhan and Lee (2008). Hazard-related causative layers can be fed into the neural network as "input nodes", which are considered independent variables, and the weightage computation of the input data was processed in "hidden layer". Input nodes receive the variables independently, while hidden nodes run the learning process before passing on to output node (a dependent variable) for prediction. Each network carries weights. These weighted connections act as coefficients to the input nodes. The hidden nodes (sum of the weighted inputs) compute network output (feed-forward) through a non-linear activation function, determining the difference (error) to the expected output (actual output). This error is distributed among the connection weights (back-propagation) in order to progressively reduce the error. In the network, the hidden and output layer neurons process their inputs by multiplying each input by a corresponding weight, summing the product, and then processing the sum using a nonlinear transfer function to produce a result. There are two stages involved in using neural networks for multi-source classification: the training stage, in which the internal weights are adjusted; and the classifying stage. Typically, the back-propagation algorithm trains the network until some targeted minimal error is achieved between the desired and actual output values of the network. Once the training is complete, the network is used as a feed-forward structure to produce a classification for the entire data (Paola and Schowengerdt 1995).

The first layer of the network, or input layer, contains a node for each of l input variables (Figure 2). The l input variables are analogous to the independent variables in multiple regressions. When a given set of l input values for one of the n samples in the training data set is presented to the input nodes, we say that the network is presented with an input pattern ($x^p_{i,l} = x_{i,1}, x_{i,2}, ..., x_{i,3}$, where $i = 1$ to $n$). The superscript $p$ indicates terms that consists of or refer to a given pattern of values (Moody and Katz 2003).

The last layer of the network, or output layer, contains nodes, one for each output type (Figure 2). In this case, there are 11 input nodes (one each for slope, aspect, curvature, distance from drainage, distance from lineaments, geology, land use, vegetation index, soil texture, soil material, and topography), and the output layer will have 11 nodes, one for each input layer. Sandwiched between the input and output layers is one "hidden" layer, which will allow complexities to develop in the mapping functions. In this case, a three-tiered ANN architecture model is used. The hidden layer, like the input and output layers, consists of nodes.

Pradhan and Lee (2010a, b, c) have applied ANN model for landslide susceptibility mapping in Penang Island. Before running the ANN programme, the training site was selected. So, the hazard-prone (occurrence) area and the area not prone to hazards can be selected as training sites. Cells from each of the two classes were randomly selected as training cells, with number of hazard cells denoting areas where hazard occurred or did not occur. In Pradhan and Lee (2010a, b, c), first, areas where the landslide did not occur were classified as "areas not prone to landslide" and areas where landslide was known to occur were assigned to an "areas prone to landslide" training set. The back-propagation algorithm was then applied to calculate the weights between the input layer and the hidden layer, and between the hidden layer and the output layer, by modifying the number of hidden node and the learning rate. Three-layered feed-forward network was implemented using the MATLAB software package. Here, "feed-forward" denotes that the interconnections between the layers propagate forward to the next layer. The number of hidden layers and the number of nodes in a hidden layer required for a particular classification problem are not easy to deduce. In their study, a structure was selected for the network, with input data normalized in the range 0.1–0.9. The nominal and interval class group data were converted into continuous values ranging between 0.1 and 0.9. Therefore, the continuous values were not ordinal data, but nominal data, and the numbers denote the classification of the input data.

Pradhan and Lee (2010a, b, c) have set the learning rate set to 0.01, and the initial weights were randomly selected to range between 0.1 and 0.3. The weights calculated from 10 test cases were compared to determine whether the variation in the final weights was dependent on the selection of the initial weights. The back-propagation algorithm was used to minimize the error between the predicted output values and the calculated output values.

## NATURAL HAZARD MAPPING EXAMPLES: LANDSLIDES

Damages and losses due to landslides are regular features in Malaysia because, historically, there has been too little consideration of the potential problems in land use planning and slope management. In recent years, greater awareness of landslide problems has led to significant changes in the control of development on unstable slope areas. Recently, the Malaysian government and highway authorities are stressing on the need for local planning authorities to take landslide into account at all stages of the landslide hazard mapping process. So far, few attempts have been made to predict these landslides or prevent the damage caused by them.

The value of a landslide risk assessment has always remained primarily dependent on the extent that the hazards are recognized and understood (Powell 2000). Introduction of risk management scheme as the major strength of the landslide management system can be understood by answering the following questions.

- What might happen? (Recognition)
- How likely is it? (Probability)
- What damage or injury might result? (Consequence)
- How important is it? (Assessment)
- What can be done about it? (Recovery)

The above questions can be answered through the following scheme components: zonation, assessment, and monitoring.

Through scientific analysis of landslides, we can assess and predict landslide-susceptible areas, and thus decrease landslide damage through

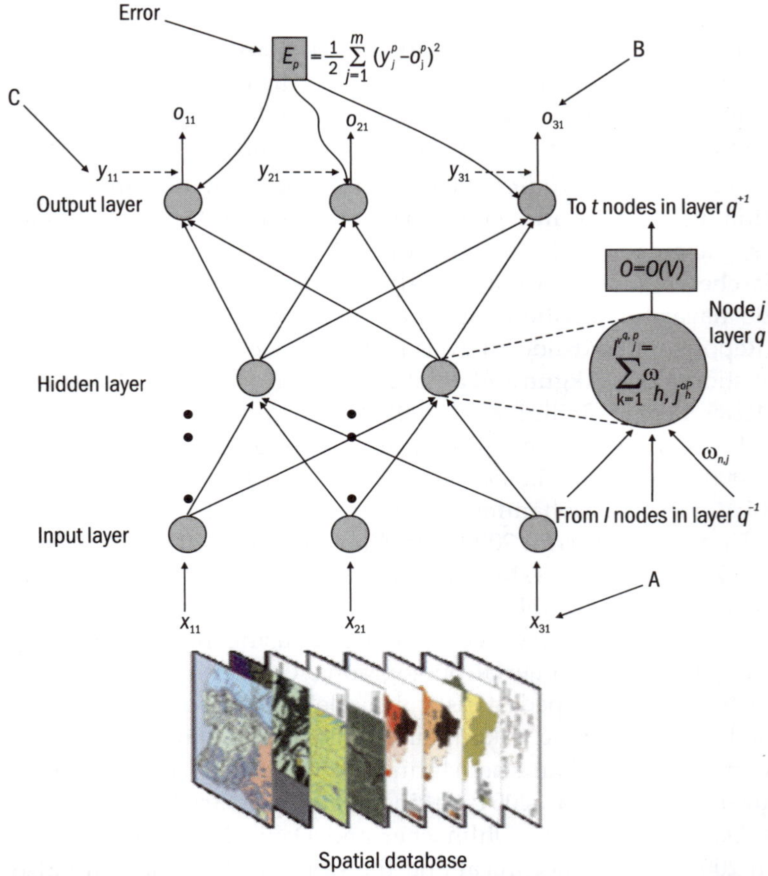


**Figure 2** Three-tiered architecture of feed-forward, back-propagation neural network (multilayer perception). (A) The presentation of training input data layer pattern 1 with the values of $x_1$, $x_2$ and $x_3$. (B) The output values of the network ($o_1$, $o_2$ and $o_3$). (C) The desired output pattern for the first samples of the training data ($y_1$, $y_2$ and $y_3$)

**Source** Modified from Moody and Katz (2003)

proper preparation. To achieve this aim, landslide hazard analysis techniques have been applied by Pradhan and Lee (2010a, b, c) using a combined frequency ratio, logistic regression, and neural network model, and verified using landslide location data. Through this kind of prediction model, landslide damages could be greatly decreased to certain extent. In addition, landslide-related factors were also assessed, and weightage for each of the thematic layers was calculated. Further, landslide risk analysis was performed using the landslide hazard map and other socio-economic related factors. However, to be able to spatialize these statistical prediction models (or any other one-dimensional model based on measurements made on small plots), few assumptions are normally made. Under this assumption, we have to imagine that the landscape under study is a continuum of small units on which the models are valid. This can be true for homogeneous land cover or soil types, but it is less evident for fragmented landscapes composed of a mosaic of units behaving differently in terms of landslides.

Historically, many studies have been carried out on landslide hazard evaluation using GIS and geoinformation-related techniques. Guzzetti, Carrarra, Cardinali, *et al.* (1999) conducted GIS-based study in the Umbria and Marche regions of Central Italy and have also summarized many landslide hazard evaluation case studies along the Apennines Mountain Chain. Reports on landslide analysis have also been prepared using GIS and probabilistic models (Akgün and Bulut 2007; Dahal, Hasegawa, Nonomura, *et al.* 2007; Clerici, Perego, Tellini, *et al.* 2006; Cevik and Topal 2003; Rowbotham and Dudycha 1998; Jibson, Edwin, and John 2000; Luzi, Pergalani, and Terlien 2000; Parise and Jibson 2000; Baeza and Corominas 2001; Lee and Min 2001; Temesgen, Mohammed, and Korme 2001; Clerici, Perego, Tellini, *et al.* 2002; Donati and Turrini 2000; Lee, Choi, and Min 2002; Lee, Chwae, and Min 2002; Zhou G, Esaki T, Mitani, *et al.* 2002; Lee and Choi 2003; Lee, Choi, and Min 2004; Lee and Talib 2005; Lee and Dan 2005; Lee and Lee 2006). Most of these studies have been conducted using the regional landslide inventories derived from aerial photographs. Guzzetti, Carrarra, Cardinali, *et al.* (1999) also developed statistical models using logistic regression for landslide hazard mapping (Akgün and Bulut 2007; Tunusluoglu, Gokceoglu, Nefeslioglu, *et al.* 2008; Lamelas, Marinoni, Hoppe, *et al.* 2008; Wang and Sasa 2005; Süzen and Doyuran 2004; Atkinson and Massari 1998; Dai, Lee, Li, *et al.* 2001; Dai and Lee 2002; Ohlmacher and Davis 2003; Lee 2005; Lee and Sambath 2006). The geotechnical and the safety factor model are also good tools to spatialize landslide hazard analysis, and these have the ability to develop scenarios by changing the input parameters (Gokceoglu, Sonmez, and Ercanoglu 2000; Romeo 2000; Refice and Capolongo 2002; Carro, De Amicis, Luzi, *et al.* 2003; Shou and Wang 2003; Zhou, Esaki, Mitani, *et al.* 2003; Lee 2007a; Lee and Pradhan 2007). All these models provide solutions for integrating information levels and mapping the outputs. Recently, other new methods have also been applied for landslide hazard evaluation using data mining, fuzzy logic, safety factor, and ANN models (Xie, Esaki, and Cai 2004; Ercanoglu and Gokceoglu 2002; Pistocchi, Luzi, and Napolitano 2002; Lee, Ryu, Min, *et al.* 2003; Lee, Ryu, Lee, *et al.* 2003; Lee, Ryu, Won, *et al.* 2004; Tangestani 2004; Lee 2007b; Mansor, Pradhan, Daud, *et al.* 2007a, b; Pradhan and Buchroithner 2009; Pradhan, Buchroithner, and Mansor 2009; Pradhan, Lee, and Buchroithner 2009a, b; Pradhan, Lee, Buchroithner, *et al.* 2009; Pradhan, Lee, and Mansor 2008; Pradhan, Lee, Mansor, *et al.* 2008; Pradhan, Mansor, Lee, *et al.* 2008a, b; Pradhan and Mansor 2007; Pradhan, Singh, and Buchroithner 2006). The cartographic results of these approaches are generally appealing, and they give rise to qualitative and quantitative landslide-susceptible and hazard areas. However, more study needs to be validated simply because of the spatio-temporal occurrence of landslides over large areas during a long period of time. Therefore, the utility and the accuracy of these maps are often questioned.

Pradhan and Lee (2010a, b, c) have applied frequency ratio, logistic regression, ANN, and fuzzy logic model to landslide susceptibility mapping at Penang Island, Malaysia. Different rock types, with different soil textures, and a long history of land use have led to a variety of landslides. Pradhan and Lee have developed a complete landslide inventory map using the archived aerial photographs and satellite images, which are well supported with field data.

## Penang Island Landslide Susceptibility and Hazard and Risk Mapping

Scientifically, there are three steps in landslide-susceptibility analysis: susceptibility, possibility (often termed as hazard), and risk (Varnes and IAEG Commission on Landslides 1984). In this study, hazard and risk analysis was performed using a combination of statistical models. Historical rainfall distribution data was combined with the susceptibility map to prepare the hazard map. Then landslide risk analysis was performed by taking into account the potential vulnerability of settlements.

$$\text{Susceptibility} = f\,(\text{landslide, landslide-related factors}) \tag{9}$$

$$\text{Possibility/hazard} = f\,(\text{susceptibility, impact factors}) \tag{10}$$

$$\text{Risk} = f\,(\text{possibility/hazard, damageable objects}) \tag{11}$$

Landslide hazard zonation (LHZ) refers to "the division of a land surface into homogeneous areas or domains and their ranking according to degrees of actual or potential hazard caused by mass movement" (Varnes and IAEG Commission on Landslides 1984).

The first task is to identify various terrain factors that rule the stability of the slope. Under the present study, an attempt has been made to prepare an LHZ map based on the combination of data acquired from various geological, geo-morphological, and land use/cover thematic maps, which includes tectonic setting, geological setting, slope material (weather ability and homogeneous), discontinuity (faults, joints, number of major joint sets, and spacing), slope dimensions (slope, aspect, aperture, persistence, and height), hydrologic conditions (rainfall and water content), previous instability, land use, land cover, and drainage pattern. Most of the previous parameters can be extracted from remote sensing data with field check. These parameters can be converted to digital format in GIS environment. Using special algorithm, an LHZ map is created.

Remote sensing data and GIS techniques have been used extensively to create thematic maps, and by overlapping them, final hazard map is generated which, based on the instability in the region, classifies the region into three categories by using the integration of GIS with remotely sensed data at Penang Island, Peninsular Malaysia: low, medium, and high risk (Pradhan and Lee 2010a, b, c).

### *Spatial Database Used*

Accurate detection of the landslide location is very important for probabilistic landslide susceptibility and risk analysis. Remote sensing methods, using aerial photographs and satellite images, are employed to obtain significant and cost-effective information on landslides. In the study of Pradhan and Lee (2010, a, b, c), 1:10 000–1:50 000-scale aerial photographs were used to detect the landslide locations. These aerial photographs were taken from the archives during the period 1981–2003. Landslide locations were detected by aerial photo interpretation and further verified by field work. These landslides can be seen in aerial photographs by interpreting breaks in the forest canopy, bare soil, and other typical geomorphic

**Figure 3** Landslide scar shown on the IKONOS satellite image (1 m) (landslide occurred in May 1999)

characteristics of landslide scars (Figure 3). A total of 324 landslides were mapped within 285 km$^2$ to assemble a database to assess the surface area and the number of landslides in the study area.

Identification and mapping of a suitable set of instability factors having a relationship with the slope failures require a priori knowledge of the main causes of landslides (Guzzetti, Carrarra, Cardinali, *et al.* 1999). These instability factors include surface and bedrock lithology and structure, seismicity, slope steepness and morphology, stream evolution, groundwater conditions, climate, vegetation cover, land use, and human activity. The availability of thematic data varies widely, depending on the type, scale, and method of data acquisition. To apply the probabilistic model, a spatial database that considers landslide-related factors was designed and constructed. The spatial database constructed by Pradhan and Lee (2010a, b, c) and used for landslide susceptibility modelling is presented in Table 2 and Figure 4.

**Table 2** List of thematic data layer used in the landslide hazard analysis

| Classification | | Geographical information system data type | | Scale or resolution | |
|---|---|---|---|---|---|
| **Spatial database** | **Factor** | **Spatial database** | **Factor** | **Spatial database** | **Factor** |
| Landslide | Landslide | ARC/INFO polygon coverage | ARC/INFO GRID | 1:25 000 | 10 m × 10 m |
| Topographic map | Slope<br>Aspect<br>Coverage | ARC/INFO line and point Curvature | | 1:25 000 | |
| Drainage map | Distance from drainage | ARC/INFO line coverage | | | |
| Soil map | Types | ARC/INFO polygon coverage | | 1:100,000 | |
| Geology map | Litho types<br>Distance from lineaments | ARC/INFO polygon, line coverage | | 1:63 300 | |
| Land cover | Land cover | ARC/INFO GRID | | 30 m × 30 m | |
| Normalized Difference Vegetation Index (NDVI) | NDVI | ARC/INFO GRID | | 10 m × 10 m | |
| Precipitation | Precipitation | GRID | | 10 m × 10 m | |

Nine landslide-inducing factors were considered in calculating the probability. These factors were transformed into a vector-type spatial

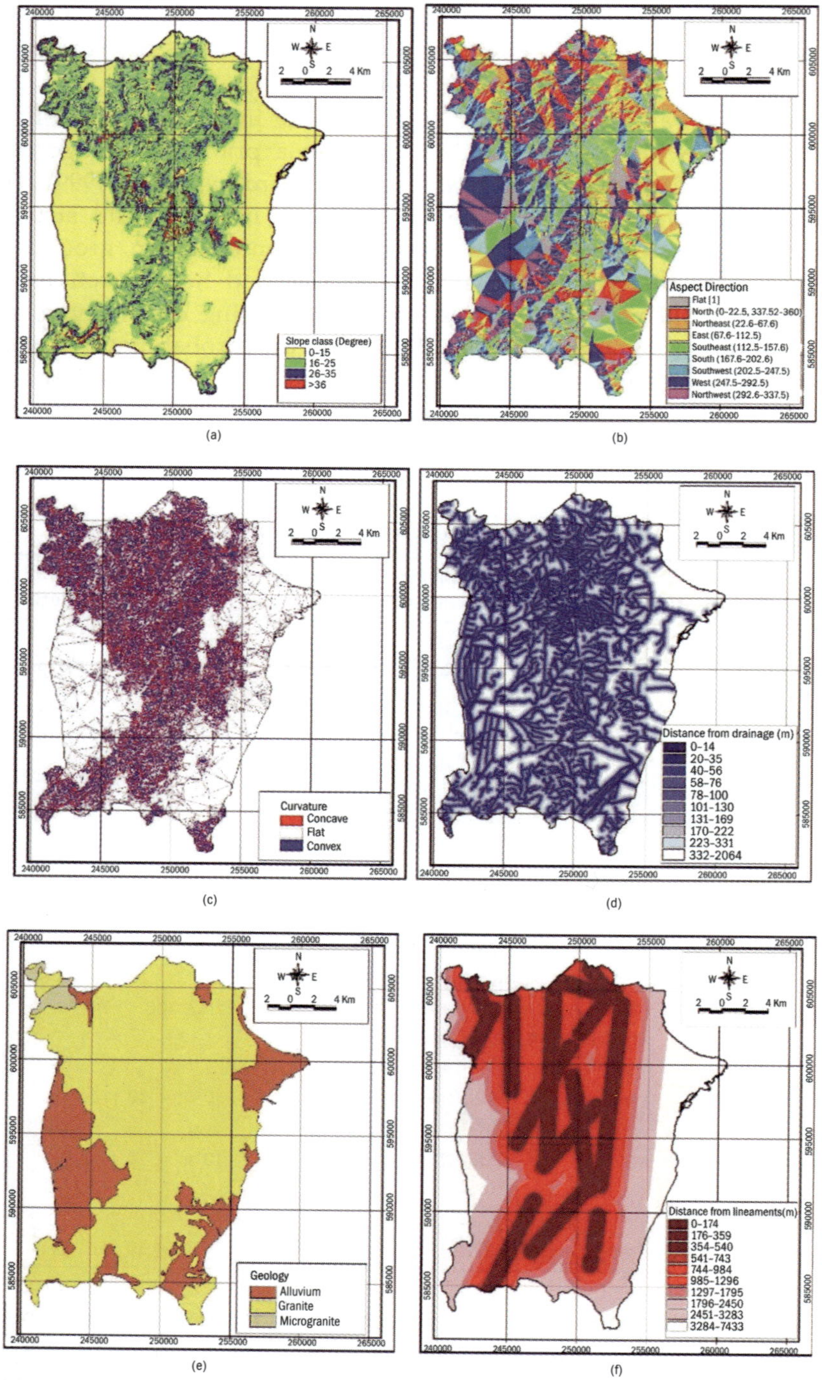


**Figure 4** Input thematic layers: (a) slope; (b) aspect; (c) curvature; (d) distance from drainage; (e) geology; (f) distance from lineament; (g) soil; (h) land cover; (i) Normalized Difference Vegetation Index (NDVI); and (j) precipitation amount

*(Fig. 4 Contd..)*

*(Fig. 4 Contd..)*

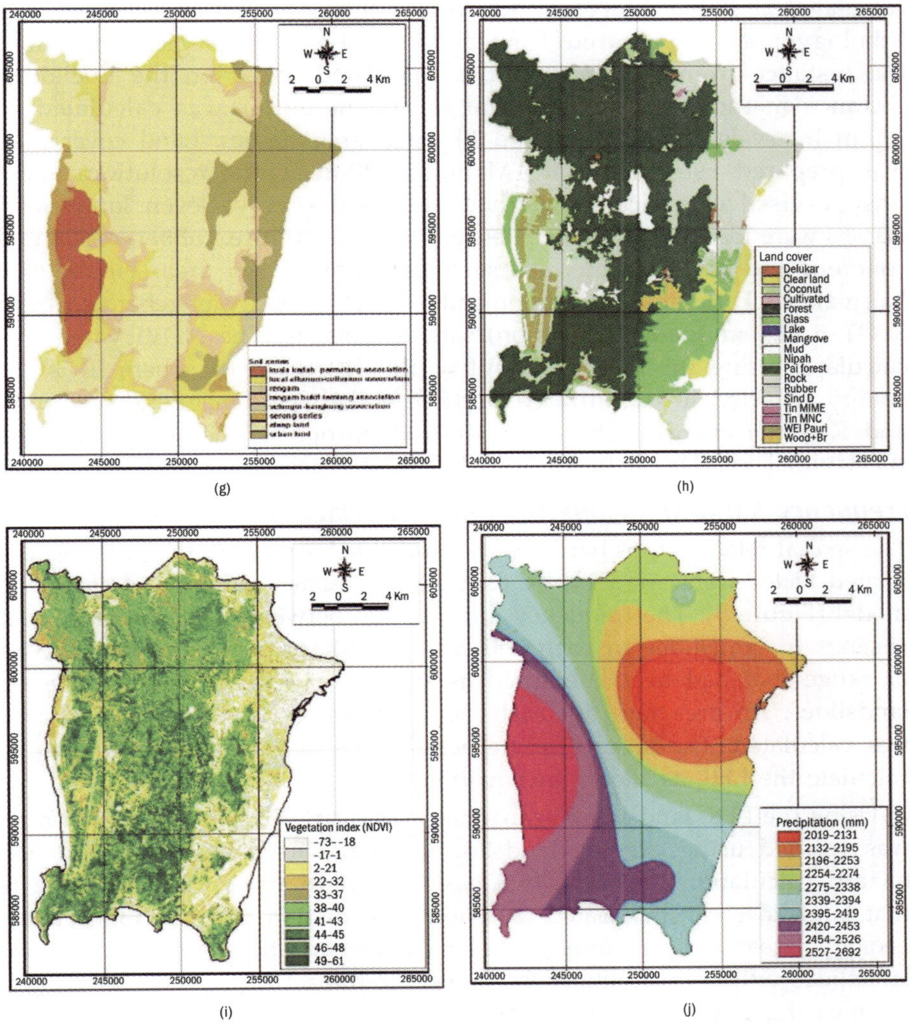


**Figure 4** Input thematic layers: (a) slope; (b) aspect; (c) curvature; (d) distance from drainage; (e) geology; (f) distance from lineament; (g) soil; (h) land cover; (i) Normalized Difference Vegetation Index (NDVI); and (j) precipitation amount

database using the GIS. For creating DEM, 10-m interval contours and survey base points showing the elevation values were extracted from the 1:25 000-scale topographic maps. Using this DEM, slope angle, slope aspect, and slope curvature were calculated. In addition, the distance from drainage was calculated using the topographic database. The drainage buffer was calculated at 100-m intervals and divided into 10 equal area classes. The lithology map is prepared from a 1:63 300-scale geological map, and the distance from the lineament is calculated based on the Euclidean distance method in ArcView. The lineament buffer was calculated in 100-m intervals and divided into 10 equal area classes. Land cover map was prepared using Landsat TM image (30-m spatial resolution) using unsupervised classification method and field survey. Seven land cover classes were identified, namely, dense forest, agriculture, rubber, residential concrete and non-concrete, sparse forest, and barren area. Finally, the Normalized Difference Vegetation Index (NDVI) map was generated from SPOT 5 (2.5 m spatial resolution) satellite images. The NDVI value was calculated using the formula NDVI = (IR – R)/(IR + R), where IR is the energy reflected in the infrared portion of the electromagnetic spectrum, and R is the energy reflected in the red portion of the electromagnetic spectrum. The NDVI is useful in delineating vegetation.

### *Frequency Ratio Modelling for Landslide Hazard Mapping*

The spatial relationships between the landslide location and each landslide-related factor were analysed by using the frequency ratio based statistical model (Figure 5). The frequency ratio, a ratio between the occurrence and absence of landslides in each cell, was calculated for each factor's type or range that had been identified as significant with respect to causing landslides. An area ratio for each factor's type or range to the total area was calculated. The ratios of each factor's type or range were summed to calculate the LHI, as shown in Equation (1).

Using the LHI, landslide hazard map was made. Finally, the hazard map was verified using existing landslide locations. To verify these landslide hazard calculation methods, two basic assumptions are needed. One is that landslides can be related to spatial information (such as topography, geology, lineament, soil, forest, and land cover) and the other is that future landslides will be precipitated by a specific impact factor, such as rainfall or an earthquake. In this study, these two assumptions are satisfied because the landslides are related to the spatial information, and all the landslides were precipitated by a single cause, namely, heavy rainfall in the study areas. For the verification, the calculated LHI values of all cells in the study area were sorted in descending order. Then, the landslides (%) were divided into classes of accumulated area ratio (%) according to the LHI value. To compare the result quantitatively, the areas under the curve (AUC) were re-calculated as the total area is 1, which means perfect prediction accuracy. So, the AUC can be used to assess the prediction accuracy qualitatively.

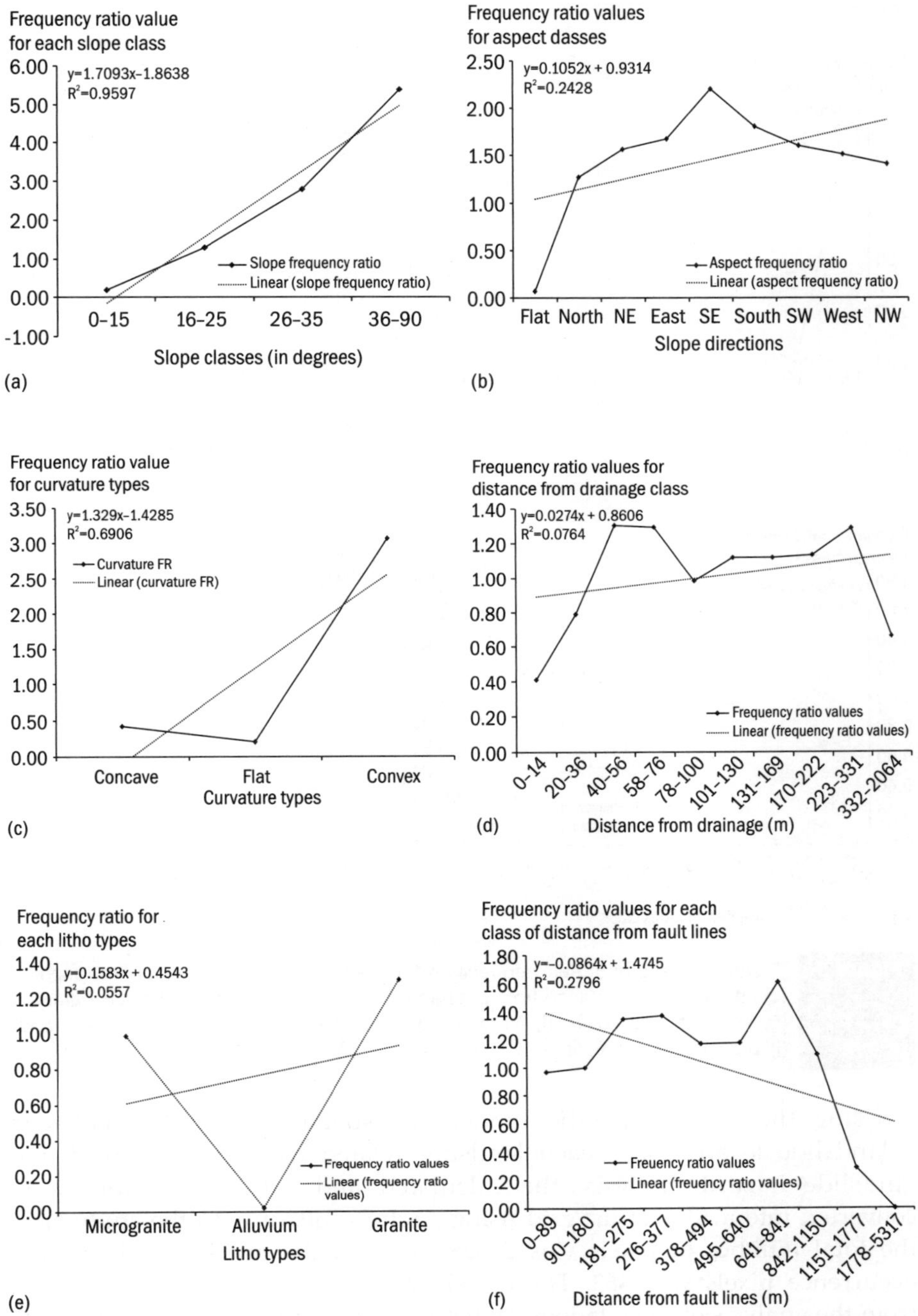


**Figure 5** Thematic layers and their frequency ratio coefficient values: (a) slope; (b) aspect; (c) curvature; (d) distance from drainage; (e) geology; (f) distance from lineament; (g) soil; (h) land cover; (i) Normalized Difference Vegetation Index (NDVI); and (j) precipitation amount

*(Fig. 5 Contd..)*

*(Fig. 5 Contd..)*

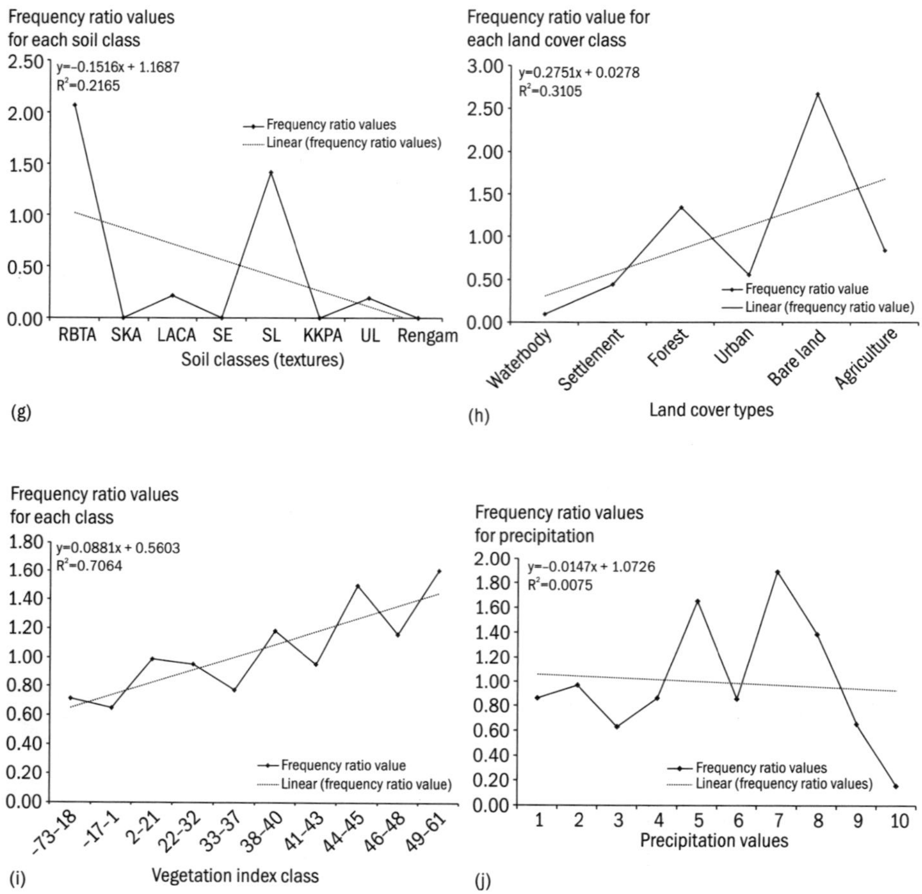


**Figure 5** Thematic layers and their frequency ratio coefficient values: (a) slope; (b) aspect; (c) curvature; (d) distance from drainage; (e) geology; (f) distance from lineament; (g) soil; (h) land cover; (i) Normalized Difference Vegetation Index (NDVI); and (j) precipitation amount

Using the frequency ratio model, the spatial relationship between a landslide location and each landslide-related factor was derived. For landslide hazard analysis, the calculated and extracted factors were converted into a 10 × 10 $m^2$ grid using ArcInfo software. In the study area, the total number of pixels was 2 928 378, and the number of landslide-occurrence pixels was 463. The correlation of coefficients was calculated from the analysis of the relationship between the landslides and the relevant factors.

### *Accuracy Assessment*

To judge or decide the potential, value, type or significance of slope stability, assessment depends on internal slope factors, while zonation is done through the outer parameters. That is why the assessment still depends more on the field work in comparison with zonation, which can be done through remote sensing data. The hazard map should be verified using existing landslide locations. In the study of Pradhan and Lee (2010a, b, c), first the landslide hazard map was created based on the frequency ratio model using map overlaying techniques (Figure 6). In the hazard map, the potential event and its probability of occurrence were combined. The hazard categories are expressed as probability in qualitative forms (for example, none, low, moderate, high). The accuracy of the hazard map was the area ratio of 0.8641 and the prediction accuracy of 86.41% (Figure 7). In a landslide hazard map, a particular surface area subjected to the same hazard can face a variety of consequences, depending on land use. Therefore, the hazard map was overlaid on damageable objects layers such as transportation network and settlement. Figure 6 shows the landslide risk map of the study area. The result shows that many settlements have been constructed on high landslide risk areas, where the probability of occurrence of landslide is very high. Those high risk areas need to be brought to the notice of the public so that people can realize the possibility of future landslides. This could save their property and life.

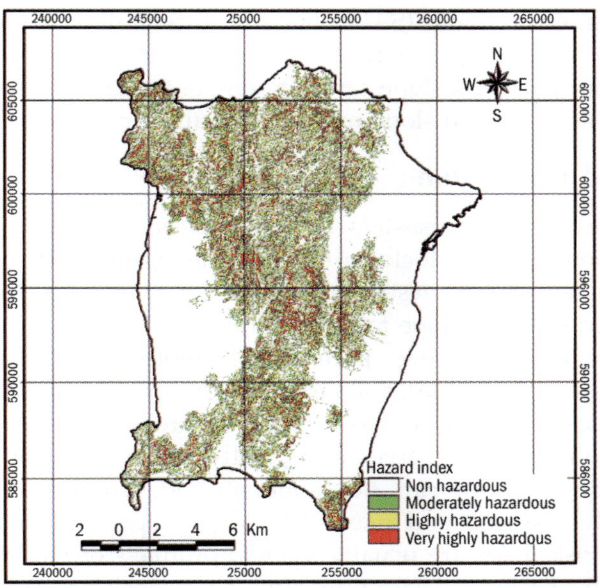


**Figure 6** Landslide hazard map of Penang Island

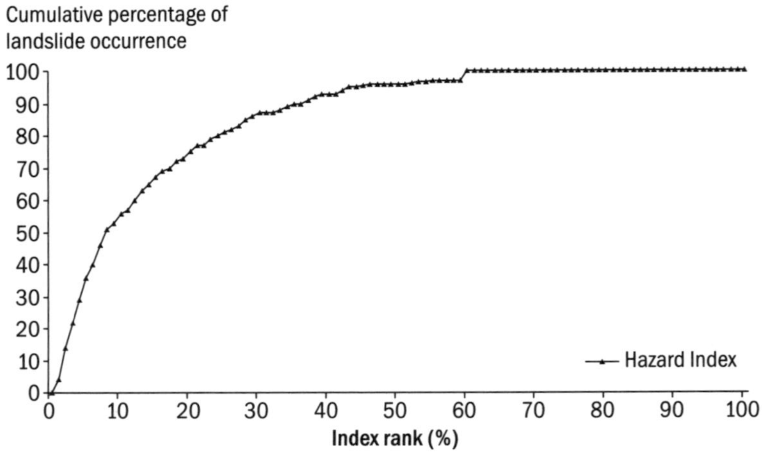


**Figure 7** Cumulative frequency diagram showing landslide hazard index rank occurring in cumulative percent of landslide occurrence

### *Monitoring*

According to pervious phases, slopes with a high potential to slide will be put under surveillance. There are two main techniques for landslide monitoring. First is on-site using geotechnical or surviving tools, while the second is implemented using remote sensing techniques.

Synthetic aperture radar (SAR) data with differential interferometry will be chosen for monitoring of highly hazardous region in Malaysia. According to Figure 8, for high risk slope category, one of the geotechnical techniques can be used according to slope morphology and significance.

## Results and Discussion

Landslides are natural disasters that have serious impacts on transport, communication, and livestock. In the study area, most often, traffic routes collapse and houses are destroyed because of landslides. Important land erosion patterns due to slope failure, with heavy impacts on soil and water losses, are witnessed. Using space-borne remote sensing data, advanced image processing techniques, and GIS tools, the spatial distribution of slides, and their scales and degree of hazard and risk prediction can be determined. A close relationship, in spatial terms, between landslide attribute factors and triggering factors can be demonstrated. Landslide risk can be systematically assessed and managed using proper mitigation measures. Hazard and risk maps are usually the end-products of landslide mapping, but they are the basis for decision-making. These decisions are usually in the form of technical countermeasures, regulatory management, or combination

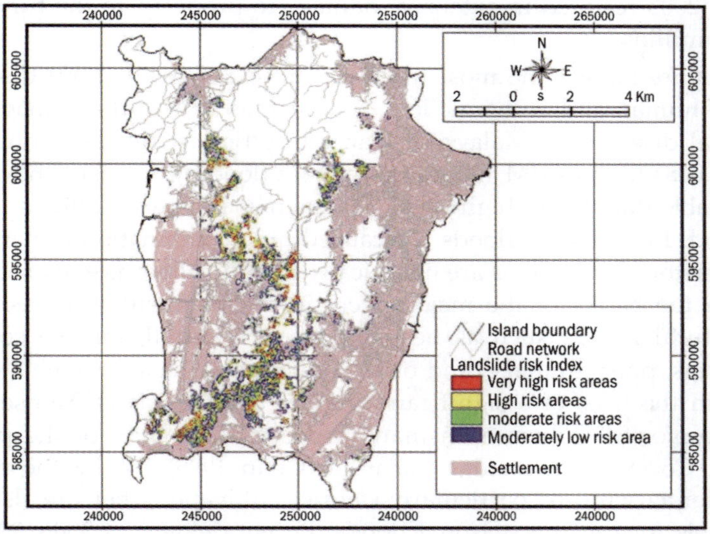


**Figure 8** Landslide risk map study area showing settlements and risk zones

of the two. Classic examples of regulatory management are zoning maps which, for instance, exclude some landslide-prone areas from habitation. The results of this study provide the scientific background for estimating the risk areas with respect to the landslides and may serve as a basis for decisions for authorities concerned. The results demonstrated here can be used as basic data to assist slope management and land-use planning in the study area. The methods used in the study are also valid for generalized planning and assessment purposes, although they may be less useful on the site-specific scale, where local geological and geographic heterogeneities may prevail.

## NATURAL HAZARD MAPPING EXAMPLES: FLOOD

The development of efficient flood management strategies involves a comparative assessment of the potential financial benefit of flood management scenarios by reducing the damage to property and infrastructure, displacement of communities, and general loss of income. Flooding is a natural phenomenon that regularly causes disruption and damage to different parts of the world, including Malaysia. Floods vary in scale and range from water running off a saturated hillside to large rivers bursting their banks. The impacts of floods range from waterlogged fields and blocked roads to widespread inundation of houses and commercial property and, occasionally, loss of life. The consequences can be enormous. Recent flash floods in Selangor and Kuala Lumpur in Malaysia in 2002 and again in 2003 resulted in extensive damage to property, to the tune of over RM 300 million.

Floods are among the most frequent and expensive natural disasters in terms of human and economic loss. As much as 90% of the damage related to natural disasters in Malaysia is caused by floods. Average annual flood damage is as high as RM 100 million. These flooding incidents have caused considerable damage to highways, settlements, and agriculture, and loss of livelihood. In Malaysia, floods are caused by a combination of natural and human factors. Malaysians are historically river dwellers as early settlements grew on the banks of the major rivers in the peninsula. Floods, coupled with natural factors such as heavy monsoon rainfall, intense convection rain storms, poor drainage, and other local factors, have become a common feature in the lives of a significant number of Malaysians. Monsoon rains have a profound influence on many aspects of the lives of the people in the East Coast of Peninsular Malaysia (Chan 1995). While the rains are needed for agriculture, particularly wet rice cultivation, they are also largely responsible for causing seasonal floods. Recent deluges of 2006, 2007, 2008, and 2009 have triggered floods along Malaysia's East Coast as well as in the southern state of Johor. The hardest hit areas are along the East Coast of peninsular Malaysia in the states of Kelantan, Terengganu, and Pahang. The city of Johor was particularly hard hit in southern side. The flood cost nearly billion Ringgit of property and many lives. The extent of damage could have been reduced or minimized if an early warning system would have been in place.

## Kelantan River Flood Mapping

The study area (Figure 9) is part of Kelantan state, which is one of the 13 states of Malaysia. The Kelantan river emerges at the confluence of the Galas river and Lebir river near Kuala Krai and meanders over the coastal plain until it finally drains into the South China Sea, about 12 km north of Kota Bahru. The main reach of the Kelantan river has some further larger tributaries downstream. However, the Galas and Lebir rivers themselves have many tributaries, which account for the majority of the flow in the main Kelantan river. Kota Bharu is the main city and centre of commercial trade and administration in the Kelantan state. Its ideal geographical location also makes it a gateway to the neighbouring countries of southern Thailand. It is thus a city with high tourism potential for domestic and foreign tourists who visit throughout the year. Due to its geographical characteristics, unplanned urbanization, and proximity to the South China Sea, Kota Bharu has become extremely vulnerable to monsoon floods every year. The city's unique characteristics like its population density and it being a commercial centre can magnify the impacts of flooding to which it is prone.

The unprecedented flooding of November 2005, triggered by monsoon rains, has been described as one of the worst natural floods in the history of Kota Bharu.

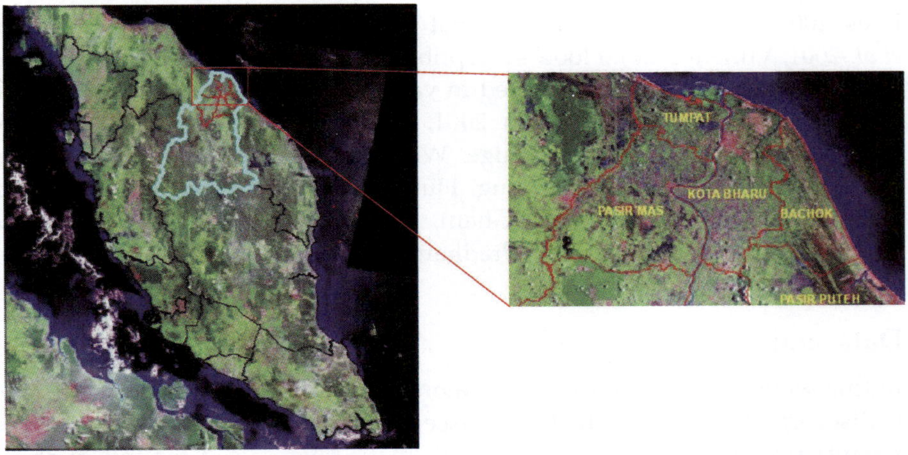


**Figure 9** Study area and Kelantan river basin

Annual rainfall is very high, averaging between 2500 mm and 3500 mm per year. Two pronounced wet seasons range from September to December and February to May. Rainfall peaks between November and December and between March and May. Peninsular Malaysia experiences a hot, wet, and humid equatorial climate regime in which the most distinguishing feature is the heavy year round rainfall, ranging from 1500 mm to more than 3500 mm annually. More significant, however, is the occurrence of sustained heavy rain spells (sometimes for several weeks) during the monsoon season, and a total rainfall of 610 mm within a matter of 24 hours is not uncommon. The seasonal floods in the East Coast are, therefore, a natural consequence of these heavy rains occurring over a short period of time. During these north-east monsoon months, a total monthly rainfall of 500 mm is not uncommon.

## Flood Susceptibility Mapping

There are many studies on flood susceptibility mapping based on remote sensing data and GIS tools. Radar remote sensing data has been extensively used for flood monitoring across the world (Hess, Melack, Filoso, *et al.* 1995; Hess, Melack, and Simonett 1990; Le Toan, Ribbes, Wange, *et al.* 1997), and many of these studies have used probabilistic methods (Landau, Mitchell, Barnett, *et al.* 2000; Farajzadeh 2001, 2002; Horritt and Bates 2002). Logistic regression model has also been applied to other natural hazard modelling such as landslide hazard and susceptibility mapping (Atkinson and Massari 1998; Dai, Lee, Li, *et al.* 2001; Dai and Lee 2002; Ohlmacher and Davis 2003). Hydrological and stochastic rainfall methods for flood susceptibility mapping have been employed in other areas (Blazkova and Beven 1997; Cunderlik and Burn 2002; Ebisemiju 1986; Haeng, Salas, and Boes 2001; Nageshwar and Bhagabat 1997; Yakoo, Kazama, Sawamoto, *et al.* 2001; Villiers 1986). Flood susceptibility mapping using GIS and neural network methods has been applied in various case studies (Honda, Francis, and Sah 1997; Islam and Sadu 2001, 2002; Sanyal and Lu 2004, 2005; Townsend and Walsh 2005; Wadge, Wislocki, Pearson, *et al.* 1993; Profeti and Machintosh 1997; Knebl, Yang, Hutchison, *et al.* 2005; Masmoudi and Habajeb 1993; Sinnakaudan, Ab Ghani, Ahmad, *et al.* 2003; Merwade, Cook, and Coonrod 2008; Zerger 2001; Pradhan, Pirasteh, and Shafie 2009; Pradhan 2009; Pradhan and Shafie 2009).

## Data and Materials Used

In this section, remote sensing data, along with other tabular and meta data, is discussed to delineate the flood susceptibility mapping for the part of the Kelantan river basin. The details on thematic layers used for this analysis are presented in Table 3. Terrain information such as historical flooded areas extracted from RADARSAT images, DEM, slope, aspect, curvature, distance from drainage, flow direction, flow accumulation, soil, land cover, soil texture, and precipitation has been updated to enable the quantification of attributes associated with floods. Flood susceptibility mapping has been applied using logistic regression model. To apply the multiple logistic regression model, a spatial database that considers flood-related factors should be first designed and constructed (Townsend and Walsh 1998). Some studies indicate that various factors are effective in flood susceptibility mapping. Slope gradient, basin shape, aspect, curvature, geological conditions, rainfall regime, and vegetation cover are some of the factors that control run-off volume (Wadge, Wislocki, Pearson, *et al.* 1993). Some of this data was available in Malaysia either as a paper or as digital maps. The remote sensing data and the thematic layers used in the analysis are highlighted in Table 3 and Figure 10, respectively. In this study, nine factors (slope, DEM, curvature, flow direction, flow accumulation, distance from drainage, land cover, soil, and precipitation) were considered in calculating the probability, and the factors were extracted from the constructed spatial database. The factors were transformed into a grid spatial database, and flood-related factors were extracted using the database. A key assumption using the multiple logistic regression approach is that the potential (inundated areas due to future flood) of future flooding areas will be comparable to the actual frequency and extent of previous historical floods. Historical flooded areas were detected from RADARSAT images for the years 1998, 2003, and 2004. A historical flooded map was prepared from

RADARSAT images coupled with the field data and existing reports, and GIS data, and this map was used to evaluate the frequency and extent of future floods in the area. Then, the calculated and extracted factors were converted into a 10 m × 10 m grid (ARC/INFO GRID type). Statistical multiple logistic regressions were applied using the database. Further, the spatial relationships between the historic flooded areas and each flood-related factor were analysed. Using the logistic regression model, the relationship was used, as each factor's rating in the overlay analysis and a formula of flood extent possibility were extracted using the relationships. This formula, as shown in Equation (4), was used to calculate the Flood Susceptibility Index and the index was mapped to represent flood-prone areas. The qualitative flood susceptibility analysis has been carried out using the map overlying techniques in the GIS environment.

**Table 3** List of thematic data layers used for flood susceptibility mapping

| Classification | Sub-classification | Geographical information system data type | Scale |
|---|---|---|---|
| Historical flooded areas | Flood extent | Polygon coverage (derived from RADARSAT images) | 10 m × 10 m |
| Basic map | Topographic map | Line and point coverage (digital elevation models) | 1:25 000 |
| | Slope | GRID | 10 m × 10 m |
| | Curvature | GRID | 10 m × 10 m |
| | Flow direction | GRID | 10 m × 10 m |
| | Flow accumulation | GRID | 10 m × 10 m |
| | Land cover | GRID | 10 m × 10 m |
| | Soil | GRID | 10 m × 10 m |
| | Precipitation | GRID | 100 m × 100 m |

## Bivariate Logistic Regression Modelling Approach

A key assumption using the statistical approach is that the potential (occurrence possibility) of future flooding areas will be comparable to the actual frequency and extent of previous historical floods. Historical flooded areas were detected from RADARSAT images and field surveys. A historical flooded map was prepared from RADARSAT images, in combination with the GIS, and this was used to evaluate the frequency and extent of future floods in the area. Topography database was constructed and land cover and vegetation index values were extracted from SPOT 5 satellite image, and precipitation distribution was obtained from the meteorological data, for analysis. Then, the calculated and extracted factors were converted into a 10 m × 10 m grid (ARC/INFO GRID type). Statistical logistic regressions were applied using the database. Further, the spatial relationships between

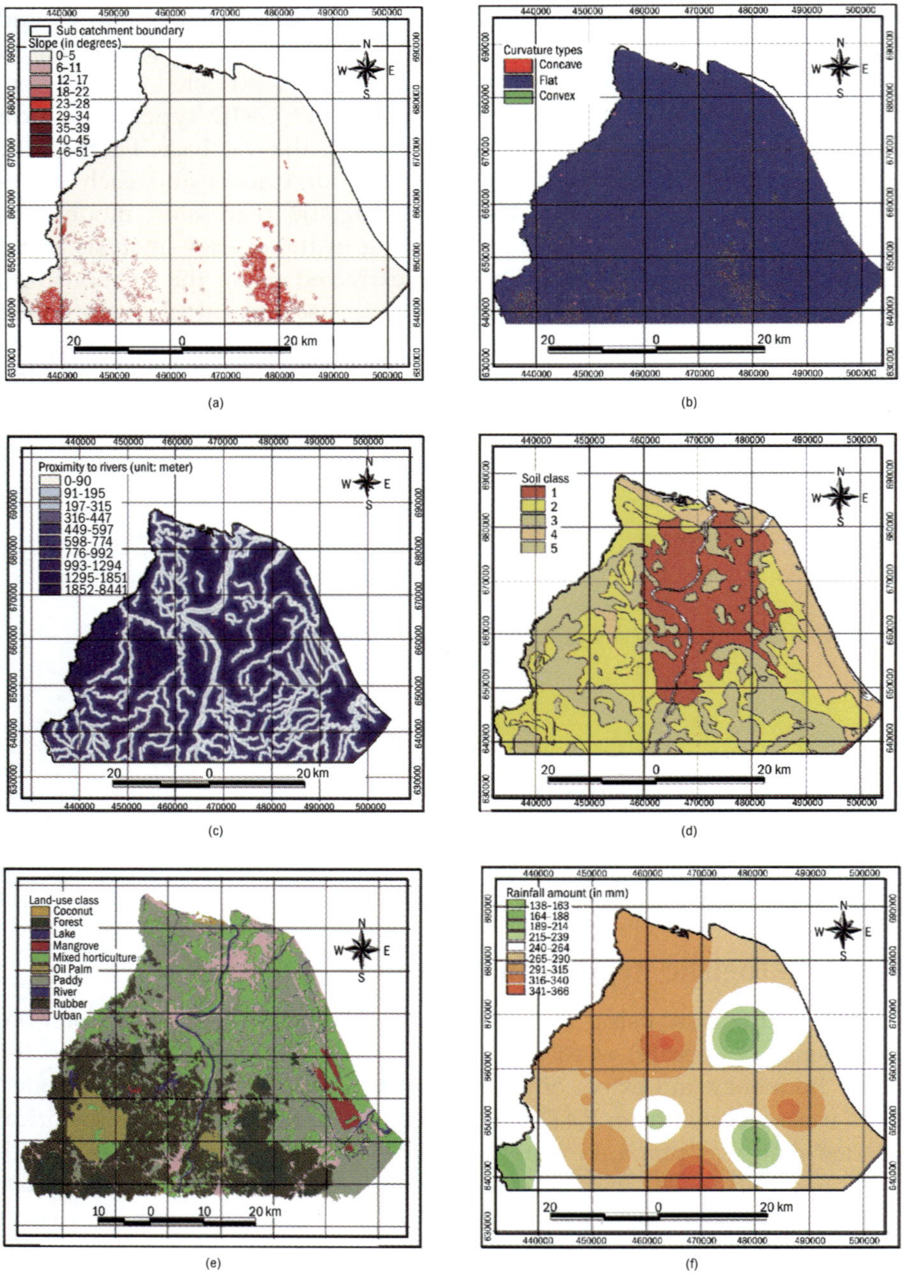


**Figure 10** Input data layers: (a) slope; (b) curvature; (c) distance from drainage; (d) soil; (e) land cover; (f) precipitation amount; (g) flow accumulation amount; (h) digital elevation models, and (i) historical flood extent derived from RADARSAT images

*(Fig. 10 Contd..)*

*(Fig. 10 Contd..)*

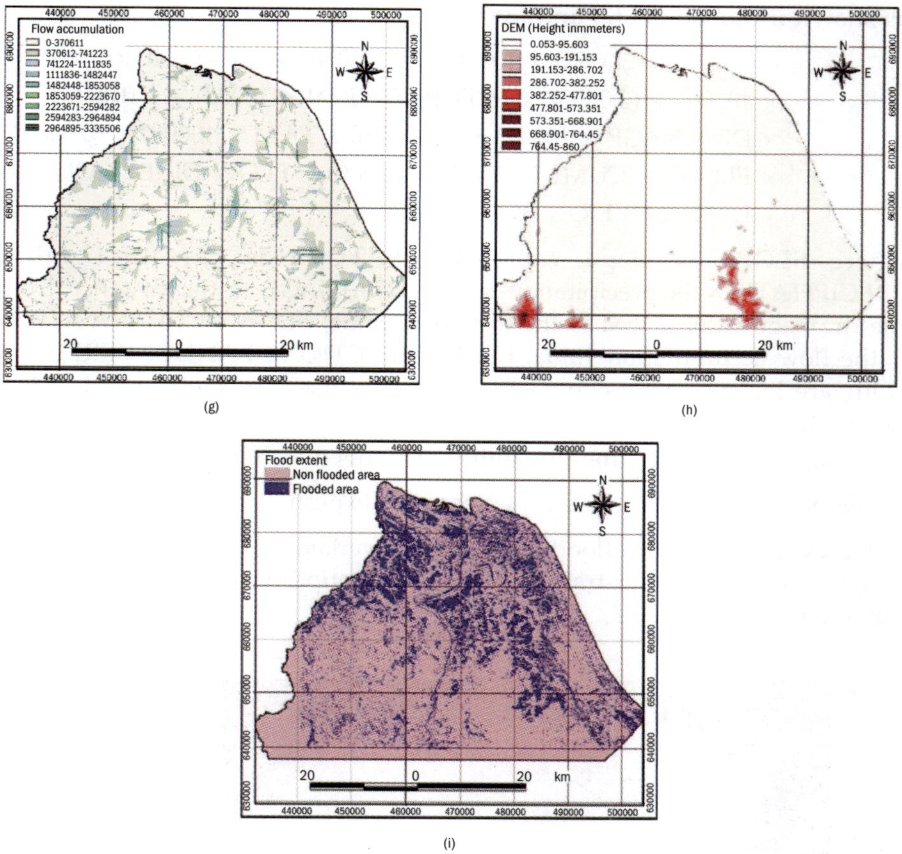


**Figure 10** Input data layers: (a) slope; (b) curvature; (c) distance from drainage; (d) soil; (e) land cover; (f) precipitation amount; (g) flow accumulation amount; (h) digital elevation models, and (i) historical flood extent derived from RADARSAT images

the historic flooded areas and each flood-related factor were analysed. Using the logistic regression model, the relationship was used, as each factor's rating in the overlay analysis and a formula for flood extent possibility were extracted using the relationships. This formula was used to calculate the Flood Susceptibility Index, and the Index was mapped to represent flood-prone areas. The susceptibility map was made using the map overlaying techniques using Equation (4). In the study, GIS software, ArcView 3.2, and ARC/INFO 9.0 version software packages, along with SPSS 12.0 statistical program, were used as the basic analysis tools for spatial management and data manipulation.

In this case, logistic regression mathematical equations were formulated as shown in Equations (2) and (3) for each case.

$$
\begin{aligned}
z_n = {} & (-0.00179 \times \text{SLOPE} \times 10000) + (-0.00562 \times \text{CURVATURE} \\
& \times 10000) + (0.00537 \times \text{PRECIPITATION} \times 10000) + (-0.00002 \\
& \times \text{DRAINAGE} \times 10000) + (-0.00080 \times \text{DEM} \times 10000) + (0.00001 \\
& \times \text{FLOW ACCUMULATION} \times 10000) + \text{FLOW DIRECTION}_c \\
& + \text{LANDCOVER}_c + \text{SOIL}_c - 3.98050
\end{aligned}
\tag{12}
$$

where SLOPE is slope value; CURVATURE is curvature value; PRECIPITATION is precipitation value; DRAINAGE is the distance from the drainage value; DEM is the elevation value; FLOW ACCUMULATION is the flow accumulation value; FLOW $\text{DIRECTION}_c$, $\text{LANDCOVERE}_c$, and $\text{SOIL}_c$ are logistic regression coefficient values listed in Table 2; and $z_n$ is a parameter.

Using Equation (12), the possibility of flooded areas was calculated.

$$\text{Flood Susceptibility Index} = \exp(z)/(1 + \exp(z)) \tag{13}$$

Figure 11 shows the flood extent map overlaid on the SPOT 5 satellite image and Figure 12 illustrates the flood susceptibility map created by using Equation (10).

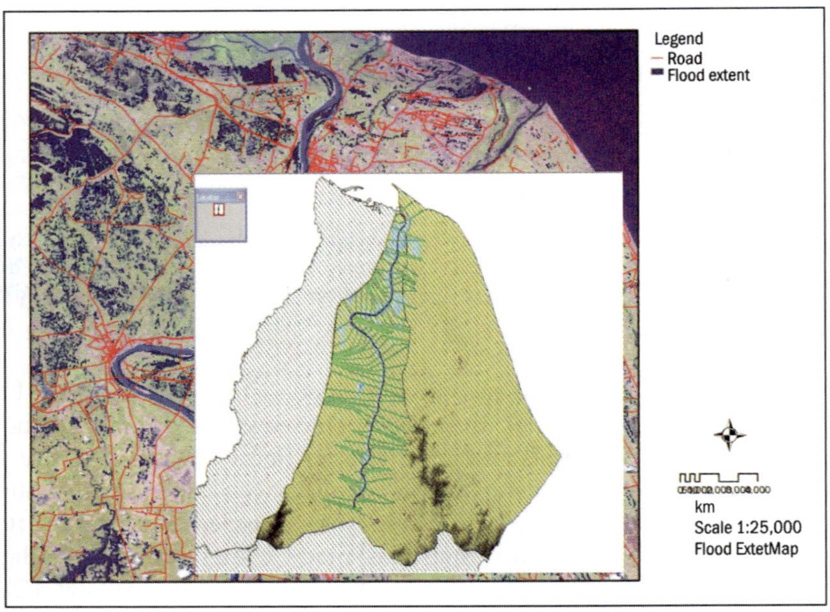


**Figure 11** SPOT 5 satellite image showing the flood extent area for the Kelentan river basin

## Results and Discussion

In the present study, logistic regression model was applied for the flood susceptibility mapping for part of Kelantan river basin. In this research, a statistical approach for estimating the susceptible flow-flood area using remote sensing technique and the GIS was adopted. For the flood susceptibility analysis, the detected historical flooded areas and the flood-related database were constructed for Kelantan river basin. Using the constructed database, flood susceptibility analysis was performed using logistic regression model (Figure 12). It was found that the probability method is somewhat simplistic, and the process of input, calculation, and output could be understood easily. Moreover, no simple conversion of database from GIS to ASCII is required, as a large quantity of data can be processed in the GIS environment quickly and easily.

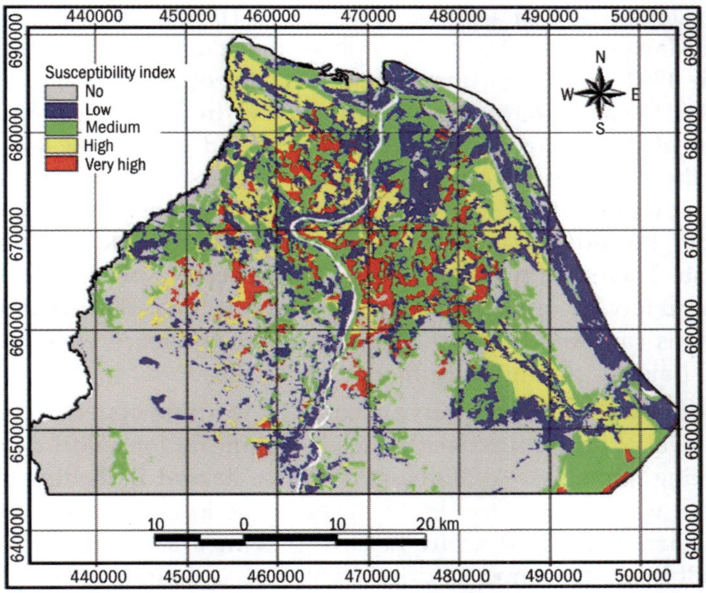


Figure 12 Flood susceptibility map based on logistic regression model

The logistic regression model is simple; the process of input, calculation, and output can be easily understood. The large quantity of data can be processed in the GIS environment quickly and easily. The logistic regression model requires the conversion of the data into ASCII or other formats for use in the statistical package, and later re-conversion to incorporate it into the GIS database. Moreover, it is hard to process the large quantity of data in the statistical package. In the case of a similar statistical model (discriminant analysis), the factors must have a normal distribution, and in the case of multi-regression analysis, the factors must be numerical. However, for logistical regression, the dependent variable must be input as 0 or 1; therefore, the model applies well to flood susceptibility analysis.

## NATURAL HAZARD MAPPING EXAMPLES: FOREST FIRE

Forest fire is defined as an uncontained and freely spreading combustion, which consumes the natural fuels of either a forest, a plantation, a grassland or a shifting cultivation area, that is, duff litter, grass, dead branch wood, snags, logs, stumps, weeds, brush foliage, and to a limited degree, green trees (Ketpraneet 1991). Fire has been identified as one of the major threats causing the loss of forests in several states in Malaysia. According to the Forestry Department of Peninsular Malaysia (JPSM) and Forest Research Institute Malaysia (FRIM), statistics show that during the last seven years (between 1992 and 1998), more than 1600 ha of peat swamp forests of Peninsular Malaysia have been destroyed by fire (Wan Ahmad 2002). In the event of a prolonged spell without rain and a lowering of the water table in the peat swamp forest, the organic layers become completely dry and more prone to fire. Fires in these peat swamp forest create much more smoke per hectare than other types of forest fires and are difficult to extinguish. Therefore, the areas at risk to fire need to be studied closely in peat swamp forests. A precise evaluation of forest fire problems and decision on solutions can only be satisfactory when a fire hazard zone mapping is available (Jaiswal, Saumitra, Kumaran, *et al.* 2002).

A wildfire risk model can serve as an early warning system to predict the severity level of future fire risks, which is significant for wildfire prevention and fighting strategies. A GIS-based wildfire hazard modelling should be developed, with an aim to identify geographic locations with the highest wildfire hazards and risks. Fire hazard is defined as a fuel condition that may result in an undesired wildfire event. Risk is defined as the probability of an event occurring. For example, dense housing within a high wildfire hazard area may have a higher probability or risk of burning than homes within a patchy fuel complex (Sampson, Neuenschwander, and Harkins 1998). In literature, the words "hazard" and "requirement" have different meanings. Hazard includes both risk and danger component (risk is associated with prevention and ignition, danger corresponds to spread and fighting actions) (Wybo, Gouma, Guarnieri, *et al.* 1995).

A fire risk model can be developed by combining the point data such as weather data and the surface data such as satellite data. We introduce the fire risk model to the forest fire early detection system, which presently offers information that is useful for deciding on the fire prevention planning in the targeted area. For disaster mitigation, information on the present condition and prediction of the situation is necessary.

## Study Area

The study area is located approximately between upper left (3°23′53.6″ E and 101°3′36.3″ N) and lower right (3°45′18.05″ E and 101°30′55.33″ N). The area is located within the Kuala Selangor district, in the northern part of Selangor. The land use at the study area comprises mainly peat swamp forest, plantation forest, inland forest, scrub, grassland, and ex-mining area. The landform of the area ranges from very flat terrain, especially for the peat swamp forest, ex mining, grassland, and scrub area, to hilly area, with the natural forest ranging between 0 and 420 m above sea level. Based on the data of Malaysian Meteorological Services Department, the temperature of the northern part of Selangor is between 29°C and 32°C, and mean relative humidity is in the range 65%–70%. The highest temperature occurs between April and June, while the relative humidity is lowest in June, July, and September. About 58.6–240 mm rainfall per month was recorded in the study area (Tanjung Karang weather station provided by Malaysian Meteorological Services Department).

There is a high potential of danger of fire in the dry season, especially in the peat swamp forest and plantation forest. Most of the fires are caused by human activities, due to either carelessness or burning activities undertaken during crop plantations. During 1995 and 1999, fire incidents were reported in the peat swamp forest area within the study area. Figure 13 shows Raja Muda Musa and Sungai Karang Reserve Forest, Selangor.

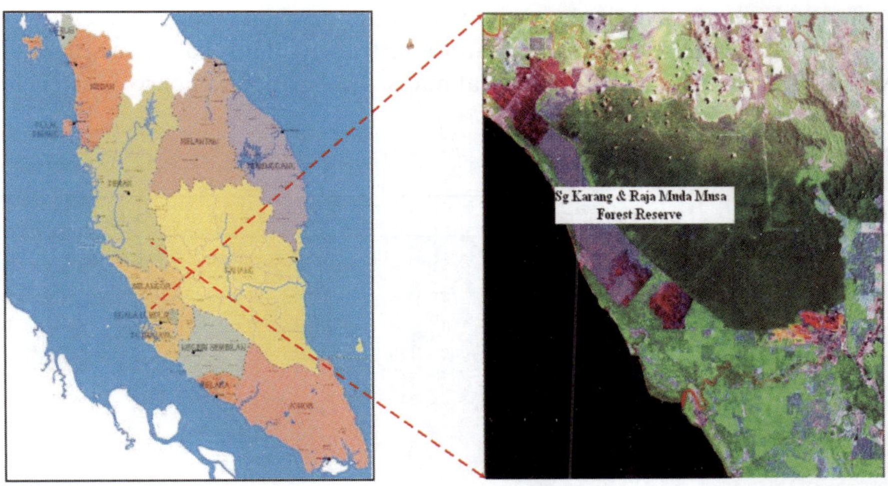


**Figure 13** Sungai Karang and Raja Muda Musa Forest Reserve, Selangor

## Data and Materials Used

Accurate detection of the location of hotspots is very important for probabilistic forest fire susceptibility analysis. Recent advances in remote sensing, GIS, and computer technologies provided an opportunity to assess and monitor the land cover changes on a near real time basis. NOAA AVHRR satellite data with a spatial resolution of 1.1 km at nadir was found to be extremely useful for national-scale assessment and monitoring of major land cover types (Giri and Shrestha 1996). Historical forest fire data for the past five years was collected from satellite remote sensing NOAA AVHRR 12 and NOAA 14 sensors. To assemble a database to assess the surface area and number of hotspots in the study area, a total of 112 hotspots were mapped in a mapped area of 616 km$^2$. The imagery from Landsat-7 ETM of path 157 and row 058, acquired on 21 September 2001, was used in this study. The spatial resolution for Landsat-7 ETM was 30 m × 30 m. Fuel map was extracted from the satellite imagery.

GIS and ancillary data consists of biophysical and socio-economic variable on 1:25 000 scale. Contour, administrative boundaries, water resources, settlement, and transportation infrastructure are based on the topographic map from the Survey Department (JUPEM). Forest fire reports have been collected from the Forest Department Peninsular Malaysia (JPSM). Fire-prone areas, fire occurrence map, peat swamp map, and soil map were acquired from MACRES. Socio-economic data such as population data was obtained from the statistical department. Meteorological data such as temperature and relative humidity and fire danger rating system (FDRS) map were obtained from the Malaysian Meteorological Services Department. Image processing was carried out using ERDAS Imagine 8.7 and PCI Geomatica 9.0.

To apply the probabilistic method, a spatial database that considers forest fire related factors was designed and constructed. This data is available in Malaysia either as paper or as digital maps. The spatial database constructed is shown in Table 4 and Figure 14.

**Table 4** Data layer of study area

| Classification | Sub-classification | GIS data type | Scale |
|---|---|---|---|
| Forest fire events | Hotspot location | Point coverage | 1:25 000 |
| Basic map | Land use | GRID | 30 m × 30 m |
| | Soil | GRID | 30 m × 30 m |
| | Normalized Difference Vegetation Index (NDVI) | GRID | 30 m × 30 m |
| | Slope | GRID | 30 m × 30 m |
| | Aspect | GRID | 30 m × 30 m |
| | Agroclimate | GRID | 30 m × 30 m |

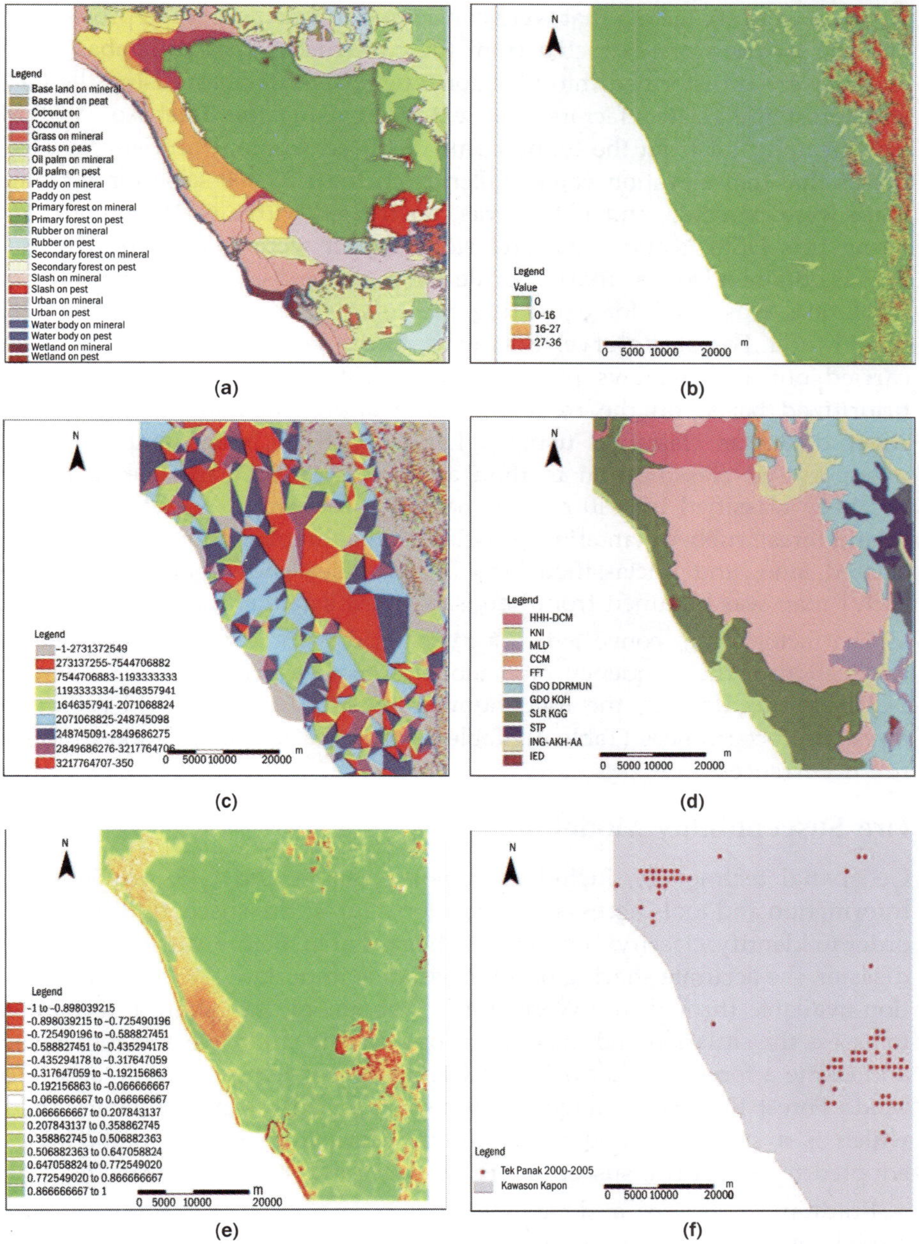


**Figure 14** Input data layers: (a) land use/land cover; (b) slope; (c) aspect classes; (d) soil types; (e) Normalized Difference Vegetation Index (NDVI); and (f) historical fire (hotspot) event derived from NOAA-AVHRR and MODIS images

There were six factors that were considered in calculating the probability, and the factors were extracted from the constructed spatial database. The factors were transformed into a vector-type spatial database using the GIS, and forest fire related factors were extracted using the database. A DEM was created first from the topographic database. Contour and survey base points that had elevation values taken from the 1:25 000-scale topographic maps were extracted, and a DEM was constructed with a resolution of 20 m. Using this DEM, slope angle and slope aspect were calculated. The soil map is obtained from a 1:100 000-scale soil map. Landsat-7 ETM, 30 m × 30 m resolution, was used for extracting fuel map in the Sungai Karang and Raja Muda Forest Reserve, Selangor. Multi-parametric analyses were carried out and overlays prepared using GIS, and severity zones were prioritized based on the frequency ratio approach. Land use and land cover data was classified using a Landsat-7 ETM image employing an unsupervised classification method and topographic map. The fuel type has been classified into 10 classes, namely, peat swamp forest, mangrove, inland forest, rubber plantation, grassland, oil palm plantation, paddy, shrub, cleared land, and unclassified land (waterbodies and urban). Finally, the NDVI map was obtained from Landsat-7 ETM satellite images.

The factors were converted to a raster grid with 30 m × 30 m cells for application of the frequency ratio model. The area grid was 2418 rows by 1490 columns (that is, the total number is 3 033 610), and 112 cells had forest fire occurrences (Table 4). Table 5 gives the frequency ratio to forest fire occurrences.

## Fire Susceptibility Modelling

Geospatial technology, including remote sensing and GIS, provides the information and tools necessary to develop a forest fire susceptibility map in order to identify, classify, and map fire hazard area. Before, during, and after disaster, the accurate sharing of information is important. With the information available through the World Wide Web, people can share information to assess the situation and make decisions. In this study, we aim to develop and create a forest fire hazard model and map for Sungai Karang and Raja Muda Forest Reserve, Selangor (peat swamp forest), using frequency ratio, which is a statistical model. The overall flow chart of the methodology adopted for forest fire susceptibility mapping is shown in Figure 15.

Pradhan, Suliman, and Awang (2007) have used a frequency ratio model for forest fire susceptibility mapping. To calculate the Forest Fire Susceptibility Index (FFSI), each factor's frequency ratio values were summed to the training area, as in Equation (14). The hotspot susceptibility value represents the relative susceptible to forest fire occurrence. So the greater the

**Table 5** Frequency ratio to forest fire occurrences

| Factors | Class | Forest fire occurrence points | Forest fire occurrence points (%) | Pixels in domain | Pixel (%) | Frequency ratio |
|---|---|---|---|---|---|---|
| Land use | Oil palm | 882 330 | 29.09 | 33 | 29.46 | 1.01 |
| | Rubber | 173 028 | 5.70 | 1 | 0.89 | 0.16 |
| | Inland forest | 251 786 | 8.30 | 1 | 0.89 | 0.11 |
| | Wetland | 47 900 | 1.58 | 0 | 0.00 | 0.00 |
| | Coconut | 267 027 | 8.80 | 2 | 1.79 | 0.20 |
| | Tin mining areas | 70 524 | 2.32 | 11 | 9.82 | 4.22 |
| | Peat swamp forest | 978 612 | 32.26 | 42 | 37.50 | 1.16 |
| | Cocoa | 21 710 | 0.72 | 0 | 0.00 | 0.00 |
| | Grassland | 11 453 | 0.38 | 0 | 0.00 | 0.00 |
| | Urban and associated areas | 21 067 | 0.69 | 0 | 0.00 | 0.00 |
| | Paddy | 237 293 | 7.82 | 22 | 19.64 | 2.51 |
| | Mixed horticulture | 14 600 | 0.48 | 0 | 0.00 | 0.00 |
| | Coffee | 6966 | 0.23 | 0 | 0.00 | 0.00 |
| | Shrub | 6381 | 0.21 | 0 | 0.00 | 0.00 |
| | Market garden | 1391 | 0.05 | 0 | 0.00 | 0.00 |
| | Orchards | 547 | 0.02 | 0 | 0.00 | 0.00 |
| | Cleared land | 791 | 0.03 | 0 | 0.00 | 0.00 |
| | Mangrove | 32 737 | 1.08 | 0 | 0.00 | 0.00 |
| | Waterbodies | 7467 | 0.25 | 0 | 0.00 | 0.00 |
| Soil | KNJ | 116 467 | 4 | 0 | 0.00 | 0.00 |
| | SLR-KGG | 570 646 | 19 | 19 | 16.96 | 0.90 |
| | BRH-OCM | 197 953 | 7 | 7 | 6.25 | 0.96 |
| | OCM | 22 676 | 1 | 1 | 0.89 | 1.19 |
| | PET | 1 233 182 | 41 | 58 | 51.79 | 1.27 |
| | TMG-AKB-LAA | 149 523 | 5 | 4 | 3.57 | 0.72 |
| | SDG-BGR-MUN | 508 282 | 17 | 12 | 10.71 | 0.64 |
| | SDG-KDH | 58 860 | 2 | 0 | 0.00 | 0.00 |
| | STP | 83 663 | 3 | 0 | 0.00 | 0.00 |
| | ULD | 6224 | 0 | 0 | 0.00 | 0.00 |
| | MLD | 86 131 | 3 | 11 | 9.82 | 3.46 |
| Slope | 0–15 m | 2 688 943 | 89 | 109 | 97.32 | 1.10 |
| | 16–26 m | 119 360 | 4 | 3 | 2.68 | 0.68 |
| | 27–35 m | 120 067 | 4 | 0 | 0.00 | 0.00 |
| | 36–89 m | 101 330 | 3 | 0 | 0.00 | 0.00 |
| Aspect | Flat | 840 143 | 27.73 | 25 | 22.32 | 0.80 |
| | North | 271 677 | 8.97 | 16 | 14.29 | 1.59 |
| | North East | 135 333 | 4.47 | 5 | 4.46 | 1.00 |
| | East | 221 514 | 7.31 | 11 | 9.82 | 1.34 |

*(Table 5 Contd..)*

*(Table 5 Contd..)*

| Factors | Class | Forest fire occurrence points | Forest fire occurrence points (%) | Pixels in domain | Pixel (%) | Frequency ratio |
|---|---|---|---|---|---|---|
| | South East | 291 939 | 9.64 | 17 | 15.18 | 1.57 |
| | South | 395 470 | 13.05 | 8 | 7.14 | 0.55 |
| | South West | 452 322 | 14.93 | 21 | 18.75 | 1.26 |
| | West | 236 762 | 7.82 | 5 | 4.46 | 0.57 |
| | North West | 184 133 | 6.08 | 4 | 3.57 | 0.59 |
| Normalized Difference Vegetation Index (NDVI) | 1 | 313 831 | 10.34 | 18 | 16.07 | 1.55 |
| | 2 | 315 165 | 10.39 | 18 | 16.07 | 1.55 |
| | 3 | 303 956 | 10.02 | 9 | 8.04 | 0.80 |
| | 4 | 422 030 | 13.91 | 14 | 12.50 | 0.90 |
| | 5 | 220 812 | 7.28 | 9 | 8.04 | 1.10 |
| | 6 | 288 182 | 9.50 | 7 | 6.25 | 0.66 |
| | 7 | 303 253 | 10.00 | 15 | 13.39 | 1.34 |
| | 8 | 510 397 | 16.82 | 18 | 16.07 | 0.96 |
| | 9 | 190 298 | 6.27 | 4 | 3.57 | 0.57 |
| | 10 | 165 816 | 5.47 | 0 | 0.00 | 0.00 |
| Agro-climate | 2 | 207 061 | 6.83 | 13 | 11.61 | 1.70 |
| | 3 | 1 309 686 | 43.17 | 53 | 47.32 | 1.10 |
| | 4 | 1 342 888 | 44.26 | 41 | 36.61 | 0.83 |
| | 5 | 77 843 | 2.57 | 2 | 1.79 | 0.70 |
| | 6 | 2137 | 0.07 | 1 | 0.89 | 12.68 |
| | 14 | 71 658 | 2.36 | 0 | 0.00 | 0.00 |
| | 16 | 22 475 | 0.74 | 2 | 1.79 | 2.41 |

**Note** A number of total cells in study area – 3 033 610 (without no data); number of forest fire occurrence points – 121

value, the higher the susceptibility to forest fire occurrence, and the lower the value, the lower the susceptibility to forest fire occurrence. The forest fire susceptibility map was developed using the FFSI values, and the map and its interpretation are shown in Figure 16.

$$\text{FFSI} = \text{Fr1} + \text{Fr2} + \ldots + \text{Frn} \tag{14}$$

where Fr is the rating of each factor's type or range.

Fire hazard model is developed based on Chuvieco and Congalton (1989), who consider the influence of several factors on forest fire. The methodology can be modified for classifying the fire hazard area as assessed for each pixel according to the following variables: forest species, land use type, elevation, slope, aspect, and distance to roads.

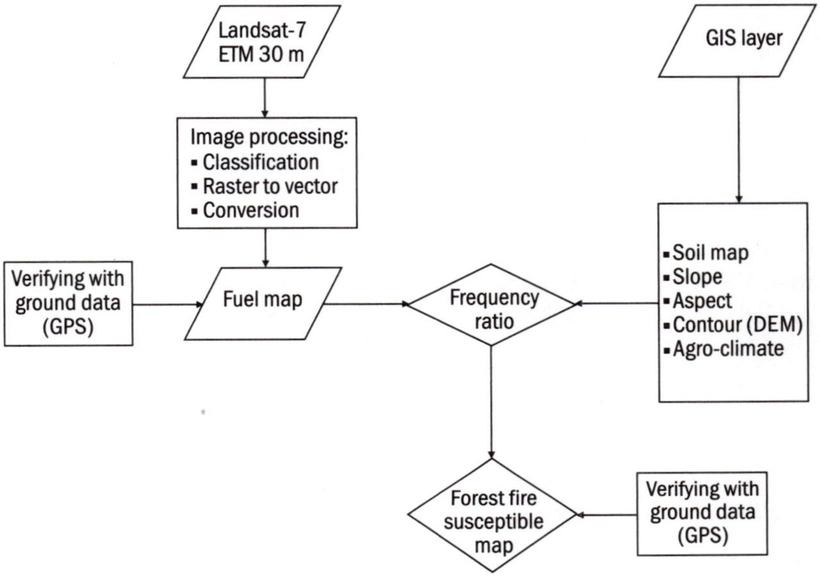


**Figure 15** Forest fire modelling flow chart methodology

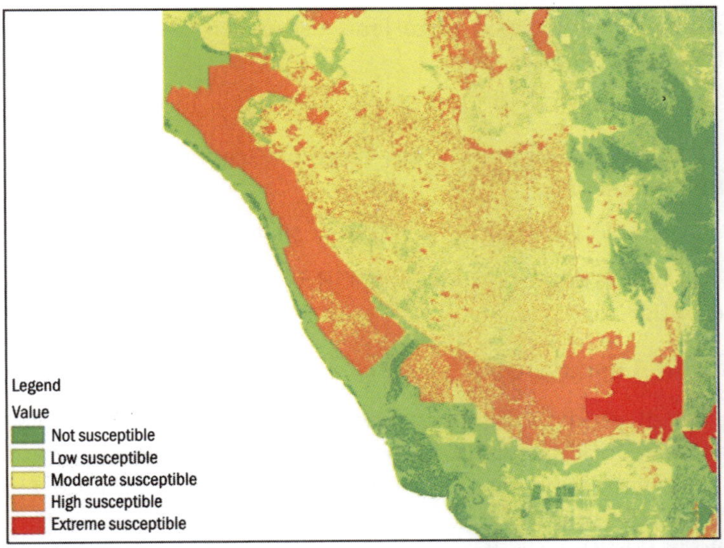


**Figure 16** Forest fire susceptibility map for Sungai Karang and Raja Muda Musa forest area

The following is the model used to identify the fire risk area based on factors affecting fire based on logistic regression (Myung-Hee, Myung-Bo, Si-Young, *et al.*1999).

$$P = 0.1 \text{ AVI} + 0.046 \text{ AVII} + 0.022 \text{ AVIII} + 0.010 \text{ AVIV} + 0.004 \text{ AAI} + 0.01 \text{ AAII} + 0.024 \text{ AIII} + 0.156 \text{ VF.I} + 0.094 \text{ VF.II} + 0.078 \text{ VF III} + 0.062 \text{ VF.IV} + 0.047 \text{ VF.V} + 0.039 \text{ VF.VI} + 0.000 \text{ VF.VII} + 0.04 \text{ DI} + 0.014 \text{ DII} + 0.014 \text{ DII} + 0.027 \text{ DIII} + 0.082 \text{ DIV} + 0.123 \text{ DV} + 0.004 \text{ EI} + 0.007 \text{ EII} + 0.034 \text{ EIII} + 0.014 \text{ EIV} \quad (15)$$

where AV is the distance to roads variable; AA is the distance to rivers variable; VF is the forest species variable; D is the slopes variable; and E is the aspect variable (I.....VII = subclasses of each variable)

$$Zi = 3.754 + 0.231 \text{ (slope)} + 0.324 \text{ (elevation)} + 0.165 \text{ (aspect)} + 0.328 \text{ (stream)} + 0.195 \text{ (forest type)} + -0.017 \text{ (agricultural pattern)} + -0.128 \text{ (urban)} + 0.030 \text{ (road)} + 0.872 \text{ (rainfall)} + 0.652 \text{ (sunshine)} + 0.713 \text{ (moisture)} \quad (16)$$

$$Pi = \exp(Zi)/(1 + \exp(Zi)) \quad (17)$$

**Model 1:** Darmawan Model (2000) was used to calculate the weight of the variables which have the same weight.

$$\text{FT} + \text{EL} + \text{GR} + \text{AS} + \text{BR} \quad (18)$$

**Model 2:** Fuel type derived from land use/cover has a weight twice that of the other variables.

$$20\text{FT} + 10 \text{ EL} + 10 \text{ GR} + 10 \text{ AS} + 10 \text{ BR} \quad (19)$$

**Model 3:** Fuel type derived from land use/cover has a higher weight followed by gradient.

$$30 \text{ FT} + 20 \text{ EL} + 10 \text{ GR} + 10 \text{ AS} + 10 \text{ BR} \quad (20)$$

where FT is fuel type; EL is elevation; GR is gradient; AS is the aspect; and BR is the buffer road.

## Results and Discussion

In the present study, frequency analysis method was applied for the forest fire susceptibility mapping for Sungai Karang and Raja Muda Forest Reserve. The validation results show that the frequency ratio model has prediction accuracy of 73.18%, as shown in Figure 4. Here, the authors can conclude that the results of frequency ratio model had shown the best prediction accuracy in forest fire susceptibility mapping.

The frequency ratio model is simple; the process of input, calculation, and output can be readily understood. A large quantity of data can be processed in the GIS environment quickly and easily. Moreover, it is difficult to process the large quantity of data in the statistical package. Recently, forest fire susceptibility mapping has shown a great deal of importance for haze detection and fire prevention in the forest area. The results shown in Figure 14 can help the concerned authorities in managing and mitigating forest fires.

However, one must be careful while using the models for specific mitigation. This is because of the scale of the analysis, wherein other forest fire related factors need to be considered. Therefore, the models used in the study are valid for generating awareness so that necessary prevention measures can be taken during the time of forest fire. In this article, forest fire susceptibility map was developed to determine the level of severity of forest fire hazard zones in terms of mapping susceptibility to fire by assessing the relative importance between fire factors and the location of fire ignition.

## NATURAL HAZARD MITIGATION AND MANAGEMENT PROGRAMME: MALAYSIAN PERSPECTIVE

High economic and social costs are incurred by Malaysia due to natural hazards. For this reason, the government devotes much energy for planning mitigation actions in relation to natural and human-induced disasters. Malaysian Centre for Remote Sensing (MACRES), an undertaking of the Ministry of Science, Technology and Innovation (MOSTI), like other major departments, is involved in providing relevant data and information to support the planning and monitoring of disaster events. Table 6 lists down the most commonly occurring natural hazards in Malaysia and different phases of disaster mitigation processes.

MACRES has been utilizing remote sensing, GIS, and related technologies to provide real-time value-added data and information to authorities in the areas of natural resources and environmental management. MACRES was involved in implementing various programmes such as National Resources and Environmental Management (NAREM), monitoring of environmental sensitive areas, and detecting and monitoring disasters such as oil spills, landslide, forest fires, and floods. With an aim to consolidate these various efforts in providing real-time value-added data and information on disasters commonly occurring in the country (forest fire, landslide, floods, oil spills, and "hot-installations"), MACRES had initiated the National Disaster Data and Information Management System (NADDI 2003) Project in 2002. The overall operational framework of the NADDI is shown in Figure 17.

NADDI is a technological and institutional solution to facilitate the decision-making process related to disaster detection and monitoring, early warning, and mitigation and relief. It aims at providing a higher degree of interoperability of individual subject within the scope of the whole disaster

**Table 6** Types of common natural hazards in Malaysia and various stages of disaster mitigation

| Type of disaster | Pre-disaster phase | | | Post-disaster phase |
|---|---|---|---|---|
| | **Assessment of risk** | **Mitigation** | **Preparedness** | **Recovery** |
| Flood | Specific data acquisition<br>■ Hydrologic data<br>■ Geologic data<br>■ Vegetation data<br>■ Rainfall<br>■ Water level<br>■ Soil | ■ Flood mitigation<br>■ Flood risk reduction strategies that involve engineering, settlement development, public administration, community-based strategies, and land use planning | Flood risk management with remote sensing and geographical information systems (GIS): creation of a GIS database for flood areas/ zonation | Incorporation of disaster mitigation components in reconstruction activities |
| | ■ Flood risk analysis and mapping (hazard and vulnerability maps)<br>■ Vulnerability and flood damage assessment<br>■ Capacity analysis | Multidisciplinary treatment of the flood problem/ integration of water management authorities | Flood warning system | |
| | | Geodetic supervision of protection measures (for example, banks) in the building process | | |
| | ■ Analysis with satellite digital and/or remotely sensed data | Identification of flood risk areas | Flood preparedness information network | |
| | ■ Creation of flood risk scenarios | Development of a land development priority map | | |
| | | Sustainable urban development; prevention of sealing in certain areas; avoidance of land degradation | | |
| Landslide/ Avalanche | Data acquisition: geological, topographic, and climatic condition of the area and human factors | Sustainable urban development; prevention of human settlement in certain areas; avoidance of | Landslide monitoring with surveying techniques.<br>■ Global positioning system | Incorporation of disaster mitigation components in reconstruction activities |

*(Table 6 Contd..)*

*(Table 6 Contd..)*

| Type of disaster | Pre-disaster phase | | | Post-disaster phase |
|---|---|---|---|---|
| | **Assessment of risk** | **Mitigation** | **Preparedness** | **Recovery** |
| | such as land use change and road construction | unplanned urbanization | ■ Electronic distance measurement (EDM)<br>■ Surveying triangulation<br>■ Geometrical levelling<br>■ Terrestial photogrammetry<br>■ Aerial photogrammetry | |
| | Development of hazard and vulnerability maps: analysis and prediction of slope failures in space (spatial distribution) and time (temporal distribution) | | Visualization of landslide-prone areas in 3D view for hill area planning | |
| | Creation of risk scenarios landslide hazard zonation | | | |
| | Landslide identification | | | |
| Wild Fire | Data acquisition<br>■ vegetation data<br>■ rainfall<br>■ soil | Avoidance of unplanned urbanization | ■ Near-real time monitoring of fire events<br>■ Satellite data | Incorporation of disaster mitigation components in reconstruction activities |
| | Development of fire hazard maps and vulnerability maps | Creating and implementing comprehensive urban development strategies and land use plans | Early warning (fire intelligence) systems | |
| | Fire risk modelling in expected climate change scenarios | | | |

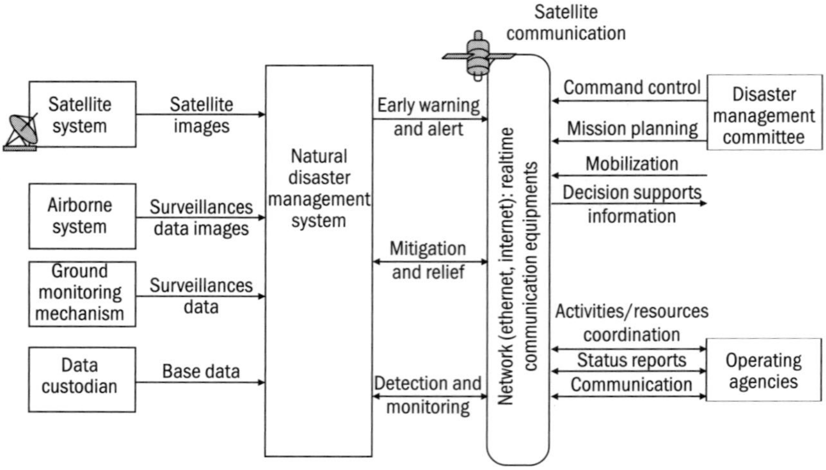


**Figure 17** Operational framework for natural hazard event management

management cycle, thereby decreasing the implication of such a situation to a tolerable limit.

Services offered by NADDI include the following.

- *Client services* Users will access the system through web client (lean client).
- *Presentation services* Allow data designed for public to be published on the Internet.
- *Application services* Enable information processing and dissemination through pre-defined tasks within the workflow, with an aim to improve the level of preparedness and state of readiness for handling hazardous events that include rescue, evacuation, and relief activities. Other very important applications are providing early warning/alert and undertaking monitoring services.
- *Centralized management database services* Providing and managing comprehensive data and information on disaster, which are easily understood, reliable, relevant, up-to-date, and easily accessible by stakeholders. Data is primarily stored in Oracle database.

NADDI collects varied data made available by various agencies, with the priority given to remote sensing as an input to disaster database, as mentioned below.

- Effective utilization of earth observation satellite data for natural hazard detection, monitoring, and modelling.
- Utilization of airborne remote sensing data (SAR and hyperspectral).

- Utilization of airborne natural hazard surveillance data.
- Ground survey and monitoring activities, for example, through telemetry stations, weather stations, reports filed by disaster teams, and so on.
- Base data from relevant authorities such as the Department of Survey and Mapping, Department of Environment, Department of Agriculture, and so on.

NADDI models and applications process collected data to generate and display critical information on map and/or report, which ensures that decision-makers take qualified decisions. The system takes into consideration the need for collective governance of resources and actions against disasters, and thus, ensures real-time interconnection and effective information exchange among relevant emergency services (Mansor and Pradhan 2007). NADDI uses the latest available information and communication technology for users to access application modules and information in real-time. Data and information can be fed into or retrieved from NADDI through an interface/ uploading utilities through the real-time communication facilities.

The overall system brings together various data and more specific system components that fully support operational disaster management activities that are currently carried out in Malaysia at the national, state, and district levels. Figure 18 gives an overview of natural hazard mitigation and management programme.

## CONCLUSION

The application of GIS technologies to natural hazards detection, modelling, and monitoring provides at least two primary advantages relative to other means of addressing these hazards. First, GIS allows efficient analysis of large, complex data sets, which may include numerous categorical and continuous variables with different degrees of spatial resolution. GIS-based analysis includes mapping of variables and derivation of new variables, such as the geometric signatures (Pike 1988). GIS also generally facilitates statistical evaluation of correlations among variables and development and calibration of pre-directive models. The statistical or data mining techniques using neural network model now applied to analysing landslide hazards would not have been implemented without the GIS technologies (Pradhan and Lee 2010a, b, c; Aleotti and Chowdhury 1999). Also, when combined with remote sensing imagery, GIS facilitates indirect mapping and analysis of widespread or hazardous conditions that are difficult to access directly during ground-based field investigations. There is an urgent need for the prediction of hill slope instability following torrential rain events. The NASA

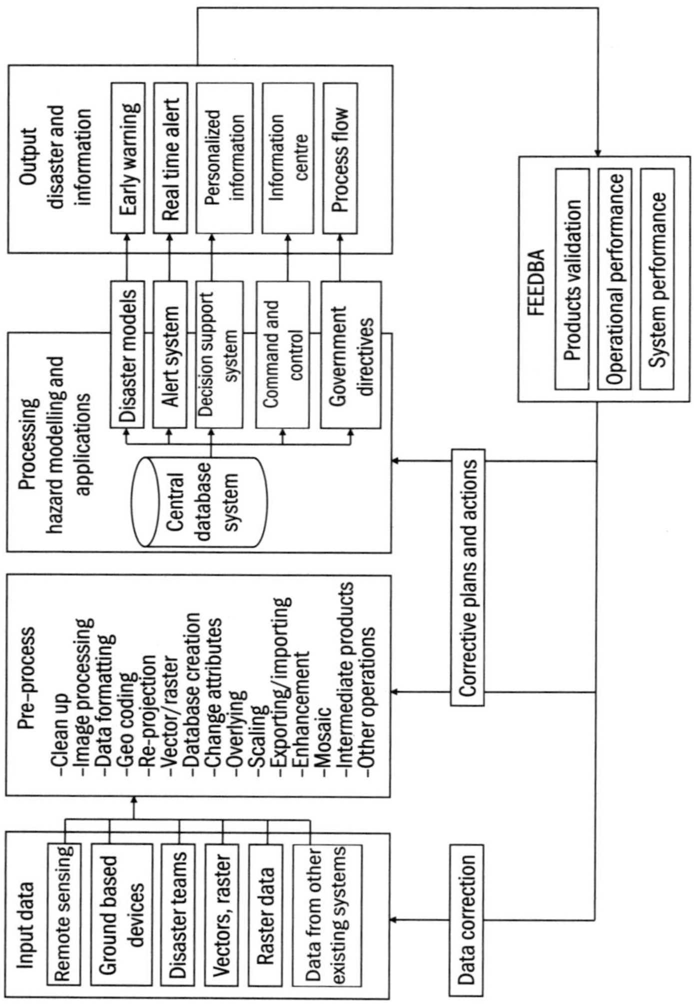


**Figure 18** Various stages of natural hazard mitigation and management processes

TRMM satellite and other satellites are providing real time precipitation information that could be used with GIS-based maps of topographic data and human settlements to issue warnings in areas where the risk of mass movement has increased as a result of precipitation.

Natural hazard modelling is an essential component in hazard management programme, wherein most of the hazardous products are generated at this level. Each natural hazard subsystem should be able to work independently to produce various types of thematic maps based on various types of scenarios using GIS techniques. Some of these products can be used in an effective decision support system (DSS) to take proper decision during crisis time, and some reports should be published for creating better awareness in public. Mitigation measurement will then be decided and carried out based on the products generated at hazard modelling phases. However, the details of the specification of hazard modelling may be varied between each type of hazards due to the different types of scenarios. The parameters for modelling are dynamics as most of the natural hazards are weather related. Therefore, understanding of the natural hazard is very important. Calibration and customization of modelling should be made available from time to time so that that the modelling output can give a very successful result.

Natural hazards can be classified into natural and man-made disasters. From a natural hazard management perspective, generally, the hazard management consists of three phases before a disaster occurs, namely, risk identification, disaster mitigation, and disaster preparedness, and two phases that occur after the occurrence of a disaster, namely, emergency response, and rehabilitation and reconstruction. Many types of information needed in disaster management have both important spatial as well as temporal components. The prime concern during any disaster is the availability of the spatial information and the dissemination of this information to all concerned. If an effective tsunami early warning system had been in place in the Indian Ocean region on 26 December 2004, thousands of lives would have been saved. The same lesson can be drawn from other disasters that have killed tens of thousands of people in the past few years. Effective early warning systems not only save lives but also help protect livelihoods and result in national development gains. Considerable progress has been made in developing the knowledge and technical tools required to generate and communicate predictions and warnings. Early warning system technologies are now available for almost all types of hazards and are in use in some parts of the world. In many cases, however, the warning systems lack the basic equipment, skills, and resources.

The development of an early warning system has been driven by the need for better decision-making. Early innovators were motivated by the belief that experts in a wide variety of fields could take better decisions if they had better tools for analysing and visualizing geographic data. Spatial-based early warning system provides assistance to decision-makers to enable them to find the best managing method in shorter time. Decision-makers need to be able to access, integrate, and share spatial data and services from multiple online sources. The variety and quantity of spatial data needed for decision-making continue to increase as do the frequency of updates. The number of applications and devices continues to grow. New hand-held, wireless, and location-based technologies are becoming essential for field personnel. The challenge is to streamline, integrate, and improve data sharing efforts, even as data proliferates and new technologies are added to the toolkit. Several case studies are presented in this article so that these can be of help at the time of the need for disaster management.

## References

Akgün A and Bulut F. 2007. **GIS-based landslide susceptibility for Arsin-Yomra (Trabzon, North Turkey) region**. *Environmental Geology* **51**: 1377–1387

Aleotti P and Chowdhury R. 1999. **Landslide hazard assessment: summary review and new perspectives**. *Bulletin of Engineering Geology and the Environment* **58**: 21–44

Atkinson P M and Massari R. 1998. **Generalized linear modeling of susceptibility to landsliding in the central Apennines, Italy**. *Computer and Geosciences* **24**: 373–385

Baeza C and Corominas J. 2001. **Assessment of shallow landslide susceptibility by means of multivariate statistical techniques**. *Earth Surface Processes and Landforms* **26**: 1251–1263

Blanchard B W. 1999. **Appendix: Hazard and disaster definitions.** Details available at <http://166.112.200.141/emi/hazards>

Blazkova S and Beven K. 1997. **Flood frequency prediction for data limited catchments in the Czech Republic using a stochastic rainfall model and TOPMODEL**. *Journal of Hydrology* **195**(1–4): 256–278

BMTPC (Building Material Technology Promotion Council). 2001. ***A Paradigm Shift from Post-disaster Reconstruction and Relief to Pre-disaster Pro-Active Approach.*** Disaster Mitigation and Vulnerability Atlas of India. BMTPC, India

Bonham-Carter G F. 1994. ***Geographic Information Systems for Geoscientists, Modeling with GIS***. Oxford: Pergamon. 398 pp.

Carro M, De Amicis M, Luzi L, Marzorati S. 2003. **The application of predictive modeling techniques to landslides induced by earthquakes: the case study of the 26 September 1997 Umbria-Marche earthquake (Italy)**. *Engineering Geology* **69**: 139–159

Cevik E and Topal T. 2003. **GIS-based landslide susceptibility mapping for a problematic segment of the natural gas pipeline, Hendek (Turkey)**. *Environmental Geology* **44**: 949–962

Chan N W. 1995. **Flood disaster management in Malaysia: an evaluation of the effectiveness of government resettlement schemes**. *Disaster Prevention and Management* **4**(4): 22–29

Chuvieco E and Congalton R G. 1989. **Application of remote sensing and geographic information systems to forest fire hazard mapping**. *Remote Sensing of the Environment* **29:** 147–159

Clerici A, Perego S, Tellini C, Vescovi P. 2002. **A procedure for landslide susceptibility zonation by the conditional analysis method**. *Geomorphology* **48**: 349–364

Clerici A, Perego S, Tellini C, Vescovi P. 2006. **A GIS-based automated procedure for landslide susceptibility mapping by the Conditional Analysis method: the Baganza valley case study (Italian Northern Apennines)**. *Environmental Geology* **50**: 941–961

Cunderlik J M and Burn D H. 2002. **Analysis of the linkage between rain and flood regime and its application to regional flood frequency estimation.** *Journal of Hydrology* **261**(1–4): 115–131

Dahal R K, Hasegawa S, Nonomura S, Yamanaka M, Masuda T, Nishino K. 2008. **GIS-based weights-of-evidence modelling of rainfall-induced landslides in small catchments for landslide susceptibility mapping.** *Environmental Geology* **54**: 311–324

Dai F C and Lee C F. 2002. **Landslide characteristics and slope instability modeling using GIS, Lantau Island, Hong Kong**. *Geomorphology* **42**: 213–228

Dai F C, Lee C F, Li J, Xu Z W. 2001. **Assessment of landslide susceptibility on the natural terrain of Lantau Island, Hong Kong**. *Environmental Geology* **40**: 381–391

Donati L and Turrini M C. 2000. **An objective method to rank the importance of the factors predisposing to landslides with the GIS methodology: application to an area of the Apennines (Valnerina; Perugia, Italy)**. *Engineering Geology* **63**: 277–289

Drabek T. 1997. ***Multi Hazard Identification and Risk Assessment***. Washington, DC: FEMA

Ebisemiju F S. 1986. **Environmental constraints of the interdependent of drainage basin morphometeric properties**. Proceedings of the First International Conference on geomorphology, Part 2, Manchester, UK

Ercanoglu M and Gokceoglu C. 2002. **Assessment of landslide susceptibility for a landslide-prone area (north of Yenice, NW Turkye) by fuzzy approach**. *Environmental Geology* **41**: 720–730

Farajzadeh M. 2001. **The flood modeling using multiple regression analysis in Zohre and Khyrabad basins**. 5th International Conference of Geomorphology, August, Tokyo, Japan

Farajzadeh M. 2002. **Flood susceptibility zonation of drainage basins using remote sensing and GIS, case study area: Gaveh rod_ Iran**. Proceedings of the International Symposium on Geographic Information Systems, Istanbul, Turkey, 23–26 September

Giri C and Shrestha S. 1996. **Land cover mapping and monitoring from NOAA AVHRR data in Bangladesh**. *International Journal of Remote Sensing* **17**(14): 2749–2759

Gokceoglu C, Sonmez H, and Ercanoglu M. 2000. **Discontinuity controlled probabilistic slope failure risk maps of the Altindag (settlement) region in Turkey**. *Engineering Geology* **55**: 277–296

Guzzetti F, Carrarra A, Cardinali M, Reichenbach P. 1999. **Landslide hazard evaluation: a review of current techniques and their application in a multi-scale study, Central Italy**. *Geomorphology* **31**: 181–216

Haeng H J, Salas J D, and Boes D C. 2001. **Regional flood frequency analysis based on a Weibull model, Part 2: Simulations and applications.** *Journal of Hydrology* **242**(3–4): 171–182

Hess L L, Melack J M, and Simonett D S. 1990. **Radar detection of flooding beneath the forest canopy: a review**. *International Journal of Remote Sensing* **11:** 1313–1325

Hess L L, Melack J, Filoso S, Wang Y. 1995. **Delineation of inundated area and vegetation along the Amazon floodplain with the SIR-C synthetic aperture radar**. *IEEE Transactions on Geoscience and Remote Sensing* **33** 896–903

Honda K C, Francis X J, and Sah V P. 1997. **Flood monitoring in central plain of Thailand using JERS-1 SAR data**. Proceedings of the 18th Asian Conference of Remote Sensing, Malaysia

Horritt M S and Bates P D. 2002. **Evaluation of 1D and 2D numerical models for predicting river flood inundation**. *Journal of Hydrology* **268:** 87–99

Islam M M and Sadu K. 2001. **Flood damage and modeling using satellite remote sensing data with GIS: case study of Bangladesh**. In *Remote Sensing and Hydrology 2000*, pp. 455–458, edited by J Ritchie, *et al.* Oxford: IAHS Publication

Islam M M and Sado K. 2002. **Development of priority map for flood countermeasures by remote sensing data with geographic information system**. *Journal of Hydrologic Engineering*: 346–355

Jaiswal R K, Saumitra M, Kumaran D R, Rajesh S. 2002. **Forest fire risk zone mapping from satellite imagery and GIS**. *International Journal of Applied Earth Observation and Geoinformation* **4:** 1–10

Jibson W R, Edwin L H, and John A M. 2000. **A method for producing digital probabilistic seismic landslide hazard maps**. *Engineering Geology* **58**: 271–289

Ketpraneet S. 1991. **Forest fire and effects of forest fire on forest system in Thailand.** Kasetsart University, Bangkok

Knebl M R, Yang Z L, Hutchison K, Maidment D R. 2005. **Regional scale flood modeling using NEXRAD rainfall, GIS, and HEC-HMS/RAS: a case study for the San Antonio River Basin Summer 2002 storm event**. *Journal of Environmental Management* **75:** 325–336

Kovach R L. 1995. ***Earth Fury: an introduction to natural hazards and disasters.*** New Jersey: Prentice Hall

Lamelas M T, Marinoni O, Hoppe A, Riva J. 2008. **Doline probability map using logistic regression and GIS technology in the central Ebro Basin (Spain).** *Environmental Geology* **54**: 963–977

Landau S, Mitchell R A C, Barnett V, Colls J J, Craigon J, Payne R W. 2000. **A parsimonious, multiple-regression model of wheat yield response to environment.** *Agricultural and Forest Meteorology* **101**(2–3): 151–166

Le Toan T, Ribbes F, Wange L F, Floury N, Ding N, Kong K H. 1997. **Rice crop mapping and monitoring using ERS-1 data based on experiment and modeling results.** *IEEE Transactions on Geoscience and Remote Sensing* **35**: 41–56

Lee S. 2005. **Application of logistic regression model and its validation for landslide susceptibility mapping using GIS and remote sensing data.** *International Journal of Remote Sensing* **26**: 1477–1491

Lee S. 2007a. **Comparison of landslide susceptibility maps generated through multiple logistic regression for three test areas in Korea.** *Earth Surface Processes and Landforms* **32**: 2133–2148

Lee S. 2007b. **Application and verification of fuzzy algebraic operators to landslide susceptibility mapping.** *Environmental Geology* **52**: 615–623

Lee S and Choi U. 2003. **Development of GIS-based geological hazard information system and its application for landslide analysis in Korea.** *Geoscience Journal* **7**: 243–252

Lee S and Dan N T. 2005. **Probabilistic landslide susceptibility mapping in the Lai Chau province of Vietnam: focus on the relationship between tectonic fractures and landslides.** *Environmental Geology* **48**: 778–787

Lee S and Lee M J. 2006. **Detecting landslide location using KOMPSAT 1 and its application to landslide-susceptibility mapping at the Gangneung area, Korea.** *Advances in Space Research* **38**: 2261–2271

Lee S and Min K. 2001. **Statistical analysis of landslide susceptibility at Youngin, Korea.** *Environmental Geology* **40**: 1095–1113

Lee S and Pradhan B. 2007. **Landslide hazard mapping at Selangor, Malaysia using frequency ratio and logistic regression models.** *Landslides* **4**: 33–41

Lee S and Sambath T. 2006. **Landslide susceptibility mapping in the Damrei Romel area, Cambodia using frequency ratio and logistic regression models.** *Environmental Geology* **50**: 847–855

Lee S and Talib J A. 2005. **Probabilistic landslide susceptibility and factor effect analysis.** *Environmental Geology* **47**: 982–990

Lee S, Choi J, and Min K. 2002. **Landslide susceptibility analysis and verification using the Bayesian probability model.** *Environmental Geology* **43**: 120–131

Lee S, Choi J, and Min K. 2004. **Probabilistic landslide hazard mapping using GIS and remote sensing data at Boeun, Korea.** *International Journal of Remote Sensing* **25**: 2037–2052

Lee S, Chwae U, and Min K. 2002. **Landslide susceptibility mapping by correlation between topography and geological structure: the Janghung area, Korea.** *Geomorphology* **46**: 149–162

Lee S, Ryu J H, Min K, Won J S. 2003. **Landslide susceptibility analysis using GIS and artificial neural network**. *Earth Surface Processes and Landforms* **27**: 1361–1376

Lee S, Ryu J H, Lee M J, Won J S. 2003. **Use of an artificial neural network model for analysis of the susceptibility to landslides at Boeun, Korea**. *Environmental Geology* **44**: 820–833

Lee S, Ryu J H, Won J S, Park H J. 2004. **Determination and application of the weights for landslide susceptibility mapping using an artificial neural network**. *Engineering Geology* **71**: 289–302

Luzi L, Pergalani F, and Terlien M T J. 2000. **Slope vulnerability to earthquakes at subregional scale, using probabilistic techniques and geographic information systems**. *Engineering Geology* **58**: 313–336

Mansor S and Pradhan B. 2007. **Image centric spatial decision support system for disaster management**. Proceedings of First International Conference on Geoinformation Technology for Natural Disaster Management *Geoinformation for Disaster Preparedness, Rehabilitation and Risk Management,* Esfahan, Iran, 8–9 May 2007,

Mansor S, Pradhan B, Daud M, Jamaludin N, Khuzaimah Z. 2007a. **Landslide susceptibility analysis using an artificial neural network**. In *Remote Sensing for Environmental Monitoring, GIS Applications, and Geology VII,* Vol. 6749, edited by M Ehlers and U Michel. Proceedings of SPIE

Mansor S, Pradhan B, Daud M, Khuzaimah Z, Lee S. 2007b. **Utilization of optical remote sensing data and GIS tools for regional landslide hazard analysis in Malaysia**. *Journal of Institution of Surveyors Malaysia* **51**: 50–55

Masmoudi M and Habaieb H. 1993. **The performance of some realtime statistical flood forecasting models seen through multicriteria analysis**. *Water Resources Management* **7**: 57–67

Merwade V, Cook A, and Coonrod J. 2008. **GIS techniques for creating river terrain models for hydrodynamic modeling and flood inundation mapping.** *Environmental Modelling and Software* **23**(10): 1300–1311

Moody A and Katz D B. 2003. **Artificial intelligence in the study of mountain landscapes.** In *Geographic Information Science and Mountain Geomorphology,* pp. 219–249, edited by M P Bishop and J F Shorder Jr. Springer

Myung-Hee J, Myung-Bo L, Si-Young L, Yun-Won J, Seong-Ryul B. 1999. **The development of forest fire forecasting system using Internet GIS and satellite remote sensing**. Details available at <http://www.gisdevelopment.net/aars/acrs/2000/ps3/ps310.shtml>

NADDI (National Disaster Data and Information Management Center). 2003. **Forest Fire Technical Specification**. Cilix Corporation Sdn. Bhd (Contract No: KSTAS/PEMB/2/2003), Project Proposal

Nageshwar R B and Bhagabat P. 1997. **Flood estimation for ungauged catchments using the GIUH**. *Journal of Water Resources Planning and Management*: 228-238

Ohlmacher G C and Davis J C. 2003. **Using multiple logistic regression and GIS technology to predict landslide hazard in northeast Kansa, USA**. *Engineering Geology* **2157**: 1–13

Paola J D and Schowengerdt R A. 1995. **A review and analysis of back propagation neural networks for classification of remotely sensed multi-spectral imagery**. *International Journal of Remote Sensing* **16**(16): 3033–3058

Parise M and Jibson W R. 2000. **A seismic landslide susceptibility rating of geologic units based on analysis of characteristics of landslides triggered by the 17 January, 1994 Northridge, California earthquake**. *Engineering Geology* **58**: 251–270

Pike R J. 1988. **The geometric signature: quantifying landslide terrain types from digital elevation models**. *Mathematical Geology* **20**(5): 491–511

Pistocchi A, Luzi L, and Napolitano P. 2002. **The use of predictive modeling techniques for optimal exploitation of spatial databases: a case study in landslide hazard mapping with expert system-like methods**. *Environmental Geology* **41**: 765–775

Powel G. 2000. **Discussion "landslide risk management concepts and guidelines"**. *Australian Geomechanics* **35** (1): 49–52

Pradhan B. 2009. **Flood susceptible mapping and risk area estimation using logistic regression, GIS and remote sensing**. *Journal of Spatial Hydrology* **9** (2): 1z8

Pradhan B and Buchroithner M F. 2009. **Cross-application of artificial neural network model for landslide susceptibility analysis**. Proceedings of 33rd International Symposium on Remote Sensing of Environment, 4–8 May 2009, Stresa, Lago Magglore, Italy

Pradhan B and Lee S. 2007. **Utilization of optical remote sensing data and GIS tools for regional landslide hazard analysis by using an artificial neural network model**. *Earth Science Frontier* **14**(6): 143–152

Pradhan B and Lee S. 2008. **Landslide risk analysis using artificial neural network model focusing on different training sites**. *International Journal of Physical Sciences* **3**(11): 1–15

Pradhan B and Lee S. 2010a. **Delineation of landslide hazard areas using frequency ratio, logistic regression and artificial neural network model at Penang Island, Malaysia**. *Environmental Earth Sciences* **60**:1037–1054

Pradhan B and Lee S. 2010b. **Regional landslide susceptibility analysis using backpropagation neural network model at Cameron Highland, Malaysia**. *Landslides* **7**: 13–30

Pradhan B and Lee S. 2010c. **Landslide susceptibility assessment and factor effect analysis: backpropagation artificial neural networks and their comparison with frequency ratio and bivariate logistic regression modeling**. *Environmental Modelling and Software* **25**:747–759

Pradhan B and Mansor S. 2007. **Application of remote sensing data and GIS tools for regional landslide hazard analysis at Cameron Highland, Malaysia by using logistic regression model**. Proceedings of IEEE International Geoscience and Remote Sensing Symposium, Barcelona, 23–27 July 2007

Pradhan B and Shafie M. 2009. **Flood hazard assessment for cloud prone rainy areas in a typical tropical environment**. *Disaster Advances* (in press)

Pradhan B, Buchroithner M F, and Mansor S. 2009. **Assessment of neural network, frequency ratio and regression models for landslide susceptibility analysis**. *Geophysical Research Abstracts* **11**, EGU2009-264-1, 2009, EGU General Assembly 2009, 19–24 April, Vienna, Austria

Pradhan B, Lee S, and Buchroithner M F. 2009a. **Landslide risk analysis using fuzzy logic based geospatial model and geoinformation techniques**. *Geophysical Research Abstracts* **11**, EGU2009-265, 2009, EGU General Assembly 2009, 19–24 April, Vienna, Austria

Pradhan B, Lee S, and Buchroithner M F. 2009b. **Use of geospatial data for the development of fuzzy algebraic operators to landslide hazard mapping: a case study in Malaysia**. *Applied Geomatics* **1**: 3–15

Pradhan B, Lee S, and Mansor S. 2008. **Application of a data mining model and its cross application for landslide hazard analysis: a case study in Malaysia**. Proceedings of COSPAR 2008, Montreal, Canada

Pradhan B, Pirasteh S, and Shafie M. 2009. **Maximum flood prone area mapping using RADARSAT images and GIS for Kelantan river basin, Malaysia**. *International Journal of Geoinformatics* **5** (2): 11–23

Pradhan B, Singh R P, and Buchroithner M F. 2006. **Estimation of stress and its use in evaluation of landslide prone regions using remote sensing data**. *Advances in Space Research* **37:** 698–709

Pradhan B, Suliman M D H, and Awang M A. 2007. **Forest fire susceptibility and risk mapping using remote sensing and geographical information systems (GIS)**. *Disaster Prevention and Management* **16**(3): 344–352

Pradhan B, Mansor S, Lee S, Buchroithner B. 2008a. **GIS and neural network model based landslide hazard analysis for Cameron highland**. Proceedings of MRSS 5th Malaysian Remote Sensing and GIS Conference and Exhibition, *Earth Observation for Climate Change, Resource Management and Industrial Applications*, 4 – 5 November 2008

Pradhan B, Mansor S, Lee S, Buchroithner M F. 2008b. ***Application of Data Mining Model for Landslide Hazard Mapping***, pp. 187–196, Vol. 37, Commission VIII, Part B8. In *Proceedings of the International Society for Photogrammetry and Remote Sensing (ISPRS) Congress*, 3–11 July 2008, Beijing, China

Pradhan B, Lee S, Mansor S, Buchroithner M F, Jallaluddin N. 2008. **Utilization of optical remote sensing data and geographic information system tools for regional landslide hazard analysis by using binomial logistic regression model**. *Journal of Applied Remote Sensing* **2**: 1–11

Pradhan B, Lee S, Buchroithner M F, Prechtel N. 2009. **Landslide hazard analysis for development of an early warning system**. Proceedings of Joint Symposium of ICA Working Group on CEWaCM and JBGIS Gi4DM Cartography and Geoinformatics for Early Warning and Emergency Management: Towards Better Solutions, 19–22 January 2009, Prague, Czech Republic

Price M F and Heywood D I (eds). 1994. ***Mountain Environments and Geographic Information Systems***. London: Taylor & Francis

Profeti G and Machintosh H. 1997. **Flood management though Landsat TM and ERS SAR data: a case study**. *Hydrological Processes* **11:** 1397–1408

Refice A and Capolongo D. 2002. **Probabilistic modeling of uncertainties in earthquake-induced landslide hazard assessment**. *Computer and Geosciences* **28**: 735–749

Romeo R. 2000. **Seismically induced landslide displacements: a predictive model**. *Engineering Geology* **58**: 337–351

Rowbotham D and Dudycha D N. 1998. **GIS modeling of slope stability in Phewa Tal watershed, Nepal**. *Geomorphology* **26**: 151–170

Sampson N, Neuenschwander L F, and Harkins K C. 1998. ***Developing a Wildfire Strategic Assessment for the North Zone of the Idaho Panhandle National Forests***. Alexandria, VA-USA: The Sampson Group, Inc.

Sanyal J and Lu X X. 2004. **Application of remote sensing in flood management with special reference to monsoon Asia – a review**. *Natural Hazards* **33**

Sanyal J and Lu X X. 2005. **Remote sensing and GIS-based flood vulnerability assessment of human settlements: a case study of Gangetic West Bengal, India**. *Hydrological Processes* **19**: 3699–3716

Sinnakaudan S K, Ab Ghani A, Ahmad M S S, Zakaria N A. 2003. **Flood risk mapping for Pari River incorporating sediment transport.** *Environmental Modelling and Software* **18**(2): 119–130

Smith K. 1996. ***Environmental Hazard***, 2nd edn. New York: Routledge

Shou K J and Wang C F. 2003. **Analysis of the Chiufengershan landslide triggered by the 1999 Chi-Chi earthquake in Taiwan**. *Engineering Geology* **68**: 237–250

Süzen M L and Doyuran V. 2004. **A comparison of the GIS based landslide susceptibility assessment methods: multivariate versus bivariate**. *Environmental Geology* **45**: 665–679

Tangang F T. 2007. ***Preliminary Assessment of Climate Variability, Climate Change, Global Warming in Malaysia and the 2006/2007 Flood in Southern Peninsular Malaysia***. A report submitted to the Malaysian Government, 70 pp.

Tangestani M H. 2004. **Landslide susceptibility mapping using the fuzzy gamma approach in a GIS, Kakan catchment area, southwest Iran**. *Australian Journal of Earth Sciences* **51**: 439–450

Temesgen B, Mohammed M U, and Korme T. 2001. **Natural hazard assessment using GIS and remote sensing methods, with particular reference to the landslides in the Wondogenet Area, Ethiopia**. *Physics and Chemistry of the Earth (C)* **26**: 665–615

Townsend P A and Walsh S J. 1998. **Modelling flood plain inundation using integrated GIS with radar and optical remote sensing**. *Geomorphology* **21**(98): 295–312

Tunusluoglu M C, Gokceoglu C, Nefeslioglu H A, Sonmez H. 2008. **Extraction of potential debris source areas by logistic regression technique: a case study from Barla, Besparmak and Kapi mountains (NW Taurids, Turkey).** *Environmental Geology* **54**: 9–22

Varnes D J and IAEG Commission on Landslides. 1984. ***Landslide Hazard Zonation – a review of principles and practice***. Paris: UNESCO. 63 pp.

Villiers A B. 1986. **A multivariate evaluation of a group of drainage basin variables, African case study,** pp. 21–33. Proceedings of the First International Conference on Geomorphology, Part 2, Manchester, UK

Wadge G, Wislocki A P, Pearson J, Whittow J B. 1993. **Mapping natural hazards with spatial modeling system**. In *Geographic Information Handling Research and Applications*, edited by P M Mather. New York: John Wiley and Sons Inc.

Wan Ahmad W S. 2002. **Forest fire situation in Malaysia,** pp. 66–74. IFFN, No. 26. Details available at <www.fire.unifreiburg.de/iffn/country/my/my_2.htm>

Wang H B and Sassa K. 2005. **Comparative evaluation of landslide susceptibility in Minamata area, Japan**. *Environmental Geology* **47**: 956–966

Wohl E and Oguchi T. 2004. **GIS and mountain hazards**. In *Geographic Information Science and Mountain Geomorphology*, pp. 309–341, edited by M Bishop and J Shroder. Berlin/Heidelberg: Springer-Verlag

Wybo J, Gouma V, Guarnieri F, Richard B. 1995. **Forest fire danger assessment: combination of methods for efficient decision making,** pp. 344–349. The International Emergency Management and Engineering Conference, TIEMEC'94, 18–21 April, Hollywood Beach, Florida. [Extended version in (1998) *IEEE Transactions on Engineering Management* **45** (2): 127–131]

Xie M, Esaki T, and Cai M. 2004. **A time-space based approach for mapping rainfall-induced shallow landslide hazard**. *Environmental Geology* **46**:840–850

Yakoo Y, Kazama S, Sawamoto M, Nishimura H. 2001. **Regionalization of lumped water balance model parameters based on multiple regression.** *Journal of Hydrology* **246**(1–4): 209–222

Zadeh L A. 1965. **Fuzzy sets**. *Infection Control* **8**: 338–253

Zerger A. 2002. **Examining GIS decision utility for natural hazard risk modeling**. *Environmental Modelling and Software* **17**(3): 287–294

Zimmerman H J. 1996. ***Fuzzy Set Theory and its Applications***. Massachusetts: Kluwer. 435 pp.

Zhou C H, Lee C F, Li J, Xu Z W. 2002. **On the spatial relationship between landslides and causative factors on Lantau Island, Hong Kong**. *Geomorphology* **43**: 197–207

Zhou G, Esaki T, Mitani Y, Xie M, Mori J. 2003. **Spatial probabilistic modeling of slope failure using an integrated GIS Monte Carlo simulation approach**. *Engineering Geology* **68**: 373–386

# Spatial Data Infrastructure Convergence: Building Spatial Data Infrastructure Bridges to Address Climate Change

**Dev Raj Paudyal, Kevin McDougall, and Armando Apan**

*Faculty of Engineering and Surveying*

*Australian Centre for Sustainable Catchments*

*University of Southern Queensland, Toowoomba, Australia*

*E-mail: paudyal@usq.edu.au; mcdougak@usq.edu.au; apana@usq.edu.au*

## INTRODUCTION

Spatial data infrastructure (SDI) is an initiative, which is defined in many different ways. However, SDI has a common intent to create an environment in which all stakeholders can cooperate with each other and interact with the technology to achieve their objectives in a better way at different political/administrative levels (Rajabifard, Feeney, and Williamson 2003). SDI is about the facilitation and coordination of the exchange and sharing of spatial data between stakeholders in the spatial data community. The ultimate objectives of these initiatives, as summarized by Masser (1998), are to promote economic development, stimulate better government, and foster environmental sustainability. Current progress in SDI initiatives shows that SDI is viewed differently by different spatial communities. Table 1 illustrates the range of SDI definitions and perspectives.

**Table 1** Differing perspectives of spatial data infrastructure (SDI)

| Source | Spatial data infrastructure definition |
|---|---|
| Brand (1998) | A global spatial data infrastructure is one that encompasses the policies, organizational remits, data technologies, standards, delivery mechanisms, and financial and human resources necessary to ensure that those working at the global or regional scale are not impeded in meeting their objectives. |
| ANZLIC (1998) | The Australian spatial data infrastructure comprises a distributed network of databases, linked by common policies, standards, and protocols to ensure compatibility. |

*(Table 1 Contd..)*

*(Table 1 Contd..)*

| Source | Spatial data infrastructure definition |
|---|---|
| Coleman and McLaughlin (1998) | A global geospatial data infrastructure encompasses the policies, technologies, standards, and human resources necessary for the effective collection, management, access, delivery, and utilization of geospatial data in a global community. |
| Executive Office of the President (1994) | National spatial data infrastructure (NSDI) means the technology, policies, standards, and human standards resources necessary to acquire, process, store, distribute, and improve the utilization of geospatial data. |
| Groot and McLaughlin (2000) | Spatial data infrastructure encompasses the networked geospatial databases and data handling facilities, the complex of institutional, organizational, technological, human, and economic resources, which interact with one another and underpin the design, implementation, and maintenance of mechanisms facilitating the sharing, access to, and responsible use of geospatial data at an affordable cost for a specific application domain or enterprise. |
| Rajabifard and Williamson (2001) | Viewing the core components of spatial data infrastructure as policy, access network, technical standards, people (including partnerships), and data, different categories can be formed based on the different nature of their interactions within the spatial data infrastructure framework. |
| UN Geospatial Information Working Group (2007) | Spatial data infrastructure is infrastructure for sharing and use of geospatial information. |
| EU (2006) | Spatial data infrastructure is defined as a coordinated framework of technologies, standards, and data, supported by policies and institutional arrangements that enable sharing and effective usage of geospatial information. |

**Source** McDougall (2006)

## EVOLUTION OF SDI AND SDI COMPONENTS

The creation of SDI is a long-term task that may take years, or even decades in some cases, before it becomes fully operational. The concept of SDI first emerged in the mid-1980s after the need was felt for cooperation and sharing of spatially related information across organizations. In Australia, national land-related information initiatives commenced with a government conference in 1984, which eventually led to the formation of a committee responsible for national SDI development. Likewise, in the US, discussion on the national SDI initiatives started during 1989, primarily among the academic community (Mapping Science Committee 1999) and progressed rapidly after the executive order from the President's Office was issued in 1994 (Gore 1998). These national SDI initiatives have been termed the first generation of national SDI initiatives, and the motivations were in reducing duplication, using resources more effectively, and creating a base from which to expand industrial productivity and the spatial information market. It was a "product based" approach, and the coordinators of the SDI developments were dominated by national mapping agencies.

The second phase of undertaking national SDI initiatives started around 2000 when some of the leading nations working on SDI development changed their development strategies and updated their SDI conceptual models, examples being the US and Australia (Williamson, Rajabifard, and Feeney 2003). This approach is "process based" and includes people as a component of SDI along with interoperability of data and resources. In this approach, spatially aware people dominate, and this approach does not support the vast majority of society, who are not spatially aware.

Now, the concept of next generation of SDI is emerging, wherein users play a vital role in information management (Budhathoki, Bruce, and Nedovic-Budic 2008; Goodchild 2008; Rajabifard, Binns, Masser, *et al.* 2006). The advent of geographical information (GI) technology and web services paves the way for more inclusive and open models of spatial services, wherein grass-root citizens and community groups with no prior experience in GI technologies can participate. Google Earth, OpenStreetMap (www.openstreetmap.org), and Wikimapia (www.wikimapia.org) are a few examples, wherein the custodianship of spatial information is no longer in the hands of mapping agencies but lies with the vast majority of people in society, who are utilizing these products.

In literature (McDougall 2006; Mohammadi, Rajabifard, and Williamson 2007; Rajabifard and Williamson 2001; Williamson, Rajabifard, and Feeney 2003), SDI components are generally defined as data, people, standards, access, technology, and policy and institutional arrangement. An SDI encompasses the policies, technologies, standards, and human resources necessary for the effective collection, management, access, delivery, and utilization of spatial data. Table 2 illustrates these SDI components.

**Table 2** Components of spatial data infrastructure

| | |
|---|---|
| Data | Fundamental data sets are themes of spatial information, regarded as primary in supporting the key functions of a country or jurisdiction, providing the common spatial reference and context, which underpin many other forms of business information. An individual agency may consider fundamental data in terms of the most important strategic spatial information that supports its business functions and processes.<br><br>Themes commonly considered fundamental can include geodetic control, cadastre, administrative boundaries, geographic names and localities, street address, transportation, elevation, hydrology, and orthophoto imagery. The list is not definitive and is dependent on the priorities of the responsible agency within each jurisdiction. |
| People | Includes the users, providers, administrators, and custodians of spatial data and also value-added re-sellers. Users can be corporates, or small or large business, public or private, or individuals.<br><br>The broad application of spatial data infrastructure beyond the traditional mapping and land administration role means that users and administrators of spatial information have very different qualifications and professional backgrounds. |

*(Table 2 Contd..)*

*(Table 2 Contd..)*

| | |
|---|---|
| Institutional framework/policy | Includes the administration, coordination, policy, and legislation components of a spatial data infrastructure. The institutional framework is reliant on successful partnerships and communication between agencies within and between jurisdictions. |
| Standards | Consistent standards and policy are required to enable the sharing, integration, and distribution of spatial data, hence, for developing standards for data models and metadata, and transfer and interoperability of storage and analysis software. Policy particularly needs to be consistent for the pricing and access to spatial data within and between jurisdictions. |
| Access and distribution technology | Consists of the access and distribution networks, clearinghouse, and other means for sourcing the spatial information or data sets by the users. Technology also involves the acquisition, storage, integration, maintenance, and enhancement of spatial data. |

**Source** Warnest (2005)

## CATCHMENT MANAGEMENT AND SDI

Catchment management refers to the practice of managing natural resources using river catchment systems as the unit of management. As an approach to managing land and water resources, catchment management involves integrating ecological, economic, and social aspects of natural resource management around an identified catchment system. It aims at integrating these considerations in a way that best ensures long-term viability while at the same time serving human needs (The Parliament of the Commonwealth of Australia 2002). Catchment management is the holistic management of natural resources within a catchment unit encompassing interrelated elements of land and water, managed on an ecological and economic basis and incorporating social systems. It is a system that favours the integration of environmental policy across governments, communities, and industry sectors through partnerships and extensive stakeholder inclusion (AFFA 2002).

Spatial data, in particular, is one of the most critical elements underpinning decision-making for many disciplines (Clinton 1994; Gore 1998; Longley, Goodchild, Maguire, *et al.* 1999; Williamson, Rajabifard, and Feeney 2003), including catchment management. The spatial data available for catchment decisions is inefficient due to different sources, scales, qualities, currencies, and so on. Furthermore, the institutional arrangements for catchment management are problematic and are led by government organizations that are responsible for many catchment decisions (Paudyal and McDougall 2008). Thus, the focus is increasingly shifting to integrate broadly sourced spatial data to create a manageable framework, which necessitates the utilization of SDI. This infrastructure facilitates access to the spatial data and services, thereby improving the existing complex and multi-stakeholder decision-making process (Feeney, Williamson, and Rajabifard 2003; McDougall and Rajabifard 2007). Moreover, it facilitates (and coordinates) the exchange and sharing of spatial data among stakeholders within the geo-information community. A preliminary step towards ensuring decision-making for catchment management has been the increasing recognition of the role of SDI to generate knowledge, provide added value to identify problems, assist in proposing alternatives, and define a course of action (Paudyal and McDougall 2008).

## ROLE OF SDI TO ADDRESS CLIMATE CHANGE AT CATCHMENT SCALE

The impact of climate change is becoming one of the greatest social, economic, and environmental challenges of the present time. Scientific evidence confirms that human activities, such as burning fossil fuels (coal, oil, and natural gas), agriculture practices, and land clearing, have increased the concentration of greenhouse gases in the atmosphere. As a consequence, the earth's average temperature is rising, and weather patterns are changing. This is affecting the rainfall patterns, water availability, sea levels, storm activity, and frequency of droughts and bushfire, subjecting the catchment communities to risk. According to the Intergovernmental Panel on Climate Change (IPCC), the world's average surface temperature is expected to increase by 1.1–6.4°C over the period 1990–2100. This temperature increase will be accompanied by a host of other environmental changes, such as an increase in global sea level, climate changes, and rise in greenhouse gas concentration (National Academy of Sciences 2008). There is a growing concern about global warming and the impact it will have on the people and the ecosystems on which they depend (Bates, Kundzewicz, Wu, *et al.* 2008). In particular, the impacts of climate change will be seen on water resources and associated catchments. This will, in turn, affect the availability of, and demand for, water resources and the way the resources are most effectively managed (Beare and Heaney 2002).

Spatial information is especially important for monitoring present and future climate change as it offers a great potential in handling climate models in both spatial and temporal dimensions. Researchers working in the field of climate science argue that the issues of regional climate change are as complicated as global issues. The task of mitigating and adapting to the impacts of climate change requires collaborative input from a wide range of experts, including physical scientists, engineers, social scientists, medical scientists, business leaders, economists, and decision-makers at all the levels of the government. To detect and assess climate change systematically, one needs to have a stable and reliable information infrastructure. By building SDI at the catchment scale, the sharing and integration of spatial data can eventually help manage and mitigate climate change impacts such as rainfall, storm patterns, and ecosystem impacts at a regional scale.

However, progress in SDI, even in developed countries, is still at an experimental stage, with emphasis particularly on building SDI model capable of supporting the monitoring, detection, and prediction of climate change on decadal time scales (Ahamed 2008). Various models, which include a spatial component, have already been developed for carbon accounting. The GLOBE, or Global Learning and Observations to Benefit the Environment, program, and the CEON, or Circumarctic Environmental Observation Network, are examples of the school-based initiatives and science projects that provide time-variant maps through a web map service. Still, the link between SDI and these initiatives needs to be explored (Bernard and Ostlander 2007). SDI offers a great potential to address climate change at both global as well as regional scales. Building SDI at a catchment scale offers the prediction and monitoring of climate change at regional scale.

## CASE STUDY OF THE MURRAY-DARLING BASIN

### Geographical Location and Climate

The study area for this research is Murray-Darling Basin (MDB), which is an area of national significance for social, cultural, economic, and environmental reasons in Australia. It covers more than one million square kilometres in south-eastern Australia and accounts for about 14% of Australia's total landmass. The basin catchment contains Australia's three largest rivers, namely, Darling (2740 km), Murray (2530 km), and Murrumbidgee (1690 km), along with their many tributaries that collectively stretch for almost 7000 km. The basin is home to over two million people, with another one million outside of the basin relying on its water resources.

Administratively, MDB falls under the four state jurisdictions and one territory, namely, Queensland, Victoria, New South Wales, South Australia, and Australian Capital Territory (Figure 1). The Basin community and the governments are partners in improving the health and productivity of the MDB. In total, there are 22 basin community committees (BCC), which work closely at local level by forming catchment authorities and sub-catchment authorities for integrated catchment management. In addition, there are various volunteer groups (like Landcare, Bushcare, Coastcare, and so on) and indigenous communities that are also working at the grass-roots level to achieve the integrated catchment management goals (ABS 2008).

The MDB experiences a range of climatic conditions and natural environments, including rainforests in the cool and humid eastern uplands,

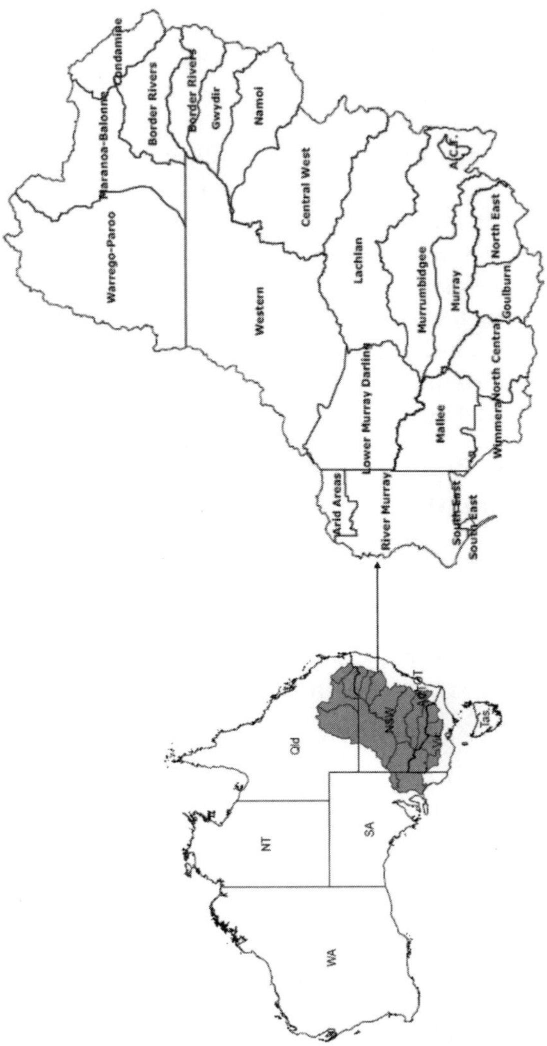


**Figure 1** Case study area: Murray-Darling Basin, Australia

**Data source** Murray-Darling Basin Authority

the temperate mallee country in the south east, and the subtropical areas in the north east, as well as the semi-arid and arid lands of the far western plains. The Basin is home to an extraordinary diversity of communities, landscapes, and natural resources. Moreover, this diversity is accentuated by the striking volatility in the Basin's weather and the economic conditions in which its industries compete (Beare and Heaney 2002).

Greenhouse gas emissions and climate change are likely to be the major issues for the future of the MDB (Hafi, Thorpe, and Foster 2009). The increasing frequency of occurrence of El Niño weather patterns and fewer La Niña patterns, which has been a trend over the last 15 years, are an indication of climate change for the future. Carbon dioxide levels in the atmosphere are rising as a result of anthropogenic activities, and there is evidence of a trend towards higher temperatures. Higher temperatures and changes in "normal" rainfall patterns (the amounts and seasonality) will inevitably result in changes in the distribution of crop and livestock production. It is also suggested that an increase in rainfall in the northern areas and a decrease in rainfall in the southern areas of the region will take place. The predicted future increase in temperature is about of 1–3°C across the Basin, by 2050 (CSIRO 2008; Hafi, Thorpe, and Foster 2009).

## Institutional Setting, Roles and Responsibilities for Catchment Management

The impact of climate change at a regional scale can be mitigated and addressed through better catchment management. The success of catchment management lies in the development of effective institutional arrangements. The organizations discussed in the following sections are working towards the sustainable catchment outcomes in the MDB.

### *Government*

Three tiers of government exist in Australia and influence catchment management of the MDB. All three levels of government play an important role in addressing climate change at a regional scale. The local government is the lowest level of government in the administrative hierarchy. It fosters community awareness and the formation of catchment care groups. It promotes the development of catchment management strategies and implements them on the relevant parts of local authority plans and procedures. There are a total of 170 local governments, which are working partly or fully within the MDB (ABS 2006).

There are four state governments and one territory government working in the MDB. Basically, the state government establishes policies, institutional arrangements, and necessary legislation to facilitate sustainable catchment management and promotes community awareness through education and services. It also provides technical and financial support to catchment coordinating committees as resources permit.

The Commonwealth Government is the highest level of government in the administrative hierarchy. It fosters catchment management strategies by participating in the strategies formulation process and provides financial support for priority catchment projects of national interest. It also implements a comprehensive strategy for tackling climate change in Australia, which is built on three pillars: reducing Australia's carbon pollution; adapting to unavoidable climate change; and helping to shape a global solution (Australian Government 2008).

### *Community Groups and Individual Landowners*

The community groups and individual landowners can play an important role in catchment management and mitigation of the effects of climate change, as they are directly concerned with land, water, and other components of nature in their locality. They are the focal nodes for catchment management at the grass-root level. The community groups involve local land care groups, catchment groups, indigenous community, conservation groups, farm/ water improvement groups, service groups, and industry associations. The Landcare program is a very successful programme involving community members and landowners for catchment management.

### *Catchment Committees/Catchment Management Authorities*

The MDB Authority (MDBA) is an independent and highest body within the environment, water, heritage, and arts portfolio. The MDBA is responsible for the integrated management of water resources in the MDB, including developing and implementing the basin plan for the long-term sustainability of diversion of surface water and groundwater. The basin plan also includes an environmental watering plan, a water quality and salinity management plan, and water trading and transfer rules (MDB 2009).

The Community Advisory Committee (CAC) of the MDB Ministerial Council is a formally appointed group of people with a wide range of expertise and networks throughout the Basin. The role of the CAC is to advise the Ministerial Council from the community viewpoint on critical natural resource management issues, including climate change issues within the Basin.

Catchment coordinating committees or catchment management authorities (CMAs) are established to address complex catchment management issues that involve many community groups and government agencies. CMAs comprise representatives of the major sectors of the community and government, which are involved in, or influenced by, the management of land and water resources in the catchment. Their major role is to provide a forum for community input and discussion, prioritize the issues, and develop and promote the adoption of catchment management strategies. There are a total of 22 CMAs in the MDB, of which four are in Queensland, nine are in New South Wales, five are in Victoria, three are in South Australia, and one is in Australian Capital Territory (ABS 2006).

### *Academia and Research*

Academia and the research community support education, research, and technology development for catchment management and for studying climate change at MDB. They help to build capacity for catchment management and aid in climate change mitigation and adaptation.

### *Industry Groups/Private Sectors*

Industry groups/private sector are the other major stakeholders that play a major role in developing management tools for ecologically sustainable catchment and in mitigating climate change at regional scale. They also rise to the challenge by pooling resources to assist in sustainable catchment management.

## BUILDING SDI TO ADDRESS CLIMATE CHANGE

Based on the case study, SDI has been proposed to address climate change at regional scale. Rajabifard, Binns, Masser, *et al.* (2006) and Warnest (2005) proposed a general SDI model having five major components, including spatial data, stakeholders, policy framework, standards, and access mechanism, as shown in Figure 2. This framework has been adopted for building SDI to address climate change.

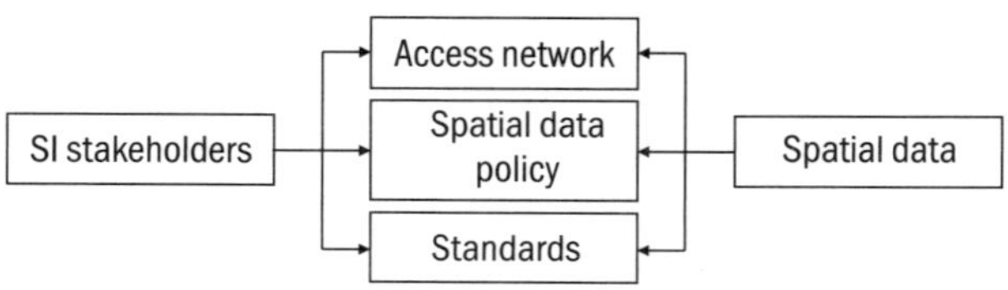


**Figure 2** Spatial data infrastructure components
**Source** Rajabifard, Binns, Masser, *et al.* (2006)

The following section describes the SDI components to address climate change.

## Spatial Data

In building SDI to address climate change, the most important issue is the identification of spatial data sets according to the user's requirement. Table

3 lists the main spatial data sets and their sources, which can be used to address climate change.

**Table 3** Main spatial data sets to address climate change

| Spatial data | Data source | | |
|---|---|---|---|
| | Local | State | National |
| Atmospheric | | | × |
| Climate | | × | × |
| Land use/land cover | | | |
| Digital elevation model | × | × | |
| Demography/population distribution | × | × | |
| Soil and geology | | × | |
| Aquifer and groundwater | × | × | |
| Surface water | | × | |
| Watershed/catchment | × | × | |
| Landownership/cadastre | | × | |
| Infrastructure | × | × | |
| Protected areas | × | × | × |
| Sources of pollution | × | × | |
| Topographic base | | × | |
| Aerial photograph/orthophoto | | × | |
| Buildings | × | | |
| Transportation | × | × | |
| Air quality | × | × | |
| Ecosystem zones | | × | × |
| Vegetation | | × | |
| Rainfall | | × | × |
| Meteorology | | × | × |

Table 3 illustrates the complexities of sources of spatial data sets at different government levels. The custodianship of these spatial data sets is not only at government levels; many high resolution spatial data sets are also freely available at the community level; however, the quality and reliability of these freely available data sets are often questionable (McDougall 2008). However, the integration and management of these multidisciplinary sources are real challenges for building SDI. The model required for accessing, sharing, and managing various kinds of data and depicting their relationships is undoubtedly complex.

## SDI Stakeholders

There are four levels of catchment hierarchy, including farm, sub-catchment, catchment, and basin, in the catchment management domain. From catchment management perspective, the main stakeholders are government organizations, community groups, individual landowners, academia, industry groups/private sectors, and catchment management authorities. However, for building SDI, they can be categorized as spatial data providers, spatial data users, and value adders. Government organizations are the key data providers, providing necessary tools, collaborations, and policy frameworks to facilitate the integration of spatial data from multidisciplinary sources and serve the user's needs. Industry groups/private sectors and academia are the value adders, which can obtain, validate, and translate data according to the user's need. They also contribute to design, development, and operations of SDI through system development, resource pooling, and research and development. In some cases, they are also the spatial data users. Community groups, including individual/indigenous landowners, are the end users.

## Standards

Consistent standards and policies are required to build SDI to address climate change. The standards include data models, metadata, and transfer and interoperability process. Specifications and translating guidelines provide the basis for the standardization of activities and facilitate the business processes. Specifications describe and identify the nature and characteristics of data sets and accompanying components, including the data model and metadata.

## Access Network

The SDI clearinghouse and user interface provide the mechanism to access metadata and the actual data sets. The clearinghouse has to have systems to authenticate data requests. The clearinghouse utilizes SDI search and access protocol engines to look for and discover data and information. The user interface is the front-end interface for user queries, as access of spatial information is one of the important components of SDI development. The user interface should be very simple so that everyone can easily access this infrastructure for mitigating and managing information related to climate change.

## Spatial Data Policy and Institutional Arrangement

As community members are the main players, community-centric institutional arrangements should be designed and developed. The appropriate spatial information policy across governments, communities, and industry sectors should be designed and developed through partnerships and extensive stakeholder inclusion. In particular, policy needs to be consistent for pricing and accessing spatial data within and between jurisdictions. The involvement of indigenous people is equally important as their indigenous knowledge could be used for sustainable natural resources management at the catchment level.

## CONCLUSION

Climate change is the greatest social, economic, and environmental challenge of the present time. Researchers working in the field of climate science argue that the issues of regional climate change are as complicated as global issues. Many climate models have been developed to monitor climate change and its future impact. Spatial information is especially important for climate models as it offers a great capability in both spatial and temporal dimensions to project climate change. Furthermore, a collaborative input from a wide range of stakeholders is needed for mitigation of and adaptation to climate change issues. Hence, the development of SDI offers great potential to study climate change at regional level.

The catchment-based approach is an effective management approach for natural resources management, including land and water. This management approach can be implemented through the creation of partnerships between different levels of government, community groups, industry groups, and academia. The current SDI development migrated from the initial top-down national approach to bottom-up sub-national and cross-jurisdictional approach. Now the concept of a more inclusive model of SDI governance is emerging, wherein grass-root-level citizens and community members can participate for environmental decisions and their sustainability. The SDI offers a great opportunity to build new institutional arrangements and governance models for better natural resource management outcomes.

The components of SDI enable practitioners to understand the complex issues regarding catchment management and regional climate change, and develop SDI. This integrated SDI model can assist in the development of a framework to integrate multi-source spatial data sets to respond to the many climate change issues.

### References

ABS (A B Statistics). 2006. ***Census DataPacks. Data and Geographic Boundaries from the 2006 Census of Population and Housing,*** edited by ABS. Basin Community Profile Release 2

AFFA (The Department of Agriculture, Fisheries and Forestry, Australia). 2002. ***Scientific Advice to the Parliament of the Commonwealth of Australia.*** AFFA

Ahamed S. 2008. **The role of spatial data infrastructure in integrating climate change information with a focus on monitoring observed climate impacts.** Conference on Climate Change and Official Statistics, Oslo, Norway

ANZLIC (The Australia New Zealand Land Information Council). 1998. **Spatial data infrastructure for Australia and New Zealand**. Discussion paper. Details available at <http://www.anzlic.org.au/infrastructure_ASDI.html>, last accessed in December 2010

Australian Government. 2008. ***Carbon Pollution Reduction Scheme: Australia's low pollution future***. Commonwealth of Australia

Bates B, Kundzewicz Z W, Wu S, Palutikof J. 2008. **Climate change and water**. Paper of the Intergovernmental Panel on Climate Change (IPCC), IPCC Secretariat, Geneva, 210 pp.

Beare S and Heaney A. 2002. **Climate change and water resources in the Murray Darling Basin, Australia**. ABARE Conference 2002 World Congress of Environmental and Resource Economists, California

Bernard L and Ostlander N. 2007. **Assessing climate change vulnerability in the Arctic using geographic information services in spatial data infrastructure**. *Climate Change* **87**: 263–281

Brand M. 1998. **Global spatial data infrastructure: policy and organisational issues**. Theme paper, Third Global Spatial Data Infrastructure Conference, 17–19 November 1998, Canberra, Australia. Details available at <http://www.gsdidocs.org/docs1998/canberra/theme.html>, last accessed in December 2010

Budhathoki N R, Bruce B, and Nedovic-Budic Z. 2008. **Reconceptualizing the role of the user of spatial data infrastructure**. *GeoJournal* **72**: 149–160

Clinton W J. 1994. **Coordinating geographic data acquisition and access to the National Spatial Data Infrastructure**. Executive Order 12096. Washington, DC, US

Coleman D J and McLaughlin J D. 1998. **Defining global geospatial data infrastructure (GGDI): components, stakeholders and interfaces**. *Geomatica* **52**(2): 129–143

CSIRO (Commonwealth Scientific and Industrial Research Organisation). 2008. ***Water Availability in the Murray-Darling Basin***. Summary of a report from CSIRO to the Australian Government

EU (European Union). 2006. ***Assessing the Impacts of Spatial Data Infrastructure***. Brussels: Joint Research Centre EU

Executive Office of the President. 1994. **The National Information Infrastructure: Agenda for Action.** Information Infrastructure Task Force, Washington, DC

Feeney M-E F, Williamson I P, and Rajabifard A. 2003. ***SDIs and Decision Support,*** pp. 195–210. London: Taylor & Francis

Goodchild M F. 2008. **Commentary: whither VGI?** *GeoJournal* **72**: 239–244

Gore A. 1998. **The digital earth: understanding our planet in the 21st century**. Digital Earth, Wikipedia. Details available at <http://en.wikipedia.org/wiki/Digital_Earth>, last accessed on 25 November 2008

Groot R and McLaughlin J. 2000. ***Geospatial Data Infrastructure: concepts, cases and good practice***. Oxford: Oxford University Press

Hafi A, Thorpe S, and Foster A. 2009. **The impact of climate change on the irrigated agricultural industries in the Murray-Darling Basin**. ABARE Conference Paper, Cairns, Queensland

Longley P A, Goodchild M F, Maguire D J, Rhind D W E. 1999. ***Geographic Information Systems***. New York: John Wiley & Sons

Mapping Science Committee. 1999. ***Distributed Geolibraries: spatial information resources***. Summary of a Workshop: Panel on Distributed Geolibraries, Mapping Science Committee (National Research Council). Washington, DC: National Academy Press

Masser I..1998.. **The first generation of national geographic information strategies.** In *Proceedings of Selected Conference Papers of the Third Global Spatial Data Infrastructure Conference,* November 1998, Canberra, Australia

McDougall K. 2006. ***A Local–State Government Spatial Data Sharing Partnership Model to Facilitate SDI Development***. Melbourne: The University of Melbourne

McDougall K. 2008. **Unlocking the potential of spatial information through data sharing – it's a two way street**. Queensland Spatial Conference, Gold Coast, Australia

McDougall K and Rajabifard A. 2007. ***A Mixed-Method Approach for Evaluating Spatial Data Sharing Partnerships for Spatial Data Infrastructure Development***. California: ESRI Press Redlands

MDB (Murray–Darling Basin). 2009. ***Governance of the Murray–Darling Basin Authority***. MDB

Mohammadi H, Rajabifard A, and Williamson I P. 2007. **Multi-source spatial data integration within the context of SDI initiatives**. *International Journal of Spatial Data Infrastructures Research* **4**

National Academy of Sciences. 2008. ***Understanding and Responding to Climate Change***. Highlights of National Academies Reports. National Academy of Sciences

Paudyal D R and McDougall K. 2008. **Building spatial data infrastructure to support sustainable catchment management**. Queensland Spatial Conference 2008, Gold Coast, Queenslands.

Rajabifard A and Williamson I. P. 2001. **Spatial data infrastructures: concepts, SDI hierarchy and future directions**. Geomatics' 80 Conference, Tehran, Iran

Rajabifard A, Feeney M-E F, and Williamson I. 2003. **Spatial data infrastructures: concept, nature and SDI hierarchy**. In *Developing Spatial Data Infrastructures: from concept to reality,* pp. 17–40, edited by I Williamson, A Rajabifard, and M E Feeney. London: Taylor & Francis

Rajabifard A, Binns A, Masser I, Williamson I. 2006. **The role of sub-national government and the private sector in future spatial data infrastructure**. *International Journal of Geographical Information Science* **20**(7): 727–741

The Parliament of the Commonwealth of Australia. 2002. **Co-ordinating catchment management: report on the inquiry into catchment management**. House of Representatives Standing Committee on Environmental Heritage

UN Geospatial Information Working Group. 2007. ***Strategy for Developing and Implementing a UN Spatial Data Infrastructure***. New York: United Nations

Warnest M. 2005. ***A Collaboration Model for National Spatial Data Infrastructure in Federated Countries***. University of Melbourne, Australia

Williamson I, Rajabifard A, and Feeney M-E F. 2003. ***Developing Spatial Data Infrastructures: from concept to reality***. London: Taylor & Francis

# Geoinformatics for Climate Change Adaptation and Disaster Risk Reduction

**P K Joshi[*], M Munsi[*, †], and A Joshi[#]**
[*]*TERI University, New Delhi, India*
[†]*Centre for Sustainable Technology, Indian Institute of Sciences, Bangalore, India*
[#]*Delhi Productivity Council Institute of Management Studies, New Delhi, India*
*E-mail: pkjoshi@teri.res.in; madhushreemunsi@gmail.com; archanapandey01@gmail.com*

## INTRODUCTION

Climate change can be defined as the variability in climatic properties, which can be attributed directly or indirectly to human activity that alters the composition of the global atmosphere, observed over comparable time periods, typically decades or longer. The change in global climatic trend has been attributed to two major factors: natural (climatic and environmental variability, for example, in ocean currents, solar activity, volcanic eruptions, and so on) and anthropogenic (infrastructure development and land use land cover change) (ISDR 2008). In recent times, increasing anthropogenic pressure is leading to the emission of greenhouse gases (GHGs) and aerosols, which, in turn, are increasing the rate of global warming and snow melting, thus resulting in increased sea level and increased frequency and intensity of cyclones, floods, droughts, and other disasters (IPCC 2001, 2007; Sperling and Szekely 2005; ISDR 2008). Climate change can have an adverse impact on humans, which can be witnessed in the form of increased frequency of natural disasters, reduced environmental and societal resilience, and, thus, increased vulnerability (O'Brien, O'Keefe, Rose, *et al.* 2008). Some of the impacts of extreme climatic variability on some of the key sectors are given in Table 1.

The United Nations International Strategy for Disaster Reduction (UNISDR) terminology on Disaster Risk Reduction (DRR) defines *disaster* as "a serious disruption of the functioning of a community or a society involving widespread human, material, economic or environmental losses and impacts, which exceeds the ability of the affected community or society to cope using its own resources", and *disaster risk* as "the potential disaster losses, in lives, health status, livelihoods, assets and services, which could occur to a particular community or a society over some specified future time period" (ISDR 2009). In recent decades, the hydro-meteorological

**Table 1** Possible impacts of climate change on some key sectors

| Phenomenon/event | Likelihood of future trends in absence of any adaptive capacity | Major projected impacts on key sectors | | | |
|---|---|---|---|---|---|
| | | Ecosystems (agriculture/forest) | Water resources | Human health | Industry, settlement, and society |
| Warmer and fewer cold days, warmer and more frequent hot days and nights | Virtually certain (>99%) | Increased yield in colder environments, decreased yields in warmer environments, increased insect infestation | Effects on water resources relying on snowmelt and some water supplies | Reduced human mortality due to decreased exposure to cold | Reduced energy demand for heating, increased demand for cooling, declining air quality in cities, reduced disruption to transport due to snow and ice, effects on winter tourism |
| Increased frequency of warm spells/heat waves | Very likely (>90%) | Effects on winter tourism due to heat stress, reduced yield due to heat stress, increased risk of wildfire | Increased water demand and water quality problems like eutrophication | Increased heat-related mortality | Reduction in quality of life for people in warm areas without appropriate housing; impacts on the poor |
| Increased frequency of heavy precipitation events | Very likely (>90%) | Crop damage, soil erosion, waterlogging | Contaminated surface water and groundwater; no water scarcity problem | Increased risk of infectious, respiratory, and skin diseases, and deaths | Disruption of commerce, transport, and effection settlements and societies due to flooding; loss of property and life |
| Increase in drought-prone areas | Likely (>66%) | Land degradation, lower crop yield, higher livestock mortality, increased risk of wildfire | Widespread water stress | Food and water shortage and related diseases; increased cases of malnutrition | Water shortage for settlements, industry, and societies; reduced hydro-power generation potentials; potential for population migration |

| Phenomenon/event | Likelihood of future trends in absence of any adaptive capacity | Major projected impacts on key sectors: Ecosystems (agriculture/forest) | Water resources | Human health | Industry, settlement, and society |
|---|---|---|---|---|---|
| Increase in intense tropical cyclone events | Likely (>66%) | Damage to crops, wind throw (uprooting) of trees, damage to coral reefs | Power outages causing disruption of public water supply | Increased risk of deaths, injuries, and water- and food-borne diseases; post-traumatic stress disorders | Disruption by flood and high winds; withdrawal of risk coverage in vulner able areas by private insurers; potential for population migration; potential loss of property |
| Increased incidence of extreme high sea level (excluding tsunamis) | Likely (>66%) | Salinization of irrigation water, estuaries, and freshwater systems | Saltwater intrusion and decreased freshwater availability | Increased risk of deaths and injuries by drowning in floods; migration-related health effects | Costs of coastal protection versus costs of land-use relocation; potential for movement of populations and infrastructure; disruption by flood and high winds; withdrawal of risk coverage in vulnerable areas by private insurers; potential for population migration; potential of loss of property |

**Source** Adapted from IPCC (2007)

disasters (that is, disasters associated with climatic change and variability) have increased manifold (>40% during 1900–09 to 76% during 2000–05). Figures 1a and 1b show the regional distribution of natural disasters by origin. Figure 1a shows the distribution of natural disasters by origin during 1991–2005 for the major regions of the world. Asia experiences the maximum incidences of disasters, followed by America, Africa, and Europe. Figure 1b shows the global distribution of natural disasters by origin, between 1900 and 2005. It identifies hydro-meteorological disaster as the most dominant one followed by the disasters that have geological and biological origins.

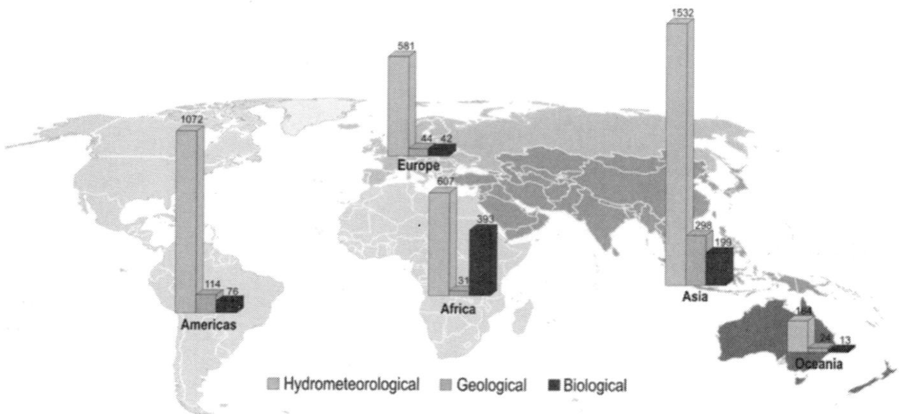


**Figure 1a** Regional distribution of natural disasters by origin during 1991–2005 (bars represent number of disasters)

**Source** Adapted from <http://www.unisdr.org/>

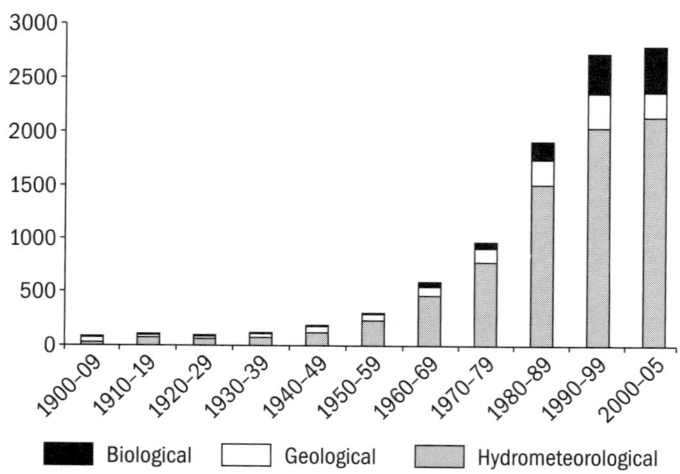


**Figure 1b** Global distribution of natural disasters by origin between 1900 and 2005 (Y-axis – number of disasters)

**Source** <http://www.unisdr.org/>

The climate-related stressors have affected the lives of many, particularly the poor, throughout the world. Climate change not only increases the magnitude and frequency of natural disasters but also generates new threats by affecting the climatic conditions and climatic variability (IPCC 2007). The current trend of increasing disasters indicates a future with extreme climate variability, which will reduce the coping capacity of the human society, making it more vulnerable to the existing risks (ISDR 2008). According to the IPCC (2007) reports, the global surface temperature is expected to increase by 1.1–6.4°C, which will lead to sea level rise in the range 18–59 cm and ocean acidification. Extreme temperatures, heat waves, heavy precipitation, and tropical cyclones will become more frequent and intense, while precipitation in subtropical lands will decrease.

The incidences of disasters, especially in Asia and the Pacific region, have increased in the past few decades, which have affected the infrastructure and livelihoods at a large scale (UNESCAP 2005). Figure 2 shows the distribution of natural disaster occurrence in 2008. At present, nearly 634 million people (one-tenth of the global population) live in coastal areas that are a few metres above the existing sea levels and are at risk from disaster. Three-quarters of these areas are located in the Asian flood-prone river deltas or in low-lying small island states, and urbanization is still increasing in those areas, thus making the area even more vulnerable to natural hazards (McGranahan, Balk, and Anderson 2007; ISDR 2008).

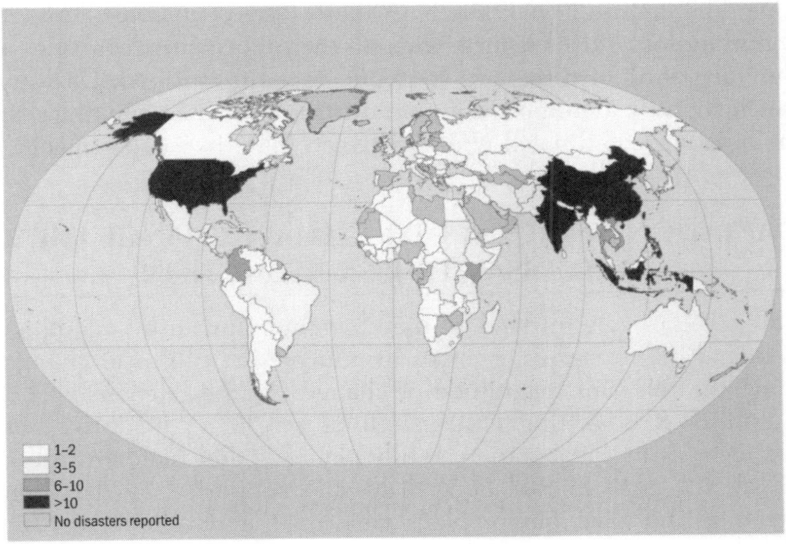


**Figure 2** Distribution of the most disaster-prone areas, as per the occurrence of natural disasters in 2008

**Source** Adapted from EM-DAT

## CLIMATE CHANGE ADAPTATION AND DISASTER RISK REDUCTION

Over the years, a positive trend has been observed between climatic change and occurrence of natural disasters. The understanding of this trend has brought to the fore the need to incorporate DRR strategies and mechanisms as part of climate change adaptation (CCA) (Mitchell and van Aalst 2007; ISDR 2008; O'Brien, O'Keefe, Rose, *et al.* 2008).

CCA is defined as adjustments in natural or human systems in response to actual or expected climatic stimuli or their effects, which moderate harm or exploit beneficial opportunities (IPCC 2007). DRR is defined as the conceptual framework of elements considered with the possibilities to minimize vulnerability and disaster risks throughout a society, to avoid or to limit the adverse impacts of hazards, within the broad context of sustainable development (ISDR 2008).

The concept of CCA emerged in the UNFCCC (United Nation Framework Convention on Climate Change) and was signed in 1992. Hyogo Framework for Action (HFA) in 2005 was adopted by 168 countries, and this framework provides strategic and systematic approaches to reduce risks to hazards. Over the years, CCA and DRR have developed through separate framework and have different views and concepts. But, in recent years, more emphasis has been placed on reducing the gaps between these two concepts by working in collaboration and discussing tools, techniques, and methods for adaptation (O'Brien, O'Keefe, Rose, *et al.* 2008; ISDR 2008; Helmer and Hilhorst 2006; Schipper and Pelling 2006; O'Brien, O'Keefe, Rose, *et al.* 2006). BAP (2007) is an important tool in this regard, wherein the participating countries agreed on the integration of these two concepts. Towards this end, DRR and risk management were identified as important aspects of CCA and, thus, stressed on the need for coordinated efforts between these two departments.

## ADAPTATION STRATEGIES TO CLIMATE CHANGE IMPACTS AND DISASTER RISK REDUCTION

Human societies can respond to the changing climate by adapting to its impacts and taking necessary steps to mitigate the causes of change, thus reducing the rate and magnitude of change. As the relationship between CCA and DRR is getting clearer, efforts are being made by both the sectors to work in collaboration. While planning and mainstreaming CCA and DRR, the goals should be realistic and supported by developmental programmes and rehabilitation plans, ensuring that developmental activities do not increase the vulnerability of human society. These activities should be well funded and well coordinated at all stages. Reducing hazard vulnerability and identifying a common approach towards CCA and DRR are the most appropriate methods adopted. To achieve this goal, characteristics of CCA and DRR need to be assessed, evaluated, and understood. Table 2 provides some key characteristics of CCA and DRR.

**Table 2** Characterization of climate change adaptation and disaster risk reduction

| Climate change adaptation | Disaster risk reduction |
|---|---|
| **Approach** | |
| • Risk management<br>• Strong scientific basis<br>• Environmental science perspective<br>• Vulnerability perspective<br>• Long-term perspective<br>• Global scale | • Measure, assess, evaluate, and manage<br>• Engineering and natural science basis<br>• Traditional focus on events and exposure and technological solutions<br>• Response and recovery, awareness<br>• Short term but increasingly long term<br>• Community based |
| **Organizations** | |
| • United Nations Framework Convention on Climate Change (UNFCCC)<br>• Inter-governmental Panel on Climate Change (IPCC)<br>• Academic research<br>• National environment and energy authorities | • United Nations<br>• World Bank<br>• International Federation of Red Cross and Red Crescent Societies (IFRC)<br>• International, national, and local civil society organizations<br>• National civil defence authorities |
| **International conference** | |
| • Conference of the Parties (COP) | • World Conference on Disaster Reduction<br>• International Disasters and Risk Conference-DAVOS |
| **Assessment** | |
| • IPCC assessment reports | • IFRC Vulnerability and Capacity Assessment (VCA)<br>• IFRC World Disaster Report<br>• International disasters databases: Emergency Events Database (EM-DAT), NatCatSERVICE (Munich Re), Sigma (Swiss Re) |
| **Strategies** | |
| • National communications to the UNFCCC<br>• National Adaptation Plans of Action (NAPA) for Least Developed Countries | • UN International Decade for Natural Disaster Reduction (IDNDR)<br>• Yokohama Strategy and Plan of Action for a Safer World<br>• UN International Strategy for Disaster Reduction (ISDR)<br>• Hyogo Framework for Action 2005–15 |

*(Table 2 Contd..)*

(Table 2 Contd..)

| Climate change adaptation | Disaster risk reduction |
|---|---|
| **Funding** | |
| • Special Climate Change Fund<br>• Least Developed Countries Fund | • National civil defence/emergency response<br>• International humanitarian funding, for example, UN Office for the Coordination of Humanitarian Affairs (OCHA) |
| • Kyoto Protocol Adaptation Fund | • Multilateral banks<br>• Bilateral aid |

**Source** Thomalla, Downing, Spanger-Siegfried, *et al.* (2006)

In recent years, there is an increasing focus on taking effective measures to cope up with the problems. Adaptation Policy Framework (developed by the United Nations Development Programme [UNDP]), UNFCCC guidelines, Caribbean Risk Management Guidelines for Climate Change Decision Making, and International Decade for Natural Disaster Reduction (IDNDR) talk about developing measures for adaptation to climate change (Sperling and Szekely 2005; Jayaraman, Chandrasekhar, and Rao 1997). Adaptation and disaster management reduce the negative impacts, both in terms of money and lives; for example, China spent about $3.15 billion on flood control between 1960 and 2000, which helped in reducing the losses to the tune of $12 billion. Similarly, disaster management and preparedness in Andhra Pradesh, India, yielded an estimated benefit–cost ratio of 13.38. Vietnam started a mangrove plantation project, which aimed at protecting coastal population from typhoons and storms, yielding an estimated benefit–cost ratio of 52 over the period 1994 to 2001 (ISDR 2008). The capacity of adaptation, however, depends on socio-economic conditions and the availability of information and technology (IPCC 2007). In Hyogo Framework (2005), gaps in disaster management were identified in areas related to governance issue (institutional and policy frameworks), risk management (risk identification, assessment, monitoring, and early warning), knowledge dissemination, reducing underlying risks, and preparedness for response and recovery. Some of the practical actions that can be followed for CCA and DRR are listed below.

## Institutional Frameworks and Funding for Adaptation

DRR is gaining importance in various UN programmes like Economic and Social Council (ECOSOC), United Nations Office of the High Representative for the Least Developed Countries, Landlocked Developing Countries and Small Island Developing States (UNOHRLLS), UN Development Group, and World Trade Organization (WTO) (ISDR 2008). International organizations like UNDP, Institute for Security and Development Policy (ISDP), Asian Emergency Preparedness and Response (AEPR) Initiative, and international non-governmental organizations also work towards enhancing disaster preparedness and risk reduction. Apart from developing adaptation policies at national and international levels, communities and local government should be engaged to assess and develop adaptation policies at local and individual levels (APEC 2008). Lack of funding and resources is also a major concern as governments often believe that these should not be considered within regular developmental financing as the disaster risks cannot be quantified in exact figures due to their unpredictable nature, thus, both the sectors lack funding (Mitchell and van Aalst 2008; Hyogo Framework 2005). Various regional, bilateral, and multilateral organizations exist, which work together to generate funds and enhance DRR programmes (ISDR 2008; Schipper and Pelling 2006). In Asia, organizations like Association of South East Asian Nations (ASEAN), Asian Disaster Preparedness Centre (ADPC), and Asian Disaster Risk Reduction Centre (ADRC) help in establishing policies to enhance risk reduction. Bilateral and multilateral organizations (World Bank, Asian Development Bank, European Commission, and so on) provide funding for disaster preparedness and related developmental planning. Task Force on Emergency Preparedness (TFEP) was established for coordination between these organizations (APEC 2008; Schipper and Pelling 2006).

## Integration of Climate Change Adaptation and Disaster Risk Reduction in Development Plans

Reduction in environmental and socio-economic impacts of climate change related disaster risks requires effective coordination between different levels of governance at national level (IPCC 2007). Figure 3 shows a schematic representation of the relationship between DRR, CCA, and national development policy. CCA and DRR need to be integrated and included with all developmental processes and budgets and macro-economic projections, for example, in city planning, land use planning, infrastructure and coastal zone development, forest use, engineering programmes, and building designs, so as to reduce the vulnerability of human society and facilitate better adaptation and sustainable development (Hyogo Framework 2005; Schipper and Pelling 2006; BAP 2007; ISDR 2008). Risk transfer mechanism, social safety nets, and livelihood supporting programmes should be incorporated in development plans (ISDR 2008).

## Disaster Preparedness Planning

Disaster preparedness planning is important for efficient response (that is, the activities followed by a disaster to provide emergency assistance to the victims, which will reduce the secondary impacts of the disaster) and recovery (this phase continues till all the systems return to normal) in case of emergency (Montoya 2003; Hyogo Framework 2005), including

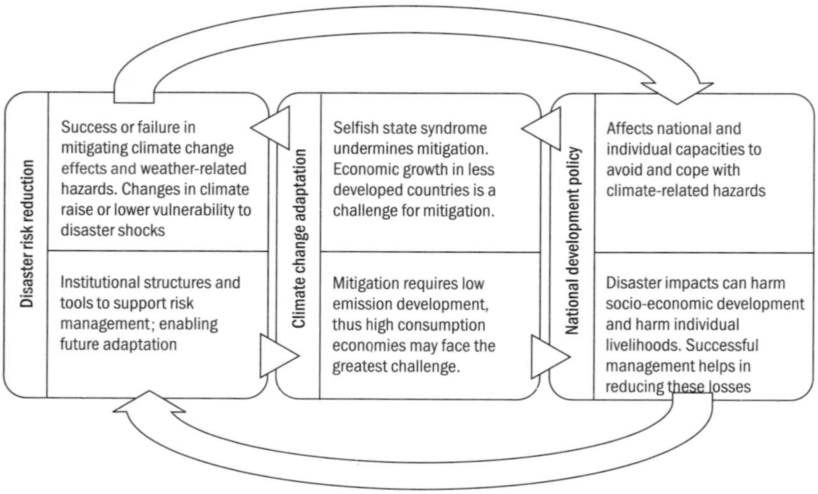


**Figure 3** Relationship among disaster risk reduction, climate change adaptation, and national development policy

**Source** Modified from Schipper and Pelling (2006)

climate change. The most important step for disaster preparedness is identifying gaps, and regional and international organizations should work in coordination with the bilateral, multilateral, and other financing organizations. Disaster preparedness should be promoted at regional, national, local, and individual levels through seminars, exhibitions, and developing new projects. Business communities at all levels should develop resilience capacity, and long-term economy recovery planning should be there for the affected sectors (APEC 2008; Hyogo Framework 2005). Experiences from various disasters and projections of climate change emphasize on the importance of joint effort of preparedness and adaptation; however, these need proper planning and team work. Trainings, including computer-supported programmes, should be promoted.

## Reducing the Underlying Risks

Environmental management is very important for CCA and DRR. A healthy ecosystem not only benefits the livelihood of the local people, but also provides resilience to disaster and increases adaptive capacity. Measures should be taken to protect ecosystems that help in alleviating hazards and related risks (for example, coral reefs and mangroves for coastal hazards) (ISDR 2008; Hyogo Framework 2005). Apart from that, hazard-resistant buildings and hospitals are also required. Building evacuation mechanisms are also required to be implemented. Preparedness and contingency plans should be revised from time to time to minimize the impacts of existing as well as new hazards (ISDR 2008; Hyogo Framework 2005).

### Enhanced Knowledge Sharing

Knowledge dissemination and education are important to establish a culture of safety and resilience. Public information programmes at local level are important for transferring information on climate-related hazards and issues. Schools and educational institutes should have basic curriculum on disaster management and CCA, and research programmes should be encouraged in these sectors. Radio and television programmes and poster campaigns can also be used for knowledge transfer and sharing on these issues (ISDR 2008). Internet has played an important role in knowledge sharing at community level. Geoportals also play an important role in knowledge dissemination by providing information and data on disasters (Rajabifard, Feeney, and Williamson 2002; Mansourian, Rajabifard, Zoej, *et al.* 2006). In the present era of mobile phone networks, mobile phone, cell phone, short message services, and even access to Internet via these devices can be of great help.

### Risk Management and Early Warning Systems

Risk identification, assessment, transfer, and reduction are the major elements of risk management and are desirable for managing disasters through emergency management (ISDR 2008; Hunt 2002; Dilley, Chen, and Deichmann 2005). Risk factor identification helps not only in mitigation (Montoya 2003), relief, and reconstruction, but also towards planning preparedness to reduce recovery time (Dilley, Chen, and Deichmann 2005), which thus aids in better policy formulation and implementation by policy-makers and increases effectiveness on warning systems (ISDR 2008). Information and communication technologies (ICTs), especially space technology, have been useful in risk management. Regular monitoring and using remote sensing techniques and GIS help in enhancing early warning systems, thus enhancing risk management technology (UNESCAP 2005).

## GEOINFORMATICS—A TOOL

Geoinformatics—an umbrella team covering a wide gamut of GIS and related technologies cremote sensing, global positioning system [GPS], image processing, high bandwidth communications—is widely believed to play an important role in climate change research and disaster risk reduction. It comprises collection, processing, analysis, storage, and dissemination of geospatial data for efficient resource management and decision-making. It uses real-time spatial databases updated regularly by means of telecommunication systems and aims at improving accessibility of geospatial information and enhancing the overall quality of problem solving and decision-making at affordable costs. The main purpose of geospatial tools is to provide information on the earth surface and document the impacts of natural and anthropogenic events on the ongoing changes (Lillesand, Kiefer, and Chipman 2007).

Information technologies, in general, and communication technologies, in particular, have seen massive growth in the past years and have been introduced in disaster management and relief processes in various degrees. Advances in communication, networking, image processing, database management, and artificial intelligence have led to the development of new technologies aimed at disaster relief. The importance of ICTs in disaster relief has been acknowledged in many international initiatives like the Declaration of Principles, and the Geneva Plan of Action of the World Summit on the Information Society (UNESCAP 2008). Geoprocessing reduces the time and expense of travel required for data collection and improves the quality of decision-making by providing the required services in a timely fashion and expanding the pool of specialists required for handling different cases. Thus, the role of geographic information providers in ICTs needs to be enhanced for better and economically viable management system. It, however, requires high coordination between the practitioners and information technologists to ensure that the technologies satisfy the needs of the current users and anticipate the future needs. Internet, media, GIS, and satellite-based communication are important forms of ICTs that play a vital role in all stages of disaster management cycle (UNESCAP 2008). These assist in information collection and processing, early warnings, and emergency communication, thus playing an important role in planning and policy formulation for devising better adaptation strategies (UNESCAP 2008; Mohanty, Karelia, and Issar Unpublished). Thus, the geoprocessing tools can be classified into five categories, namely, (1) spatial information systems, (2) remote sensing, (3) Internet, website, and portals, (4) communication systems, and (5) ICT applications (UNESCAP 2008). Figure 4 gives a brief description of these categories. Table 3 lists some of the case studies from different parts of the world, which used ICTs for disaster management.

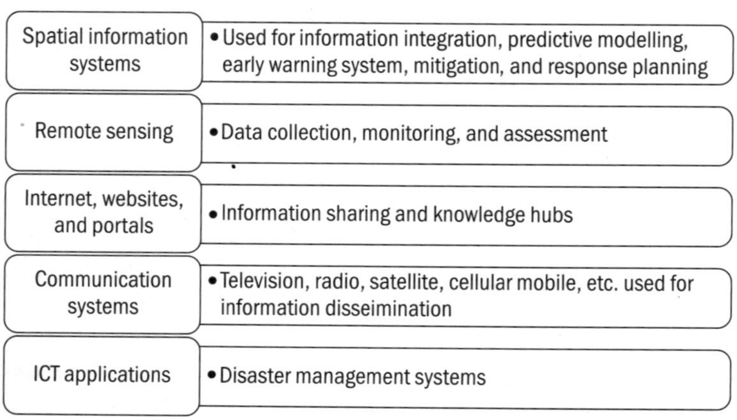


**Figure 4** Geoprocessing tools for disaster risk reduction and climate change adaptation

**Table 3** Examples of information and communication technologies used for disaster management

| Place of study | Applications | Author |
|---|---|---|
| Bangladesh | Flood hazard maps of Bangladesh were developed using National Oceanic and Atmospheric Administration (NOAA)–Advanced Very High Resolution Radiometer (AVHRR) images and geographic information system (GIS) | Islam and Sado (2000) |
| Mackay (Queensland) | GIS application for cyclone disaster management was examined and evaluated | Zerger and Smith (2003) |
| Korea | Spatial database (SDB), a GIS-based geological hazard information system, has been developed for landslide analysis in Korea | Lee and Choi (2003) |
| Latin America | National-level DesInventor disaster databases (30 years of data) were used to cover major and localized disasters | Wattegama (2007) |
| South Africa | Monitoring, Mapping, and Analysis Incidents in South Africa (MANDISA) was used for tracking disaster events in the western province of South Africa | Wattegama (2007) |
| Vietnam | GIS was integrated with traditional knowledge for flood risk mapping | Tran, Shaw, Chantry, *et al.* (2009) |
| Antarctic Peninsula | Spatial and temporal data of ERS-1/2 synthetic aperture radar (SAR) and Envisat advanced SAR (ASAR) were used to study the impact of climate warming | Arigony-Neto, Saurer, Simoes, *et al.* (2009) |

## Earth Observation System

Earth observation (EO, or remote sensing) technology can provide repeat and continuous data coverage for the entire earth's surface (Asrar and Dozier 1994). Remote sensing techniques, integrated with GIS and GPS technologies, are widely used for studying the earth system, risk and hazard factor identification and prediction, decision-making, and developing adaptation strategies (Jayaraman, Chandrasekhar, and Rao 1997; Dilley, Chen, and Deichmann 2005; Tralli, Blom, ZlotNicki *et al.* 2005; ISDR 2008; UNESCAP 2008).

Remote sensing is the science of identifying different earth surface features by estimating their geo-biophysical properties using electromagnetic radiation as a medium of interaction (Navalgund, Jayaraman, and Roy 2007). Remotely sensed data provides continuous information for any geographical region at both global and local levels, and thus helps in understanding the dynamics of an area (Jayaraman, Chandrasekhar, and Rao 1997; Holdaway 2001; Tralli, Blom, ZlotNicki *et al.* 2005). Disaster management basically uses *optical monitoring* (simplest in concept, but subject to environmental factors such as weather and cloud, and works only in daylight), *IR monitoring* (detects the thermal radiation either emitted by the body or radiation emitted from another source, which is subsequently scattered by the object under study), and *microwave/radio monitoring* (the radiation is caused either by emission or scattering; all time monitoring) (Holdaway 2001). *Geostationary satellites* are capable of detecting storms and other weather-related events/disasters, and *polar-orbiting satellites* are capable of detecting floods, volcanic eruptions, and so on, whereas *EO satellites* help in pre-disaster preparedness, disaster responses, and monitoring activities. Thus, the coordinated use of these satellites can help in better disaster risk identification and adaptation (Jayaraman, Chandrasekhar, and Rao 1997; Holdaway 2001). The appropriateness, however, depends on the resolution of the data. Some of the satellite data that is being used increasingly for disaster management is IKONOS, OrbView, MODIS, and Quickbird (Montoya 2003; Tralli, Blom, ZlotNicki *et al.* 2005). Table 4 lists some of the satellite missions developed for disaster management.

**Table 4** Some of the global satellite missions for climate change studies

| Satellite mission | Status | Applications |
|---|---|---|
| Advanced Land Observing Satellite (ALOS) | Approved | Digital terrain models, environmental monitoring, disaster monitoring, civil planning, agriculture and forestry, earth resources, land surface |
| Aqua (formerly EOS PM-1) | Currently orbiting | Atmospheric water and energy cycles, cloud formation, precipitation and radiative properties, air/sea fluxes of energy and moisture, sea ice extent and heat exchange with the atmosphere |
| China Brazil Earth Resources Satellite-2 (CBERS-2) | Currently orbiting | Earth resources, environmental monitoring |
| China Brazil Earth Resources Satellite (CBERS-2B, 3,4) | Approved | Earth resources, environmental monitoring, land surface |
| Constellation of small Satellites for Mediterranean basin Observation (COSMO-SkyMed) | Planned | Environmental monitoring, surveillance and risk management applications, environmental resources management, maritime management, earth topographic mapping, law enforcement, informative/science application |
| CRYOSAT (Earth Explorer Opportunity Mission) | Approved | Primary objective is to test the prediction of thinning arctic ice due to global warming |
| Detection of Electro-Magnetic Emissions Transmitted from Earth-quake Regions (DEMETER) | Approved | Ionospheric disturbances related to seismic activity, ionospheric disturbances related to human activity, pre- and post-seismic effects in the ionosphere, global information on the earth's electromagnetic environment |

*(Table 4 Contd..)*

*(Table 4 Contd..)*

| Satellite mission | Status | Applications |
|---|---|---|
| Defense Meteorological Satellite Program (DMSP) F-13, 15, 16, 17, 18, 19, 20 | F-13, F-15: currently orbiting, F-16, F-17, F-18, F-19, F-20: approved | The long-term meteorological programme of the US Department of Defense (DoD)—with the objective to collect and disseminate worldwide cloud cover data on a daily basis |
| Elektro-L (Geostationary Operational Meteorological Satellite - 2) | Approved | Hydrometeorology, climatology, disaster management, space environment, ice and snow, land surface, space environment, data collection and communication |
| FY-2C (Geostationary Meteorological Satellite) | Currently orbiting | Meteorology and environmental monitoring, data collection, and redistribution |
| Geostationary Operational Environmental Satellite (GOES)-9, 10, 11, 12 | Currently orbiting | Meteorology (primary mission), search and rescue, space environment monitoring, data collection platform, data gathering |
| HJ-1A, 1B, 1C (Disaster and Environment Monitoring and Forecast Small Satellite) | Planned | Disaster and environment monitoring and forecast |
| National Oceanic and Atmospheric Administration (NOAA) | Currently orbiting | Meteorology, agriculture and forestry, environmental monitoring, climatology, physical oceanography, volcanic eruption monitoring, ice and snow cover, space environment, solar flux analysis, search and rescue |

**Source** Adapted from <http://www.docstoc.com/docs/2193069/List-of-satellite-missions>

## Geographic Information System

GIS is a computer-based database management system, which is designed to store and manipulate large volume of geo-referenced spatial data (locational as well as attribute data). Thus, GIS allows the mapping or display of the location of features, and recording and analysing descriptive characteristics thereof. It can integrate and relate any data with a spatial component, regardless of the source of the data (Lillesand, Kiefer, and Chipman 2007; Levin 1999). GIS works with static data, whereas remote sensing provides dynamic information, and thus, remote sensing and GIS should be integrated together to improve the ways of hazard modelling and analysis (Chen, Blong, and Jacobson 2001; Montoya 2003; Zerger and Smith 2003; Tralli, Blom, ZlotNicki *et al.* 2005). Apart from planning, preparedness, education, and so on, GIS is also used for preparing evacuation model, which is an important part of disaster preparedness and response (Zerger and Smith 2003). The services provided by space-based ICTs are given in Table 5.

With the advancement of geospatial techniques, various multi-criteria evaluation (MCE)–GIS programmes like MCE-RISK, Weighted Linear Combination (WLC), Technique for Order Preference by Similarity to Ideal

**Table 5** Services obtained from space technology in disaster reduction

| Space-based ICTs | Services provided |
|---|---|
| Early warning system | Early detection of events, and information dissemination |
| Risk information | Information on terrain, hydrologic, climatic, socio-economic and ecological aspects of community vulnerability |
| Impact assessment | Pre- and post-event change detection, and damage identification |
| Preparedness | Creation of vulnerability information for developmental planning and reconstruction |
| Emergency communication | Satellite radio, broadcasting, HAM radio, VSAT, WLLVSAT, Satphone |

ICT – information and communication technology

**Source** Adapted from UNESCAP (2005)

Solution (TOPSIS), Compromise Programming (CP), and so on are being developed to support DRR and CCA (Chen, Blong, and Jacobson 2001). Spatial Decision Support System (SDSS) has emerged as a Decision Support System (DSS), which takes into account the spatial and temporal data for decision-making to solve various problems with spatial components (Sugumaran and Sugumaran 2005). It helps in integrating analytical modelling potentials, spatial and non-spatial data management, domain knowledge, spatial display capabilities, and reporting capabilities under a single framework (Sugumaran and Sugumaran 2005).

Integration of GIS and GPS (to determine the ground location where the user is standing) with digital video (DV) has proved to be a valuable alternative for urban disaster management. DV uses charge couple devices (CCDs) for capturing moving images and stores them in high-quality, reproducible, editable, and digital format (Montoya 2003). It helps in generic data collection, which allows the users to extract the required attributes for a particular application. The technique is cheap and also reduces the data collection time (Montoya 2003). Mobile GIS helps in linking of various existing resources and data, rather than developing any new device, and thus, constitutes off-the-shelf technology; for example, the development of palm-top devices opens up the possibility of data collection process with higher accuracy and speed (Montoya 2003). With the development in ICTs, the effectiveness of the early warning systems has increased. In response to the recent disasters like 2004 Indian Ocean Tsunami, 2005 Kashmir Earthquake, and 2005 Katrina Hurricane, which highlighted the inadequacies in the technologies and stressed on the need for development, Grasso and Singh (2008) proposed Global Environmental Alert Service (GEAS), which will help in better coordination between scientific community and policy-makers (Figure 5).

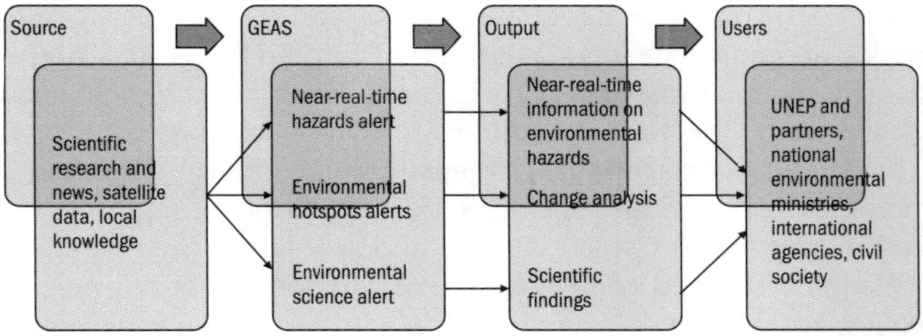


**Figure 5** Flow diagram showing the functioning of Global Environmental Alert Service
**Source** Grasso and Singh (2008)

GEAS aims at covering most of the natural hazards by streaming data from existing warning systems. GEAS will not only provide environmental information and help in risk mitigation but will also provide online access in user-friendly format, which will help in result analysis, decision-making, disaster mitigation, and raising public awareness (Grasso and Singh 2008). Careem, De Silva, De Silva, *et al.* (2006) suggested the use of Sahana, a free and open source software (FOSS) application, for effective information transfer and coordination during disaster relief, recovery, and rehabilitation. This framework is a Web-based flexible system that can be easily accessed, managed, and synchronized with the events and tasks by the users. The presence of GIS mapping capability makes this application even more attractive. With the advancement in space-based technology, combinations of different techniques are being used to develop risk management tools.

## Spatial Data Infrastructure

Spatial data and technologies have proved to be very important in effective DRR and effective decision-making. However, for emergency situations, availability, accessibility, and usability of data are very important. Collaboration between different organizations is thus required for data collection and sharing (Mansourian, Rajabifard, Zoej, *et al.* 2006). Spatial data infrastructure (SDI) is an information infrastructure that enables data exchange and sharing in a coordinated way among different communities (Rajabifard, Feeney, and Williamson 2002; Georgiadou, Puri, and Sahay 2005; Mansourian, Rajabifard, Zoej, *et al.* 2006). The framework was initialized for creating Web-based tools that will enhance the accessibility of useful spatial data in a cost-effective way (Rajabifard, Feeney, and Williamson 2002; Georgiadou, Puri, and Sahay 2005). Availability of this input data on Internet saves the time and effort of the users. An SDI framework includes policies, access networks, online data-handling facilities, standards, and human resources (required for data collection, management, delivery, and utilization for a community) (Mansourian, Rajabifard, Zoej, *et al.* 2006).

Many countries are developing SDIs for better management and utilization of spatial data sets (Georgiadou, Puri, and Sahay 2005; Mansourian, Rajabifard, Zoej, *et al.* 2006; Crompvoets, Rajabifard, Van Loenen, *et al.* 2009). However, the implementation of SDIs is complex due to security and financial concerns, and implementation depends on a number of other technologies like remote sensing, GIS, spatial modelling, and computer networking for which trained workers and spatial information are required (Georgiadou, Puri, and Sahay 2005). The components of an integrated SDI are policy, access network, standards, people, and data. People and data are generally considered under one category, while access network, policy, and standards, constituting the technical components, form another category. The technical category is highly dynamic due to changing technologies, and there is a need for the mediation of rights and restriction between the data and users. Thus, the whole concept of SDI is very dynamic (Rajabifard, Feeney, and Williamson 2002).

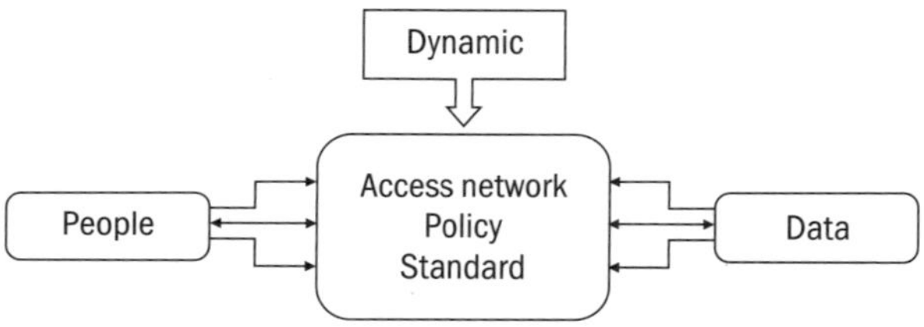


**Figure 6** Components of product-based spatial data infrastructure
**Source** Mansourian, Rajabifard, Zoej, *et al.* (2006)

On the basis of the aims, objectives, and political levels, SDI development has been grouped as process based and product based. Figure 6 represents a product-based SDI, wherein the SDI links the available database (Rajabifard, Feeney, and Williamson 2002). The process-based SDI model (Figure 7), on the other hand, helps in the management of information through communication channels for efficient data sharing and usage by community (Rajabifard, Feeney, and Williamson 2002).

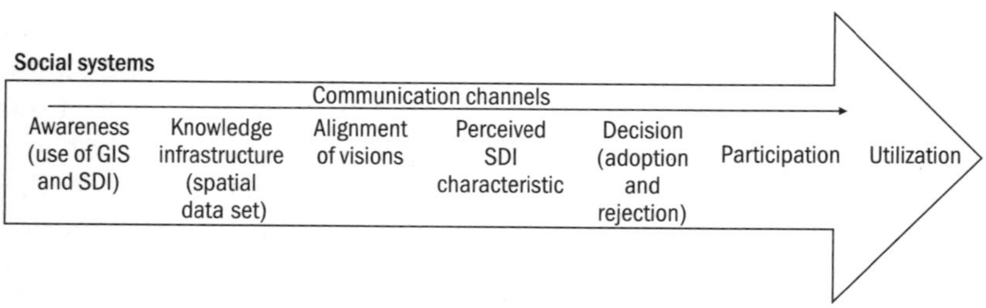


**Figure 7** Components of process-based spatial data infrastructure
**Source** Rajabifard, Feeney, and Williamson (2002)

Most of the SDI organizations/hierarchy (Figure 8) have similar organizational structures with three organizational tiers, which gives rise to different SDI development models. This structure is generally referred to as organizational pyramid (Rajabifard, Feeney, and Williamson 2002).

**Figure 8** Relationships between spatial data infrastructure hierarchy and different models of spatial data infrastructure development

**Source** Rajabifard, Feeney, and Williamson (2002)

The three organizational tiers are operational organizational tiers, management organizational tiers, and strategic organizational tiers. In the operational level (state and community SDI), production processes take place. The management organizational level (national SDI) consists of managers, researchers, and administrators. They carry out research, monitor activities, generate data, and prepare policies for the highest level, that is, strategic level (global and regional SDI), which is composed of decision-makers. The management and operational organizational tiers follow product-based approach as they play an important role in data production. Process-based model can be followed at strategic organizational tier and in nations with federated systems only, as SDI participation is voluntary in nature (Rajabifard, Feeney, and Williamson 2002).

While SDI is the framework that sets the rule of data sharing, Web-based GIS is the platform where these rules are implemented. Geoportals are Web gateways for collecting information and data in an organized manner. The contents are organized in the form of search tools, directories, support resources, community information, data, and applications. This Web environment allows online data sharing and queries, and data purchasing. It allows interactive mapping analysis and enhances the processing capabilities (ESPON 2006). SDI, along with Web-based GIS, can play an important role in disaster management, as it will bridge the gap between scientific communities, decision-makers, and disaster management communities (Mansourian, Rajabifard, Zoej, *et al.* 2006; Crompvoets, Rajabifard, Van Loenen, *et al.* 2009).

Thus, WebGIS as a tool and SDI framework will facilitate partnership between different organizations and ensure that all the required data is available, and thus, will facilitate better disaster management (Mansourian, Rajabifard, Zoej, *et al.* 2006). Some of the well-known geoportals are the British Geological Survey, which covers geoscience resources, US Geospatial One Stop and EU INSPIRE, which deal with national government data, and the Geography Network and Global Spatial Data Infrastructure (GSDI) portal that lists a wide variety of global geographic data (Maguire and Longley 2005). GSDI association includes many organizations throughout the world and supports SDI development at local, national, and international levels for the benefit of the society (<http://www.gsdi.org/>).

## Geoprocessing for Disaster Risk Reduction in India

Asia is highly vulnerable to disasters (Figure 9) and accounts for 83% of the population affected by disasters globally, between 1991 and 2000. Within Asia, in India, 24% of the deaths are caused by disasters and floods, and cyclones and storms account for 60% of all the disasters in India (NDMD 2004). Thus, India is one of the most disaster-prone countries, and effective steps should be taken to mitigate disasters. Due to the unique geo-climatic conditions in India, about 59% of the landmass is prone to earthquakes, 5% (40 million hectares) of the landmass is vulnerable to floods, 8.5% of the land (8000 km of coastline) is prone to cyclones and storms, and almost 68% of the total geographical area is vulnerable to droughts (BMTPC 1997; Mohanty, Karelia, and Issar Unpublished; Bhandari Unpublished; NDMD 2004).

Realizing the concern on disaster management, the Government of India, in 1999, set up a high-powered committee for disaster management in India. This committee further recommended the formation of the Cabinet Committee on Disaster Management, the National Institute of Disaster Management, and the National Centre for Calamity Management for these to assist in the operation of disaster management plans (Bhandari Unpublished). Disaster management was also included in the Tenth Five-year Plan and Twelfth Finance Commission.

After the Gujarat Earthquake of 2001, the disaster management responsibility was transferred from the Ministry of Agriculture to the Ministry of Home Affairs, which was an effort for mainstreaming disaster management, and shifting the emphasis from response to risk management and preparedness (World Bank Good Practice Notes 2008). The Ministry of Home Affairs drew up National Disaster Management Framework,

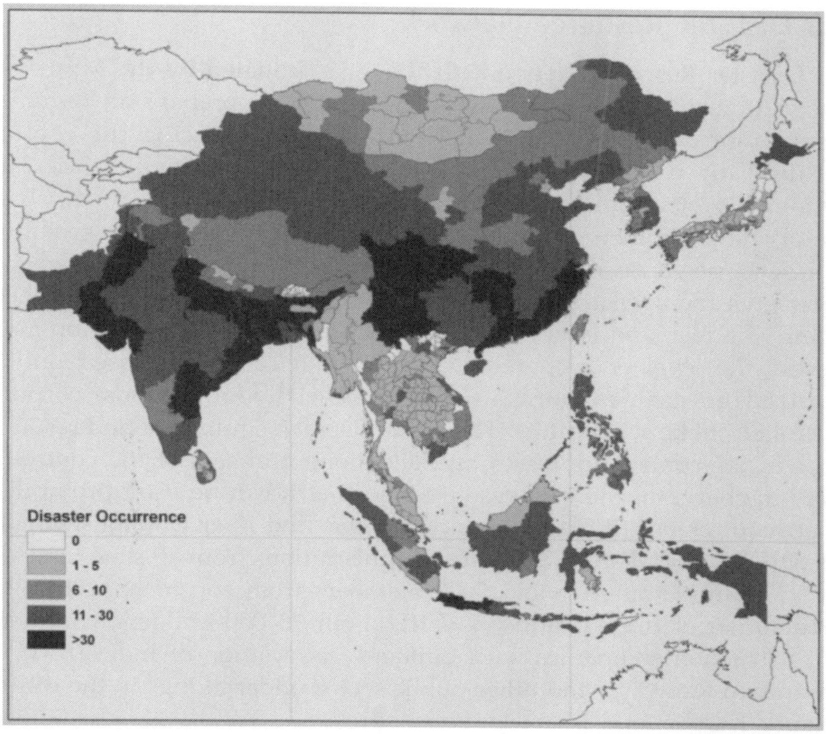


**Figure 9** Number of disaster occurrence in Asian countries, between 1975 and 2004
**Source** Adapted from <www.emdat.be>

which changed the approach of disaster management and shifted the focus on institutional mechanisms, preparedness, prevention, planning, and human resource development to minimize the impacts of natural disasters (Bhandari Unpublished; Mohanty, Karelia, and Issar Unpublished; NDMD 2004). For disaster mitigation and prevention, the Government of India has shifted the focus to disaster prevention, mitigation, and preparedness, and disaster management programmes are being integrated with development programmes. Mitigation programmes for different disasters have been developed, and building bye-laws have also been reviewed (NDMD 2004). The Government of India, along with the UNDP, United States Agency for International Development (USAID), and European Union, is implementing Disaster Risk Management (DRM) Program in the most vulnerable 169 districts of 17 multi-disaster-prone states in India with the help of ICT tools (Mohanty, Karelia, and Issar Unpublished; NDMD 2004). A number of ICT-based tools and initiatives (including the space-based technologies) have been developed for effective disaster management.

## India Disaster Resource Network

India Disaster Resource Network (IDRN) was initiated by the Ministry of Home Affairs in collaboration with the UNDP to prepare an organized disaster information system, which will help in reducing the response time after any disaster (Mohanty, Karelia, and Issar Unpublished). It is a nationwide electronic application with controlled access to database/ inventory of equipment, human resources, and critical supplies required for disaster response collated from state- and national-level government departments and agencies and entered into the portal at district level (IDRN; Mohanty, Karelia, and Issar Unpublished). This enables the decision-maker to assess the level of preparedness (IDRN). It can be accessed only by authorized government officials, district-level nodal persons, corporate bodies, and public sector units. The portal lists the equipment and resources and gives the contact addresses and telephone numbers of the controlling officers in-charge of the said resources so that it can be used promptly in case of requirement (IDRN; Mohanty, Karelia, and Issar Unpublished). The users of IDRN portal are 602 district administrations from all states and UTs, state/UT administration, about 5000 members from corporate bodies with Confederation of Indian Industry (CII), about 33 000 builders, contractors and construction companies with Builders' Association of India (BAI), the entire Indian Railways, and other public sector undertakings in the country (Mohanty, Karelia, and Issar Unpublished).

## Space-based Technology for Disaster Management

EO plays an important role in observing and assessing the phenomena and vulnerability, which leads to the development of disaster management framework (Hegde, Jayaraman, and Srivastava 2009). The Indian EO series satellites are moving towards thematic constellation (characterized with multi-platform, multi-resolution, and multi-parameter), and along with the contemporary satellites, these are helping in reducing the gaps in terms of capturing catastrophic events and their attributes. The IRS – P6, IRS – P5, OCEANSAT, and the forthcoming radar imaging satellite (RISAT) represent the thematic characteristics of India's EO mission (Hegde, Jayaraman, and Srivastava 2009). The Department of Space (DOS) plans to launch Disaster Management Support (DMS) programme, which will provide services that will use satellites and air-borne systems for disaster management (Hegde, Jayaraman, and Srivastava 2009).

The Bureau of Indian Standards Technical Committees on Earthquake Engineering and Wind Engineering has a Seismic Zoning Map and a Wind Velocity Map, which takes into account the cyclonic winds for the country. The Seismic Zoning Map was revised in 2002. The Central Water Commission has a Flood Atlas of India. The Group has used these hazard maps to prepare 1:2 million scale maps by superposing the above available data on the digitized Survey of India map as the base map. The earthquake, wind storm, and flood hazard maps are drawn for each state and UT separately. Seismic occurrence and cyclones are monitored by the India Meteorological Department (IMD), whereas the flood monitoring is done by the Central Water Commission. The Geological Survey of India and the Department of Earthquake Engineering are the other important organizations. The Bureau of Indian Standards Technical Committees on Earthquake Engineering and Wind Engineering has prepared hazard maps for the country, which have been used by these organizations for preparing earthquake, wind storm, and flood hazard maps for each state and UT. Building Materials Technology Promotion Council (BMTPC) of the Government of India, on the basis of information and data present on Indian landslide and thematic factors map, prepared the first small-scale landscape hazard map for India. For cyclone detection and tracking, INSAT satellite data and RADAR have been installed at Kolkata, Paradeep, Vishakhapatnam, Machhilipatnam, Chennai, Karaikal, Mumbai, Goa, Cochin, and Bhuj. The cyclone detection mechanism is very successful and has been used for issuing timely cyclone warning (Bhandari Unpublished). For monitoring drought, various weather parameters are monitored and are integrated with GIS. The International Water Management Institute (IWMI) has developed South East Asia Drought Monitor (based on remote sensing data and related ancillary data), which covers only the western India. This provides drought data at regional, district, and pixel levels and thus helps in drought monitoring and mitigation at local level. This system is still under development and can be a good beginning (Prabhakar and Shaw 2008). Table 6 lists the EO satellites of India that are used in disaster management.

GIS, in conjunction with satellite data, provides a versatile platform for decision support by providing crucial information regarding the location of the public facilities, communication links, and transportation network at national, state, and district levels (Mohanty, Karelia, and Issar Unpublished; NDMD 2004). Remote sensing and GIS have not only made it possible to extract more information by editing and updating the data, but also helped in efficient and faster data collection, handling, data management, and storage (Rao, Pandey, Ahuja, *et al.* 2002). Thus, GIS-based database ensures better planning and implementation of preparedness and mitigation activities (Mohanty, Karelia, and Issar Unpublished; NDMD 2004). Recognizing the importance of GIS, the Ministry of Home Affairs, in collaboration with other agencies like the Department of Space, Department of Science and Technology, and Ministry of Communications and IT, has initiated a GIS-based National Database for Emergency Management (NDEM). The ministry has also collaborated with the UNDP for the development of GIS-based tools (Mohanty, Karelia, and Issar Unpublished). The ministry

**Table 6** India's earth observation satellites for disaster management

| Satellite | Spatial resolution (m) | Swath (km) | Temporal resolution (days) | Applications in disaster management |
|---|---|---|---|---|
| IRS - 1C | PAN: 5.8<br>LISS III: 23.5<br>70.8<br>WiFS: 188 | 70<br>141<br>148<br>810 | 24 | Agriculture, forestry, urban, land use, soil, geology, terrain, water resources, digital elevation models (DEMs), environment disasters (damage assessment/relief) |
| IRS - P3 | WiFS: 188<br>MOS A: 1569×1395<br>MOS B: 523×52<br>MOS C: 523×644 | 770<br>195<br>200<br>192 | 24 | Remote sensing of earth resources, study of X-ray astronomy, periodic calibration of PSLV radar located at tracking stations |
| IRS - 1D | PAN: 5.2 - 5.8<br>LISS III: 21.2 - 23.5<br>63.6 - 70.5<br><br><br>WiFS: 169 - 188 | 63 - 70<br>127<br>141<br>133<br>148<br>728 - 812 | 25<br>—<br><br>— | Agriculture, forestry, urban, land use, soil, geology, terrain, water resources, DEMs, environment disasters (damage assessment/relief) |
| Resourcesat (IRS - P6) | LISS IV: 5.8<br>LISS III: 23.5<br>AWiFS: 56 | 23.9<br>141<br>740 | 24 | Agriculture, forestry, urban, land use, soil, geology, terrain, water resources, DEMs, environment disasters (damage assessment/relief) |
| IRS - P4 | OCM: 360<br>MSMR: 120<br>80<br>40<br>40 | 1420<br>1360 | 2 | Systematic data for oceanography, coastal, atmospheric applications, disasters (weather system assessment) |
| Cartosat 1 (IRS - P5) | Two panchromatic camera: 2.5 | 30 | 5 | Cartographic applications, DEMs useful for large-scale hazard zonation (disaster risk assessment) |
| Cartosat 2 | Panchromatic camera: < 1 | 10 | 3–4 | Large-scale cartographic applications: DEMs useful for hazard micro-zonation (disaster risk assessment) |
| INSAT - 1 | Visible and TIR | | Continuous | Clouds, sea surface temperature, precipitation, disasters (weather system assessment) |
| INSAT - 2 | Visible, TIR, WV, and CCD | | Continuous | Telecommunications, weather forecasting, disaster warning |
| INSAT - 3 | Visible, TIR, WV, sounder, and CCD | | Continuous | Clouds, disasters (weather system assessment) |
| KALPANA | Visible/thermal, water vapour | 2 ×2 | | Sea surface temperature, cloud, CMVs, WV, mesoscale, temperature/humidity profiles disasters (weather system assessment) |

**Source** Modified from <www.nrsa.gov.in>; Hegde, Jayaraman, and Srivastava (2009)

is also planning to establish National Database for Disaster Management (NDDM), which, in conjunction with satellite images, will help in hazard zonation, risk assessment, preparedness, and emergency response management (NDMD 2004).

## National Emergency Communication Plan

For effective disaster management and response, coordination between decision-makers and disaster response teams is very important. However, at the time of the occurrence of disasters, telecommunication infrastructure gets damaged. So, the Ministry of Home Affairs has planned to set up a reliable information and communication network using both terrestrial and space-based technologies for emergency communication during disasters (Mohanty, Karelia, and Issar Unpublished). In the first phase of the plan, the VSAT network will use resources from POLNET and will provide communication between National Emergency Operation Centre (NEOC), remote disaster sites, and state emergency operation centres (SEOC). In the second phase, the government plans to integrate all EOCs of national, state, and district levels, and the transportable terminals at disaster sites, using satellite and terrestrial communication networks (Mohanty, Karelia, and Issar Unpublished).

### *Spatial Data Infrastructure in India*

In recent years, India has produced spatial data sets with high accuracy, using remotely sensed images, GPS, GIS, topographic surveys, soil surveys, geological surveys, and various other natural resources inventory programmes. This spatial data is vital for planning, infrastructure development, disaster management, and so on at local, regional, and national scales (Rao, Pandey, Ahuja, *et al.* 2002). Organizations like Geological Survey of India, Survey of India, Forest Survey of India, Department of Space, National Informatics Centre, Central Ground Water Board, Census Department, National Atlas and Thematic Mapping Organization, National Bureau of Soil Survey and Land Use Planning, National Remote Sensing Agency, Central Water Commission, Department of Ocean Developments, National Council for Applied Economics Research, Ministry of Environment and Forests, Central Pollution Control Board, and Indian Meteorological Department are important repository of spatial data; however, sharing of data by these institutes is done on a very limited scale (Georgiadou, Puri, and Sahay 2005), which often leads to the duplication of efforts. Thus, there is a need for the establishment of National Spatial Data Infrastructure (NSDI), which can serve as one-step source of spatial and non-spatial data and will ensure proper data collection, efficient data analysis, easy accessibility of spatial and non-spatial data, and increased awareness (Rao, Pandey, Ahuja, *et al.* 2002; Georgiadou, Puri, and Sahay 2005). Table 7 lists the characteristics of Indian NSDI.

**Table 7** Characteristics of Indian National Spatial Data Infrastructure (NSDI)

| Learning from the domain of information infrastructure | As applied to Indian NSDI |
|---|---|
| Installed base and lock-in effect | Installed base recognized, but no clear strategy exists of how it would form the basis of the NSDI. Lock-in created by diversity of maps, other spatial sources, and institutional issues not addressed |
| Reflexive standardization | Standardization process mired in scientific thinking. Developing a "hierarchy" of standards reflexively and as a negotiated process not in evidence. |
| Cultivation approach to design | Top-down, "constructionist" approach evident at present, data-centric focus, end users not involved in determining their perceived needs, inculcation of bottom-up approaches also not considered; visions of "grand" design. |

**Source** Adapted from Georgiadou, Puri, and Sahay (2005)

## LOOKING FORWARD

The relationship between disaster risk occurrence and climate change is becoming clearer. Thus, CCA policies and negotiations should be included in DRR policies and frameworks. The most important step towards CCA and DRR is to integrate these two concepts. CCA and DRR should be of national priority with strong institutional framework for funding and better implementation and should be included in various development activities to prepare disaster-resilient communities (Hyogo Framework 2005; Mitchell and van Aalst 2008; APEC 2008; ISDR 2008). Adaptation and mitigation activities need to be enhanced. Geospatial tools with better spatial and temporal coverage of climate data should be developed for risk and vulnerability assessment studies. Continuous monitoring and climate change modelling techniques can be used for preparing disaster inventories and enhancing early warning systems (Hyogo Framework 2005; BAP 2007; APEC 2008). Risk reduction can be resolved completely only through improving local resilience and developing the capacity to prepare for and respond to emergency events at local level and thus requires humanitarian assistance. Effective response and recovery mechanisms should be devised and implemented (ISDR 2008). Spatial data should be made available to users and communities for knowledge dissemination, data accessibility, and better management decisions. This will also support capacity building and technology advancement for disaster reduction (ISDR 2008).

The Bali Conference of 2007 has played an important role in creating awareness to make CCA and DRR a local, national, and regional priority. Now, the association between CCA and DRR is well understood. Though both the concepts have developed from different frameworks, initiatives are being taken to integrate these two concepts into developmental plans for better management. One of the important aspects for CCA and DRR is continuous monitoring of climatic factors. The development of ICTs, especially the EO satellites, has helped in timely monitoring and assessment, and thus, enhancing early warning systems. Timely assessment and monitoring will help in prediction modelling, damage estimation, and thus, enhanced early warning system, and better preparedness for response and recovery. The integration of these geospatial tools with Internet has helped in creating awareness, strengthening the decision support mechanism.

## References


APEC (Asia-Pacific Economic Cooperation). 2008. ***Strategy for Disaster Risk Reduction and Emergency Preparedness and Response in the Asia Pacific Region: 2009 to 2015***. Report prepared for Task Force on Emergency Preparedness Meeting. Lima, Peru. Details available at <http://aimp.apec.org/Documents/2008/TFEP/TFEP1/08_tfep1_012.pdf>, last accessed on 1 July 2009

Arigony-Neto J, Saurer H, Simoes J C, Rau F, Jana R, Vogt S, Gossmann H. 2009. **Spatial and temporal changes in dry-snow line altitude on the Antarctic Peninsula**. *Climate Change* **94**: 19–33

Asrar G and Dozier J. 1994. ***EOS Science Strategy for the Earth Observing System***, pp. 1–119. AIP Press

BAP (Bali Action Plan). 2007. **Bali Action Plan**. Details available at <http://unfccc.int/files/meetings/cop_13/application/pdf/cp_bali_action.pdf>, last accessed on 20 June 2009

Bhandari R K. Unpublished. **Disaster management in India: a new awakening**. Details available at <http://www.vitcdmm.org/journal_on_disasters.pdf>, last accessed on 10 June 2009

BMTPC (Building Materials Technology Promotion Council). 1997. **An introduction to the vulnerability atlas of India.** Details available at <www.bmtpc.org>, last accessed on 23 June 2009

Careem M, De Silva C, De Silva R, Raschid L, Weerawarana S. 2006. **Sahana: overview of a disaster management system**. In *Proceedings of the International Conference on Information and Automation*, 15–17 December 2006, Colombo, Sri Lanka

Chen K, Blong R, and Jacobson C. 2001. **MCE-RISK: integrating multicriteria evaluation and GIS for risk decision-making in natural hazards**. *Environmental Modeling and Software* **16**: 387–397

Crompvoets J, Rajabifard A, Van Loenen B, Fernandez T D. 2009. **Future directions for SDI assessment**. GSDI 11 World Conference Spatial Data Infrastructure Convergence: Building SDI Bridges to address Global Challenges, 15–19 June 2009, Rotterdam, The Netherlands. Details available at <http://www.gsdi.org/gsdi11/papers/pdf/93.pdf>, last accessed on 19 June 2009

Dilley M, Chen R S, and Deichmann U. 2005. ***Natural Disaster Hotspots: a global risk analysis***. Washington, DC: World Bank

ESPON (European Spatial Planning Observation Network). 2006. **ESPON Web-GIS**. Details available at <http://www.espon.eu/mmp/online/website/content/tools/913/index_EN.html>, last accessed on 11 July 2009

Georgiadou Y, Puri S K, and Sahay S. 2005. **Towards a potential research agenda to guide the implementation of spatial data infrastructures – a case study from India**. *International Journal of Geographical Information Science* **19**(10): 1113–1130

Grasso V F and Singh A. 2008. **Global Environmental Alert Service (GEAS)**. *Advances in Space Research* **41**: 1836–1852

Hegde V S, Jayaraman V, and Srivastava S K. 2009. **India's EO infrastructure for disaster reduction: lessons and perspectives**. *Acta Astronautica* **65**(9–10): 1471–1478

Helmer M and Hilhorst D. 2006. **Natural disasters and climate change**. *Disasters* **30**(1): 1–4

Holdaway R. 2001. **Is space global disaster warning and monitoring now nearing reality?** *Space Policy* **17**: 127–132

Hunt J C R. 2002. **Floods in changing climate: a review**. *Philosophical Transactions of the Royal Society* **360**: 1531–1543

Hyogo Framework. 2005. **Hyogo framework for action 2005–2015: building the resilience of nations and communities to disasters**. *World Conference on Disaster Reduction*, 18–22 January 2005, Kobe, Hyogo, Japan

IPCC (Intergovernmental Panel on Climate change). 2001. ***Climate Change 2001: the scientific basis***. Contribution of Working Group I to the Third Assessment Report of the IPCC, edited by J T Houghton, Y Ding, D J Griggs, M Noguer, P J van der Linden, X Dai, K Maskell, C A Johnson. Cambridge, UK, and New York, NY: Cambridge University Press. 881 pp.

IPCC (Intergovernmental Panel on Climate change). 2007. **Climate change 2007: synthesis report** [online]. Details available at <http://www.ipcc.ch/pdf/assessment-report/ar4/syr/ar4_syr.pdf>, last accessed on 12 June 2009

ISDR (International Strategy for Disaster Reduction). 2008. **Climate change and disaster risk reduction** [online]. Details available at <http://www.unisdr.org/eng/risk-reduction/climate-change/docs/Climate-Change-DRR.pdf>, last accessed on 10 June 2009

ISDR (International Strategy for Disaster Reduction). 2009. **UNISDR terminology on disaster risk reduction** [online]. Details available at <http://www.unisdr.org/eng/terminology/terminology-2009-eng.html>, last accessed on 29 July 2009

Islam M M and Sado K. 2000. **Development of flood hazard maps of Bangladesh using NOAA–AVHRR images with GIS.** *Hydrological Sciences—Journal-des Sciences Hydrologiques* **45**(3): 337–355

Jayaraman V, Chandrasekhar M G, and Rao U R. 1997. **Managing the natural disasters from space technology inputs**. *Acta Astronautica* **40**(2–8): 291–325

Lee S and Choi U. 2003. **Development of GIS-based geological hazard information system and its application for landslide analysis in Korea**. *Geosciences Journal* **7**(3): 243–252

Levin N. 1999. **Fundamentals of remote sensing** [e-book]. Details available at <http://atar.mscc.huji.ac.il/~geo/personal/Noam%20 Levin/1999-fundamentals-of-remote-sensing.pdf>, last accessed on 7 June 2009

Lillesand T M, Kiefer R W, and Chipman J W. 2007. ***Remote Sensing and Image Interpretation***, 5th edn. New Delhi: Wiley India Pvt. Ltd

Maguire D J and Longley P A. 2005. **The emergence of geoportals and their role in spatial data infrastructures**. *Computers, Environment and Urban Systems* **29**: 3–14

Mansourian A, Rajabifard A, Zoej M J V, Williamson I. 2006. **Using SDI and web-based system to facilitate disaster management**. *Computers and Geosciences* **32**(3): 303–315

McGranahan G, Balk D, and Anderson B. 2007. **The rising tide: assessing the risks of climate change and human settlements in low elevation coastal zones**. *Environment and Urbanisation* **19**(1): 17–37

Mitchell T and van Aalst M. 2008. **Disaster risk reduction and climate change adaptation – closing the gap** [online]. Details available at <www.id21.org>, last accessed on 3 June 2009

Mohanty S, Karelia H, and Issar R. Unpublished. ***ICT for Disaster Risk Reduction: the Indian express***. Prepared for GoI–UNDP (Government of India–United Nations Development Programme) Disaster Risk Management Programme [online]. Details available at <http://www.ndmindia.nic.in/WCDRDOCS/ICT%20for%20Disaster%20Risk%20Reduction.pdf>, last accessed on 7 June 2009

Montoya L. 2003. **Geo-data acquisition through mobile GIS and digital video: an urban disaster management perspective**. *Environmental Modeling and Software* **18**: 869–876

Navalgund R R, Jayaraman V, and Roy P S. 2007. **Remote sensing applications: an overview**. *Current Science* **93**(12): 1747–1766

NDMD (National Disaster Management Division). 2004. ***Disaster Management in India: a status report***. Government of India, Ministry of Home Affairs, National Disaster Management Division [online]. Details available at <http://www.ndmindia.nic.in/GoIUNDP/ReportPub/DM-Statu-%20Report.pdf>, last accessed on 9 June 2009

O'Brien G, O'Keefe P, Rose J, Wisner B. 2006. **Climate change and disaster management**. *Disasters* **30**(1): 64–80

Prabhakar S V R K and Shaw R. 2008. **Climate change adaptation implications for drought risk mitigation: a perspective for India**. *Climate Change* **88**: 113–130

Rajabifard A, Feeney M F, and Williamson I P. 2002. **Future directions for SDI development**. *International Journal of Applied Earth Observation and Geoinformation* **4**: 11–22

Rao M, Pandey A, Ahuja A K, Ramamurthy V S, and Kasturirangan K. 2002. **National Spatial Data Infrastructure – coming together of GIS and EO in India**. *Acta Astronautica* **51**(1–9): 527–535

Schipper L and Pelling M. 2006. **Disaster risk, climate change and international development: scope for, and challenges to, integration.** *Disasters* **30**(1): 19–38

Sperling F and Szekely F. 2005. **Disaster risk management in a changing climate.** In *The World Conference on Disaster Reduction*, on behalf of the Vulnerability and Adaptation Resource Group (VARG). Reprint with Addendum on Conference outcomes. Washington, DC

Sugumaran V and Sugumaran R. 2005. **Web-based spatial decision support systems (WebSDSS): evolution, architecture, and challenges**. In *Third Annual SIGDSS pre-ICIS Workshop, Designing Complex Decision Support: discovery and presentation of information and knowledge*, 11 December 2005, Las Vegas, Nevada

Thomalla F, Downing T, Spanger-Siegfried E, Han G, Rockstrom J. 2006. **Reducing hazard vulnerability: towards a common approach between disaster risk reduction and climate adaptation**. *Disasters* **30**(1): 39–48

Tralli D M, Blom R G, ZlotNicki V, Donnellan A, Evans D L. 2005. **Satellite remote sensing of earthquake, volcano, flood, landslide, and coastal inundation hazards**. *Photogrammetry and Remote Sensing* **59**: 185–198

Tran P, Shaw R, Chantry G, Norton J. 2009. **GIS and local knowledge in disaster management: a case study of flood risk mapping in Vietnam**. *Disasters* **33**(1): 152–169

UNESCAP (United Nations Economic and Social Commission for Asia and the Pacific). 2005. **Framework for regional cooperation on space technology supported disaster reduction strategies in Asia and the Pacific**. Study report prepared for the Meeting of Experts on Space Applications for Disaster Management, Chiang Mai, 25–28 July 2005

UNESCAP (United Nations Economic and Social Commission for Asia and the Pacific). 2008. **Enhancing regional cooperation on disaster risk reduction in Asia and the Pacific: information, communications, and space technologies for disaster risk reduction** [online]. Details available at <http://www.unescap.org/idd/events/cdrr-2009/CDR_5E.pdf>, last accessed on 16 June 2009

Wattegama C. 2007. **ICT for disaster management** [online]. Details available at <http://www.apdip.net/publications/iespprimers/eprimer-dm.pdf>, last accessed on 9 June 2009

World Bank Good Practice Notes 2008. **Disaster risk reduction** [online]. Details available at <http://siteresources.worldbank.org/CHINAEXTN/Resources/318949-1217387111415/Disaster_Risk_en.pdf>, last accessed on 28 June 2009

Zerger A and Smith D I. 2003. **Impediments to using GIS for real-time disaster decision support**. *Computers, Environment and Urban Systems* **27**: 123–141

## Websites

<www.docstoc.com/docs/2193069/List-of-satellite-missions>
<www.emdat.be>

<www.gsdi.org>

<www.nrsa.gov.in>

<www.unisdr.org>

GSDI. Global spatial data infrastructure [online]. Details available at <http://www.gsdi.org/>, last accessed on 22 July 2009

# Geoinformatics and Communication Technologies for Climate Change Hazards: Musing beyond the Technical Issues

**Fausto Marincioni*** **and Luca Abeti**†, #

**Department of Marine Sciences, Marche Polytechnic University, Via Brecce Bianche, Ancona, Italy*

*†IMT Institute for Advanced Studies of Lucca, Piazza S. Ponziano, Lucca, Italy*

*#Marche Region, Civil Protection and integrated Security Policies Department, Stada Cameranense, Varano (AN), Italy*

*E-mail: f.marincioni@univpm.it; luca.abeti@regione.marche.it*

## INTRODUCTION

Information and communication technologies (ICTs), including geospatial analysis tools, are part of both the triggering causes of and the possible solutions for the climate change issue. In 2007, it was estimated that the operation of ICT equipment accounts for about 2% to 2.5% of the global greenhouse gas emissions annually (Kumar and Mieritz 2007). Yet, at the same time, ICTs are substantially contributing to reduction in $CO_2$ and other greenhouse gas emissions, for example, by replacing paper-based technology with online publishing. In addition, the use of these technologies permits better understanding of climate change and facilitates the transfer of the acquired knowledge. ICTs hold immense potential in the area of natural hazard prediction and detection. The ability to access any information on climate change hazards from practically anywhere in the world is fostering more effective links among the various disaster management communities, as well as among the affected populations and rescuers. Overall, geoinformatics and communication technologies are impacting all types of disaster communications, and great advancements are expected in disaster education and knowledge transfers (van Oosterom, Zlatanova, and Fendel 2005). For better or worse, these technologies have changed the hierarchical structure of information flow. By lowering the barriers of ethnicity, gender, geographic location or time zone, ICTs are providing new means and a new location, the cyberspace, to access, exchange, and translate information, which only a few years ago were confined to conferences, professional meetings, workshops or university classrooms. Nonetheless, the great potential of these communication tools is not yet supplemented with a clear understanding of how and to what degree these technologies are going to affect the decision-making and knowledge transfer processes, especially within the realm of climate change hazards. It appears, in fact, that effective applications and benefits of ICTs not only depend on their technical features, but also, and perhaps more importantly, on the users' perception of their applicability. The available literature (see, for example, Quarantelli 1997; Marx 2003; Marincioni 2007) links the effectiveness of ICTs in transferring disaster knowledge with the users' professional culture and modus operandi. At the same time, exceedingly high expectations on the availability of connections and communication infrastructure negatively influence the perceptions of emergency operators and victims (Sheetz, Quek, Kavanaugh, *et al.* 2009). Indeed, emerging spatial analysis systems and network communication architecture are improving the reliability of the infrastructure and services needed to face the new emergencies due to climate alterations. In the following sections, three major segments are reviewed, wherein ICTs can contribute to the management of climate change hazards. These three segments are (1) decision support systems, (2) disaster communications, and (3) education and training. The non-technical issues connected with the application of these technologies with climate change hazards and disaster mitigation are also discussed.

## APPLYING ICTS TO CLIMATE CHANGE HAZARDS

Emergency management and climate change risk assessment and mitigation share a common technological framework based on geoinformatics. Since more and more emergencies originate from or depend on climate change, these two fields of study are progressively becoming blurred disciplines (Lemmen, Warren, Lacroix, *et al.* 2008). The ICTs employed for emergency management are usually similar to the ones used for assessing climate change. There are many activities that are both adaptive to climate change and mitigative to potential disasters. Examples of such activities include hazard mapping and risk assessment, identification of vulnerable communities or developing database of critical infrastructure to enhance the decision-making process, and increasing public understanding of risk through education.

### Decision Support Systems

Decision support systems are defined in this chapter as all those information tools that enable the decision-makers to develop effective disaster mitigation strategies and improve rescue and recovery operations. The outputs of these systems are traditionally in the form of reports, maps, raw data or statistics. Modern ICTs enable the integration of various data sources, thus allowing instantaneous links to be created among databases, spatial simulation models, and management tools. Integrated decision support systems have great potential in the mitigation of climate change hazards as they can project scenarios of impact on the pre-existing economic and social structures. Hence, the decision-makers can see the effects of today's decisions in terms of future impacts and losses, and estimate possible alternatives. Four major components are necessary in a decision support system: control systems, database systems, models, and output systems. These components can be actualized by a variety of software. Let us focus on the geoinformatics core of such applications: geographic information systems (GIS), remote sensing, and global positioning systems (GPS).

### *Geographic Information System*

GIS provides an extremely important human–computer interaction paradigm in the management of emergency and climate change hazards. It allows a rapid interaction with computer-based services in a decision support system and provides the decision-makers with a more holistic view of the territory and its physical processes. Two important developments significantly enhanced the effectiveness of GIS in the field of emergency management, namely, (1) the online availability of basic geographical mapping functionalities, such as Web GIS applications, Google Map, Google Earth, Microsoft Virtual Earth, and so on (Boulos 2005) and (2) the easy and free-of-charge access to high definition satellite images and three-dimensional images. The pervasiveness of communication and computing capabilities makes GIS work in emergency situations also, for instance, by means of palmtops and cellular phones. These applications further improve the rapid exchange of data between the emergency operators in the field and the coordination centres, thus enabling informed decisions with a better view of the evolving emergency. Indeed, spatial analysis is pivotal to almost every aspect of climate change hazard and risk management (Thomas and Humenik-Sappington 2009). For example, in mitigation and preparation, it can be used to define escape routes, identify shelter locations, and calculate their capacity. During rescue and recovery, GIS can quickly pinpoint critically damaged areas, the number of affected people, the disruption of transportation networks or public health emergencies, and so forth. Likewise, GIS can be used proficiently during reconstruction to evaluate hazards and estimate the vulnerability of the newly developed areas. Linking the physical attributes of risk with the social, economic, and political aspects, GIS can help improve community's resilience and adaptation to climate change. The principal problem hindering such application is the difficulty of accessing detailed local-scale data. While there is a relatively well-standardized variety of national-scale data sets, availability and access to local-scale data sets can be problematic. Furthermore, the increasing involvement of the private sector in the development of local and regional coverage has resulted in non-standard data sets. This problem is not technical. It rather results from the lack of a common understanding of the various organizational tasks involved in disaster management, and of the possible interactions between multi-layered government activities, the public, non-governmental organizations, and the private sector. Various private national and international organizations are working on this very problem by defining sets of standards and regulations for geospatial data processing and exchange (Matthews 2005). A good example is the Open Geospatial Consortium, which includes participation from the government and industry and is working to develop global standards to enable sharing of geographical data. The goal is to create standardized cross-platform information banks with sufficient regional and local detail.

### *Remote Sensing*

Acquiring images of the earth from space has added a global dimension to the study of the planet's environment and its processes. With the help of sensors placed either on aircraft or satellites, remote sensing provides broad views of the earth's changes that may be missed by local-scale monitoring on the ground. As mentioned before, remote sensing is a very important data source for GIS, digital elevation model (DEM), and decision support systems in general (Bhatta 2008; Thomas and Humenik-Sappington 2009), and is widely used in emergency management and climate change monitoring (Billa, Mansor, and Mahmud 2004; Chen, Blong, and Jacobson 2003). Two remote-sensing technologies—LIght Detection and Ranging (LIDAR) and InterFerometric Synthetic Aperture Radar (IFSAR)—are methodically used by emergency management agencies across the world, and are considered all-time detection tools. LIDAR is an airborne laser system that can be combined with a GPS receiver to enable precise measurement of the elevation of ground points at the earth's surface. The resulting DCMs can reach very high accuracy levels (ranging between centimetres) along narrow corridors such as shorelines, waterways, and levees. IFSAR is a technology that uses airborne or space-borne radar antennae to obtain terrain data over larger geographic areas. It is an all-weather system for generating elevation data to an accuracy level of 1.5–3 m, depending on the size of the area surveyed. Remote sensing has been applied to disaster management in each of the different phases of the disaster cycle. For example, during the pre-impact and mitigation phases, it has been used to monitor and forecast hurricane tracks; during the post-impact assessment, to define the change in the coastline; and in long-term reconstruction, remote sensing has been used to monitor urban growth and land use changes. Indeed, remote sensing is particularly effective for environmental monitoring activities, such as, for example, continuous real time acquiring of data on rainfall, hydrological levels, temperature, snow levels, and so on. Beside real-time data acquisition, remote sensing networks are also employed to create historical databases (very important to understand the long-term variation and effects of climate change hazards). Recent advancements in sensing hardware are improving the infrastructure of remote sensing networks. A promising trend in environmental remote sensing is the use of small sensor nodes in wireless sensor networks (WSN) (Mainwaring, Culler, Polastre, *et al.* 2002). They are low cost and tiny sensor devices that are able to communicate data to Wi-Fi access points. Small sensor nodes can thus be spread in critical areas in order to duly monitor a particular phenomenon. For instance, such sensor networks can be employed in monitoring forest fires in order to alert operative centres in the event a forest fire takes place. New algorithms are also enhancing the efficiency of such WSN, extending battery endurance and multi-hop routing of data.

### Global Positioning System and Global Navigation Satellite System

These satellite-based radio navigation systems originally developed by the US Department of Defense are able to determine the three-dimensional location (latitude, longitude, and elevation) and velocity of a receiver on, or in proximity to, the earth's surface. GPS was primarily developed to improve land, sea, and air navigation, but since the first satellite launch in 1978, it has found wide applications, especially in environmental monitoring and disaster management. GPS and global navigation satellite system (GNSS), with GIS and remote sensing systems, provide decision support system with real-time positioning (georeferencing) of the acquired data/knowledge. A significant use in emergency management is in tracing system. The European Union Galileo GNSS project includes dedicated services for safety and public management of emergencies (Issler and Hein 2003). GPS can also be used to detect global and regional climate changes by means of the radio occultation techniques (Yuan, Anthes, Ware, *et al.* 1993). By measuring differences of wet and total delays and vertical profiles of refractivity, it is possible to model temperature and water vapour in the lower troposphere (increased temperature tends to decrease the delay, while increased water vapour increases the delay). More recently, a project called Constellation Observing System for Meteorology, Ionosphere and Climate (COSMIC) exploits the radio occultation techniques by means of a dedicated satellite network of six satellites (Anthes, Bernhardt, Chen, *et al.* 2008). The data collected by COSMIC, in particular temperature and water vapour profiles, helps observing and forecasting hurricanes, typhoons, and other storm patterns over the oceans and facilitates weather prediction. Furthermore, the measurement of electron density in the ionosphere could improve forecasting of geomagnetic storms interrupting satellite and communications systems and affecting power grids. The European Organisation for the Exploitation of Meteorological Satellites launched a radio occultation satellite in October

2006 aboard the Metop-A environmental satellite. It is operationally similar to COSMIC satellites, but provides fewer atmospheric profiles per day as only one satellite is in the orbit. In addition, various commercial projects based on the radio occultation technique are in the pipeline. For instance, the GeoOptics Company is planning to launch a cluster of 24 tiny 30-kg satellites by 2011 (hoping to increase the number to 100 satellites by 2016). GeoOptics plans to invest about $100 million and hopes returns by selling meteorological and climate change data. More fine-tuned usage of GPS is being envisaged from the integration of GIS, GPS, and mobile phones, increasing the potential application to emergency management. Because modern mobile phones are usually provided with a GPS receiver, and the capacity to transmit information over the Internet (by means of GPRS, EDGE or HSDPA technologies) has increased, it has become possible to track and find people in a disaster area. This will be very useful for search and rescue operations. A similar, and already existing, service is the Google Latitude (see <http://www.google.com/latitude>), which is a web-based application that integrates both GIS and GPS/GPRS functionalities to share the location of people in a social network. In long-term studies, it can help in monitoring patterns of movements of people, or even relocation, after a disaster.

## Disaster Communications

As with almost all extreme events, the management of climate change hazards and disasters requires a broad variety of communications, each with specific functions and purposes. They are generically called disaster communications (Table 1).

**Table 1** Communications in disasters

| | |
|---|---|
| Technical communications | Through satellites and computer networks to collect and analyse data for hazard forecasting and emergency planning. |
| Disaster site communications | To link affected populations, emergency response officials, and sources of relief supplies. |
| Organizational communications | To link private, governmental, non-governmental, and multi-national disaster relief organizations. |
| Communications for scientific development and policy formulation | Government to develop and exchange knowledge on natural hazards among scientists, emergency managers, planners, decision-makers and legislators. |
| Public education communication | To educate the public, at large, about natural hazards and disaster issues through electronic and print media |

**Source** Modified from Cate (1995)

Two broad groups of communication technologies can be defined: (1) satellite-based communications and (2) computer network communications. Systematic experiments to use satellites in disaster communications have been carried out since 1973 by the Office of the United Nations Disaster Relief Coordinator (now United Nations Office for the Coordination of Humanitarian Affairs). Their actual use during emergencies was initially limited, as the satellite technology of the 1970s was not suitable for mobile transmissions (Zimmermann 1995). Improvements were made during the 1980s with the introduction of the Intelsat, ATS, and Argos satellites. Eventually, at the beginning of the 1990s, Inmarsat, the International Maritime Satellite Program, started in 1982, made satellite links a regular component of disaster communications (Zimmermann 1995). At present, the number of satellite systems suitable for disaster communications is growing, and with it are growing the options and trade-offs between the cost of communications, continuity of transmission, reduction in voice delays, and size of portable terminals. The principal application of satellites is during the response and rescue phase to support emergency communications, and during the recovery and reconstruction phases to provide an interim solution for public telecommunications until the ground networks are rehabilitated (an operation that could take weeks or even months). One of the early examples of the successful use of satellite communications in disasters can be taken with reference to Hurricane Marylin in 1995. Before the hurricane hit the US Virgin Islands in September 1995, the US Army sent in a 12-person "early assessment team" equipped with a satellite-capable phone, a laptop with a commercial crisis-orientated software database, and a modem that linked up with the Inmarsat satellite system (Hazard Technology 1995). During the first 24 hours after the storm, this was the only working communication system on the island (National Research Council 1996). Nowadays, hurricane tracking and related civil protection activities would be simply inconceivable without satellite technology. The challenge has, in fact, moved to the ability of integrating other communication technologies with satellite channels (Thomasson, Greet Verelst, Deprey, *et al.* 2008). Readers can refer to, for example, the IEEE 802.1x[1] set of standard for local and metropolitan area network, in particular wireless protocols like Wi-Fi, TErrestrial Trunked RAdio (TETRA), Worldwide Interoperability for Microwave Access (Wi-MAX), and Global System for Mobile (GSM). Current efforts on satellite communication integration are not only limited to physical and network levels (ISO 1997), but other important communication standards, based on the data level of networks, are also required to be integrated with disaster communications. In particular, Voice Over IP (VOIP) and videoconferencing applications can operate because of common set of well accepted standard protocols such as H.323, SIP, and SCCP, which enable the definition of voice and video communications whenever an IP network is present. More sophisticated frameworks enable a multiplexed mush-up of communications, wherein

---

[1] IEEE 802 Groups of standards, <http://www.ieee802.org/11/>

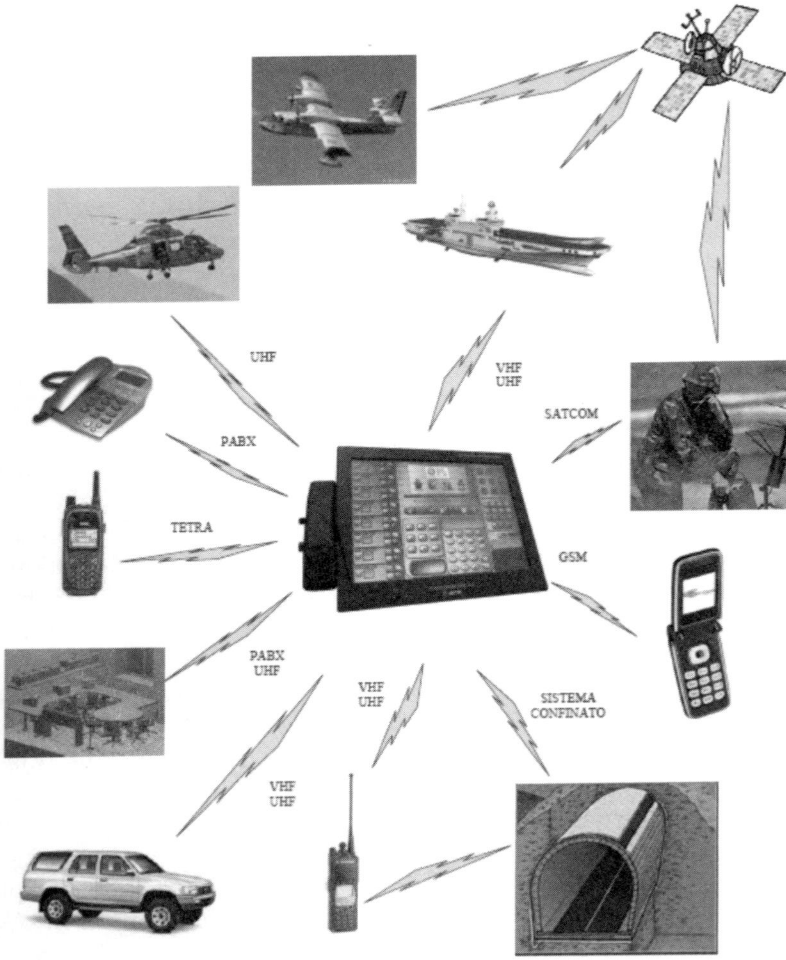


**Figure 1** An example of a multi-communication system for the disaster communications

applications can interact without the limitations of the physical devices and transmission protocols. The resulting scenario of a multi-communication system developed by MPM[2] is showed in Figure 1. The application of computer network technologies to natural disasters started as early as the beginning of the 1970s (Stephenson and Anderson 1997), but it was only during the 1990s that researchers and practitioners of this and related disciplines started consistently using electronic means such as electronic mail, File Transfer Protocol, and news groups to exchange information on disasters. These independent initiatives of individuals, who operated

[2] MPM Società Telecomunicazioni, <http://www.mpm-tlc.it/ita/prodotti3.htm>

without pre-defined methodologies, procedures, or regulations, to efficiently apply such technologies to disaster management, immediately yielded good results. With the development of the World Wide Web, the Internet emerged as a powerful and convenient tool for collecting and disseminating situation reports, ground and aerial photos, maps, and even movie clips of a disaster area. The feasibility of computer network communications in disasters was tested on a large scale during the 1994 Northridge (California) and 1995 Kobe (Japan) earthquakes. Emani and Kasperson (1996), studying the coverage of the 1995 hurricane season in the US, found that in six months, the US Federal Emergency Management Agency (FEMA) distributed 138 hurricane-related e-mail messages describing preparedness and response activities during impacts. In the second half of the 1990s, universities, research institutes, emergency management agencies, and international relief organizations started using computer networks more and more systematically in disaster communications. The general outcome was the development of online disaster databases aimed at providing reliable and objective information for vulnerability assessment, policy formulation, and research. Nowadays, computer network technologies are used during all phases of the disaster response cycle. Easy access to remote sensing images, remote GIS databases, or any other relevant database has enhanced the effectiveness of the response plans both before and after the event. Clearly, system configurations vary with each individual phase of the disaster response cycle. For example, in the mitigation phase, the activities to reduce the impact of disaster when it occurs are widespread, and timeliness is generally not a critical issue. The ICT requirements to support the mitigation phases are broad connectivity among diverse groups of organizations and the ability to move large amount of information. Most of the data is GIS-based risk assessments, claim histories, and facilities and resources lists. This data is archive-based and accessible through search engines. The preparation phases range from logistical support to supply and resource distribution preceding impacts. Preparation also includes public education and community training. Distance learning and other training activities are part of this phase. Public awareness and alerts through broadcast announcements are key activities during the preparation phases. The process of giving out warnings through the network during this phase is intense, and timeliness becomes critical. The response and rescue phases range from bringing relief to victims, to conducting damage surveys, preparing baseline maps, and undertaking logistical operations. Effective communications among response teams and the general public become time-critical. Hence, the network configuration must be rapid, reliable, and amenable to continuous re-configuration. Controlled-access and data security also become vital to efficient response and rescue operations. Intranets are the most common network configurations for these phases. The major problems faced during emergency communications are system incompatibility, high cost, and complexity of use. For the recovery phases, the information needs include on-site data collection related to rebuilding, claim processing, and reports on lessons learnt. Feedback on the mitigation process and historical databases is important in order to document the experience and prevent the occurrence of same mistakes in the future. The network configuration for this phase reflects this change of focus, as timeliness is no longer a first priority, but effective data transfer becomes important. The Internet is often ideal for such transfers. Emerging "Internet standards" help in defining a common basis for an interoperable and lightweight environment for emergency management. Social networking systems that enable the so-called Web 2.0 are becoming useful instruments in emergency management. For instance, systems such as Wiki platforms (Leuf and Cunningham 2001) allow an open and easy-to-use collaborative environment wherein institutions can collaborate in order to collect and share information about an emergency. Wikis allow everyone to "write the Web" and hence represent a lightweight alternative to challenging architectural integration for information sharing. Other interoperability standards and architecture are promoting the usage of the Internet, which goes beyond the simple information sharing. The Web is evolving in a middleware technology framework, wherein full featured applications can be developed and used with an effective interaction. In particular, the Service Oriented Architecture (SOA) (Tsai 2005; Paulson 2006) enables applications to work together by sharing the functions. Other specifications like Portlets (Trujillo, Batory, and Dıaz 2007) allow the plug-and-play of functionalities also at a presentation level, that is, to assemble a Web portal integrating features from different sources. More sophisticated and challenging approaches also attempt to integrate hardware resources like processing and storing resources. Grid systems are able to define a logical computing cluster for data mining and modelling of climatic information (Scharl 2004). A promising field of Web 2.0 application to climate change hazards is also the opening of social collaboration to a wider audience affected by an emergency. The architecture and instruments that we have described so far help institutions and citizens to collaborate together in order to face an emergency. For instance, Wikis can be used by citizens to enrich their knowledge about an emergency. Moreover, the usage of the Web technologies in an emergency is not to be necessarily controlled by the emergency management institutions; in many cases, it is a spontaneous phenomenon. A first noticeable experience of this approach was during the Hurricane Katrina in 2005. Personal blogs, social network websites, and on-line forums reported a plethora of information on the emergency and its evolution. Among all, the *Scipionus.com* project (Singel 2005) furnished a way for people to report and find damage assessments on a geographical Wiki. The site was developed by two 24-year-old computer programmers, who were frustrated combing message boards, trying to find out if their family home had been destroyed. A more recent example concerning the social usage of pre-existing Web portal in an emergency is the sharing of videos on YouTube (a video broadcast website where all users can upload files) during the Italian flooding in October 2008. Pervasiveness and ubiquity of computing, which means the ability to communicate and share knowledge potentially everywhere with the available computing device, point towards a trend of technological evolution, which will help in managing risks and rapid environmental changes. The more the communication infrastructure is distributed over the territory (with some providing redundancies), the less such infrastructure can be completely damaged or destroyed during a disaster. Finally, the flexibility of a ubiquitous and highly interactive technological infrastructure helps in achieving rapid changes even when they involve thousands or millions of people, such as, for instance, the climate change adaptation process.

## Disaster Education and Training

Scholars and practitioners have shown that vulnerability to disaster is strongly related to the way risks are identified and managed. They have also demonstrated that education is the key to reducing losses associated with the catastrophic impact of natural hazards (Alexander 1991). In this respect, the ability of geoinformatics and communication technologies to provide easy access to multimedia information offers great potential for online education on climate change hazards (Table 2).

**Table 2** Advantages in the application of information and communication technologies to climate change education

- Constrictions of time and space for communications have been greatly reduced.
- Remote access to library systems and databases, in general, facilitates links between users and information sources.
- Multimedia and interactive features help users to better visualize problems and solutions.
- Online courses and scheduled satellite transmissions enable long distance education programmes.
- Groupware or workgroup support systems sustain non-disciplinary learning pathways and increase the share of expertise, implementing cooperative learning and multicultural education.
- Cyberspace communications are forcing users to critically analyse the information retrieved, thus increasing their personal information screening capability.
- Within the Internet, ethnicity, sex, and socio-economic and cultural inequities are greatly reduced.

Contrary to traditional and formal education that assigns to the teacher an almost complete control of the type of learning strategies to be used in the classroom, the key principle of computerized interactive education is that the student is able to determine his or her own pathways to learning. By choosing among multiple options, users can effectively decide what to learn and at what rate of progress. Two types of online study material may be distinguished: (1) general information, including scientific articles and (2) independent study programmes. The former is used to convey basic or advanced knowledge of climate change hazards to the public, whereas the latter is intended to increase the analytical and practical skills of practitioners and scholars focused on the emergency response operations. With regard to the online independent study packages, the multimedia features (embedded images, videos, and sounds) have also greatly increased the level of interest and appeal of certain topics. Furthermore, the interactive features of certain programmes enable the educational process to be focused on the needs of learners (Clark and Craig 1992; Soloway, Jackson, Klein, *et al.* 1995). A wide variety of disaster study packages are presently accessible through the online archives of universities, research centres, governmental emergency agencies, and international relief organizations, such as, for instance, the United Nations and the International Red Cross. Scholars, professionals, or laymen can access countless archives containing scientific articles and books, raw data, maps, satellite imagery, and even footage accompanied with full sounds with respect to extreme events such as tropical cyclones, flooding, drought or ice melting. Furthermore, the increasing availability of geospatial information enables users to model scenarios and facilitate interaction with the spatial dimension of climate change hazards. Since most disasters can be mapped, web GIS technology provides a formidable learning tool to simulate different disaster situations and formulate effective responses. Lately, groupware services are increasingly used to exchange disaster information within specific social networks. Interaction with other people from across the world, who have been the victims of climate-change-triggered disasters, can bring about great benefit sharing through their experience and knowledge on such events. Certainly, this educational potential of online reference material and archives is still hampered by information overload and quality control issues. Another important technology for climate change education is virtual training. Increasingly sophisticated simulation software enables diverse interactive environments. By changing the basic physical parameters (such as, for example, wind speed and direction, temperature or soil moisture), these systems dynamically produce different hazard/disaster scenarios. These systems offer unprecedented opportunities to train the emergency managers (and lay people as well) for disaster events without exposing individuals to real risk. Although, three-dimensional scenarios of hazards, emergency or recovery situations are being created for virtual reality exercises, technical problems still occur. For example, to achieve a credible virtual reality, simulation does not necessarily need to appear photographically perfect, but rather must respond promptly to the user's actions (thus the need for very fast computing systems). The aim is to reach a level of credibility for which the users perceive an actual hazard threat.

In fact, if the simulation appears unrealistic, the experience will not be educative. Indeed, these technologies should not be used alone, as the reality of, and psychological involvement in, real hazard events is more complex than the virtual representation: losses of relatives, property or income are immediate examples (Mitchell 1997). Nonetheless, by complementing field-based simulation exercises, virtual reality can become a very important educational tool in promoting the adoption of existing climate change mitigation strategies. A game such as "Stop Disasters",[3] created by Playerthree for UN/International Strategy for Disaster Reduction (ISDR), is a good example of a virtual reality tool created to support disaster education. The game is aimed at 9–16-year-old children and helps them understand disaster events while they play the video game. Another online simulation application that is completely devoted to emergency training is the "Red Cross Emergency Response Unit (ERU)".[4] This game can be played online or downloaded and played on a personal computer. The player is in control of a Red Cross base camp and is responsible for its management. A promising environment to experiment emergency and climate change hazard simulations is "Second Life", a virtual word system with realistic 3D sceneries. Users must orient themselves, socialize, connect, and interact with others by using voice and text chat. Second Life is considered a large-scale multiplayer game, which in 2007 had about 1.3 million active subscribers (ComScore 2007). A two-year-old project called "Play2Train", developed at the Idaho State University at Pocatello, is using tools like Second Life to enable distance learning and simulation for the emergency response training (Hewitt, Spencer, Ramloll, *et al.* 2008a,b). These large-scale social network simulations are also useful for studying personal or group behaviour in a critical situation. Indeed, more research is necessary to evaluate whether individuals with virtual hazard experience will act efficiently in a real hazard situation.

## NON-TECHNICAL LIMITS OF THE APPLICABILITY OF ICT TO MANAGE CLIMATE CHANGE DISASTERS

Undeniably, ICTs enabled humans to expand the reach of communications beyond their physical limits, increasing the quantity of information that a single person can access and share. Yet, faster and more efficient technology does not seem to automatically trigger knowledge transfer. In an early study on the applicability of ICTs to disaster management, Quarantelli (1997) pointed out several non-technical questions for disaster communications, many of which are still unanswered even to date, some 12 years later (Table 3).

[3] <http://www.stopdisastersgame.org>

[4] <http://redcrossthegame.nl/site_en/>

**Table 3** Ten non-technical issues and questions in the application of ICTs to disaster management

| |
|---|
| 1. ICTs contain inherent potentials of disaster: the cascade effects of systems failure may have serious and unpredictable consequences for activities and infrastructure maintenance. |
| 2. ICTs may increase the gap between the rich and the poor (who cannot afford it). |
| 3. ICTs from being a "means" may turn into an "end" in itself. |
| 4. ICTs will inevitably bring information overload. |
| 5. ICTs only offer transient storage for information as opposed to the permanent archival offered by paper. |
| 6. Through ICTs, inappropriate and erroneous information has an increased likelihood of diffusion. |
| 7. ICTs will produce even more diminution of nonverbal communication such as gestures, inflection, body languages, or tones. |
| 8. ICTs will complicate the delicate intra- and inter-level communication systems existing in a chain of command. |
| 9. ICTs will bring fad and fashion practices into disaster management (often pushed for individual rather than collective interests). |
| 10. There is a lack of adequate social and cultural infrastructure that can be used in ICTs during disaster management. |

ICTs – information and communication technologies
**Source** Modified from Quarantelli (1997)

To become truly effective, technical innovations must be supported by a culture that favours sharing of information. In a study comparing the knowledge transfer within disaster management communities in Italy and in the US, Marincioni (2007) pointed out that the factors found to be influencing the transfer of disaster knowledge can be generalized into four conceptual categories: (1) professional culture; (2) economic and environmental context; (3) available and used technology; and (4) interaction (Figure 2). The professional culture (consisting of values, principles, traditions, education, and experience of the local disaster management community) appears to influence the scope of disaster management and the definition of disaster knowledge. Context, including social organization, the natural setting, and the amount of resources available represent the external conditions under which the disaster manager operates. Technology, providing the means to support information exchange, set the stage for interaction among the members of the disaster management community. Indeed, as the disaster manager community lacks homogeneity, these categories are not mutually exclusive, and their influence can vary from case to case. However, challenges to improve the information transfer processes do not occur only at the technology–user interface, but also on the data management side. Perhaps, the most insidious problem faced by the developers of geoinformatics and information technologies, in general, concerns standards for data quality and format. Especially with government agencies, large amount of disaster information of varying quality is being collected and

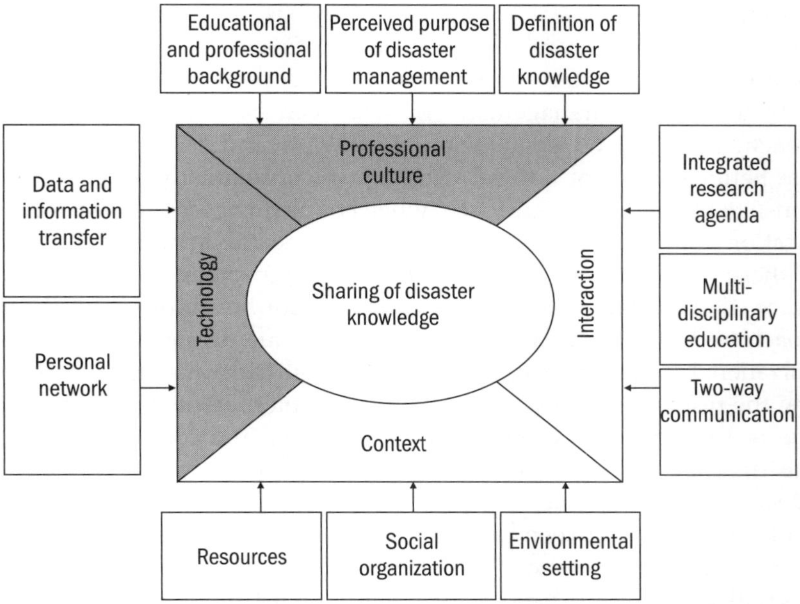


**Figure 2** The transfer of disaster knowledge
**Source** Marincioni (2007)

made available online. The integration and comparison of this information across different national and international organizations are paramount tasks; challenges arise with validation, integration, and distribution of data sets. Furthermore, requirements for integrating data are not uniform across the disaster phases, particularly during the emergency phases when the uniqueness of certain conditions often requires ad hoc strategies. Various initiatives concerning metadata standards and format are under way, as numerous ontologies used to classify information already exist, and new ones are created all the time. Hence, new information technologies must be capable of supporting the evolution of metadata, both in terms of new entries or the merging of old taxonomies, as well as for supporting multiple document and data set standards (Matthews 2005). Especially with real-time data, it is critical that data repositories be accessed and operated during a crisis by different analysis systems

## ICTS AND DISASTER MANAGEMENT: THE CULTURAL DIMENSION

As mentioned previously, the degree of application of geospatial analysis and communication technologies to the management of climate change hazards and disasters does not appear to be rooted in the technology itself, but in the diverse cultural approaches of the users. Cross-cultural studies on this subject (Marincioni 2007; Sokura and Cosby 2007; Stolzenburg 2007; Marx 2003) showed interesting patterns in the perception and application of ICTs to disaster management. In what can be defined a *technocentric* approach, more advanced geospatial analysis and ICTs are believed to enable better control of natural cycles, and consequently, improve human response to impacts. This approach, when coupled with less experience in the application of such information technologies, generates very high expectations from these technologies; ICTs are attributed high potential to significantly enhance the exchange of disaster information. A more *anthropocentric* approach to disaster management appears to value personal experience and organizational and communication skills more than the availability of technology. Basing the resolution of problems on effective interpersonal and professional networks, this approach values the organizational and communication skills of the disaster managers more than their technological expertise. This perspective, on the one hand, appears to value ICTs for having increased the level of interaction between professionals across specialties and disciplines, while on the other, questions the overall ability of technology to transmit non-verbal information needed to create a trustful relationship between interlocutors. Furthermore, the learning process is based on personal predisposition to accept new ideas and experiences, and no technology can trigger such interest. Another approach, which can be labelled as *ecocentric*, seeks relief from natural disasters through a combination of new ICTs and public education programmes aimed at improving social resilience to natural cycles. Thus, technologies able to increase environmental education are perceived to have potential to enhance disaster mitigation. Furthermore, because of the multidisciplinary and transnational nature of ICTs, the expectation is an increased sharing of disaster information across professional and cultural boundaries. Yet, despite recognizing improvements in the circulation of disaster information, disaster managers with an ecocentric perspective claim that disaster knowledge is the result of the unique and direct interaction between the individual and the surrounding landscape; thus, no ICTs can enhance the sharing of such knowledge, which must be learnt first-hand (Marincioni 2007).

## ENHANCING THE EXCHANGE OF CLIMATE CHANGE KNOWLEDGE

Nowadays, geoinformatics and communication technologies have reached very high level of effectiveness and sophistication. They allow forecast, monitoring, and mitigation of climate change hazards, providing scientists, emergency managers, decision–makers, and legislators with very detailed maps and models. In spite of these supports, effective exchange of climate hazard knowledge is still a weak link in the emergency management. As for many other technological progresses, their actual outcome depends also on the cultural context in which they are used. Indeed fine-tunings of both the technology and the user perception can be performed to maximize benefits. Further discussion outlines some general recommendations to improve the exchange of climate change knowledge through ICTs (Table 4). It has become evident that effective communication and information exchange are essential, but not sufficient, for knowledge transfer. Under the current overwhelming information barrage, sending a report, data set, or a map to someone does not ensure knowledge exchange with that person; to achieve a true interaction, it is necessary to establish a dialogue with the interlocutor. Effective exchange of knowledge takes place when various parties engage in a dialogue in which factual information is mixed with personal experience, inventiveness, and values (Michaels 1992; Popkin and Rubin 2000). When individuals develop working relationships with others, they not only exchange ideas, information, and experience, but also evaluate the similarities between their contexts. Thus, whether an individual acquires knowledge or not does not depend only on the means of transfer, namely the technology, but also, and perhaps more importantly, on the context in which such transfers take place. In addition, the transfer of knowledge is not a linear, but an interactive and dynamic process. According to the "issue-network model" of Michaels (1992, p. 162), who studied the knowledge exchange between disaster practitioners and academics, information does not flow from one group to another, but "reverberates among their various points of contact". It is during these reverberations that the information gets translated and interpreted within the specific framework of the issue-network community. The means through which information is transferred within, between, and beyond such a community is through what Michaels calls "information webs". Linking together experts with different levels of commitment and different professional approach means recognizing their diverse agendas, interests, level of understanding, and skills. This diversity is the main engine that supports knowledge transfer; in such a dynamic system, people who have a particular problem to solve seek out other people with different expertise, different links, and maybe even different goals. ICTs can help in the facilitation of these interactions, thus strengthening the multidisciplinary approach. Creating and sustaining rich links across disciplines could help develop cross-disciplinary knowledge and encourage a holistic approach to manage climate change hazards. Tailoring ICTs to the cultural setting of climate change research organizations could significantly improve the processes discussed above. Climate change knowledge can be captured and shared by using the same tools that are employed to create it.

**Table 4** Key recommendations to improve the exchange of disaster knowledge

| | |
|---|---|
| Dialogue | Effective communication and information exchange are essential, but not sufficient, for knowledge sharing. To achieve a true interaction, it is necessary to engage in a dialogue through which factual information is mixed with personal experience, inventiveness, and values. |
| Face-to-face interaction | Direct personal interaction, validating and sharing the dialogue context, is the most effective channel through which knowledge can be exchanged. |
| Professional culture | To foster a dialogue among multidisciplinary groups (for example, those involved in the management of natural disasters) is necessary to understand the differences in their professional culture. |
| Dynamic process | The transfer of knowledge is not a linear process, but an interactive and dynamic process. |
| Multidisciplinary approach | Interdisciplinary education is a necessary step to approach environmental issues from a holistic perspective and to learn how to bridge the different challenges they pose. |
| Flexible ICTs | ICTs, if properly customized to the needs of the practitioners and researcher communities, can serve as an excellent linking tool for climate change knowledge exchange. |

ICTs – information and communication technologies

## CONCLUSION

Geoinformatics and communication technologies have become very important tools to support research in environmental sciences. Their application has greatly advanced the understanding of climate change phenomena, and has also improved the management of natural disasters and emergencies. Continuous advances and progresses with ICTs are expected in the years to come, and their speed of change makes any attempt to forecast possible future applications almost futile. Surely, any new development will increase the quality and quantity of accessible data, and thus will improve the understanding of environmental processes. In spite of the great potential that these technologies have for the prevention and mitigation of climate change disasters, if improperly used, their outcomes may turn out to be modest. The reasons for this are rooted in the very source of current climate change hazards, namely individual and collective behaviour. This is to say that the present climate crisis is a social and not a technological problem. Thus, as Henry Quarantelli has been explaining for decades, with his work at the Disaster Research Center at the University of Delaware, technology cannot solve inherently social issues, such as disasters. Despite this, the use of ICTs for their very ability to support social interaction and foster the exchange of knowledge may influence societal behaviour on climate change. But to be effective, these technologies must be attuned to the unique cultural context of the local communities. Once more, it is not the intrinsic features of the ICTs, but the way these are used, which can make these innovations an important part of the solutions for the contemporary climate change crisis.

## References


Alexander D E. 1991. **Natural disaster: a framework for research and teaching.** *Disaster* **15**(3): 209–226

Anthes R A, Bernhardt P A, Chen Y, Cucurull L, Dymond K F, Ector D, Healy S B, Ho S P, Hunt D C, Kuo Y H, Liu H, Manning K, McCormick C, Meehan T K, Randel W J, Rocken C, Schreiner W S, Sokolovskiy S V, Syndergaard S, Thompson D C, Trenberth K E, Wee T K, Yen N L, Zeng Z. 2008. **The COSMIC/FORMOSAT-3 Mission: early results**. *Bulletin of the American Meteorological Society* **89** (3)

Bhatta B. 2008. ***Remote Sensing and Geographic Information Systems.*** Oxford University Press. 872 pp.

Billa L, Mansor S, and Mahmud A. R. 2004. **Spatial information technology in flood early warning systems: an overview of theory, application and latest developments in Malaysia**. *Disaster Prevention and Management* **13**(5):356–363

Boulos M. 2005. **GIS in practice III: creating a simple interactive map of England's Strategic Health Authorities using Google Maps API, Google Earth KML, and MSN Virtual Earth Map Control**. *International Journal of Health Geographics* **4**(1): 4–22

Cate F H. 1995. ***Communication and Disaster Mitigation.*** Scientific and Technical Committee. International Decade for Natural Disaster Reduction. National Academy of Sciences, Washington, DC

Chen K, Blong R, and Jacobson C. 2003. **Towards an integrated approach to natural hazards risk assessment using GIS: with reference to bushfires.** *Environmental Management* **31**(4): 546–560

Clark R and Craig T. 1992. **Research and theory on multi-media learning effects.** In *Interactive Learning Environments: human factors and technical considerations on design issues,* edited by M Giardina. Berlin: Springer

ComScore. 2007. **Geographical location of Second Life residents who logged-in during January and March 2007.** Details available at <http://www.comscore.com/press/release.asp?press=1425>

Emani S and Kasperson J X. 1996. **Disaster communication via the information superhighway: data and observations on the 1995 hurricane season.** *International Journal of Mass Emergency and Disasters* **14**(3): 321–342

Hazard Technology. 1995. **Fast action from FEMA speeds hurricane relief to Virgin Island. XV**(2): 11

Hewitt A, Spencer S, Ramloll R, Twall R. 2008a. **Learning in a virtual environment: managing emergency preparedness and health security using Second Life as a teaching tool.** In *Proceedings of the TCC 2008 World Wide Online Conference.* Details available at <http://pirate.shu.edu/~twalriad/presentations/neconnect_2008/>, last accessed on 3 March 2009

Hewitt A, Spencer S, Ramloll R, Trotta H. 2008b. **Expanding CERC beyond public health: sharing best practices with healthcare managers via virtual learning**. *Health Promotion Practice* **9**(4): 83–87

ISO, Open Systems Interconnection. 1997. **Information and documentation interlibrary loan application protocol specification**. Details available at <http://www.iso.org/iso/catalogue_detail?csnumber=22248>

Issler J L and Hein G W. 2003. **Galileo frequency and signal design**. *GPS World* **14**(6): 30–37

Kumar R and Mieritz L. 2007. **Conceptualising green IT and data centre power and cooling issues**. Gartner Research Paper No.G00150322

Lemmen D S, Warren F J, Lacroix J, Bush E. 2008. ***From Impacts to Adaptation: Canada in a changing climate***. Government of Canada, Ottawa, ON, 448 pp. Details available at <http://adaptation.nrcan.gc.ca/assess/2007/pdf/full-complet_e.pdf>

Leuf B and Cunningham B. 2001. ***The Wiki Way: quick collaboration on the Web***. Boston, MA: Addison Wesley Longman Publishing

Mainwaring A, Culler D, Polastre J, Szewczyk R, Anderson J. 2002. **Wireless sensor networks for habitat monitoring**. In *Proceedings of the 1st ACM international Workshop on Wireless Sensor Networks and Applications*, Atlanta, Georgia, USA, 28 September 2002. Details available at <http://doi.acm.org/10.1145/570738.570751>

Marincioni F. 2007. **Information technologies and the sharing of disaster knowledge: the critical role of professional culture**. *Disasters* **31**(4): 459–476.

Marx G. 2003. **Some information age technofallacies**. *Journal of Contingencies and Crisis Management* **11**(1): 25–31

Matthews G. 2005. **Disaster management. Sharing experience, working together across the sector**. *Journal of Librarianship and Information Science* **37**(2): 63–74

Michaels S. 1992. **New perspectives on diffusion of earthquake knowledge**. *Earthquake Spectra* **8**(1): 159–175

Mitchell J T. 1997. **Can hazard risk be communicated through a virtual experience?** *Disasters* **21**(3): 258–266

National Research Council. 1996. ***Computing and Communications in the Extreme; research for crisis management and other applications***. Washington, DC: National Academy Press

Paulson L D. 2006. **Services science: a new field for today's economy**. *IEEE Computer* **39**(8): 18–21

Popkin R and Rubin C B. 2000. **Practitioners' views of the Natural Hazard Center**. *Natural Hazard Review*, November

Quarantelli E L. 1997. **Problematic aspect of the information/communication revolution for disaster planning and disaster research: ten non technical issues and questions**. *Disaster Prevention and Management* **2**(6): 94–106

Scharl A. 2004. ***Environmental Online Communication, Advanced Information and Knowledge Processing Series***. London: Springer

Sheetz S D, Quek F, Kavanaugh A, Byoung Joon K. 2009. **Expectation of connectedness and cell phone use in crisis**. In *Proceedings of ISCRAM2009 6th International Conference on Information Systems for Crisis Response and Management*, 10–13 May 2009, Göteborg, Sweden

Singel R. 2005. **A disaster map 'Wiki' is born**. *WIRED*. Details available at <http://www.wired.com/software/coolapps/news/2005/09/68743>

Soloway E, Jackson S L, Klein J, Quintana C, Reed J, Spitulnik J, Stratford S J, Studer S. 1995. ***Learning Theory in Practice: case studies of learner-centered design***. Technical Report. Highly Interactive Computing Group. Ann Arbor, MI: University of Michigan. Details available at <http://www.umich.edu/~spit/Hi-C/DIS.html>

Sokura B and Cosby A. G. 2007. **The role of information and communication technology (ICT) in the resilience of educational institutions in the wake of Hurricane Katrina**. *Journal of the Mississippi Academy of Sciences* **52**(4): 243–261

Stephenson R and Anderson P. S. 1997. **Disasters and the information technology revolution**. *Disasters* **21**(4): 305–334

Stolzenburg K. 2007. **Regional Perspectives on Digital Disaster Management in Latin America and the Caribbean**. ECLAC – Project Documents Collection. United Nations Publication. LC/W.128. Santiago, Chile. 46 pp.

Thomas C and Humenik-Sappington N. 2009. ***GIS for Decision Support and Public Policy Making***. ESRI Press

Thomasson L, Greet Verelst G, Deprey S, Boutry P. 2008. **Hybrid satellite-terrestrial based solutions for rapid deployement of wireless telecommunication networks in emergency situations.** In *Proceedings of the 15th Annual Conference of the International Emergency Management Society* (TIEMS), 17–19 June 2008, Prague, Czech Republic

Trujillo S, Batory D S, and Dıaz O. 2007. **Feature oriented model driven development: a case study for portlets**. ICSE. *IEEE Computer Society*, pp. 44–53

Tsai W T. 2005. **Service-oriented system engineering: a new paradigm**. In *Service-oriented System Engineering. IEEE International Workshop*, pp. 3–8

van Oosterom P, Zlatanova S, and Fendel M (eds). 2005. ***Geo-Information for Disaster Management***. Berlin: Springer-Verlag

Yuan L L, Anthes R A, Ware R H, Rocken C, Bonner W D, Bevis M G, Businger S. 1993. **Sensing climate change using the global positioning system**. *Journal of Geophysical Research* **98**(D8): 14 925–14 937

Zimmermann H. 1995. **The use of satellite telecommunications in disaster relief operations**. *Acta Astronautica* **37**: 437–446

# Index

**D**

# About the Editors

**Dr P K Joshi** has held the positions of Associate Professor and Head of the Department of Natural Resources at TERI University, New Delhi, India, since 2007. Prior to joining TERI University, Dr Joshi spent a decade with the Indian Space Research Organisation (ISRO) on secondment from the Indian Institute of Remote Sensing (IIRS) and jointly as Adjunct Faculty to the Centre for Space Science and Technology Education in Asia and the Pacific (CSSTEAP), Dehradun. He was also Visiting Faculty at the Department of Geography, University of Helsinki (Spring 2009), and Yale School of Forestry and Environmental Studies (Spring 2011).

His peer recognitions include the National Academy of Sciences India (NASI) Young Scientist Platinum Jubilee Award (2008), Indian National Science Academy (INSA) Young Scientist Medal (2006), Gold Medal of Indian Academy of Environmental Science (IAES) (2006), and Fellow of Indian Academy of Environmental Science (FIAES) (2004). He has published research papers in peer reviewed journals, written monographs/books and is regularly invited to attend conferences in related fields. He is also on the Editorial Board of many national/international journals.

**Dr T P Singh** is Associate Professor, Symbiosis Institute of Geoinformatics (SIG), Symbiosis International University, Pune, India. Prior to joining SIG, he was remote sensing (RS) and GIS scientist with National Resources Data Management System (NRDMS), Department of Science and Technology, Nainital, and also with Salim Ali Centre for Ornithology and Natural History (SACON), Dehradun. He spent around four years with the IIRS, while working for national programme on biodiversity characterization at landscape level.

He holds an MSc degree in Environmental Sciences, MPhil in Geo-engineering (RS and GIS), MS RS (Pierre and Marie Curie University, Paris VI), and a PhD in Botany (specialization in RS and GIS). His area of research is application of geoinformatics for natural resource management. He is specialized in environmental application of geospatial tools and LiDAR technology. He has supervized about 80 MSc theses and project works and has published more than 25 papers in peer-reviewed journals.

# About the Editors

**Dr P K Joshi** has held the positions of Associate Professor and Head of the Department of Natural Resources at TERI University, New Delhi, India, since 2007. Prior to joining TERI University, Dr Joshi spent a decade with the Indian Space Research Organisation (ISRO) on secondment from the Indian Institute of Remote Sensing (IIRS) and jointly as Adjunct Faculty to the Centre for Space Science and Technology Education in Asia and the Pacific (CSSTEAP), Dehradun. He was also Visiting Faculty at the Department of Geography, University of Helsinki (Spring 2009), and Yale School of Forestry and Environmental Studies (Spring 2011).

His peer recognitions include the National Academy of Sciences India (NASI) Young Scientist Platinum Jubilee Award (2008), Indian National Science Academy (INSA) Young Scientist Medal (2006), Gold Medal of Indian Academy of Environmental Science (IAES) (2006), and Fellow of Indian Academy of Environmental Science (FIAES) (2004). He has published research papers in peer reviewed journals, written monographs/books and is regularly invited to attend conferences in related fields. He is also on the Editorial Board of many national/international journals.

**Dr T P Singh** is Associate Professor, Symbiosis Institute of Geoinformatics (SIG), Symbiosis International University, Pune, India. Prior to joining SIG, he was remote sensing (RS) and GIS scientist with National Resources Data Management System (NRDMS), Department of Science and Technology, Nainital, and also with Salim Ali Centre for Ornithology and Natural History (SACON), Dehradun. He spent around four years with the IIRS, while working for national programme on biodiversity characterization at landscape level.

He holds an MSc degree in Environmental Sciences, MPhil in Geo-engineering (RS and GIS), MS RS (Pierre and Marie Curie University, Paris VI), and a PhD in Botany (specialization in RS and GIS). His area of research is application of geoinformatics for natural resource management. He is specialized in environmental application of geospatial tools and LiDAR technology. He has supervized about 80 MSc theses and project works and has published more than 25 papers in peer-reviewed journals.